Philosophy of Engineering and Technology

Volume 36

D1941552

The *Philosophy of Engineering and Technology* book series provides the multifaceted and rapidly growing discipline of philosophy of technology with a central overarching and integrative platform. Specifically it publishes edited volumes and monographs in: the phenomenology, anthropology and socio-politics of technology and engineering the emergent fields of the ontology and epistemology of artifacts, design, knowledge bases, and instrumentation engineering ethics and the ethics of specific technologies ranging from nuclear technologies to the converging nano-, bio-, information and cognitive technologies written from philosophical and practitioners' perspectives and authored by philosophers and practitioners. The series also welcomes proposals that bring these fields together or advance philosophy of engineering and technology in other integrative ways. Proposals should include: A short synopsis of the work or the introduction chapter. The proposed Table of Contents. The CV of the lead author(s). If available: one sample chapter. We aim to make a first decision within 1 month of submission. In case of a positive first decision the work will be provisionally contracted: the final decision about publication will depend upon the result of the anonymous peer review of the complete manuscript. We aim to have the complete work peer-reviewed within 3 months of submission. The series discourages the submission of manuscripts that contain reprints of previous published material and/or manuscripts that are below 150 pages/75,000 words. For inquiries and submission of proposals authors can contact the editor-in-chief Pieter Vermaas via: p.e.vermaas@tudelft.nl, or contact one of the associate editors.

More information about this series at http://www.springer.com/series/8657

Michael Nagenborg • Taylor Stone
Margoth González Woge • Pieter E. Vermaas
Editors

Technology and the City

Towards a Philosophy of Urban Technologies

Springer

Editors
Michael Nagenborg
Department of Philosophy
University of Twente
Enschede, The Netherlands

Margoth González Woge
Department of Philosophy
University of Twente
Enschede, The Netherlands

Taylor Stone
Ethics and Philosophy of Technology Section
Delft University of Technology
Delft, The Netherlands

Pieter E. Vermaas
Department of Philosophy
Delft University of Technology
Delft, The Netherlands

ISSN 1879-7202 ISSN 1879-7210 (electronic)
Philosophy of Engineering and Technology
ISBN 978-3-030-52312-1 ISBN 978-3-030-52313-8 (eBook)
https://doi.org/10.1007/978-3-030-52313-8

This Springer imprint is published by the registered company Springer Nature Switzerland AG
The registered company address is: Gewerbestrasse 11, 6330 Cham, Switzerland

Contents

Part II Responsible Design of Urban Technologies

Part III Urban Futures and "Smart" Cities

Chapter 1
Introduction

Michael Nagenborg, Taylor Stone, and Pieter E. Vermaas

Abstract Technology is no stranger to the city. Cities are planned, built, maintained, governed, demolished, and destroyed by technical means. Yet, the city has yet to receive much attention within the philosophy of technology. This volume addresses this gap, and in doing so contributes to the much-needed discussion on technology-enabled urban futures from the perspective of the philosophy of technology. In this introductory chapter, the larger volume is introduced by reflecting on the rationale and need for such a collection, sketching the main themes analyzed throughout, and providing an overview of the contributions.

Technology is no stranger to the city. Cities are planned, built, maintained, governed, demolished, and destroyed by technical means. Technologies can play a central role in making future cities more sustainable, making urban governance more transparent and just, and increasing the efficiency in various sectors ranging from transport to education. Conversely, poor planning and short-sited policy and design choices for urban technologies can exacerbate social, political, and ecological challenges. It should come as little surprise that questions of *why* and *how* to design and manage urban environments in the twenty-first century has emerged as a transdisciplinary topic of inquiry (e.g., Batty, 2018; Graham & Marvin, 2001; Sennett, 2019). Yet, neither does the city receive much attention by Philosophers of Technology, nor do Philosophers of the City consider technology to be a central topic. Recent volumes on Philosophy of the City offer important contributions to the burgeoning field via historical, ethical, aesthetic, ecological, and political analyses (e.g., Jacobs & Malpas, 2019; Meagher, 2008; Meagher, Noll, & Biehl, 2020; Stefanovic &

M. Nagenborg (✉)
Department of Philosophy, University of Twente, Enschede, The Netherlands
e-mail: m.h.nagenborg@utwente.nl

T. Stone
Ethics and Philosophy of Technology Section, Delft University of Technology, Delft, The Netherlands

P. E. Vermaas
Department of Philosophy, Delft University of Technology, Delft, The Netherlands

M. Nagenborg et al. (eds.), *Technology and the City*, Philosophy of Engineering and Technology 36, https://doi.org/10.1007/978-3-030-52313-8_1

Scharper, 2012). However, explicit and focused reflections on the technological nature of urban environments have occupied a peripheral role. Indeed, the last widely recognized contribution from a prominent representative of Philosophy of Technology on the city is Lewis Mumford's *The City in History* (1961).

The notable exception is discussions of "smart cities" and "smart urban technologies." All over the world massive investments are being made to realize visions of the smart city. This has spurred a growing body of (largely critical) scholarship on the ethical and political ramifications of said visions from adjacent disciplines (Cardullo, Di Feliciantonio, & Kitchin, 2019; Kitchin, 2016a, 2016b; Mitchell, 1996; Sadowski & Bendor, 2019; Sadowski & Pasquale, 2015). However, discourse on smart cities or smart urban technologies from *within* Philosophy of Technology has only begun to take shape (e.g., Epting, 2019; Martens, 2017; Nagenborg, 2018; Nagenborg, Albrechtslund, Klamt, & Murakami Wood, 2010; Ryan & Gregory, 2019; Stone, Santoni de Sio, & Vermaas, 2019; Voordijk & Dorrestijn, 2019).[1] Yet, even in these writings, the fundamental role of technologies in the urban lifeworld is taken for granted. Technologies of all scales—artefacts, buildings, systems, and infrastructures—are inexorably intertwined with the very concept of "city." To discuss social and political facets without appreciating and analysing the interrelated technological components is therefore to omit a defining feature of cities. Rather than framing smart city technologies as invaders into the urban lifeworld, we need to understand their role in discourse and praxis from within the context of an already technologically constituted urban lifeworld. We may still remain critical and ultimately reject these developments, but the reason should not be that we are addressing a *technological* development.

Philosophy of Technology can help to understand the role of urban technologies, and in doing so contribute to a more sophisticated and comprehensive approach to developing smart cities. As in Philosophy of Technology in general, we need to move from Capital-T Technology, which appears as a monolithic threat to Capital-C Culture (Franssen, Lokhorst, & van de Poel, 2018). Rather, we need to recognize the ambivalence and ubiquity of technologies in the life of city users, towards a more nuanced understanding of *how* different technologies foster or disrupt perceptions, behaviours, values, and politics. The contributions in this volume demonstrate how urban technologies shape—and are shaped by—fundamental concepts and principles, such as citizenship, publicness, democracy, and nature. And, they explore how to think about technologically mediated urban space as part of the human condition. Through these explorations, "smart" technologies come to be understood via critical and constructive examinations of philosophical concepts (e.g., authenticity), human experience (e.g., soundscapes), socio-technical practices (e.g., mainte-

[1] Within Philosophy of Technology there is also a large body of literature that focuses on technologies and technology-related issues that can be categorized as "smart" (e.g., cell phones and apps, sensors, wearable devices, drones, AI, big data, etc.). However, the scope of these investigations remains at the level of individual artefacts or specific use cases, and are to a lesser extent contributions to Philosophy of the City.

nance), socio-political institutions (e.g., policing), urban form and planning (e.g., public space), and specific domains of application (e.g., streetlights).

The volume consists of three parts. The first part is dedicated to exploring technologies of the city. The chapters in this part are aimed at answering questions of how we can understand and know the city by means of technological mediation, technology-enabled practices, and technology-city relations. The second part features contributions to the responsible design of urban technologies. The focus here is on applying insights from the ethics (and politics) of technology to contemporary developments and concrete projects in urban planning and design. The final part broadens the scope and addresses potential urban futures, especially the underlying ideas of so-called "smart cities." In doing so, timely questions regarding, for example, the future of public space are brought to the fore.

1.1 Part 1: Ontological Foundations: City-Technology Relations

The first three chapters study the city through the lens of aesthetics and artistic practices. Sanna Lehtinen and Vesa Vihanninjoki (Chap. 2) address the challenges of new and emerging urban technologies from the perspective of everyday aesthetics. They argue that we need to understand the impact of such technologies on our experience in order to improve the skills and capacities of city dwellers to adequately deal with these changes. The authors demonstrate how aesthetics and Philosophy of Technology can be brought together to better understand currently under-researched phenomena.

Tea Lobo (Chap. 3) highlights the value of contemporary literature as a "strange tool" to grasp the interplay between technology and city life. She argues that urban technologies also bring about new forms of life and are therefore also poetic in the sense of *poiesis*. Going back to the seminal writings of Georg Simmel and Walter Benjamin she first shows how the process of industrialization has been reflected and captured in the works of Charles Baudelaire and others. She then turns to three prominent examples of contemporary works, namely Zadie Smith, Rana Dasgupta, and Tom McCarthy to analyse how the process of digitalization can be explored through twenty-first century writings. She further argues that literature allows us to gain insights into what she calls "subjective modes of urban experience" which are not accessible via quantitative data analytics.

EL Putnam (Chap. 4) invites us to pay closer attention to the sound of urban spaces through a critical engagement with artistic works. In the first part, she discusses various projects that aim to map urban soundscapes, and thereby mediate the way that urban dwellers perceive such soundscapes. However, she also shows how such projects may not be sufficient to challenge dominant perceptions of the city. The case in point here is the reliance on Google Maps as the unchallenged base map. Therefore, she turns to different projects that challenge the base map by making use of re-designed listening technologies.

Vlad Niculescu-Dincă (Chap. 5) investigates an important facet of cities that, alongside many other practices and professions, is changing with new smart technologies: policing. For this, insights from Postphenomenology are applied to the role and effects of new digital technologies on the strategies used by law enforcement, combined with a detailed ethnographic study of policing practices in England and The Netherlands. Through this combination, Niculescu-Dincă shows how new profiling techniques can have certain biases layered into the system itself—an issue highlighted via drawing a comparison to *sedimentation* in geology. As smart technologies continue to mediate the perceptions and actions shaping policing practices, explains Niculescu-Dincă, there is an imperative to actively "dig up" these layers and examine the ethical issues at stake therein.

Mark Thomas Young (Chap. 6) and Marcel Müller (Chap. 7) offer philosophical explorations of a prominent subject in urban research: infrastructures. Young (Chap. 6) challenges the prominent role of infrastructure failure in the philosophical discussion. His main concern is that the sole focus on failures, which reveal what is otherwise hidden, stops us from paying due attention to other important aspects, namely the everyday work of maintenance, repair, and modification. Without denying the epistemic value of breakdowns, he argues that we will not be able to fully grasp the nature and role of infrastructures unless we also understand these other practices.

Marcel Müller (Chap. 7) combines critical infrastructure research, Philosophy of Technology, Phenomenology, and the work of Jean-Paul Sartre to study what it means to live within infrastructures. By going back to Sartre, Müller succeeds in addressing a blind spot in current works of the Philosophy of Technology caused by the emphasis on technological artefacts in use. As Müller shows, living within infrastructures has the desirable consequence of living in a stable and, thus, secure environment. Yet, infrastructures can also trap city dwellers in their routines.

In the final chapter of the first part, Jaana Parviainen and Seija Ridell (Chap. 8) bring together the topics of infrastructure and human bodies. While resonating with Müller's contribution, they offer a specific tool to understand the multiple ways in which power is enacted through infrastructures: choreography. Through analysing the technologically mediated kinaesthesia and movement trajectories of lived bodies, the authors demonstrate how city dwellers become enrolled in global cybernetic feedback loops on various levels.

1.2 Part 2: Responsible Design of Urban Technologies

The chapters of part two explore the needs and the opportunities for the responsible design of urban technologies. Rockwell F. Clancy and Aline Chevalier (Chap. 9) offer a rich case study on dockless app-based bicycle-sharing systems in China. The lessons to be learned from the particular and localized case go beyond the single technology. While the systems have been widely adapted, specific design choices in combination with lack of public policy-making resulted in contested spaces and significant waste issues rather than a viable addition to existing transportation systems. The chapter, thus, makes a significant contribution by raising questions about design, the privatization of transport services, and the challenges of integrating new alternatives into existing transportation infrastructures. Further, it offers an in-depth case study exploring the impacts of emergent technologies and the important role that cultural factors play in their ultimate use patterns.

Kevin Mintz (Chap. 10) addresses the need for universal design to ensure that people with disabilities have equal access to a city. He also addresses the tension between assistive technologies and universal design by arguing that technologies assisting people with disabilities are an important element in realizing universal design principles. Mintz not only provides illustrative examples of disabling urban design, but also grounds his work in the ongoing debate about disabilities and distributive justice.

The next two contributions investigate the opportunities for the value-sensitive design of particular technologies. Taylor Stone (Chap. 11) focuses on a seemingly mundane technology system, street lighting, and recent developments in making streetlights more sustainable by using LEDs and "smart" technologies. He proposes the value-sensitive framework of *designing for darkness*, which aims at incorporating substantive environmental values and making room for meaningful nighttime experiences. The chapter includes three design proposals to show the alternatives available and to highlight the differences responsible design can make to future urban nightscapes.

Pieter E. Vermaas and Sara Eloy (Chap. 12) provide us with a critical reflection on digital tools for urban and architectural design. They consider 'shape grammar' design systems that allow inhabitants to generate new designs and adjustments of existing designs of apartments on the basis of design rules. On the one hand, such systems may allow for a more participatory approach to architecture while respecting the structure of the existing built environment. On the other hand, the approach raises interesting questions about the role of expert knowledge. The chapter's findings are illustrated by presenting two cases. First the authors look into the potential integration of a shape grammar system into the open-source software The HouseMaker© by the Dutch architectural office MVRDV. The second case is the Rabo-de-Bacalhau Transformation Grammar developed for the refurbishment of apartments built in the mid-twentieth century in Lisbon.

Ryan Mitchell Wittingslow (Chap. 13) makes the case that the design of smart cities needs to allow for an authentic experience. In order to outline the design requirements to facilitate an authentic city, he turns to Albert Borgmann's device paradigm and the idea of focal practices. In contrast to Borgmann, he considers "authenticity" as an aesthetic rather than an ontological category. In turn, the experience of inauthenticity becomes an indicator for design failures, which allows for an exploration of how to design for an authentic experiences in—and of—the smart city.

Finally, Henry Dicks (Chap. 14) examines the implications of biomimicry for urban design and planning through an in-depth look at the idea of modelling the city like a forest. In the first part of the chapter, the author presents three arguments (about "fittingness", "scale", and "complexity") to make the overall idea plausible.

In the second part, he explores the implications of the forest-model in the domains of urban planning, urban water systems, urban energy and transport systems, and urban agriculture.

1.3 Part 3: Urban Futures and "Smart" Cities

Udo Pesch (Chap. 15) proposes a historically informed normative understanding of urban spaces. He argues that the introduction of smart systems leads to the emergence of the experimental city, which allows for new forms of civic engagement. However, the experimental city needs to accommodate for the normative understanding of public spaces developed in the context of the ideals of the liberal, sanitary, and rational city.

Bart van der Sloot and Marjolein Lanzing (Chap. 16) offer a critical evaluation of a prominent element of the smart city: the Living Lab. By studying three contemporary cases (Singapore, Toronto, and Eindhoven), the authors show how the idea of testing various "smart" products in public urban spaces changes the meaning of the public sphere, thereby undermining the foundation of a deliberative democratic society. Thus, they support Pesch's (Chap. 15) argument that the experimental city needs to accommodate the liberal city. At the same time, they offer insights into how we can safeguard the experimental use of technologies in urban space.

In a similar spirit, Germán Bula (Chap. 17) mobilizes Jane Jacobs' writings, notably her 1961 book *The Death and Life of Great American Cities*, to counter what he calls *control cartesianism* rooted in the *planner's itch*, that is: the "drive to anticipate and control every future state of a system" (Bula, this volume). By contrasting the Cartesian tradition with Spinoza's approach, Bula works towards the idea of planning for self-organization, which is underlying Jacobs' work. Taking this idea as a starting point, he arrives at three conditions for the smart city, namely: the political power of cities and neighborhoods, the willingness to engage in participatory democracy, and the possibility to reclaim urban spaces by communities.

Stefano Borgo, Dino Borri, Domenico Camarda, and Maria Rosaria Stufano Melone (Chap. 18) enter the discussion on the smart city by demonstrating the lack of a clear understanding of what a city is. They propose an ontological-grounded and historically informed analysis of the city as a complex system consisting of three components: *city-place*, *city-agency*, and *city-knowledge*. The outcome is a unifying framework for understanding cities, which does justice to the traditional ("non-smart") city as well as the smart city, which is marked by the dominant role of the *city-knowledge* component.

In contrast, Brandt Dainow (Chap. 19) builds on insights from Actor Network Theory and system-theory. He takes as a starting point that human beings and the (smart) city do not directly encounter each other but interact through the medium of digital systems. Persons and their digital devices are thus fused to nodes, which he refers to as "Integrated Personages." The communication between various nodes gives rise to an "Integrated Domain," such as a smart city. This process-oriented framework allows Dainow to show that an analysis of ethical issues within a given Integrated Domain must focus on connectivity and processing. He explicates his findings in a discussion on how autonomy may come into play within such an Integrated Personage.

Wang Qian and Yu Xue (Chap. 20) offer an analysis of the smart city from the perspective of the Philosophy of Organism and the Tao of the city. The vocabulary in this chapter may deviate at points from the one that comes with the European or American tradition of Philosophy. Yet, Wang and Yu provide useful links to Western Philosophy, and reveal a similarity between Chinese Philosophy and Western discourses about the city as a system of systems that invites a comparison with Dainow's account of the integrated domain.

We conclude the third part and our volume with Robert Seddon's reflections on the city of tomorrow after tomorrow (Chap. 21). Echoing Vlad Niculescu-Dincă's contribution (Chap. 5), Seddon argues that the city is made up of various historical layers, which form the memory of a city. In the process of digitalisation, data not only represents the past in the form of VR and AR application and data records. On the long run, data captured today and in the future will become the next layer. Seddon thus leaves us wondering: what will be discovered in and under that layer in a more distant future?

1.4 Conclusion: Towards a Philosophy of Urban Technologies

As a new direction for the Philosophy of Technology, the chapters present a multiplicity of methodological approaches, and many incorporate insights from adjacent fields (e.g., STS, environmental philosophy, political philosophy, geography, and urban design). However, they are unified in the common starting point and final output: providing conceptual and practical insights into how technologies can—and in many chapters *should*—interact with cities and city life. The volume thus contributes to the much-needed discussion on technology-enabled urban futures from the perspective of Philosophy of Technology. In doing so, it opens new directions for inquiry at the intersection of Philosophy of Technology and Philosophy of the City.

Acknowledgements The volume we present you has been initiated by three events in 2017 and 2018. In 2017, Michael Nagenborg and Shane Epting organized a session on "ITC and the City" at the Ethicomp/CEPE 2017 conference in Turin. Also in 2017 the editors (together with Remmon Barbaza) organized a special track on "Technology and the City" at the Society for Philosophy and Technology's bi-annual conference in Darmstadt. In 2018, the first Summer Colloquium of the Philosophy of the City Research Group was held at the University of Twente. Various authors presented the first draft of their chapters at these three meetings and we are grateful for all the encouragement and valuable critique provided by the participants. We also express out gratitude to the 4TU.Center for Ethics and Technology for providing additional support for the Philosophy of the City Summer Colloquium, as well as covering some of the costs connected to the previous events. Finally, we would like to thank Brandt Dainow for his support in proof-reading and copy-editing, as well as many of the volume's authors for providing written feedback on earlier drafts of their colleague's chapters.

References

Batty, M. (2018). *Inventing future cities*. Cambridge: MIT Press.

Cardullo, P., Di Feliciantonio, C., & Kitchin, R. (2019). *The right to the smart city*. Melbourne: Emerland Publishing.

Epting, S. (2019). Automated vehicles and transportation justice. *Philosophy & Technology, 32*(3), 389–403.

Franssen, M., Lokhorst, G.-J., & van de Poel, I. (2018). Philosophy of technology. In E. N. Zalta (Ed.), *The Stanford encyclopedia of philosophy*. Stanford, CA: Stanford University. https://plato.stanford.edu/archives/fall2018/entries/technology/

Graham, S., & Marvin, S. (2001). *Splintering urbanism: Networked infrastructures, technological mobilities and the urban condition*. New York: Routledge.

Jacobs, K., & Malpas, J. (2019). *Philosophy and the city: Interdisciplinary and transcultural perspectives*. London: Rowman & Littlefield.

Kitchin, R. (2016a). *Reframing, reimagining and remaking smart cities* (The Programmable City Working Paper 20).

Kitchin, R. (2016b). The ethics of smart cities and urban science. *Philosophical Transactions of the Royal Society, 374*, 1–15.

Martens, K. (2017). *Transport justice: Designing fair transportation systems*. New York: Routledge.

Meagher, S. (2008). *Philosophy and the City: Classic to contemporary writings*. Albany: SUNY Press.

Meagher, S., Noll, S., & Biehl, J. (2020). *The Routledge handbook of philosophy of the city*. New York: Routledge.

Mitchell, W. (1996). *City of bits: Space, place, and the infobahn*. Cambridge: MIT Press.

Mumford, L. (1961). *The city in history: Its origins, its transformations, and its prospects*. New York: Harcourt, Brace & World.

Nagenborg, M. (2018). Urban robotics and responsible urban innovation. *Ethics and Information Technology*, 1–11. https://doi.org/10.1007/s10676-018-9446-8

Nagenborg, M., Albrechtslund, A., Klamt, M., & Murakami Wood, D. (2010). On "ICT & the city". *International Review of Information Ethics, 3*, 2–6.

Ryan, M., & Gregory, A. (2019). Ethics of using Smart City AI and big data: The case of four large European cities. *ORBIT Journal, 2*, 2. https://doi.org/10.29297/orbit.v2i2.110

Sadowski, J., & Bendor, R. (2019). Selling smartness: Corporate narratives and the smart city as a sociotechnical Imaginary. *Science, Technology & Human Values, 44*(3), 540–563.

Sadowski, J., & Pasquale, F. (2015). The spectrum of control: A social theory of the smart city. *First Monday, 20*(7), 1–22.

Sennett, R. (2019). *Building and dwelling: Ethics for the city*. London: Penguin Books.

Stefanovic, I., & Scharper, S. (2012). *The natural city: Re-envisioning the built environment*. Toronto: University of Toronto Press.

Stone, T., Santoni de Sio, F., & Vermaas, P. E. (2019). Driving in the dark: Designing autonomous vehicles for reducing light pollution. *Science and Engineering Ethics, 26*(1), 387–403.

Voordijk, H., & Dorrestijn, S. (2019). Smart city technologies and figures of technical mediation. *Urban Research & Practice, 26*, 1. https://doi.org/10.1080/17535069.2019.1634141

Michael Nagenborg is an Assistant Professor of Philosophy of Technology at the Philosophy department of the University of Twente. Michael holds a doctoral degree in philosophy (University Karlsruhe, Germany). His doctoral thesis has been on Privacy and ICT. From 2007 till 2013 he has been a member of the Research Group for Security Ethics at the International Centre for Ethics in the Sciences and Humanities (University of Tuebingen, Germany). In 2016, he joined the board of the "Philosophy of the City Research Group." His current research is on the interplay between cities, technologies, and human self-understanding. Since 2018, he is the co-leader of the NWO funded research project BRIDE (Bridging Data in the Build Environment, https://www.utwente.nl/en/bms/bride/).

Taylor Stone is currently cross-appointed as a Lecturer and Postdoctoral Researcher in Ethics of Technology at Delft University of Technology. He received his PhD in Ethics of Technology from Delft University of Technology and the 4TU.Centre for Ethics and Technology, and was previously a Postdoctoral Researcher in TU Delft's Department of Industrial Design. His research explores the convergence of philosophy and design, with an emphasis on cities and urban infrastructure. Within this larger field, he specializes in the ethics and aesthetics of nighttime lighting.

Pieter E. Vermaas is associate professor at the Philosophy Department of TU Delft, the Netherlands. One of his research lines concerns the structure of design processes and their applications in engineering, architecture and society at large. A second line focusses on the emerging quantum technologies and ways to explore their future impact on society. Vermaas is editor in chief of the book series "Philosophy of Engineering and Technology" and "Design Research Foundations", and President of the Society for Philosophy and Technology.

Part I
Ontological Foundations: City-Technology Relations

Chapter 2
Aesthetic Perspectives on Urban Technologies: Conceptualizing and Evaluating the Technology-Driven Changes in the Urban Everyday Experience

Sanna Lehtinen and Vesa Vihanninjoki

Abstract The pervasiveness of technology has changed the way urban everyday is structured and experienced. An understanding of the deep impact of this development on everyday experience and its foundational aesthetic components is necessary in order to determine how skills and capacities can be improved in coping with such change, as well as managing it. Urban technology solutions—how they are defined, applied and used—are changing the sphere of everyday experience for urban dwellers. Philosophical and applied approaches to urban aesthetics offer perspectives on understanding technologically mediated sensory experiences within the urban realm. This chapter shows how new urban technologies act as *an agent of change* within the familiar urban environment. We outline how the perspective of philosophical aesthetics can be used to understand urban technologies and their role in the constitution of everyday urban lifeworlds.

Keywords Urban technologies · Urban aesthetics · Everyday experience · Urban lifeworld · Wayfinding · Affordances

2.1 Introduction

In order to understand life in contemporary cities, one must develop a refined understanding of how technology contributes to it. The urban everyday lifeworld consists increasingly of objects, activities and relations that combine advances in technology and design in complex ways. Technology has already profoundly changed the way urban environments are perceived and experienced. Our aim in this chapter is to

S. Lehtinen (✉) · V. Vihanninjoki
University of Helsinki, Helsinki, Finland
e-mail: sanna.t.lehtinen@helsinki.fi

M. Nagenborg et al. (eds.), *Technology and the City*, Philosophy of Engineering and Technology 36, https://doi.org/10.1007/978-3-030-52313-8_2

describe how this development and process of change within the experiential sphere is affecting urban dwellers and their relationship with the urban environment in which they live their everyday lives.

One of the aims is to introduce *urban aesthetics*, a relatively new strand of philosophical aesthetics, into the philosophical discussion about urban technology, and the various ways it affects the experience and use of contemporary cities. Even though aesthetics is still most commonly considered to concern only the form, appearance or even just the visual look within the extraordinary sphere of art or other creative human pursuits, we emphasize that it is also useful in understanding multi-sensory, embodied and much more basal levels of everyday experience. This comprehensive understanding of aesthetics offers valuable insight into the experiential repercussions that incremental change through adaptation of technologies has caused, and will continue to cause, as cities change in the future. Urban aesthetics is thus used as a framework for discerning and assessing not only the qualitative changes that urban technologies set in motion but also beyond their most immediate implications for the formation of the experience. This is done more specifically through selected approaches from urban and everyday aesthetics and reflecting them through current theories about affordances of urban technologies.

An aesthetic perspective on technologies emphasizes the view that everyday experiences are to a great extent grounded in the materiality and sensory formation of the surrounding conditions, even though these become altered by imaginary and interpretative extensions that point beyond the most immediate experiential qualities. A central question is, how does each new urban technology affect this sensory basis and the socio-cultural interpretations of it? The intention on a larger scale is to reveal the complex interdependencies between applications of new urban technologies and the human agents within the urban sphere.

Our hypothesis is that technology often acts as *an agent of change* in urban environments. Since many elements in urban environments are relatively long-lasting, new technologies are added to an already existing framework of meanings and functionalities. Technological solutions can thus be experienced as a mere addition to previously existing conditions, even though their effect would actually be more fundamental. However, in most cases, we argue, they initiate a process of deeper adaptation which transforms the everyday experience of urban dwellers by changing their behavior, habits, expectations, and preferences. What this actually entails is beyond the scope of this article, but we are raising some relevant points for future debate. We see clear value, for example, for planning and design processes in discussing urban experience as technologically mediated, and in doing this more specifically through the lens of philosophical aesthetics. So far this has been done fairly little and even then, the notion of aesthetics is often reduced to refer only to artistic or other kinds of creative practices.

Our interest in this topic stems from the realization that technology significantly alters not only how the city is experienced directly but also regarding the everyday habits and activities beyond the most immediate effects of how a city is perceived. In this chapter, first, we take a look at urban aesthetics in general. This is important since it is still a fairly unrecognized area of study, yet its attention to the experiential quality of everyday life is particularly apt for discussing the effects of contemporary uses of technologies. These effects are analyzed in more depth in the third part which focuses on technologization of the urban everyday. Throughout the chapter, GPS-based navigation technologies are used as an example of new urban technologies in everyday use: the focus is on how their use via mobile phones in, for example, bicycle and pedestrian wayfinding (map apps, route planners) is affecting how even the familiar and otherwise unaltered urban environment is perceived and experienced. This is studied more precisely through the notion of affordance, which is central to understanding how advanced technologies have acquired a fundamental and formative role in the contemporary urban lifeworld.

2.2 Urban Aesthetics

Since the look and feel of urban environments affect their inhabitants in both explicit and implicit ways, aesthetic factors play a central role in urban lifeworld and its constitution (cf. Madsen & Plunz, 2002). It is thus reasonable to assume that philosophical aesthetics offers a relevant framework for assessing both *what* the urban context is and *how* it becomes processed in human experience and the sphere of human action. One has different types of relationships with many different cities, especially in the contemporary globalised world, but it is also fair to state that urban life is about experiencing urban environments—their material, social and symbolic dimensions—mostly in the context of the *everyday*.

Even though the everyday is "an essentially contested concept" (Saito, 2017), it does provide an obvious framework for looking at how different functions of cities have been traditionally planned. It thus makes sense to focus on the concept of the everyday in order to discuss how cities are used and experienced as well. The everyday attitude of an urban dweller is pronouncedly "colored with routines, familiarity, continuity, normalcy, habits, the slow process of acclimatization, even superficiality and a sort of half-consciousness" (Naukkarinen, 2013). Through everyday actions many elements of the urban environment 'make sense' that might not even exist otherwise. These elements include everything from the systematically planned temporal rhythms of urban transportation to how different functions and services are located geographically, for example.

What, then, is specifically *aesthetic* about the everyday urban experience and what makes aesthetics in general a relevant approach in studying the relationship we necessarily have with the urban surroundings? According to some views in the field of everyday and urban aesthetics, there are two ways of relating aesthetically to one's environment. The first is what is most often understood by aesthetic experiences: the extraordinary experiences of beauty or otherwise exquisite and outstanding experiential qualities of the environment. This view emphasises the highlights, those moments that truly make one pay attention to the perceivable features of the city. The "intentional attention to aesthetics" (Saito, 2017, p. 9) marks a break in one's everyday life. It speaks of peak moments that make one admire or detest something for the sake of its appearance (for the role of "negative aesthetics", see e.g. Berleant, 2010). This idea is linked to the traditional view in the philosophy of art which regards *disinterestedness* as a core feature of aesthetic judgment. In the context of the built environment, this is also the traditional perspective on architectural aesthetics, conceiving buildings and other structures as something to be evaluated primarily through their visual and spatial appearance. Even though the multisensory aspects of experience are acknowledged, the emphasis tends to be on visual qualities, experienced even statically and as detached from the actual use of these spaces.

The concept of the "tourist gaze" (Urry & Larsen, 2011) offers one way of characterising the aesthetic relationship to a city that one is only visiting or assessing from an otherwise distanced position. In this case the experience is defined by the very lack of most everyday associations with places. The place is not "our place", and we are able to examine it in a way that is detached from how it is used on a daily basis. One of the main motivations for observing the features of a city as a tourist is to satisfy our interest in things that are new to us. The camera is the emblematic technology of the tourist, and is used in an attempt to *record* the experience of our visit. This kind of relationship risks reducing urban environments to aloofly observed landscapes and fixed backdrops for taking selfies. When interested in the appearance of the city in this way, one does not need knowledge about the more or less hidden structures that make everyday life possible within it (Latour & Hermant, 1998). Neither does one have access to how familiarity with a certain place will emphasise some aesthetic features and suppress others. The tourist gaze is thus emblematic of one end of the scale of aesthetic interest in urban environments.

The other end of the scale of aesthetic interest points towards understanding the aesthetics of the urban environment as it emerges from one's everyday experience. This kind of experience is defined by *familiarity* and takes place in the everyday engagement with the city (Haapala, 1998, 2005, 2017). Routine, repetition and temporally-based habits characterise everyday life in the urban environment. Aesthetic value is traditionally ascribed to the new (Saito, 2007) or the extraordinary (Leddy, 2012) as something that piques interest, but recent views in the philosophical study of everyday aesthetics also find fundamental aesthetic value in that which is most familiar to us (e.g. Haapala, 2005; Saito, 2007). Aesthetic pleasure (or displeasure) thus forms one significant or even fundamental part of this basal level of experience that defines our relationship to our particular everyday environment. This relationship is further defined by individual variations in using and interpreting the particular environment in question.

Making this distinction between two modes of aesthetic attitude towards urban environments is based on the experience of *strangeness* and *familiarity* (Haapala, 1998; see also Tuan, 1974). It also emphasises that either the environment and its physical features are in one's direct focus of attention—and subjected as such to aesthetic judgment—or they are understood to be experienced through their functioning as the wider, experiential and enabling context of everyday life. Implementing new urban technologies also adds elements of strangeness to the familiar urban lifeworld. The manifold and even contested effects of such implementation are the focus here. It is important to bear in mind, though, that any actual separation of technology from its field of operation would be illusory. Neither is it possible to separate technology from the urban lifeform, however the co-evolvement of urban experience and technology can be studied through the themes of the everyday, familiarity, and strangeness.

The two aforementioned ways of relating to one's environment aesthetically are not contradictory or mutually exclusive, even if they become easily presented in a way that emphasizes the difference in experience. On the contrary, both elements co-exist in the urban experience, and they support each other to a considerable degree (Lehtinen, 2015). In this chapter, however, we are more interested in the latter perspective—how the everyday as such constitutes the urban lifeworld as aesthetic. The everyday as characterised by repetition, routines and familiarity emphasizes that one reacts to and thus also experiences in a different way that which has with time become most familiar and mundane to oneself (Saito, 2017). Unintuitive, new, or otherwise abrupt phenomena in the urban sphere might result in rupture or even collapse of experiential familiarity, which together with a smooth seamlessness created by a certain continuity in experience is characteristic of our individual set of quotidian activities (Haapala, 2005). In abrupt situations, the relationship with the familiar environment becomes cracked, unbalanced and even unsatisfactory, and the normalised reliance on the presence of technology may become problematic as a consequence.

The role of technology within the framework of familiarity in the urban environment is indiscernible from the very goals that one has within one's everyday life. New technologies, when successfully adopted, merge with other ways of doing and thinking in inseparable ways. Different types of user interface are generally designed with intuitiveness and ease of use in mind, even though it is a difficult and highly case-specific goal to reach. Home automation, for example, aims at facilitating the use of the most immediately experienced and familiar surroundings, therefore interfering with a very intimate sphere of personal life. Achieving the desired level of intuitiveness through the design process is a complex task, and failing to do so will risk leaving the end-user discontented with and estranged from the new technological component.

New technologies that are implemented into existing structures and uses of the urban environment build new experiential layers that are more or less based on already existing networks of affordances. These layers refer directly to how affordances build up and become experienced and interpreted. The intertwining of experiential layers affects and constitutes the current conditions of the urban lifeworld as it is. These layers are essentially related to the physical features of objects in the environment, but they are also more or less mediated by the technologies used. GPS and location-based or—aware mobile applications are a good example of this: besides the universally valid spatial co-ordinates, many uses of these technologies are firmly based on the material features of the environment, such as buildings, conventional route patterns and different types of natural and man-made objects that create the detailed urban landscape. But these technologies also add a significantly different layer to the perceived elements: by making relations visible and by giving new, advanced form to previously intuitive modes of wayfinding, they contribute to the experiential fabric of the urban sphere even more than might have been thought. Besides this actionable layer, the skilled use of mobile applications gives them other dimensions too: for example, by stimulating memories and imagination tied to the familiarity of places, or to the technologies themselves, as well as a possibility of creative and strategic variations in the use of urban space.

The description of experiential layers is not used in this context to prove the existence of some type of original, basal level of experience beneath the socially conditioned skills. On the contrary, this points towards understanding how human behaviour, the ensuing actions and especially the prevailing preferences, are the product of a deep and multi-level experiential engagement with the environment. This perspective also emphasises that, in fact, most technologies are based on older technologies and that their overlap is also reflected in the process of learning, unlearning and re-learning the skills and habits required to use them. New technologies emerge at a fast pace and these changes affect the human experiential sphere. Change as such, however, is difficult to grasp conceptually, even though it is a central part of the urban everyday experience. Technology-induced change—whether incremental, disruptive or transformative—is a key factor in understanding how the urban everyday is experienced, and how the urban environment is perceived and evaluated as a part of this experience.

2.3 Technologisation of the Urban Everyday

As a starting point for our experiential analysis of urban technologies, we acknowledge the fundamental role of practices and routines in the constitution of the urban experience. We thus rely on the Heideggerian (1978) approach to the human condition as the "focal point" of various and practically innumerable functional relationships between human beings and their surroundings. What is important here is that the essence of technological things—their functionality or "toolness"—can never be completely understood via a traditional, instrumental view of technology. According to this instrumentalism, technologies are basically neutral connectors between intentions and effects; that is, they are mere means-to-an-end. On the contrary, as postphenomenological philosophy of technology acknowledges (see, e.g. Ihde, 1990, 1993, 2010; Verbeek, 2005, 2011), tools and technologies open up and make comprehensible new possibilities of use and action: they mediate our experience of reality, and it is often the various tools and technologies we make use of in our everyday lives that eventually allow us to conceive our pursued ends as ends in the first place.

As Ingold (2000) has convincingly pointed out, technologies are always embedded at various levels in the structures of society, and thus the effects that the changes in technological realities have on society turn out to be pervasive and widespread. For example, thinking of the technologically induced changes in the forms and structures of production in Western societies, "the development of [production] forces has transformed the entire system of relations between worker, tool and raw material, replacing subject-centred knowledge and skills with objective principles of mechanical functioning" (p. 319).

What is more, this embeddedness of technologies in the structural foundations of any society is opaque by nature: the technology-laden societal structures and mechanisms do not reveal themselves, but they remain implicit in the practices and conventions that constitute the socio-material realities of a culture. The exact role that various technologies have in a society is seldom (if ever) observable as such, but may be addressed through anthropological study: only by examining in more detail the concrete practices and the related socio-cultural ways of experiencing reality, can we access the role in the mediation process that commonly used technologies have.

In order to understand more thoroughly the experiential implications that present-day urban technologies might have, certain conceptual clarifications are required and, above all, the relationship between mere tools and "full-blown" technologies has to be elaborated. The central point seems to be that tools compose a very general and open-ended category of practical items (the usage of tools is most likely not even restricted to mankind), whereas technologies pertain to particular types of socio-material forms of life—that is, to the so-called modernised societies. In other words, "tool" is a more general category, and "technology" is a sub-category of tools, presumably characteristic of modernity. Hence the question concerning the essence of technology turns out to be a question concerning the essence of modernity.

However, as Verbeek (2005) has argued, such approaches to the essence of technology tend to be overly abstract and monolithic, thus losing sight of the significance that particular, concrete technologies have. According to Verbeek's view, the fundamental problem in such a line of thought lies in focusing on Technology as the determining condition of modernity, instead of examining various technologies as practical and useful entities that open up new possibilities for acting in the world—as well as experiencing it through a course of action. Moreover, the coupling of modernity and the technological seems to rely on a circular mode of argumentation, for the technological mode of thinking is already assumed as a necessary precondition for the occurrence of concrete modern technological practices. The question of where, exactly, the technological thinking itself originates remains unanswered.

Contrary to the traditional approaches, the supposed "technologization" of our worldly relations—that is, how modern technologies substitute for more traditional tools—ought to be seen as an essentially socio-material process. Such a process, in turn, consists of the lengthy and laborious development of concrete engineering and design practices, taking place in certain particular societal contexts, thus exemplifying the prevailing ideologies and values of the time. Indeed, it is the gradual emergence of an established engineering industry that is of central importance here, for such a development enables—and perhaps even forces—viewing the everyday and its contents externally, from an external point of view. This, in turn, has far-reaching implications regarding our relationship to the various devices we encounter in our everyday life.

For example, Ingold has analyzed the emergence of proper technologies as a process of externalisation, of "a progressive cutting out of technical from social relations" (Ingold, 2000, p. 314), which removes the productive force of tools and devices from the user's everyday experience of them. In short, the subject-based skills and techniques that the adequate use of tools requires are thus replaced by objective technological knowledge that lies "elsewhere"—outside the context of application—altering and potentially impoverishing the relationship between user and device. Though such an externalisation may be a phenomenon that primarily affects society on the "objective", structural level, it also has profound consequences for the "subjective", experiential level. The experiential implications, which are the proper scope of this chapter, can in turn be approached through an analysis of particular devices and their materiality that shape people's everyday lives and experience.

The Ingoldian (2000, p. 316) concept of "machine" is helpful in understanding the essence of a technological device, since at the level of material instruments, it is precisely "the machine [that] has come to signify the independence of technical operations from human sensibility". The difference between a tool and a machine has to be viewed from the perspective of goal-oriented everyday practices: a tool is essentially an integral part of a user's day-to-day activities, withdrawing from attention in order to enable its user more means of worldly actions for achieving various objectives; a machine, in turn, has an objective logic of its own, demanding a user's attention and forcing them to concentrate on the operation of the machine itself, thus leaving other daily tasks aside (Raudaskoski, 2009).

It has to be emphasised that the distinction between tools and machines refers to the two modes of user experience that often overlap, and in any case do not exclude each other; the distinction merely points out the two different dimensions of a user's relation to a single material device. Considering, for instance, mobile phones, it is clear that "on the one hand [a mobile phone] is a tool that extends our 'bodies' by giving us more means for mutual communication, on the other hand it is a machine, a technological device that operates under technical rules and processes of its own" (Raudaskoski, 2009, p. 44). A mobile phone—as well as any other technological device—thus has a kind of "dual identity" as part tool and part machine, and the development of a device's identity is, in principle, a matter of contingency: a technological device might either become a handy tool, or it might remain as a mere disruptive machine.

Now the interesting question is how, and under what circumstances, do complex machines actually become tools, as they seem to do: how, exactly, does a machine become involved in practical activities in such a manner that the machine-side of a device eventually yields to the tool-side? As applied and underlying technologies are becoming more and more fine-tuned and complex, and thus more distant from the user's everyday experience, this is a topical and urgent question. In any case, technologies simply have to gain a "tool-identity" by various means of familiarisation in order to enter the sphere of the everyday—and to have practical value within it.

Despite the fact that complex technological devices become integral parts of our everyday life, this is merely a partial truth. It is due to their dynamic dual identity as part tool and part machine that technological devices tend to inject a certain instability or even vulnerability into the everyday sphere of uses and actions. This means that even though various devices have become ordinary parts of our activity systems, and thus experientially familiar, the machine-side of their identity and the related experiential strangeness has not completely disappeared.

A comparison of a more "traditional" tool with a technological device illustrates the idea of vulnerability, and helps to understand the existential condition of the far-reaching technologisation of our everyday lives. Here we may, in part, rely on Heidegger's (1978) classic analysis of the broken tool: the breakdown of the hammer halts the everyday procedures of the workshop and thus reveals the normally hidden functional connections (the referential in-order-to structures) that are based on the handiness of the various interrelated tools and utensils. In short, when the hammer does not fulfil its task, it no longer exists as a genuine tool but as a mere object that is present to us, demanding our attention.

Such a "standard interpretation" of a broken tool also applies to technological devices: just as the breakdown of the hammer paralyses the operations of the workshop, the breakdown of a smartphone prevents us from doing what we normally do with it—whether that "normality" consists of being in touch with our friends, answering work-related e-mails on the go, or getting to places with the aid of a navigation application. However, despite these similarities, there are still notable differences between the "toolness" of a hammer and that of a smartphone; this becomes apparent when one asks what, exactly, will happen next, if (and when) the tool de facto breaks down? To be more precise: what will the user do in order to restore and maintain the normal functions that the tool affords?

It is likely that the blacksmith—who in Heidegger's original example runs the workshop—is capable of either fixing the hammer on their own, or at least replacing it with an alternative tool from the workshop. The smartphone user, on the contrary, is helpless: they may try to switch the device on and off, or remove the battery for a while and hope for the best. Probably they will have to rely on the expertise of a repair service or replace the device altogether by purchasing a new one.

Here the machine-side of the smartphone is apparent: despite the seeming familiarity of a technological device, even a relatively minor event of hardship allows the fundamental strangeness of the machine reappear. Indeed, even if we learn to use technologies as tools—that is, if we manage to integrate various technologies into our daily lives quite seamlessly—we do not necessarily familiarise ourselves "with the functioning of the technology as a machine", as Raudaskoski (2009, p. 45) has suggested. For example, not many smartphone users actually learn to use the device as a tool by getting acquainted with the internal logic of the software or the details of the hardware, though it is precisely these dimensions that eventually constitute the functioning of the technology as a machine.

Indeed, while integrating various devices into our everyday, we remain on the superficial level of *interface*, and it is these different interfaces that practically constitute our relation to technology in general (see Hookway, 2014, especially pp. 44–46).

Thus one can argue that the "tool-interpretation" of a machine or a technological device in itself remains necessarily at the level of interface: "the machine as a tool" then means technology as conditioned by the human perspective of various intentional uses and actions. Even further, technological devices may present themselves as mere tools, though there is always more to them, for the "objective logic of the machine" is typically inaccessible through the ordinary user interface.

Design also typically operates at the level of interface, rather than the level of the machine that lies "behind" the interface, potentially blurring our understanding of the complexity of various technologies. This is why the excessive refining of design processes—the pursuit of ever "easier" and "more intuitive" user experience—might even foster the above-mentioned vulnerability inherent in the technological dimension of various devices. As the everyday use of a technological device does not call for any kind of understanding of the machine-side of it, we are rather defenseless against its inevitable malfunctions. Indeed, it is not too exceptional that every now and then our "smart" devices can leave us quite helpless in an everyday situation—whether due to a run-down battery, an unnoticeably outdated application, or some other kind of temporary jam.

The ever-increasing amount of "superficial" interfaces in our everyday life has, along with the apparent political and power-related implications (see e.g. Galloway, 2012), certain experiential consequences that have so far been barely addressed, if at all. If the penetrating technologisation of our everyday truly gives rise to an emerging vulnerability, this may eventually pose a threat to the experiential quality of the everyday, and even to its *everydayness* itself. This has to do with the fundamental relationship between tools and the everyday or, to be more exact, the role that tools have in the constitution of everydayness as a specific mode of experience. According to the Heideggerian argument, it is precisely the unobtrusiveness of tools (their ability to withdraw from experience) that enables the peculiar smoothness characteristic of an (valuable) everyday experience. The unobtrusiveness of tools is ultimately based on their essence as being something *reliable*—i.e., *not* vulnerable (Heidegger, 2002). The overall reliability of tools (that is apparently at stake here) can thus be seen as a ground or essential prerequisite for the everyday experience of smoothness, and the related aesthetics of familiarity.

When discussing in general the reliability of the tools that afford our everyday uses and actions, we are dealing with the normally tacit meaning-structures that eventually make our everyday sensible to us, and thus comprise the substantive basis of everyday aesthetics as a whole. In addition, the smoothness of the everyday can be seen as a necessary prerequisite for experiencing other, perhaps more traditional types of aesthetic values that stand out and pique our conscious interest. Hence a reliability-based experiential smoothness has aesthetic relevance in two different senses: as an aesthetically valuable phenomenon in itself, and as a "preaesthetic" condition for further-refined and conceptually analysable aesthetic experiences (Lehtinen, 2015). These fundamental-level remarks have to be taken into consideration when evaluating the various aesthetic potentialities—both towards positive or negative outcomes—that are necessarily involved in introducing new technologies to our urban everyday.

2.4 New and Emerging Technologies as Agents of Change

Urban environments are by no means homogenous or stable, and the particular aesthetic fascination of cities can be linked to their diversity (von Bonsdorff, 2007) and even to some extent to their fast pace of change. With a focus on urban technologies, futurity and change in general are central themes to be addressed. The aesthetic dimension of experience related to urban technologies necessarily affects and is affected by the responsibilities and prospects of urban planning and understanding the experiential impact of particular technologies would help in determining what exactly is changing and in what type of time frame.

Implementing new urban technologies realigns various functions in urban environments according to their current norms and projected technological level of the city in question. How new technologies are brought into use also raises questions regarding the status of the elements that are most affected by the implementation and adoption of these technologies. "Conservatism" versus "futurism" are two approaches regarding whether and to what extent the existing features of the urban environment should, or could, be changed. Traditionalist perspectives emphasise the "precious quality of human continuity" (Berleant, 2007, p. 81) even going as far as preserving urban environments for the sake of their "museum value". Innovation-oriented paradigms such as the smart city ideology lean towards fascination for change for its own sake and emphasise assessing the old through its relationship to that which is new or emerging. Interestingly, both of these value discourses tend to be rather limited in the way they focus on the given conditions or some strong interpretation of the current situation and thus neglect the inherent potentiality of the environment. In other words, a conservation perspective is overly suspicious of change, whereas a futurist perspective idealises change led by new and emerging technologies.

When adopting new technologies, small features can have relatively large experiential consequences. The "clumsiness" resulting from early steps in learning new technology-mediated habits is a reminder of this. These inescapable side-effects of change have fuelled suspicions towards new technologies. Change can be feared as generating more change that would affect everyday routines and habit patterns in negative ways. If technologically-induced change becomes naturalised, it might also be easy to lose track of the actual drivers directing the development. Some of these change-related blind spots of implementing new technologies might thus be avoided by focusing more systematically on the *quality* of experiences (Sanakulov & Karjaluoto, 2015).

The aesthetics of the city is also of concern in the process of making as accurate as possible different technological representations of an urban environment. It is possible to generate digital representations of urban environments, for example, through building or city information modelling (BIM & CIM). The focus of these techniques is often on making the functional features of a city visible, but the aim is to make the representation as realistic as possible. Existing and chosen technologies dictate many of the parameters, but choices are also made as to which perceptual features are interpreted as being close to the average everyday experience of the actual conditions. The fascination of these virtual representations is that they can be used to replace the real city: for example, making it possible to perceive the city at one glance, something that is not physically possible for an individual without some kind of technology. Applications of information modelling technologies are increasingly finding their way to the urban everyday: they are used for example in route planning, urban game design and people-finding. These applications represent a paradigm that naturalises technology-induced change in the development of the urban lifeform.

If technology is seen as an agent of change in the context of urban environments, technological development or technologicality in general could be understood to drive a "culture of change" even more widely. The desired smoothness of the everyday urban experience is subject to variation in quantity and quality, depending on the quality and quantity of technological mediation. This development is driven further by overlapping and interlacing of various technologies. Traditional objects in the urban environment are not fixed either, as technologies have expanded the range and amount of these changes and objects significantly. Also the pace of technology-induced change is different compared to, for example, many natural processes that take place in the timescale of the urban environment.

The implementation of a technological innovation is a paradigmatic example of how change actually takes place. Many contingent elements in conditions determine the logic by which technologies come into use. For an analysis of experiential change, it is crucial to focus on the actual everyday experience instead of the idealised experience of a certain technology. This distinction is often difficult to pinpoint, especially because designing technology understandably relies on an imagined and streamlined idea of how the everyday experience will unfold. In reality, everyday experiences are marked by interruptions, abrupt changes, failed attempts and other irregularities that are difficult to anticipate (Naukkarinen, 2013).

Technologies direct and fix attention to certain features of the environment. They are thus affecting how and what of the urban everyday realm is distributed to the sensory perception and thus categorisable for further evaluation (on applying the Rancièrean notion of the distribution of the sensible into everyday aesthetics, see e.g. Highmore, 2011). Some technologies enlarge perceptual possibilities within the urban sphere: drones, for example, record visual data from a height above the regular level of human visual perception. These images stimulate interest and elicit reactions by making everyday urban environments visible from previously unseen or rare angles.

Emerging technologies challenge the Heideggerian notion of the pre-reflexive familiarity of the world in everyday life. This is linked to how the new becomes interpreted in terms of and related to that which is familiar. This is true even in cases when there is little of no difference in the resemblance between the new technology and the technology it is replacing. Advancing the implementation of new technologies through relatively small adjustments makes maintaining the prevailing ways of doing things easier, but on the other hand, some unnecessarily demanding and antiquated action patterns might also be retained long past their actual usefulness. This can be seen, for example, in how many of the initial uses and behaviours related to mobile phone use derived directly from using landline telephones, ending up diminishing or even hindering the development of the actual application possibilities and perceived affordances of the portable new technology. These types of anachronisms in user habits might be slight, but their impact grows as they are repeated routinely on a daily basis.

What then, are the stable, long-lasting elements in urban environments and what are the most prone to change? How are the lasting and changing elements determined, what is their relationship and how are they experienced? The theme of "fixed vs. flux" is central in understanding urban change and how it is experienced. There is no uniform, fits-all-types-of-city answer to these questions. The fixed elements are often associated with built elements and spaces, and those in constant change with natural—biotic but also abiotic—elements in an urban environment. However, it is clear that the increase in urban technologies adds more changing elements to the urban sphere. It is therefore important to understand what reactions and patterns are central in coping with change, whether incremental or more radical.

By acknowledging the fundamental role of change, we question the validity of any arguments that rely on the existence of an "original state" or stable equilibrium of human–environment-relations: urban experience (as a systemic whole) has more or less dynamic states, and the experience of stability regarding such a state is always relative. From the viewpoint of everyday urban life this means that, though some experience of stability is necessary for the constitution of an everyday mode of urban life (that is, of its everydayness itself), the exact content of a "particular everyday" (as a contingent constellation of conventions) varies de facto continuously. To exaggerate this point, we live our lives under an "illusion of stability", for the practices we take part in are constantly changing, but this illusion is nevertheless of crucial importance from a practical point of view. In our everyday life, we simply have to be able to rely on certain things—for example that our regular bus route is still operating, that our smartphones still work adequately, that fresh water still is available on tap—though we may even be aware that most of these particular examples probably might not form a part of our everyday life in the far future.

The technological artefact "can give rise to 'emergent' mediations" (Verbeek, 2005, p. 259). These emergent mediations can be interpreted through new, unintentional affordances that become embedded into the existing urban environment. In this case, a particular technology or a version of it is the driver of change as it unleashes some previously unactualised or even non-existing potentiality in the environment. These affordances that embed the horizon for human activity, become perceived and interpreted in active engagement with the urban environment in question.

2.5 Affordances and the Normativity of Everyday Practices

The Gibsonian theory of affordances (see, e.g. Gibson, 1979) aims to explain how human activity is necessarily linked to the existing physical, social, and above all functional action-enabling features in the environment. The concept of affordance has come a long way from the context of Gibson's original formulations and, in fact, an up-to-date understanding of affordances reminds us rather closely of the post-phenomenological perspective on technologies and other tools as a constitutive part of human-world relationships. For example, Kiran and Verbeek (2010, p. 417) conclude that "a piece of equipment influences the world to appear for us in a manner that is in accordance with its affordances. A claw hammer [...] affords rock hard hits and pulling things, which points it toward a context that includes nails and spikes, items that in turn afford hitting or pulling. The nails and the spikes stick out from a background, the world, because of the hammer."

What is particularly important in the passage above is that affordances always belong to certain socio-culturally defined sets of practices, and that they essentially affect our experience of the world. In short, this means that affordances constitute lifeform-specific ways of both acting in the environment, and perceiving it in accordance with these actions (see Rietveld & Kiverstein, 2014). Just as there is no "reality-in-itself" but mere "reality-to-us" (that is, reality as mediated by the various uses and actions characteristic of our culture, see Verbeek, 2005), there are no absolute or objective affordances that an environment could provide but merely affordances to a specific lifeform.

Despite this, some affordances may, indeed, be more obvious to a "skilled agent" than others. A "skilled agent" here refers to an actor who has gained expertise in certain socio-cultural practices—that is, who has internalised a lifeform-specific "normal way" of doing things. It is important to notice that such an internalisation takes place also, or even primarily, on the level of embodiment and perception, so that the expertise has essentially become integrated in the agent's *bodily experience* of reality. As Rietveld and Kiverstein (2014, p. 341, emphasis added) put it, "[skilled agents] do not need to select *reflectively* from the possible actions they can perform, [but] they *perceive* what action the specific situation demands. [...] What the skilled person has learned to do over the years feeds back into the way the meaningful world appears to [them] *in perception*."

The possibility of experiencing certain affordances as "obvious", gives rise to the emergence of a lifeform-specific normativity of practices. Indeed, the normativity of possible uses and actions is closely related to the "normality" that is constitutive of practices: a certain regularity and relative stability of behavioral patterns is the basis for experiencing diverse uses and actions as both meaningful and evaluable. In short, the normative evaluation of behavior is essentially bound to a specific situation in which socio-cultural conventions apply, and the notion of normativity that is "applicable to a skilled individual's engagement with affordances comes from the individual's ability to distinguish correct from incorrect, better from worse, optimal from suboptimal, or adequate from inadequate activities in a specific, concrete material setting" (Rietveld & Kiverstein, 2014, p. 332). Not all possible actions are thus equally correct or adequate, but the potential realisation of available affordances is always subject to context-dependent evaluation, which takes into account subtle social nuances, as well as the material conditions of the situation.

However, such "situated normativity" (see Rietveld & Kiverstein, 2014) pertains merely to certain types of affordances, namely those that already have an established role in a particular lifeform. Indeed, the notion of situated normativity provides an adequate description of how certain uses and actions become "self-evident" and "normal", and also of how the experience of such "self-evidence" and "normality" regarding these uses and actions is carried over to the following generations. Costall's (2012, p. 91) account of "canonical affordances" is particularly illustrative of this matter:

> The meanings of things can indeed become objectified and normative. Artifacts embody human intentions. Indeed, it is through the tacit, embodied understanding of the "canonical affordances" of things, as much as through explicit representations, that young children enter our cultural world. [...] The concept of "canonical affordances" itself alerts us to those important cases where the affordances of some thing are not simply shared between people but also normatively predefined.

The concept of canonical affordances however falls short of explaining how, exactly, the process of meanings becoming "objectified and normative" could or should *itself* be evaluated. In other words, it seems that the dimension of normativity comes along only if an affordance becomes canonised, but the meta-level question of whether or not it is appropriate for a certain lifeform to have a particular affordance as "canonical" cannot be answered.

For example, let's say that it is entirely "normal" for the people within a particular lifeform to make use of an affordance, provided by an artefact called an airplane, in order to travel long distances on a regular basis for the sake of recreation. Such a canonical affordance itself appears, in a sense, to be situated outside the scope of genuine normative assessment, despite the fact that the excessive realisation of such an affordance essentially contributes to larger processes that may eventually be fatal to the lifeform itself as a whole, and to the entire planet.

This is mainly due to certain notable limitations concerning the concept of "situated normativity", for it essentially operates at the "micro-level normativity" of individual behavior *in* a particular socio-materially conditioned situation, but it does not provide an adequate basis for addressing the "macro-level normativity" of the varying socio-material situational *conditions themselves*. This apparent shortcoming regarding the notion of affordance may have something to do with its origin in ecological psychology, where normativity has, after all, a rather limited scope of validity, as the ecological system is supposed to be self-organising and to steer itself accordingly.

Such questions related to replacing existing, perhaps detrimental canonical affordances with new, more appropriate ones illustrate the fundamental challenge with the theory of affordances as a whole: how is it possible to evaluate and compare different affordances themselves, and the consequences that their full-scale utilisation might have? In particular, how is it possible to evaluate unforeseen affordances that involve a potential to change the lifeform-specific conceptions of normality altogether? The possibility for carrying out such an evaluation may well be very beneficial and also rather urgent, since ever-increasing technologicality will for its part drastically affect which particular affordances will be utilised on a large scale, and thus gain "canonical status".

Indeed, from the viewpoint of change and innovation, it is particularly interesting to ponder how the emergence of novel technologies gives rise to certain unforeseen environmental affordances and even promotes them, but at the same time might displace or erode some existing ones. For example, one can think of GPS-based navigation applications that enable us to find our way and move freely without the fear of getting lost both in familiar and strange environments—both in our home town and in the cities we have never visited before. Indeed, wayfinding has never been easier, and as practically everyone carries a localizable smartphone with them nearly all the time, the ease and everydayness of using such an application provides its user with a comforting sense of security.

From the viewpoint of affordances, the rise of technological navigation aids has opened up numerous new environmental affordances, as previously out-of-reach places are now effortlessly and safely accessible to more people. However, the triumph of navigation applications has not resulted in a mere increase in the number of available affordances, but it has already caused a major change in which environmental affordances are de facto utilised. This has to do with the fact that the applications are transforming the way we experience our environment remarkably—that is, the way we distribute our attention and the way we construct various "mental maps" in order to orientate ourselves. In short, they even make us "blind" to many such environmental qualities and properties that we would otherwise notice and pay attention to, as Grabar (2014) has noted: "With their small screens and egocentric perspectives, mobile navigation systems function like blinders, reducing the landscape to the width of a street. They narrow the world."

2.6 Opening and Closing Affordances

From the aesthetic point of view this may be worrisome, for these particular environmental qualities and properties can be exactly those elements that make the surroundings distinguishable and identifiable by comprising the character and the idiosyncratic experiential quality of the local environment. If such elements remain more and more "invisible" to us—that is, due to a change in the way we experience our surroundings—the related urban aesthetic values are also increasingly ignored. In other words, continuous and excessive use of mobile navigators may, in fact, "narrow the world" in such a way that it seemingly includes less aesthetic value than before. Of course, the material basis of the aesthetically valuable qualities and properties remains unaltered in the environment, but there is no point in discussing these potential values without an experiencer who is actually capable of experiencing them.

As became clear with the analysis of a "skilled agent", the way people make use of affordances in their everyday life essentially affects their skills that eventually define their relation to their environment—"the way the meaningful world appears to [them] in perception" (Rietveld & Kiverstein, 2014, p. 341). In other words, people are able to learn new skills, but they are likewise able to forget their existing ones: if traditional navigation skills are no longer needed and thus used, such skills will fade away. Such a tendency has already been verified in an empirical research, the outcome being that the "users of navigation tools have poorer memory of surrounding scenes and less accurate configurational knowledge of travelled routes, compared with people who use maps or directly experience the routes" (Ishikawa, 2016, p. 124).

The message here is not that the emergence of present-day navigation technology and its related applications would be either good or bad, but that it has rather far-reaching and perhaps irreversible consequences concerning our relationship to the environment, and the way we experience reality in general. For example, as Ishikawa (2016, p. 133) points out, many important skills other than mere wayfinding are at risk: "As long as spatial thinking has pedagogical and practical importance for our society, the issue of geospatial literacy in the age of satellite navigation is a topic that calls for continued discussions." Additionally, if the ability of navigating in the urban environment by directly experiencing it slowly but surely perishes, this development renders quite useless a whole branch of urban design that has focused precisely on the questions of *legibility* and *continuity* in cities—that is, on the question related to their *understandability* in general (see, e.g. King & de Jong, 2016).

The features of legibility and continuity are also of central significance regarding the aesthetic quality of urban environments. This connection can, in part, be traced back to the overall functionality of the urban environment, and to its related aesthetic values and meanings. For example, the environmental feature of "being orientable" presupposes the feature of being legible, and being orientable in itself can be seen as a central function of any urban environment. Then, at least according to the "functional beauty" argument (see Parsons & Carlson, 2009), part of the aesthetic value of a city stems precisely from its appropriateness concerning its function of being orientable. Hence if we let go of the requirement for cities to be orientable (on the basis of direct experience and legibility), we will also most likely lose those aesthetic values that originate in this particular function.

The legibility of an environment, in turn, can be divided into two "subspecies", according to the role of conscious attention and action when finding one's way in the city. The point here is that one can orient in the city either by referring to maps, verbal instructions, sets of landmarks, vistas and other visual cues, or by the means of a more direct and straightforward method of simply being acquainted with one's surroundings. Eventually this division comes back to the question of what, exactly, does it mean to "know" a route in an environment, for "knowing" can signify multiple ways of relating to an environment. For example, on the basis of an empirical study, Gale, Golledge, Pellegrino, and Doherty (1990) call for a clear distinction between knowledge *about* a route and knowledge of *how* to navigate the route.

This kind of division in knowing one's movement in an environment bears a clear resemblance to the two modes of relating to it aesthetically—the tourist's gaze, and engagement based on experiential strangeness and familiarity, respectively. For example, Tuan (1977, p. 183) describes the two branches of environmental knowledge, and also their aesthetic relevance, as follows:

> Abstract knowledge about a place can be acquired in short order if one is diligent. The visual quality of an environment is quickly tallied if one has the artist's eye. But the "feel" of a place takes longer to acquire. It is made up of experiences, mostly fleeting and undra-

> matic, repeated day after day and over the span of years. [...] The feel of a place is registered in one's muscles and bones. [...] Knowing a place, in the above senses, clearly takes time. It is a subconscious kind of knowing

On the basis of this, there is a clear difference between the domain of visual or mental images and the bodily engagement with the materiality of an environment. This kind of analysis also provides an improved ground for assessing the potential experiential repercussions that widespread usage of navigation applications might eventually have. For example, if navigation applications rely on visual representations of the surroundings—as by and large they now seem to do—does this emphasise in a detrimental way the domain of abstract knowledge about the environment and finding one's way in it? And further: is knowing an environment through bodily engagement now under threat, compared to other ways that are based on objectifying images of the environment?

These are, in part, questions of contingency that call for empirical analysis. It is, for instance, important to empirically investigate how people de facto become acquainted with their surroundings: do they use navigation applications merely as "early stage" tools, or do they rely on them more continuously in their everyday life? If one uses them in order to familiarize oneself with a strange environment, this can even remarkably enhance the bodily engagement with one's surroundings, but if one keeps on using them without more direct experiential contact with the environment, the level of engagement is likely to remain minimal.

To be sure, the fact that navigation applications have become very common does not merely pose a threat to our environmental relationship, but also opens up various new possibilities of using and experiencing our everyday surroundings. These unforeseen environmental affordances can certainly include notable aesthetic potentialities, so that the canonisation of such affordances might improve the quality of our urban experience even quite drastically.

The aforementioned freedom of movement may, for example, give rise to a new kind of aesthetic sensitivity or openness to aesthetic potentialities inherent in our urban environments. In particular, when one no longer depends solely on the information provided by roads, buildings and other structures to orientate oneself in a city, the urban landscape and its components (understood widely as the entire sphere of perceivable entities) can cease to function as mere guiding features. Consequently, they can be approached with increased enthusiasm and curiosity, providing an enhanced possibility of experiencing urban aesthetics anew, both at the level of detail and in terms of conceiving the city as a systemic whole, with certain dynamics of its own.

Furthermore, thanks to the possibility of "getting lost safely" provided by the navigator in one's pocket, one is perhaps more prone to experiencing aesthetically relevant features that are characteristic of *urbanity* as a phenomenon. These kinds of features include, for example, a certain perceptual inexhaustibility in the form of a "surplus of meaning", suggesting that there is always more to a city than a single human being could possibly experience and know thoroughly; hence, "in some respects the city remains a perpetual mystery, just as great art remains a mystery—extending beyond our capabilities" (Haapala, 2003, p. 21). Moreover, this kind of vastness and radical otherness associated with the city may give rise to an experience of *urban sublime*, in the sense that "the city's human aggregates [...] inspire ambivalent feelings, mingling exhilaration with a threat to selfhood" (Den Tandt, 2014, p. 127). Such an experience of urban sublimity has its base in conceiving the city primarily as a *human context* of massive complexity, not as a mere collection of impressive physical structures, for as Den Tandt (2014, p. 127, emphasis added) points out, "just as sublime landscapes hint at a divine presence in nature, cityscapes spark off epiphanies about *multitudinous humanity*".

Summarising the argument so far, we should not remain passive or helpless in the face of emerging technologies, silently accepting all possible implications, but we should have more and better means to evaluate the pros and cons of each technological device or application more comprehensively. In particular, the experiential and aesthetic repercussions of various technologies have so far escaped further examination, perhaps due to the difficulty of indicating the exact mechanisms in action and the hardships in verifying the concrete effects that a particular technological solution might have.

These related difficulties do not, however, in any way decrease the importance and the urgency of such a task, but on the contrary, despite the apparent speculativeness inherent in the philosophical analysis of urban experience, it is precisely the sphere of experience that eventually "determines" individual behavior in the urban context, and thus the urban future as a whole. The theory of affordances has proven to be somewhat insufficient regarding this matter, and it needs to be further developed or supplemented with an alternative theoretical point of view.

In short, aesthetics has a sort of "double role" in the formation and the evaluation of our everyday practices that, in a way, can be seen comprising a very diverse set of "canonical affordances". First, aesthetics has much to do with the "constitution of normalcy", understood as the capability of perceiving various environmental affordances "correctly", and acting accordingly. This means that the everyday dimension of aesthetics has a remarkable role in the process of conditioning an agent to perceive *certain* affordances and *not* others; perceiving affordances is in this sense always "aesthetically conditioned". This is the relevance of aesthetics at the micro-level of situated normativity.

Secondly, examining the wider aesthetic and experiential repercussions of different technology-laden or -induced affordances provides a way to address the questions of the macro-level of normativity. By asking what kind of aesthetic and experiential implications and outcomes particular "normal practices" might have, one is able to gain a renewed perspective that is *external* in relation to the concrete situation in which the socio-materially conditioned individual necessarily assesses the normativity of their actions. Such an aesthetically and experientially-informed point of view regarding technology-related affordances essentially contributes to evaluating "the politics of artefacts", thus complementing and consummating the critical perspective that Winner (1980) called for in his seminal article decades ago.

For example, the above-mentioned navigation applications are used in a particular situation in which certain social norms apply. The application may, indeed, be used in a way that is assessed as "correct" or "incorrect": the usage of the application can be potentially insulting, if one ends up ignoring one's companion due to an excessive focus on the device. In addition to this, the application may have harmful experiential effects even though it is used according the prevailing norms: one can eventually lose one's ability to read the environment and its cues, if one relies solely on aided means of navigation. These kinds of repercussions cannot be accessed through the conditioned situation itself, but from a wider perspective that examines the socio-material conditions themselves, and their foundation in a particular life-form.

2.7 Conclusions

In this chapter, we set out to study the experiential consequences of the reliance on and use of new urban technologies. One aspect is related to whether technologies ultimately widen or limit possibilities or affordances in the everyday urban environment. This means understanding in different contexts, whether technology manifests as an "enabler" or a "filter" for the everyday urban experience. We have been building a general theoretical framework, but the focus in our examples has been on technological solutions that take part in contemporary wayfinding practices.

Experiential factors are proving to be increasingly important in understanding what types of urban environments enable human flourishing. The experiential consequences of change in the urban environment necessarily affect relationships with the city. Adaptation skills and suitable coping mechanisms become more central and they are developed partly out of existing, established and familiar habits and more intuitive modes of action based on perception and directly engaging with the environment. Urban life as such requires an impressive amount of different types of skills that are learned in practice.

Naturally, there are still elements relating to urban wayfinding that require direct—or, less mediated—attention to the features and cues of the environment: these are often linked to reacting to the movements of other human or non-human users of the urban environment. This can include, for example, assessing visible distances, co-ordination of movements when passing by others or reacting to traffic lights as a pedestrian. Freedom to choose where to direct attention in the urban space is illusory or exists only to a limited extent, since one is inevitably bound to pay attention to the cues of the environment. Over-reliance on mediated knowledge about the environment has been considered to be a risk for the safety of an individual, if it takes attention away from observing threats in the environment.

One further question is to consider how technological convergence in general affects affordances and urban experience. Convergence refers here to the deep integration of multiple technologies and can be seen as one prominent sign of the densification of technological development as more interlinked features are covered by fewer interfaces. Converged technologies also seem to centralise experiences through the use of fewer interfaces, and thus make processes less explicit from the user's perspective.

Stable and static elements are complemented by changing and dynamic elements within the urban lifeworld. Adapting to change requires flexibility but it also affects preferences, which often change far more slowly than the actual pace of change in urban environments. How this type of "acclimatisation" takes place, is a topic for further study. What we propose, is however, that aesthetically based and manifested preferences and their relation to urban life should be taken into account when implementing new urban technologies. Adding new features to urban systems is not only an interesting problem from the design and planning perspective, but also from an experiential point of view: one becomes aware of many old habits and preferences when new technological elements are introduced into existing conditions.

Acknowledgments We are thankful for close reading and insightful comments to the editors and the anonymous reviewers, and wish to thank especially Michael Nagenborg and David Murakami Wood.

References

Berleant, A. (2007). Cultivating an urban aesthetic. In A. Berleant & A. Carlson (Eds.), *The aesthetics of human environments* (pp. 79–91). Toronto: Broadview.

Berleant, A. (2010). *Sensibility and sense: The transformation of the human world.* Exeter: Academic Press.

Costall, A. (2012). Canonical affordances in context. *Avant, 3*(2), 85–93.

Den Tandt, C. (2014). Masses, forces, and the urban sublime. In K. R. McNamara (Ed.), *The Cambridge companion to the city in literature*. Cambridge: Cambridge University Press.

Gale, N., Golledge, R. G., Pellegrino, J. W., & Doherty, S. (1990). The acquisition and integration of route knowledge in an unfamiliar neighborhood. *Journal of Environmental Psychology, 10*(1), 3–25.

Galloway, A. R. (2012). *The interface effect*. Cambridge: Polity.

Gibson, J. J. (1979). *The ecological approach to visual perception*. Boston: Houghton-Mifflin.

Grabar, H. (2014). *Smartphones and the uncertain future of 'spatial thinking'*. Retrieved March 15, 2018, from https://www.citylab.com/life/2014/09/smartphones-and-the-uncertain-future-of-spatial-thinking/379796/

Haapala, A. (1998). Strangeness and familiarity in the urban environment. In A. Haapala (Ed.), *City as cultural metaphor. Studies in urban aesthetics* (pp. 108–125). Lahti: International Institute of Applied Aesthetics.

Haapala, A. (2003). The urban identity. The city as a place to dwell. In V. Sarapik & K. Tüür (Eds.), *Place and location*. Tallinn: Estonian Academy of Arts.

Haapala, A. (2005). On the aesthetics of everyday. familiarity, strangeness, and the meaning of place. In A. Light & J. M. Smith (Eds.), *The aesthetics of everyday life* (pp. 39–55). New York: Columbia University Press.

Haapala, A. (2017). The everyday, building, and architecture. Reflections on the ethos and beauty of our built surroundings. In E. Führ (Ed.), *Ethics in architecture: Festschrift for Karsten Harries* (Vol. 22, pp. 169–182).

Heidegger, M. (1978). *Being and time* (J. Macquarrie & E. Robinson, Trans.). 4th edn. Oxford: Basil Blackwell.

Heidegger, M. (2002). [1935–37] The origin of the work of art (J. Young, Trans.) In *Off the beaten track*. Cambridge: Cambridge University Press.
Highmore, B. (2011). *Ordinary lives. Studies in the everyday*. New York: Routledge.
Hookway, B. (2014). *Interface*. Cambridge: MIT Press.
Ihde, D. (1990). *Technology and the lifeworld*. Bloomington: Indiana University Press.
Ihde, D. (1993). *Postphenomenology*. Evanston: Northwestern University Press.
Ihde, D. (2010). *Heidegger's technologies. postphenomenological perspectives*. New York: Fordham University Press.
Ingold, T. (2000). *The perception of the environment. Essays on livelihood, dwelling and skill*. London: Routledge.
Ishikawa, T. (2016). Maps in the head and tools in the hand: Wayfinding and navigation in a spatially enabled society. In R. H. Hunter, L. A. Anderson, & B. L. Belza (Eds.), *Community wayfinding. Pathways to understanding*. Cham: Springer.
King, M. R., & de Jong, E. (2016). Legibility and continuity in the built environment. In R. H. Hunter, L. A. Anderson, & B. L. Belza (Eds.), *Community wayfinding: Pathways to understanding*. Cham: Springer.
Kiran, A. H., & Verbeek, P.-P. (2010). Trusting our selves to technology. *Knowledge, Technology & Policy, 23*(3), 409–427.
Latour, B., & Hermant, E. (1998). *Paris ville invisible*. Paris: La Découverte-Les Empêcheurs de penser en rond.
Leddy, T. (2012). *The extraordinary in the ordinary: The aesthetics of everyday life*. Peterborough: Broadview Press.
Lehtinen, S. (2015). *Excursions into everyday spaces. Mapping aesthetic potentiality of urban environments through preaesthetic sensitivities*. Helsinki: University of Helsinki.
Madsen, P., & Plunz, R. (Eds.). (2002). *The urban lifeworld: Formation, perception, representation*. London & New York: Routledge.
Naukkarinen, O. (2013). What Is 'everyday' in everyday aesthetics? *Contemporary Aesthetics, 11*, 25–36.
Parsons, G., & Carlson, A. (2009). *Functional beauty*. Oxford: Oxford University Press.
Raudaskoski, S. (2009). *Tool and machine. The affordances of the mobile phone*. Tampere: University of Tampere.
Rietveld, E., & Kiverstein, J. (2014). A rich landscape of affordances. *Ecological Psychology, 26*(4), 325–352.
Saito, Y. (2007). *Everyday aesthetics*. Oxford: Oxford University Press.
Saito, Y. (2017). *Aesthetics of the familiar: Everyday life and world-making*. Oxford: Oxford University Press.
Sanakulov, N., & Karjaluoto, H. (2015). Consumer adoption of mobile technologies: A literature review. *International Journal of Mobile Communications, 13*(3), 244–275. https://doi.org/10.1504/IJMC.2015.069120
Tuan, Y.-F. (1974). *Topophilia. A study of attitudes, and values*. London: Prentice-Hall International.
Tuan, Y.-F. (1977). *Space and place. The perspective of experience*. Minneapolis: University of Minnesota Press.
Urry, J., & Larsen, J. (2011). *The tourist gaze 3.0*. London: SAGE.
Verbeek, P.-P. (2005). *What things do. Philosophical reflections on technology, agency, and design*. University Park, PA: Pennsylvania State University Press.
Verbeek, P.-P. (2011). *Moralizing technology. Understanding and designing the morality of things*. Chicago: University of Chicago Press.
von Bonsdorff, P. (2007). Urban richness and the art of building. In A. Berleant & A. Carlson (Eds.), *The aesthetics of human environments* (pp. 66–78). Toronto: Broadview.
Winner, L. (1980). Do artifacts have politics? *Daedalus, 109*(1), 121–136.

Sanna Lehtinen (PhD in Aesthetics, 2015, University of Helsinki) is a Postdoctoral Researcher at the Helsinki Institute of Sustainability Science HELSUS of the University of Helsinki. Her current research interests include developing philosophical approaches to urban aesthetics, philosophy of urban technologies and updating environmental aesthetics. She has collaborated with architects and urban planners and also teaches aesthetics at Aalto University, Helsinki. Lehtinen was a recipient of the Young Scholar Award by the International Association of Aesthetics (IAA) in 2013. She is a Codirector in the Philosophy of the City Research Group (PotC) and the President of the Finnish Society for Aesthetics.

Vesa Vihanninjoki (MA in Aesthetics, 2015, University of Helsinki) is a salaried Doctoral Candidate at the University of Helsinki. His doctoral research examines the role that aesthetics plays both in urban everyday life and in the diverse forms of present-day urban development. The emphasis is on the questions of sustainable urban infill and a revised notion of urban place, combining the relative stability of the material environment and the dynamism inherent in all experience. With an academic background that also covers various fields of technology (e.g. civil engineering and urban planning), Vihanninjoki aims at providing a multidisciplinary approach to urban aesthetics.

Chapter 3
The Techne and Poiesis of Urban Life-Forms

Tea Lobo

Abstract Technology extends human perception and it intervenes in relations to the environment. Life in cities is particularly affected by newest technological developments, and city dwellers are most shielded and disconnected from the natural world by these very same technologies. The term technology stems from the Greek *techne*, and it refers to an instrumental relation to the world—a manipulation and adaptation of the environment to human needs. However, by intervening in everyday life and modifying relations to the environment, technology also produces new forms of life and is therefore also poetic in the sense of *poiesis*. It is clearly distinguished from literary works because it opens worlds of possibilities that are not fictional. This makes a reflection on its poeticity all the more urgent, because the large-scale advent of new technologies such as the industrialization or the digitization radically changes the modes of perception and relations to urban environments. Art works like literature absorb and articulate these changes and are therefore useful "strange tools" for putting on display the changing modes of experiencing cities. For instance, industrialization, explosive urban growth and the rapid modernization of cities overwhelmed the senses. In its very form and rhythm, modernist literature performs the fragmented and brisk-paced experience of modernity. The digital age ostensibly promises intelligibility and order, for digital technologies can survey and manage vast amounts of information. Smart cities are designed for this purpose and for offering tailored solutions to the inhabitants' individual needs. However, data analytics cannot capture the phenomenological quality of *lived* experience. Art works can, and contemporary literary works put on display urban life-forms characterized by disembodiment and by exhaustive information on any subject that on its own does not translate into any sense of meaning or belonging.

Keywords Techne · Poiesis · Art · Literature · Modernism · New urban imaginary · Industrialization · Digitization · Smart cities

---

T. Lobo (✉)
Swiss Federal Institute of Technology Zurich, Basel, Switzerland

M. Nagenborg et al. (eds.), *Technology and the City*, Philosophy of Engineering and Technology 36, https://doi.org/10.1007/978-3-030-52313-8_3

3.1 Introduction

Technology is not only a process of rendering one's surroundings more convenient and efficient. It is also an intervention in a form of life and a manner of experiencing the world. It is a commonplace to point out how technology extends human senses and agencies: the microscope and the telescope extend our ability to see parts of the world hitherto too small or too far; the bicycle extends the human capacity for locomotion to greater speeds. With the increasingly salient dependencies on technologies (like our smart phones), some might proclaim today to be a transhuman age of the cyborg. However, the use of technology to modify our form of life seems to be constitutive for human life—think only of the manner in which the domestication of fire and the invention of cooking has modified our digestive tract (Sterelny, 2010)—there is nothing "trans" human about it. As Oliver Müller argues by reference to Ernst Cassirer and Hannah Arendt, technology is a manner of shaping one's reality, and at the same time one's very self (2014, pp. 102–107, p. 223). In this sense technology is not only an instrumental *techne,* but it can also be considered as *poiesis*—a production of something that did not exist before. Like poetry, technology produces new worlds. By contrast to poetry, these worlds are not fictional. However, art and literary works can be used as "strange tools" (Noë, 2015) to critically study the technologically inflicted changes to our life form through time, for as Walter Benjamin has argued in relation to the industrialization and the explosive growth of modern cities, literary forms often absorb and reflect the technological changes in the age in which they were produced (1969). What Benjamin has argued about in relation to the industrial age can and should be applied and transformed to address the rapid changes in the digital age and the advent of smart or responsive cities.

The application of digital technologies and data analytics is becoming increasingly common in today's urban design. For instance, in a study within the project "Big-Data Informed Urban Design" at the ETH Singapore, the researchers measured urban walkers' emotional responses via their skin conductance and temperature and correlated them with geospatial data from GPS tracking devices (Hijazi et al., 2016). This data could be used for a new approach to urban design, which is more responsive to people's experience of the city. In *The Responsive City,* Stephen Goldsmith and Susan Crawford focus on how analyzing real-time data from the social media can efficiently replace cumbersome and unresponsive bureaucratic machinery. It could provide a more flexible, democratic and quicker means of participatory urban planning by accessing "the community's collective knowledge" (2014, p. 13), for instance by analyzing social media feeds. A critical question that should be asked is whether and how digital technology in itself alters our perceptions of the city in the first place. This question should be kept in mind when discussing large scale 'smartification' of cities using digital technology. If it is the case that digital technology alters our perception of cities—and my first thesis in this chapter is that this is indeed the case (1)—then such projects will not simply result in cleaner, more efficient and sustainable cities, but will contribute (for better or worse) to a radical change in how we relate to cities in the first place.

My second thesis is that art works in general and literature in particular can communicate these cultural changes (2).[1] As Noë (2015) has pointed out, knowledge is not only about amassing data: it's "a matter of seeing how the data you already have—your own experiences, observations, beliefs—hang together." (Noë, 2015, p. 16). Here, Noë draws on Ludwig Wittgenstein's understanding of philosophical activity as working on a perspicuous or a synoptic presentation (*übersichtliche Darstellung*). Noë's main example is dance, and how dance, as an art form, is a way of putting-on-display. It can serve "to fashion for us a representation of ourselves as dancers; to make clear what is otherwise concealed and poorly understood" (2015, p. 17). According to Noë, art in general "exhibits our manner of organization to ourselves" (2015, p. 205). As Wolfgang Huemer has argued about literary texts, they too "display social practices" (2007, p. 237). Huemer takes up Franz Kafka's *Trial* as an example and points out that the text shows us the workings of a blind and opaque bureaucratic system, but also how that system only functions if individuals play by its rules (ibid.). Even though neither Noë nor Huemer directly address urban design, or the aesthetic presentation of cities, their considerations are applicable to these issues: urban literary works can be read as affording a different kind of perspicuity of how we organize our social practices in their most plastic form, namely the city.

In the following, I will summarize how my thesis that technology alters our perception of cities (1) has been argued for in relation to the processes of industrialization, and how this was reflected in twentieth-century literature (2). Then I will consider contemporary literature and the manner in which it engages with and displays the experience of current digitization processes in urban environments (1 and 2). I focus on literary works that prominently operate with technological motifs and that can contribute to a philosophical reflection on the manner technology changes our way of being situated in and moving through (urban) landscapes.

3.2 The Industrial and Phantasmatic Aesthetics of Twentieth-Century Urban Literature

In his essay from 1903 "The Metropolis and Mental Life", the sociologist Georg Simmel postulates a new variety of alienation in the conditions of industrialized and rapidly growing big cities. He describes the danger for the individual to be "leveled, swallowed up in the social-technological mechanism" (2002, p. 11). The sheer mass of people moving in the streets of turn-of-the-century Berlin, London and Paris and the number of vehicles crowding the roads overwhelmed the senses. Simmel thus explains the necessity for "metropolitan types" to develop a "protective organ", a way of filtering out excess stimuli—an unemotional and impersonal "matter of fact

[1] Art works arguably even shape these changes in the consciousness to some extent. For instance the sense of modernity being exciting and joyous has been shaped by Virginia Woolf's literary works, whereas its tragic side by T. S. Eliot.

attitude in the treatment of persons and things" (2002, p. 12) and a certain cool, "blasé outlook" (2002, p. 14). Simmel does not only refer to the simple difficulty of staying afloat in a sea of stimulation of the big city, but also the existential necessity of keeping up to date with the *right* sort of information. The metropolitan type forms his or her personality in relation to the current fashions, intellectual debates etc. in the city, either by signaling their endorsement, their rejection or quirky modification of these trends. According to Simmel, by contrast to big city dwellers, the rural inhabitants are free to be more sincere and to allow more time to get to know others' unique personalities. The sheer size of the city, mass of people and brevity of time to meet them breeds in the people the feeling of necessity "to appear to-the-point, clear-cut and individual" (2002, p. 18), but always in relation to "impersonal cultural elements and existing goods and values" (2002, p. 19) of the metropolis. This creates a brand of individuality unique to the metropolitan type, one that distinguishes itself not based on one's own irreplaceability, but on one's position in the complex coordinates of the pre-existing cultural codes battling for dominance in the city.

Walter Benjamin refers to Simmel's observations in his essay from 1939 "On Some Motifs in Baudelaire" (1969, p. 341). The essay is on the late nineteenth-century Parisian poet Charles Baudelaire and it deals with changes in aesthetic and poetic forms correlated with the conditions of industrialized big city life. Benjamin takes up Simmel's postulate that big city inhabitants have had to develop a way of filtering out the "series of shocks and collisions" inherent in the over-stimulation of the senses (1969, p. 328). Gone were the days of the nineteenth-century flâneurs,[2] who contemplated city life in their own leisurely strolling pace. Benjamin explicitly names technology as one of the main factors that influence this modification in perception, namely the traffic, electric lights and advertising. He states, "[T]echnology has subjected the human sensorium to a complex kind of training" (1969, p. 328). As his equivalent to that which Simmel terms "a protective organ", Benjamin more specifically refers to the "protective eye" (1969, p. 341) that city-dwellers cultivate to shield themselves from the excess of stimuli. The resulting tunnel vision renders impossible the kind of lyrical and leisurely poetic mode still possible in the countryside, and robs the urban poet-flâneur of his dreamy, distant gaze, forcing him to focus on the mundane task of managing to navigate the city (1969, pp. 340–341).

Neither Baudelaire nor Benjamin appear particularly mournful about the lost lyrical mode in literature, rather they are gesturing towards new poetic possibilities opened up by the industrialized urban experience. In accordance to Benjamin's observation in "The Storyteller" that the genre of the novel flourishes in modernity, the fragmentation of urban experience and the sensory overstimulation had become one of the main leitmotifs of the twentieth-century novel (1969, pp. 83–84). For instance, Andrei Bely's *Petersburg,* a novel published in 1913 presents the experience

[2] Gregory Shaya defines the flâneur as follows: "He was a common figure of the nineteenth century, essential to any picture of the streets of Paris. The flaneur was the man of leisure who went into the street in search of some satisfaction of his overdeveloped sensibilities. He was, by various accounts, a gastronome, a connoisseur, an idler, an artist [...]" (2004, p. 47).

of this large and rapidly growing Russian city built from scratch and almost overnight as if the city itself offered resistance to comprehensive observation (Alter, 2005, p. 91). The characters' visual fields are therefore highly fragmented and phantasmagoric, dominated by flashes of color and light as opposed by any attempts to provide an intelligible and reliable account of life in the city of Petersburg on the whole. The novel's style is correspondingly not at all realistic and linear, but highly experimental, full of repetitions apparently mimicking the traffic rhythm of the city. Better-known examples of modernist urban novels are Virginia Woolf's *Mrs Dalloway* and James Joyce's *Ulysses.* The latter, which appeared serially from 1918 to 1920 is perhaps the most famous modernist urban novel that stages fragmented, seemingly unconnected glimpses of early twentieth-century Dublin. In Woolf's *Mrs Dalloway* from 1925, the industrial metaphor is expressed in Mrs. Dalloway's self-understanding of her role as a hostess for London's society—she must "assemble" the various personages together into a successful party in order to actualize the multifarious and kaleidoscopic experience of London's collective consciousness (Alter, 2005, p. 115).

Ezra Pound is one example of a poet who explores the new aesthetic possibilities opened up by the urban form of life. This is one of his poems from 1913, "In a Station of the Metro":

> The apparition of these faces in the crowd;
> Petals on a wet, black bough.
> (1957, p. 35)

The poem is compact and precise, evoking the rhythmic motion of trains and other machines in the industrialized age, the necessity of moving briskly with the times, of not wasting words. As Alex Goody argues, modernist poetry performs the economical and precise rhythms of modernity (2011, p. 25). This poem moves the way a pedestrian needs to move in modernized and technologically advanced large cities: without the leisure of strolling, lingering or contemplating the view lest he be overrun by a tram or jostled by the crowds. The modernist literary aesthetic expressed in Pound's poem is emblematic of the kind of fragmented and fast-paced experience marked by the mechanical rhythm of traffic. Therefore, in Noë's words, modernist literature puts on display and makes tangible the overwhelming quality of explosive urban growth and modernization.

The latter half of the twentieth century is characterized by the massively spread use of communication media partially blurring the line between mediality and reality. In the 1980s Jean Baudrillard claimed that the television and the telephone have introduced the element of simulacra into our everyday experience. Decades before the advent of social media, he noted "the ecstasy of communication" (1983) which has dissolved the boundaries between the private and the public and between individuals, rendering any talk of a surveyable public sphere meaningless. W.G. Sebald's novels reflect this change in the media landscape and its influence on urban experience and aesthetic possibilities, however, as I will argue, without buying into Baudrillard's resigned diagnosis of unreality. For instance, in the 2001 novel *Austerlitz,* the dome of the Lucerne train station is mentioned, and a photo of a televised image in which that very dome is burning is included within the text. There is furthermore a remark in a footnote that the fictional narrator had visited that very building in February of 1971, presenting a photograph he took of the still intact dome, alongside the photo of the real televised image of the fire which happened a day later in interdiegetic time of the novel, but also in the real world, February 5, 1971 (2013, p. 19).

In another reference to the video medium, the main protagonist of the same novel, Austerlitz, who has lost his entire family in the Holocaust, hopes to find clues about his mother by watching the Nazi propaganda video *Der Führer schenkt den Juden eine Stadt* on Theresienstadt, where he knows his mother had been. It, too, was originally a televised event, and Austerlitz watches it years later on a videotape from the Federal Archives in Berlin. He describes his anticipation while still researching on how to obtain the tape and what he thought finding it would mean to him,

> I imagined seeing her walking down the street in a summer dress and a lightweight gabardine coat, said Austerlitz: among a group of ghetto residents out for a stroll, she alone seemed to make straight for me, coming closer with every step, until at last I thought I could sense her stepping out of the frame and passing over into me.
> (2013, pp. 343–344)

In this passage, the bygone cultural image of the flâneur is evoked in most absurd and darkest circumstances, in the context of a ghetto-camp in the Potemkinian town of Theresienstadt, posing as an attractive holiday spot in a Nazi propaganda video. The life-form of unburdened urban flânerie has been definitely put an end to by the technology of mass destruction of entire cities which changed the face of warfare in World War II, all the more so in pseudo-towns such as Theresienstadt, designed to contain and control people like cattle.

And Austerlitz, too, realizes that the video technology he had placed so much hope in, secretly fantasizing that it would provide him direct acquaintance with his long dead mother turns out to only emphasize its own futility: "At first I could get none of these images into my head; they merely flickered before my eyes as the source of continual irritation or vexation […]" (2013, p. 345). He talks of the "impossibility of seeing anything more closely in these pictures, which seemed to dissolve even as they appeared" (2013, p. 346). He then plays the tape in slow motion, hoping this would help. Now "[t]he contours of [ghetto residents'] bodies were blurred and […] had dissolved at the edges […]" (2013, p. 349). The text includes still frames from the video, with pixelated, blurry pictures of almost unintelligible faces. The media which was supposed to enable him to see his mother—the video—turned out to show nothing but ghostly shadows. It was not medial in the occult sense of enabling him access to the dead, but in the sense of rendering any *im*mediate experience of the past meaningless. However, by contrast to postmodern novelists, Sebald does not simply celebrate the phantasmatic new media and the experience of simulacra in modern cities—even though his novels do reflect this obvious development. By including references to real events, people, architectural landmarks, streets and inconspicuous buildings, Sebald focuses more on the *intelligibility* of the metropolis. As Ursula Terentowicz-Fotyga has argued, W.G. Sebald, along with other contemporary novelists like Zadie Smith are representatives of a *new urban imaginary*, which defy Baudrillardian passivity in face of new media technologies and "traverse the urban simulacrum in search of a way out" (2009, p. 307).

3.3 The City on (the) Display in Contemporary Literature

In the following section, I will analyze the works of twenty-first century writers Zadie Smith, Rana Dasgupta, and Tom McCarthy with a focus on the recurrent motif of digital technologies that characterize the contemporary everyday. As a proponent of the "new urban imaginary", Smith is interested in exploring and expressing the city as it can be known and perceived as opposed to mere simulacra, yet, this does not mean that she reverts to a nineteenth-century realist mode of omnisciently presenting the city. Rather, she has been compared to the modernist author James Joyce in her stream-of-consciousness depiction of the urban experience (Strauss, 2012). Her novels *White Teeth* and *London NW* abound in rich glimpses and snippets of twenty-first-century London. However, as I will argue, she is interested in the intelligibility of contemporary cities and in the manner in which current digital technologies are apt to both deceive the senses and provide a tantalizing glimpse into the possibility of putting the city as a whole—in all its complexity—on display. This is a possibility that seemed lost with the diversification of urban life-forms in the course of industrialization and with the blurring of boundaries of the real and medially represented with the advent of mass communication media. For, digital technology, as the proponents of smart and responsive cities argue, can enable us to survey the hitherto unmanageable quantities of information produced in the cities every day that had overwhelmed or frustrated the senses in the previous century.

Zadie Smith's recent novel *Swing Time* (2016) treats on manners of displaying and making perceptually salient current socio-cultural developments by means of art. Parallel to the oscillation in setting between poor neighborhoods in both the United Kingdom and West Africa, and global nomads' hubs like New York and London, the novel reflects on the history of dance. The narrator is interested in everything from African tribal dances, coded slave dances, to contemporary pop star choreographies, which commodify multi-cultural influences and employ them to ornamental effects. As discussed above, Alva Noë has argued that dance, as an art form, displays the body as saturated in historical, cultural and communicative contexts. Here I argue that while it centers on dance and the human body, Smith's novel *at the same time* displays and reflects that collective body we call the city—in its global varieties and with regard to how current technological developments modify and influence urban experience. She includes the rooted perspective of West African locals—whose excruciatingly slow internet connection renders information-gathering an elaborate investigative process, involving fragile relationships of trust, as well as the physical traversing of spaces between villages and cities. But she also reflects on global nomads' privileged aerial view from an airplane window, and the almost omniscient gaze of a proficient internet and social networks surfer based in New York, with all the technological amenities the city affords her.

One of the novel's major themes is the differing modes of information gathering appropriate to vastly different terrains: globalized twenty-first-century cities like London and New York versus the rural landscapes of West Africa. The nameless narrator has grown up in the former world: she is a Londoner who spends most her time in New York due to her work, and who considers setting her phone to the airplane mode for several days at a time one of "the great examples of personal stoicism and moral endurance of our times" (2016, pp. 1–2). However, her mother is Afro-Caribbean, therefore in her ancestry hailing from West Africa via European slave trade to the Americas. The narrator works in the music media and when she becomes a pop star's personal assistant and accompanies her around the world, including half a dozen trips to a West African village on a dubious developmental aid project, she is in the curious position of being intensely invested and interested in the place and aware of her almost complete ignorance of the realities there. This shows particularly well on her first day at the village, where she, shocked by the poverty around her, rushes to the village school headmaster's office (with the only available power plug "run on a solar-powered generator and paid by an Italian charity years earlier" (2016, p. 176)) in order to google the country's GDP. She describes:

> frantically trying to get online, although I could, of course, have googled what I wanted to know in New York, far more quickly, with infinitely more ease, at any time in the previous six months. Here it was a laborious process. A page would half load, then crash, the energy from the solar rose and fell and sometimes cut off completely. It took more than an hour.

In the end she had the information she needed: her boss, the pop-star Aimee's net worth was a little larger than the GDP of the entire nameless West African country.

By contrast, back in New York and on her high-speed internet, the narrator has all the information she needs under her fingertips. In an increasing rebelliousness towards her boss Aimee, she does not go to one of her shows, not wanting to see the "funhouse versions of the dances" they saw in West Africa that Aimee appropriates, but spends the entire evening experiencing it via her laptop, following the social media commentary, pictures and videos released by the second: "Search Aimee, search venue, search Brooklyn dance troupe, image search, AP wire search, blog search" (2016, p. 369). She realizes that

> I could reconstruct – 140 characters at a time, image by image, blog post by blog post – the experience of having been there, until, by one a.m., nobody could have been there more than me. I was far more there than any of the people who had actually been there, they were restricted to one location and one perspective – to one stream of time – whereas I was everywhere in that room at all moments, viewing the thing from all angles, in a mighty act of collation (2016, p. 369).

She is spell-bound: "I kept refreshing and refreshing, waiting for new countries to wake up and see the images and form their own opinions or feed off opinions already voiced" (2016, p. 370). Powerless as she is, especially in her exploitative working relationship with Aimee, she is high with the power of knowledge that digital information technology lends her, allowing her to transcend her own ultimately finite spatio-temporal perspective and oversee the whole of Aimee's dance like a god, finding validation in other voices around the world criticizing Aimee's cultural appropriation of African dances and traditional clothes. She is able to access the city's venues and view them from practically any angle without ever leaving her apartment.

Another character, a Brazilian economist nicknamed Fern, is part of the team in Aimee's mission to reduce poverty, and lends the whole endeavor a shred of legitimacy and understanding of global and local dynamics. Unlike the narrator, he is well acclimatized to the conditions in rural West Africa and has the necessary patience to gather information the old fashioned way: by carefully listening and withholding assumptions and judgment. Due to the scarcity of communication media and information technology, person-to-person communication and the ability to navigate relationships in local family and village structures and hierarchies is key, and Fern has acquired these skills during his many years of field work. The narrator observes Fern's behavior:

> Whenever I spotted him in my reluctant daily walk around the village [...] Fern would be locked in intense discussion with men and women of every age and circumstance, crouching by them as they ate, jogging next to donkey-drawn carts, sitting drinking *ataya* with the old men by the market stalls, and always listening, learning, asking for more detail, assuming nothing, until he was told it (2016, pp. 248–249).

As much as she respects Fern, she is alienated by his intense sensorimotor engagement with the local culture, so different from her own disembodied, digitally enabled orientation in space.

While the narrator is not completely numbed to the world of the senses, for instance she does appreciate the quality of light in both New York and in African sunbathed landscapes ("New York was my first introduction to the possibilities of light, crashing through gaps in curtains, transforming people and sidewalks and buildings into golden icons, or black shadows [...] But [...] this light was something else again. It buzzed and held you in its heat, it was thick, alive with pollen and insects and birds, and because nothing higher than one storey interrupted its path, it gave all its gifts at once, blessing everything equally [...] (2016, pp. 254–245)), she does consider herself more at home in the city. She is accordingly blasé in her attitude, as shown in her wish to "crush" the optimism on Fern's face (2016, p. 251), trusting more her own abstract researches into the state of things in West Africa then Fern's first-hand knowledge gained from fieldwork. Her blasé attitude, in the sense Simmel introduces it as a characteristic of the metropolitan type, comes to the fore when she is scrutinizing Fern's outfit and style when he shows up in New York: "Fern, in an ill-fitting, clearly rented suit, a little too short at the ankles, was walking towards me, waving goofily" (2016, pp. 335–336). As Simmel remarks, the psychological survival in a metropolis depends on being up to date on the *right* sort of information. Committing a fashion *faux pas* and not presenting himself as being up-to-date the way New Yorkers do disqualified Fern from being taken seriously by the narrator on her own terrain, in the big city. At first she and Fern admire the view on the city together. He senses her disapproval and says, "Please don't laugh at me" and "The truth is I don't know how to dress in New York" (2016, p. 338). Later on, the narrator muses: "I knew he'd crossed Liberia, alone and on foot, aged only twenty-four [...]—but now all I could think was: *Brother, this city will eat you alive*" (2016, p. 365). While she is critical of Aimee for her flippancy and superficiality, the narrator herself is clearly very concerned with outer presentation and appearances, as a critical means of metaphorical survival in the city and holding onto one's identity and status without letting the city "eat you alive". There is a binary in the novel between the mode of being in the countryside of a third world—and correlated reliance on interpersonal experience and exchange of intelligence—and the mode of being in the city, with access to latest information technology and the correlated impersonal, disembodied access to the events and places, and with the unspoken necessity of adhering to impersonal codes and exterior markers of being-in-the-know, like fashion, characteristic of metropolitan types.

The novel shows how the reliance on smartphones and the internet has become so deeply ingrained in metropolites' navigation of space that it has become almost unnoticeable. It takes a visit to a dramatically different environment, that of a tiny West African village to properly see it. This is not in itself objectionable, but if we reflect on the use of digital technologies and their potential for designing cities, it would be fruitful to also consider how digital technology itself modifies the perception of cities. Furthermore, it is not clear whether digitally optimized cities would again afford the kind of stress-free strolling that the flâneurs of old enjoyed, or whether these cities would be technologically determined to such an extent that they would be impossible to navigate without the equivalent of today's smartphone. While Smith describes her narrator's technologically mediated experience of her surroundings neutrally, merely hinting at her alienation but presenting it as the way things are now,[3] the next novel I study—Rana Dasgupta's book draws out the more sinister and problematic aspects of digital technologies.

Rana Dasgupta's *Tokyo Cancelled* is a hybrid work. It is a collection of stories interrelated by their shared framework: they are told at the airport, by passengers whose flight to Tokyo has been cancelled because of a snowstorm, and who then have to find a way to pass the time until the next morning. As such, it is a novel that reflects on the bygone form of oral storytelling, once rendered antiquated by the advent of technologies that increased the velocity of life (Benjamin, 1969, pp. 83–84). In a situation where technology fails, however, and people are forcefully slowed down, even completely arrested in their motion by, say, the cancellation of a flight, this antiquated form re-emerges, and passengers tell each other tales from faraway lands of their origin. To mark the extent to which technology is taken for granted, the first line of the book is "There was chaos" (2005, p. 1). The disorientation of the passengers is emphasized:

[3] To be sure, Smith does criticize the standardisation of experience due to Facebook in her essay "Generation Why?", and the replacement of experience by representation due to digital technologies in "The Tattered Ruins of a Map: On Sarah Sze's *Centrifuge*" (Smith 2018). in her

> A 747 had disgorged its 323 passengers into the middle of a vacant, snow-brushed tarmac expanse, left them to trudge across it through the cold and the floodlit glare to a terminus [...] had abandoned them, in short, in the Middle of Nowhere, in a place [...] like a back corridor between two worlds, two somewhere, where people only alighted when something was seriously kaput with the normal eschatological machinery (2005, p. 1).

Some of the unlucky passengers who are unable to find accommodation for the night huddle together on uncomfortable vinyl airport furniture. Someone suggested, "You know friends I don't think we know each other well enough to sit in silence. [...] I was thinking just wondering to myself: Does anyone know any stories?" (2005, p. 7). They commence telling stories set in the cities they come from, and the fourth story told is titled "The House of the Frankfurt Mapmaker". It is a story of one named Klaus Kaufmann, who collects geospatial information "compiling them all into a complete electronic map of the planet" (2005, p. 93). There follows a list of the kind of information that interests him:

- Air routes
- Archeological sites
- Canal systems
- Climate
- Crops
- Demographics
- Electricity systems
- Housing projects
- Legal systems
- Price indices
- Rock and soil types
- Sea routes
- Telecommunication networks

This project has made him filthy rich, and he even accepts bribery to put a yet nonexistent pipeline on his map in order to put pressure on parties involved in the decision of its realization. A Turkish official suggests to Klaus, "[...] perhaps if you mark the pipeline on your map it will help to push things in our direction. Make it already a bit concrete, so to speak... [...] I think this information would make your map much more – how should I say it – prescient, no?" (2005, p. 95). Dasgupta's book was published in 2005, a year after Google has acquired the project now universally known as Google Maps (Google Earth was launched in 2001), which made the company one of the largest global players both economically and, as some suggest and fear, politically. The story reflects on the power of represented information to slowly replace subjective experience and even shape new realities, such as a pipeline.

Kaufmann clearly has power based on the information he has gathered, but he is also impoverished in his experience of the places he visits to collect this information. He considers them mere "backdrops". For instance, after picking up his rented car, he proceeds to drive through the city:

> The windows of the glitzy international stores reflected cheerful scenes of the early morning street; people read newspapers and talked on mobile phones. A man in a wheelchair was

> affixing five different posters of Atatürk to the front of an abandoned shop, with a handwritten sign saying "TL 1.5m". The single-mindedness that Klaus's vocation demanded left no time for lingering in the places he visited, but he liked the sense of momentum he got from carrying out his purposes against so many different backdrops (2005, p. 95).

Klaus does not care about the places he visits and maps, and usually does not give them a second thought. In a reference to theatrical scenography, they are merely "backdrops" to his life, not a reality to be encountered and felt. Furthermore, he does not even look at the street scene directly, his view is mediated by the reflection on the glass panes in the stores' windows.[4] This anticipates the design of his house, which, as the reader learns later on, is built with large flat screens replacing all the windows and constantly projecting scenes from beautiful places around the world (2005, p. 103), ensuring that Klaus does not have a direct view of the banal reality outside his mansion's gates. Not only is he cut off from directly perceiving the material world with his sense of vision, he is also inexperienced in handling temperature differences—for he almost exclusively approaches the world via screens, be it computer screens or that of his car's navigation device: "Next to the 21 on his dashboard was a 48 coming from outside. He realized he had not paid the journey enough thought, had not so much as put a bottle of water in the car" (2005, p. 96). His car broke down in the middle of Anatolian wilderness—close to where the new pipeline would be constructed—the literal Baudrillardian "desert of the real" (Baudrillard, 1994, p. 1).

When he comes out of the car, all his mediated information was useless and he was at the mercy of the elements: "The sun mocked his thick hair and pummeled his scalp, while gusts of dusty wind sucked mercilessly at the moisture in his body" (2005, p. 97). After some time outside, he collapses. An old woman and her daughter, who live in a cave, nurse him back into consciousness. The old woman attributes his thoughtlessness to his city provenience: "You're from the city, I suppose. You people do crazy things. Think you can just go wandering out. I don't think they teach you how your body works" (2005, p. 98). The city, usually the seat of newest technologies that serve to make life more comfortable and less helplessly exposed to nature, protects its dwellers from the rougher experiences of the elements but it also alienates them from an intuitive survival instinct and a feel for the body outside the comforts of the polis. The old lady mocks Klaus, "You are a joker, German man! [...] Did your map not tell you that summers are hot in Anatolia? That the land is without water or shade? [...] You make me laugh my pink friend. A mapmaker!" (2004, p. 100). Innocently, she asks him to take her daughter Deniz's hand in mar-

---

[4]The motif of the city as a locus of representation and dream-like effects also comes up in the second story in Dasgupta's novel/story anthology "The Memory Editor", set in London. The characters "drove down from Islington in the car, crossing over Blackfriars Bridge from where the floodlights on St. Paul's Cathedral made it look like a magnificent dead effigy of itself. The restaurant was a floating cocoon of leather and stainless steel with lighting like caresses, and their table looked down over the row of corporate places that lined the other side of the Thames". (2005, p. 29).

riage as a thank you for saving his life. Klaus, in his blasé metropolitan manner thinks it impossible that the girl would ever find her way to Frankfurt without any documents; he half-agrees in a mumbled assent and goes home.

When Deniz shows up at his doorstep, Klaus is at first dumbfounded, then he does his best to avoid her, while having his invisible house help take care of her basic needs. In a reference to Bluebeard and his wife (Klaus was listening to an opera rendition of this old tale while driving through Turkish wilderness), he forbids her to enter a room in the tower of his mansion. When she of course does enter it, she encounters his map:

> She found herself in the middle of a large circular room with high ceilings, and walls made entirely of frosted glass. Projected onto the glass, from floor to ceiling and around all 360 degrees, was a map of the world, a vast, flickering thing that kept her turning on the spot in wonder (2005, pp. 111–112).

The motif of glass continues: the map is projected on glass screen. Glass, the medium that normally allows transparence and a clear view of reality behind it is transposed with a represented reality. So much so that one is completely immersed in this representation of the world, and the contours of the actual room she finds herself in are blotted out. She approaches the set of binoculars in the center of the room and zooms in on Frankfurt (the way one can zoom on a place on Google Earth on one's computer). The glasses she wears also provide data in text and symbol form transposed over the glass:

> What a majestic place was Frankfurt! She had never imagined that human beings were engaged in so many pursuits! People made things whose purpose she could not even imagine: valves, adhesives, tubes, lenses, pumps, plastic pellets, nylon tiles, rubber seals, steel gauze, springs, washers. They sold houses and prosthetic limbs and elegant clothes. They bought books and money and cars and guitars and other people. [...] And if you did not want to buy whole people you could buy pieces: for there were hearts and kidneys on sale, and livers and slices of skin and corneas and fetuses (2005, p. 113–114).

Deniz has lived a life that is a crass antithesis to Klaus's. Her only access to mediated vision was a little fragment of a mirror she had in her cave. She is completely immersed in the electronic map of Frankfurt. So much so that she starts fantasizing about her future life based on this map—fantasies entirely unchecked by the reality of her *sans papier* existence in Frankfurt and Klaus's complete indifference to her:

> For hours she traced imaginary journeys around Klaus's majestic map, thrilled at the largeness of things and full of anticipation at what a life she would lead in this city she had come to. She made plans and resolutions about what she would have and be (2005, p. 114).

Instead of all that, Klaus finds her an employment as a maid in a shady hotel. It is suggested that the hotel is just a cover up for a more sinister business of organ and embryo trade, and that Deniz's friend, another foreign maid named Klaudia, who was pregnant, was victim to just these illegal operations. When she confronts Klaus and Karl, the person who appears to be Klaus's twin brother, Karl,[5] about Klaudia, they ignore her. In a reference to Fyodor Dostoevsky's *The Meek One,* mentioned by the two men, in which a woman in an unhappy marriage throws herself out of a window, Deniz runs and jumps through one of Klaus's projection screens that serve as his windows. She first thinks:

> *What unspeakable things are piled up at the edges of civilized people's imaginations.* And she ran across the room and leapt through the glass (2005 p. 128).

Again, the glass motif symbolizes mediation, separation from reality via represented reality by the transposition of Klaus's map on the glass in his room in the tower, or by the projection of pretty scenes from around the world in the niches where there are normally windows, providing "backdrops" for his life but not real situatedness. In an act of symbolic protest against this self-enclosed life, Deniz commits suicide by first crashing through the "backdrop" that makes up Klaus's life, the projected video of some far off beautiful place, and into the mercilessly concrete reality beyond the screen, plummeting to her death on the pavement.

Tom McCarthy's novel from *Satin Island* (2015) also commences at an airport, during a glitch in the otherwise perfectly functioning technological everyday. Unlike in Dasgupta's book, the culprit here is not nature. It is a rogue airplane, a private jet flying erratically over European airspace and therefore disabling all other airplanes from passing. The narrator, simply called U—who could be you, like any youtuber today—is one of the passengers stuck at the airport in Turin. Rather than to converse or tell stories, each stranded voyager is engrossed in his or her private little screen, a smartphone or a tablet. U notes: "Around me and my screen, more screens: of other laptops, mobiles, televisions" (2015, p. 6). And: "People who weren't clicking and scrolling their way, like me, through phones and laptops were grazing on the luxury items stacked up all about is" (2015, p. 10). These luxury items were kept behind the polished glass of store windows and U notes how the reflection of televised news played across these surfaces: "[T]he marketplace bomb-aftermath replayed across the pattern of a shawl, oil flowed and reflowed on a watch's face" (ibid.). The reality of a space designed for soliciting maximal consumption merges with that of mediated events from all over the globe.

U is an anthropologist who works for a corporate enterprise named simply the Company. It remains vague what it is that the Company does; it seems to be a consulting firm. This Company landed a deal called the Koob-Sassen Project—and

[5] At the end of the tale, another Klaus doppelgänger named Kurt appears. This suggests that Klaus, Karl and Kurt are unlikely to be triplets, but are themselves the result of the bizarre genetic experiments done at the hotel. This recalls Walter Benjamin's text *The Artwork in the Age of its Technological Reproducibility,* in which he theorizes that the technological reproducibility of art (as in photography and video) has contributed to the loss of "aura", the veneer of irreplaceability of the original (2008, p. 23). Only, in Dasgupta's tale the technological possibility of reproducing i.e. cloning people via genetic engineering has eclipsed the aura of irreplaceability of individual people. This is evident in the blasé way foreign maids' bodies are disposed of (2005, p. 122 and p. 128). Since Dasgupta's tale takes place in Frankfurt, and Benjamin was associated with the Frankfurt School of cultural critique, it is likely that this is a reference to Benjamin's famous essay.

often, simply just the Project—which, too is vague in its substance. It is supposed to be so big and all-encompassing that it is impossible to describe. One aspect of it has something to do with city-planning. The Company "advised cities how to brand and re-brand themselves [...] We dealt [...] in narratives" (2015, p. 14). They consulted other companies, cities, regions in how to sell themselves as meaningful and necessary in a post-everything reality. It is also vague how they proceed, but it is clear that their strength lies in possession of information. Similar to Klaus Kaufmann's map-making headquarters, the Company's offices are equipped with

> floor-to-ceiling glass partitions [...] creating an expansive vista in which sketches, diagrams and other such configurations of precious data, lying face-up on curved tabletops, pinned to walls or drawn on whiteboards or, occasionally (and this made the data seem all the valuable, *fragile* even), on the glass itself [...] (2015, p. 15).

The Company worked with data to create its products—narratives. They seemed to possess arcane knowledge that gave them power—this knowledge was derived from data analytics, sometimes looked upon as modern day alchemy: the Company's headquarters conveyed the impression of "a hermetic zone, a zone of alchemy, a crucible in which whole worlds were in the mix." (ibid.)

> By creating narratives for companies and cities, U's anthropological work at the Company is, as he understands it at first, a means of injecting meaning into the world:
>
> Helping a city council who were thinking of creating parks and plazas but had yet to understand the ethnographic logic driving such an act; laying out for them the history of public (as opposed to private) space, making them grasp what these zones fundamentally embody, what's at play in them from a political and structural and sacred point of view; and doing this in such a way that this whole history is injected back into the squares, sports-fields and playgrounds millions of citizens will then inhabit—same thing (2015, p. 34).

U imagines himself on par with road repair crews, bakers, milk-men and other people whose work is generally hidden, but who ensure the functioning of a city. But they, like himself, as U imagines, are those who "*put* the city there" (ibid.) He thus understands the work of creating and maintaining meaning, a *brand* for a city, its corporate identity as equally important as maintaining its physical shape—its roads, the sustenance for its populace. This idea is in line with some of his boss's slogans, such as: *A city has no 'character'; it is a schizoid headspace, filled with the cacophony of contradiction.* Or, on the end-point of design: *a state in which the world is one hundred per cent synthetic, made by man, for man, according to his desires...* (2015, pp. 42–42). In an attitude of a quaint technocratic utopianism (recalling Le Corbusier's functionalist urbanism), the Company's employees like U understand themselves as quietly and in the background working on "helping draw up blueprints for the future of the world" (2015, p. 45).

The Company's narrative within McCarthy's novel bears remarkable resemblance to IBM's marketing strategy of "smarter cities" in the real world, which was launched in 2008, in the middle of the financial crisis (Söderström, Paasche and Klauser, 2014, p. 311). This strategy, too, is a narrative, a form of storytelling that aims to sell IBM's IT services to cities around the world in order to make them "smarter". As Throgmorton has argued, storytelling or narrative has power to shape realities, and it is becoming increasingly necessary to ask, "who has the power to give meaning to things, to name others, to construct the character of collective identities, to shape the discussion of urban politics [...]?" (2003, p. 132). The answer to this question is increasingly: major corporations—both in the world of McCarthy's novel where the Company consults cities on injecting meaning back into public space by selling narratives, and in the real world as the example of IBM shows.

As Söderström, Paasche and Klauser analyze in "Smart Cities as Corporate Storytelling", this marketing strategy or "story" generally starts with problematizing cities today, diagnosing them as somehow privative and then positioning the company as an (to use Bruno Latour's term) "obligatory passage point", a crucial and indispensable actor in a socio-technical network they are seeking to create, and so tap into a new and lucrative market (Söderström et al. 2014, p. 309–311). This goes as far as to diagnose cities today as "sick" and needing therapy, utopian solutions based on a firm faith in technology and the power of data to deliver answers to all urban problems (ibid., p. 315). This double step is apparent in *Satin Island,* as well, where U's boss, the Company's CEO diagnoses cities today as sick, as "schizoid", a "cacophony of contradictions". His company seeks to iron out these contradictions, create a coherent whole that can be captured in a single catchy narrative, and design a synthetic future "made by man, for man, according to his desires...", thus positioning itself as an "obligatory passage point" for this supposedly bright future, and reaping the profits. U explains the Company's proceedings:

> Certainly, each brief the Company worked on, every pitch we made, involved an invocation of, a genuflection to, the Future: explaining how social media will become the new press-baronage [...] using the Future to confer the seal of truth on these scenarios and assertions, making them absolute and objective simply *by* placing them within this Future: that's how we won contracts (2015, p. 91).

By gesturing toward future technologies and developments, the Company gets rich without providing substantial accounts of concrete steps to be taken now.

U was hired to write a Great Report (another vague, capitalized term in the novel) that is supposed to sum up today's day and age. He procrastinates on it, noting random events and occurrences on the board in his office and seeking to connect the dots with wool threads: random events like a parachutist's death caused by a sabotaged parachute, oil spills, a garbage disposal island. In the end, he realizes that his task is impossible that an overarching meaning eludes him continuously. He imagines the Koob-Sassen Project that his company has landed as something of immense importance, but he is still vague on the details. The way he imagines the project is:

> towers rising in the desert—splendid, ornate constructions, part modern skyscraper, part sultan's palace lifted from *Arabian Nights:* steel and glass columns segueing into vaulted cupolas and stilted arches, tiled *muquarnas,* dwindling minarets that seemed, at their cloud-laced peaks, to shed their own materiality, turn into vapour (2015, pp. 67–68).

The image evokes cities rising in the deserts of the Gulf, like Dubai or Masdar, all the more so with U's mention of how his boss, tellingly named Peyman (the pay man), would be "perched on a balcony, [...] consorting with engineers and princes, architects and sheikhs and viziers" (ibid, p. 69). Peyman would have the overall view of the project, of "an entire metropolis" with the Company's logo—the Tower of Babel—rising (ibid). Meanwhile, U and the rest of the employees and workers, he calls them "ant-like" (ibid, p. 68), would be shuffling around, "delivering instructions they themselves, perhaps, did not quite understand [...] so complex was the logic governing the Project" (ibid). He imagines that upon its realization "all would become clear, to everyone, and ants would see as gods." (ibid). His fantasy is that of a future city that affords the ultimate meaning and the ultimate perspicuity of society and one's place in it, a quasi-religious vision of an all-encompassing gaze.

Instrumental for this purpose was "data itself, its pure, unfiltered content" (2015, p. 73), which would, somehow raw and uninterpreted provide the substance for the Project, its contents and the intelligence needed to realize its sublime vision. The idea behind his Great Report is to gather *all* relevant information on the current age, to "turn it *all* into data" (2015, p. 74). The impossibility of this endeavor dawns on him surprisingly late in the novel. He had become convinced that the Great Report was "unplottable, unframeable [...] *unwriteable*" (2015, p. 126). Too complex, disjointed and disparate were all the elements making up reality today. There was no single perspective from which to view the whole—except maybe that of a machine with an almost infinite memory capacity (but by definition with the corresponding incapacity to process lived experience). U notes that almost every move of almost every single person is already being mapped, not by an anthropologist, but

> by software that tabulates and cross-indexes what we buy with who we know, and what they buy, or like, and with the other objects that are bought or liked by others who we don't know but with whom we cohabit a shared buying – or liking – pattern (2015, p. 133).

He realizes that it "wasn't that the Great Report might be unwritable, but—quite the opposite—that it had *already been written* [...] by a neutral and indifferent binary system [...]" (ibid.).

U's Great Report, which he had intended to title *Satin Island,* after a phrase from a dream he had of an island off the coast of a magnificent metropolis, was a failure. Instead, we have the novel *Satin Island* in front of us. He cannot put together *all* the necessary information to draw conclusions about our age—a work only a computer could do, and it is dubious whether the information on our buying and liking patterns could truly sum up an age, or whether there *is* such a thing as a definable age or epoch we find ourselves in. Furthermore, the kind of knowledge gained by algorithms would hardly be a public good, but property of corporations like the Company in the novel, and Google, Facebook and IBM in the real world. To use the language of the novel: the ants will never see as gods. The references to data as "precious" and "valuable" in the novel already indicates the status information has acquired as the new currency, the new gold, which will not so easily be given up to public use, but is increasingly becoming a new hot political issue. However, rather than doing the work of computers, the aesthetics of McCarthy's novel *shows* the disjointedness of our times, and the difficulty to discern a coherent whole. As an artwork, it puts on display precisely these obstacles to a clear perception, which are—according to Walter Benjamin—synonymous with the advent of modernity, the acceleration and diversification of life-forms, and the role of technology therein.

3.4 Conclusion

To come back to the theses (1) and (2) posed at the beginning of this essay. As argued, technology does modify and influence perception and experience of cities (1), and literary works reflect precisely these changes (2). In the literature of the 19th and early 20th century the effect of industrialization, massive urban growth and the resulting lack of perspicuity of the city was expressed in choppy, efficient, machine like language in literature, or the description of a fragmented and even phantasmagoric perception. The rise of communication media like television in the second half of the twentieth century contributed toward the feeling of phantasmagoria and the blurring of the lines between media and reality. In the twentieth and twenty-first century, the major technological breakthroughs that most affected subjective experience were the internet, and later social media and the correlated focus on data. This development is reflected in Zadie Smith's *Swing Time,* which shows the narrator's utter epistemological reliance on the internet for the navigation of the spaces she finds herself in. Rana Dasgupta's *Tokyo Cancelled* shows how a breakdown in technology that is taken for granted—the cancellation of a flight—can draw out archaic modes of oral storytelling from times less thoroughly permeated with impersonal information. The story of the mapmaker from Frankfurt shows more sinister possible implications of the thirst for information—the replacement of reality with representations. Tom McCarthy's *Satin Island* shows that even newest developments in data analytics for city planning will not necessarily afford a new perspicuity in urban society. While data and algorithms might themselves be neutral and indifferent, their use is a highly political matter and is currently in the hands of giant corporations. This does not mean that one should advocate a retreat into aesthetics and the study of literature—available to all with a library membership—the world of literature need not be an alternative to political engagement for more participative city governance and transparency on the use of data. Rather, the study of literature provides insight into subjective modes of urban experience that cannot be captured by means of quantitative data analytics and can inspire civic engagement in real cities. The poiesis of urban life forms that affect our very experience of cities should not happen behind our backs. Taking literature and the arts seriously is just one way of sharpening one's awareness that the technological production of future cities is also a production of future selves.

References

Alter, R. (2005). *Imagined cities: Urban experience and the language of the novel*. New Haven: Yale University Press.

Baudrillard, J. (1994). *Simulacrum and simulation*. Michigan: University of Michigan Press.

Baudrillard, J. (1983). The ecstasy of communication. In H. Foster (Ed.), *The anti-aesthetic: Essays on postmodern culture*. Seattle: Bay Press.

Benjamin, W. (1969). *Illuminations*. New York: Schocken.

Benjamin, W. (2008). The work of art in the age of its technological reproducibility: Second version (Edmund Jephcott et al., Trans.). In M. W. Jennings et al. (Eds.), *The work of art in the age of its technological reproducibility and other writings on media* (pp. 19–55). Cambridge, MA: Harvard University Press.
Dasgupta, R. (2005). *Tokyo cancelled*. London and New York: Fourth Estate.
Goldsmith, S., & Crawford, S. (2014). *The responsive city: engaging communities through data-smart governance*. San Francisco: Jossey-Bass.
Goody, A. (2011). *Technology, literature and culture*. Cambridge: Polity Press.
Hijazi, I. H., et al. (2016). Geostatistical Analysis for the study of relationships between the emotional responses of urban walkers to urban spaces. *International Journal of E-planning Research, 5*(1), 1–19.
Huemer, W. (2007). Why read literature? The cognitive function of form. In J. Gibson, W. Huemer, & L. Pocci (Eds.), *A sense of the world: essays on fiction, narrative, and knowledge*. New York: Routledge.
McCarthy, T. (2015). *Satin Island*. London: Jonathan Cape.
Müller, O. (2014). *Selbst, Welt, Technik: Eine anthropologische, geistesgeschichtliche und ethische Untersuchung*. Berlin/Boston: Walter de Gruyter.
Noë, A. (2015). *Strange tools: Art and human nature*. New York: Hill and Wang.
Sebald, W. G. (2013). *Austerlitz*. London: Penguin.
Pound, E. (1957). In a station of the metro. In *Personae*. New York: New Directions.
Shaya, G. (2004). The Flaneur, the Badaud, and the making of a mass public in France, circa 1860–1910. *The American Historical Review, 109*(1), 41–77.
Simmel, G. (2002). The metropolis and the mental life. In G. Bridge & S. Watson (Eds.), *The blackwell city reader* (pp. 11–19). Oxford and Malden, MA: Wiley-Blackwell.
Smith, Z. (2018). *Feel free: Occasional essays*. London: Penguin.
Smith, Z. (2016). *Swing time*. London: Hamish Hamilton.
Söderström, O., Paasche, T., & Klauser, F. (2014). Smart cities as corporate storytelling. *City, 18*(3), 307–320.
Sterelny, K. (2010). Minds: extended or scaffolded? *Phenomenology and the Cognitive Sciences, 9*(4), 465–481.
Strauss. D. (2012). Reasons to re-joyce. *The New York Times Sunday Book Review*, December 9.
Terentowicz-Fotyga, U. (2009). Unreal city to city of referents: Urban space in contemporary London novels. *Journal of Narrative Theory: Narrating Cities, 39*(3), 305–329.
Throgmorton, J. A. (2003). Planning as a persuasive storytelling in a global-scale web of relationships. *Planning Theory, 2*(2), 125–151.

Tea Lobo, PhD is an independent researcher. She has worked at the Swiss Federal Institute of Technology (ETH Zurich) and she was a Junior Fellow at the University of Bern. Her current project is on the interfaces of ethics and aesthetics in urban landscapes. She has received a stipend from the Holcim Foundation to work in philosophy of the city. She studied philosophy and English literature at the University of Basel, Fribourg and Harvard. Her PhD, defended at the University of Fribourg, is on interrelations between literature and philosophy, with a special focus on Ludwig Wittgenstein's work.

Chapter 4
Locative Reverb: Artistic Practice, Sound Technology, and the Grammatization of the Listener in the City

EL Putnam

Abstract There are various ways that artists use technology in exploring the relation of sound to the urban environment, which has different impacts on the listener in relation to place. The rising prominence of these works is connected to a broader sonic turn in urban studies and art, underscoring a rising emphasis on the influence of sound on multisensory experience. Using Bernard Stiegler's consideration of technology as *pharmakon* (or the condition of duality in which something is both poison and cure, bringing both benefit and harm), and his definition of technological grammatization, how artistic use of technology mediates the relationship of the urban environment to the listener through sound is studied through a pharmacological approach in order to nuance the possibilities of artistic critical engagement, emphasising how this can include unintended consequences of re-enforcing certain listener behaviours. At the same time, considerations of how artistic repurposing of listening technology can provide new modes of urban engagement are taken into account, where sound offers the impetus for what Brandon LaBelle refers to as sonic agency.

Keywords Sonic ecologies · Sound art · Bernard Stiegler · Pharmakon · Sonic agency

4.1 Introduction

There are various ways that artists use technology in exploring the relation of sound to the urban environment, which has different impacts on the listener in relation to place. The rising prominence of these artworks is connected to a broader sonic turn in urban studies and art, underscoring a gaining emphasis on the influence of sound on multisensory experience. By drawing attention to the significance of the philosophy of technology in artistic practice, specifically Bernard Stiegler's (2013)

E. Putnam (✉)
School of English and Creative Arts, National University Ireland Galway, Galway, Ireland
e-mail: el.putnam@nuigalway.ie

M. Nagenborg et al. (eds.), *Technology and the City*, Philosophy of Engineering and Technology 36, https://doi.org/10.1007/978-3-030-52313-8_4

definition of technology as *pharmakon* (or the condition of duality in which something is both poison and cure, bringing both benefit and harm) and grammatization, an analytic framework of sound art and geography is constructed, which exceeds particular works of art to how listeners engage with the urban milieu as well as how technology frames the act of listening. How artists utilise technology and how this use affords particular actions from users has a significant impact on the listeners' relationship to technology, and subsequently to place, whether through a cartographic interface, as with an online sound map, or an immersive, ephemeral sound attuned mobile experience, as with locative media sound walks. Even though artistic use of sound technology may invite new perceptions of sounds, this critical analysis reveals that creative use of technology does not inherently invite new modes of participation from users. In fact, it is possible for art works to reinforce habits of listening that are complacent in the grammatization of the listener through hegemonic identities, such as private knowledge bodies like Google, and the affirmation of the base map as the primary mode of organising space. In other instances, artistic use of sound and listening technologies can potentially foster what Brandon LaBelle (2017) refers to as sonic agency, allowing for a new means of experiencing public space while cultivating different modes of listening. Beginning with an overview of the sonic turn in urban studies and the fine arts, this chapter provides an analysis of several sound-based cartographic projects, including the sound maps Radio Aporee and LimerickSoundscapes, digital locative media projects by US artist Teri Rueb, and the electrical walks of Christina Kubisch. Throughout this investigation, there is an examination of how the use of technology by artists informs audience participation, how artists contribute to multisensory experiences of city through aesthetic engagement with sound, and how this has broader implications for technological mediation in urban environments.

4.2 Technology as *Pharmakon*

How technology is able to invite control as well as agency is due to its pharmacological qualities. Philosopher Bernard Stiegler (2013) describes how technology, which is not restricted to the digital but encompasses any tool that enhances human activity, as a *pharmakon*, which has the capacity to behave as both poison and cure. These qualities are inherently interconnected and cannot be separated, meaning that a pharmacological approach to the study and use of technology accounts for its harmful and healing attributes, allowing for a more nuanced understanding of how humans engage with technology. Stiegler derives this concept from Jacques Derrida, though it originates with Plato. While Plato focuses on the poisonous qualities of technology in *Phaedrus* (Plato, 1995) where he describes how writing acts as a hindrance to true memory, Stiegler is interested in developing a positive pharmacology by exploring technology's curative properties. Even though Stiegler considers technology as integral to human evolution, he is concerned with how technologies have developed and been utilized starting in the twentieth century. Technologies once used for the sharing of knowledge through the externalisation of memory, such as writing, have become calculation technologies of Gilles Deleuze's control societies (Deleuze, 1992), where there is the control of production and consumption depriving the user of skills and knowledge. In addition, Stiegler describes the interplay of the externalisation of memory and technology as grammatization, which he derives from the linguist Sylvain Auroux who defines grammatization as the "production and discretization of structures" (Stiegler, 2014, p. 55). However, Stiegler notably diverges from Auroux who focused his understanding of grammatization in language. In contrast, Stiegler expands his framework to include bodily movements and the voice in addition to language, which he connects to grammatization through sound and image.

Stiegler's concepts of the *pharmakon* and grammatization allow for a critical exploration of the multifaceted ways that technology impacts users, avoiding sweeping generalisations in the form of condemnations or celebrations. This approach enables an examination of the unintended consequences of an artistic work, which Jethani and Leorke (2013) argue is critical to comprehending the impact of a project after its initial release. Moreover, taking a pharmacological approach to urban sound mapping and locative media affords a means of critically engaging with the paradoxical nature of these projects, which makes concrete an ephemeral quality of sound in the urban environment that can both empower people through active processes of listening while also restraining them through the often overlooked influence of institutional forces associated with digital geographic technology that is increasingly mediating how people use and understand urban space. One way in which this occurs is through web based sound maps.

4.3 Sound Maps

Artist sound maps, or online maps that allow users to listen to recordings plotted on a cartographic interface, have become increasingly popular with the opening up of GPS technology to consumer markets and increased access to cartographic platforms, such as Google Maps API mashups. Through sound maps, artists have developed means of annotating and connoting maps that offer different means of engaging with a place. While artistic engagement with maps is not a novel innovation, as there is a long history of artists challenging institutionally inscribed readings of place through the manipulation of maps and the presentation of alternative mappings (Wood, 2006), changes to mapping technology have led to an increase in artistic projects that involve collective connotation, either by a selected group or open to the general public. One of the largest sound map databases is hosted by Radio Aporee (http://aporee.org/), started by Udo Noll in 2006. The project posits itself as "a global soundmap dedicated to phonography, field recording (and related practices) and the art of listening. it [sic] connects sound recordings and places, in order to create a sonic cartography, open to the public as a collaborative project" (Noll, 2015, np). Using the Google Maps interface, the database currently hosts sounds from users all over the globe. Users are welcome to contribute recordings by simply uploading sounds to the site. Open access in terms of contribution and consumption, this net-art project exemplifies how user connotation can support the development of sound maps, exceeding the authorial parameters of a single artist. The open framework for contribution is characteristic of the grassroots-inspired ethos of web-based works, where open participation and sharing are encouraged (Cornell & Halter, 2015). Such an extensive archive celebrates the value of collective creation made possible on a mass scale with the advent of social media networks and the release of Google Maps API for free consumer use.

The use of sound as a means of mapping a place is notable as it attempts to tap into the sonic ecology of a place. As a creative medium, sound maintains an ephemeral quality. To experience sound requires listening in the moment. Sound's capacity to reconsider the public sphere is significant to its role in mapping as it posits sound as challenging the predominance of the visual in geography and cartography. At the same time, sound maps are part of a wider sonic turn in urban studies, where greater attention is paid to the acoustic characteristics of the city.

4.4 The Sonic Turn

For the past few decades, sound has increasingly been gaining attention in relation to the study and planning of urban spaces (Adams et al., 2006; Blesser & Salter, 2009; Bull, 2000; Wissmann, 2014). While there has been great attention to the visual and movement within urban design, sound has been garnering more notice. Acknowledging the qualities of sound in how a person experiences a city does not preclude attention to other senses, but as the following analysis of sound maps, locative media, and sound walks show, sound is integral to the multisensory characteristics of the present-day urban environment, working in conjunction with other senses, such as the visual and proprioception, to inform perceptions of the city. Even though this "sonic turn" (Dillane, Langlois, Power, & Bhriain, 2015) is relatively recent, the shifting nature of urban soundscapes is affiliated with the rise of the modern city in the nineteenth and twenteeth centuries as industrial sounds and noises abounded (Thompson, 2004). The sonic experience of the city is complex and varied, with the fleeting qualities of sound itself making it challenging to capture. Rowland Atkinson describes the interplay of sound and its ability to shape urban experience as sonic ecologies that consist of "a permeable, modulating, fleeting and occasionally persistent soundscape within and across different social and physical sectors of the city. These ecologies fade and grow in intensity and scale according to particular temporal and social sequences" (Atkinson, 2007, p. 1913). Sonic ecologies have connections to acoustic ecology, pioneered by Schafer (1993) and others during the 1970s, which places emphasis on acute engagement with a sound environment. When describing sonic ecologies, Atkinson highlights how sound not only identifies a place, but demarcates and regulates space that is "regularly contested at both the individual and broader political scales" (Atkinson, 2007, p. 1908). Thus, sound is not just a random by-product of urban experience, but is an unevenly distributed milieu that is both ordered and regulates city life, with social, cultural, political, and economic implications.

Sound modifies and influences human behaviours in ways that are not often recognised, as "sound affects us depending on whether we feel it is out of place, threatening to our well-being, suggestive of unpleasant experience and so on" (Atkinson, 2011, p. 19). At the same time, sound influences urban inhabitants in ways that differentiate from the visual. Michael Bull (2000) draws attention to the physical qualities of sound that enables the spacing and placing of experience through relational connection. Along similar lines, Michael Gallagher argues that sound itself functions as a kind of affect, "an oscillating difference, an intensity that moves bodies, a vibration physically pushing and pulling their material fabric" (Gallagher, 2016, p. 43). That is, sound's affective capacity is connected to its vibratory nature, or the "base level" of sound, with the human body serving as just another type of body through which sound oscillates.

Moreover, engaging with sound is not limited to just considering the physical auditory environment, but also as a way of making sense of that environment (Thompson, 2004). While sound itself may be ephemeral, the ordering of sonic ecologies emerges from their persistence and connection to place, whether they encompass the sounds of the city or attempts to subdue such noises through infrastructural and personal intervention, which can be technological. As Bull notes, the "technologies of sound … affect our relationship to the spaces we inhabit" (Bull, 2000, p. 9). Thus, sound may also be used as an explicit means of ordering space, underscoring the relationship between sound, space, and power (Gallagher, 2016).

Whether acoustic stimuli is considered sound or noise can also influence its relationship to place. According to philosopher and musician Paul Hegarty (2007), noise is not an objective term, but constitutes a judgement about sound that is a subjective assessment. In addition, Atkinson (2007, 2011) emphasises how in the context of the city, noise can take on political dimensions and even connotations of prejudice as attempts to mediate "noise" varies depending on class, cultural difference, and social clout. Urban noise abatement policies tend to focus solely on decibel levels, the quantifiable aspect of sound, while not considering how certain loud sounds contribute to the unique soundscape of a place (Adams et al., 2006).

Just as the difference between "sound" and "noise" can influence how a person responds to acoustic stimuli, there is a difference between how a person engages with it whether through "hearing" or "listening." Stephen Handel describes how "hearing" can be considered a passive process that involves the physicality of acoustic reception, while "listening" is active: "it allows age, experience, expectation, and expertise to influence perception" (Handel, 1993, p. 3). Listening, therefore, is a form of subjective perception that mediates and judges sounds, which may vary from person to person. How someone listens will depend on what sounds they are exposed to and how their judgement of sound is informed culturally and socially. The sounds a person is familiar with and the sounds that become part of an acoustic background, along with the corporeal qualities of how a person listens (i.e. does a person have sensitivity to certain sounds, use a hearing device, or have impaired hearing?) will contribute to their judgement as to what sounds are appreciated or considered unpleasant noise. This judgement can change depending on exposure. Sounds maps, like Radio Aporee, enable people to become more attuned to a soundscape through listening as they engage actively with the acoustics of a place and its distinctive sonic qualities.

The rise of artistic sound maps, like Radio Aporee, in the early twenty-first century is part of broader trend concerning aesthetic appreciation of sound, as popularised through sound art, which places sensory emphasis on the sonic and listening in fine art as opposed to the domination of the visual. Sound art emerged as an art form during the twentieth century, corresponding (though not always directly) with the increased attention to sound in urban environments. LaBelle (2015) emphasises how sound has gained recognition as an aesthetic category since the 1950s, when John Cage's experimental compositions that encouraged listening to everyday and environmental noise highlighted how sound can be treated as distinctive from music. LaBelle describes sound art as a practice that "harnesses, describes, analyses, performs, and interrogates the condition of sound and the processes by which it operates" (LaBelle, 2015, p. xi). LaBelle argues that as a result, sound has transgressive properties inherent to the medium: "sound works to unsettle and exceed areas of visibility by relating us to the unseen, the non-represented or the not-yet-apparent" (LaBelle, 2015, p. 2). This rising aesthetic appreciation for sound, which includes non-musical everyday sounds, is changing how people listen. Toby Butler points out how the advent of sound art has included a "movement to record and celebrate more contemporary, everyday sound, which has been hitherto overlooked or even resented as 'noise'" (Butler, 2006, p. 892). That is, by inviting audience members to actively listen to sounds that may have otherwise been experienced passively or with frustration, listeners are able to cultivate a different appreciation for an environment. Sonic artistic practices only became possible, however, due to innovations in sound recording, playback, and listening technologies.

4.5 Sound Technologies and Cartographic Interfaces

With the advent of recording and playback technologies in the late nineteenth and early twentieth century, sound could now be captured and re-played, revolutionising how people engage with sound personally, socially, and culturally. Ground breaking recording technological innovations of the twentieth century include the invention of the gramophone, the introduction of microphones, the advent of electromagnetic tape, the dispersal of consumer recording equipment, and the digital revolution of the twenty-first century. Through these technological advancements, the ability to record and play back sound has changed human engagement with this fleeting form, or what Mark Katz refers to as the "phonograph effect" that arises from the "interaction of three equally important and mutually influencing agents of change: the technology, the users of the technology, and society" (Katz, 2010, p. 16). LaBelle describes the relationship of recording to sound's ephemeral nature and the bodies of listeners as follows:

> To record sound, trap it on media ready for amplification, diffusion, and distribution, through systems of transport and broadcast, is to toy with the present, undo origin, and realign memory. It is also to turn sound into object, giving it weight and mass, added strength and force, a figure haunting through its continual reappearance the bodily real (LaBelle, 2015, p. 24).

While sounds can be recorded, and therefore result in identical playbacks, each listening experience may vary. As Gallagher notes, "sound constantly unfolds in difference, confounding expectations" (Gallagher, 2016, p. 45).

While the recording of sound dissociates it from its original context of presentation, the sound map repositions the recorded sound in relation to a geographic environment through cartographic inscription; transposing the sonic ecology of a place onto its visual representation. Drawing from Atkinson's description of sonic ecologies as the simultaneous nature of being in flux, yet situated in a place, sound has the ability to structure a space through the listener's experience on individual and collective levels. In this sense, there is already a process of ordering associated with listening in the city; a process that is concretised through the production of sound maps like Radio Aporee. The question of grammatization in relation to sound mapping, therefore, concerns how sound is used to condition the behaviours of listening and how these listeners subsequently engage with a place. Therefore, the grammatization of the city in relation to sound maps involves not just concretization of the fleeting ambiance of the urban milieu, but also the listener in relation to this environment.

With a web based sound map such as Radio Aporee, users upload field recordings to database, which is then connected to particular geographic locations on a map interface. Salomé Voegelin describes how sounds are inherently invisible, but may have visual counterparts, such as scores (Voegelin, 2010, loc. 40). The sound map provides visual form to sounds—making fleeting moments of urban experience (which is what a field recording captures) and makes them concrete through sonic geographies, while also emphasising the multisensory aspects of the city by bringing together the sound with the visual that a user navigates through the map's interface. The field recording is a document of a passing moment and the listener becomes attuned to qualities of an environment through sensory immersion. In that sense, the field recording is a document of a place, but it maintains an indexical relationship to the environment that is experiential while retaining an invisibility. A sound map makes this invisibility visible. Voegelin emphasises how listening is contextual, but the context is fleeting (Voegelin, 2010, loc. 30). The map interface, however, pinpoints recordings to a particular location, countering the ephemerality of sound. Through the creation of a database and the visual placement of sounds, the cartographic interface restrains the variability of listening, as moments of experience become concrete identifiers of a place. Sounds are abstracted from the context of creation, but then are associated with a particular visualisation of space.

In terms of a pharmacological analysis, sound maps can enhance appreciation of sounds through the cultivation of active listening, which is the curative aspect of the pharmakon, while at the same time reinforce listening habits that homogenise perception, which is a poisonous aspect of the pharmakon. Stiegler argues that digital discretisation of movements, which includes listening, allows for their "ordering, treatment, calculation and imitation," thus leading to an illusion of individuality through the loss of individuation (Stiegler, 2014, p. 68). Therefore, there is a grammatization of place and the listener in relation to place where sounds becomes pinned to particular locations: the fleeting urban ambiance are made concrete through the discretisation of Google Maps cultivating homogenous shared listening. These processes impact a person's relationship to sound, but also to place, emphasising a unified listening experience of that place.

Through Radio Aporee's creation of a database, the knowledge of listeners is categorised, but also shared. The concern is not the centralisation and sharing of knowledge per say, but as Stiegler (2014) points out, grammatization makes the sharing of knowledge possible as well as inviting control. Martin Crowley notes in response to Stiegler: "digital media, which might offer new creative possibilities are 'perverted' by being deployed massively to stifle differentiation and favor standardization" (Crowley, 2013, p. 124). Stephen McQuire (2017) offers such an analysis of Google Street View, where the collection of Google's image database coordinates an experience of the city through its imagery. Even though the user of Google Street View can develop their own path, this "user choice" is restricted by what the database affords. While Radio Aporee's sound database lacks the seeming seamless, integrated editing associated with Google Street View, the unacknowledged restrictions of user engagement are pertinent, which can perpetuate homogenisation in listening. Therefore, this critique of Radio Aporee is not concerning the collective sharing of field recordings through the connotation of a map per say, but how the interface of the map, as a Google platform, risks perpetuating the growth of Google's hegemonic singular knowledge base while re-inscribing a specific base map. In addition, as users contribute sounds to Radio Aporee's constantly growing database, these sounds are translated into data through Google, with no indication of how they are used beyond the surface of the interface.

Moreover, grammatization not only occurs through the shared listening of sounds, but also through the map interface and the data that is collected through the building of the sound database. As noted, Radio Aporee uses the Google Maps API as its base map. Artists may choose to use Google as their base map due to its ease of use as a seemingly neutral platform and accuracies in geographic locations. Since introducing Google maps in 2005, the company has formed an influential infrastructure of cartographic knowledge that competes with traditional institutions that relied solely on the expertise of professionals, such as the UK Ordinance Survey and the French IGN (Plantin, 2018). Jean-Christophe Plantin points out how this is not to say that experts are absent from Google's infrastructure, but that it relies heavily on its users to contribute cartographic knowledge as they actively produce geographic data, either directly through suggested changes to base map information, such as altering street names, or indirectly through the passive collection of data through use.

One of the key ways that differentiates Google from other cartographic infrastructures is that it is a private corporation driven "by its business of targeted consumer advertising" (Dalton, 2013, p. 267). Basing his analysis in part on the work of Bernard Stiegler, McQuire (2017) emphasises how Google's business model is the organisation of the physical world into data, where common resources, such as urban sounds, are attributed private values. Google creates what he refers to as an operational archive, or an archive that is "constantly 'under development' even as it is being used" (McQuire, 2017, p. 83). Its effectiveness emerges from widespread use, which is the basis of its commercial value, meaning that it functions as a source of information but also as a tool for gathering information. McQuire states: "Google Maps functions primarily as a lure, since value to Google is derived primarily from the data they can accrue through encouraging user transactions" (McQuire, 2017, pp. 84–5). The result is the mass appropriation of the urban environment as a means of extracting value, part of a wider movements of subjecting life to "value extraction through data acquisition" (McQuire, 2017, p. 86). As users upload media and listen to Radio Aporee through the Google Maps interface, they are contributing to its growing operational archive, which in turn is transforming the sounds of experience into data that have potential commercial value.

Google Maps encompasses a single body of knowledge that is algorithmically customized for our use. The perceived openness of its participatory engagement makes it more accessible to users, allowing for more diversified and individualised commercial approach, while also lending to its appeal for use in creative projects. Jethani and Leorke (2013) emphasize the influence of institutional and commercial influences in the development and impact of such projects, which can have unintended consequences. The omnipresence of Google Maps and transferable qualities of the interface to different uses through APIs do enhance accessibility of works, since the familiar format can open them up to new audiences who otherwise may not engage with sound art or field recordings. While the crowdsourcing model used in Radio Aporee have facilitated what Stiegler refers to as "bottom up" processes (Stiegler, 2012), this data is exploitable by hegemonic entities like Google. That is, the grammatization affiliated with mapping does not just involve how geographic data is categorised, structured, and shared, but influences how bodies relate to space and move within a place, and how they can be directed to do so through technological interfaces.

In addition, introducing connotations to an already existing map does not critically engage with the base map being utilised. In her analysis of collaborative mapping projects that arose during the first decade of the twenty-first century, such as *Urban Tapestries* (2004), Alison Sant (2006) describes how many of these projects replicate the datasets inherent to these cartographic interfaces. She defines how the base map consists of a "purely geographic categorisation of urban space, defined by the Cartesian coordinates, the road system, and the block plan" that perpetuate a "singular notion of urban space" (Sant, 2006, np). As the base map is replicated through interfaces such as the Google Maps API, this particular ordering of place is unquestionably replicated through cartographic rendering, thereby limiting the extent to which connoted maps contribute to the reimagining of place.

4.6 LimerickSoundscapes and the Critical (Listening) Citizen

It is important to note that grammatization extends to how sounds are collected, as this can influence the cultivation of listening behaviours. While Radio Aporee depends on an crowdsourcing model where users are invited to add sounds to its data base, LimerickSoundscapes (2014) emerged as a project that developed a sound map of the Irish city of Limerick produced through volunteer collectors who were trained and supported by the project's coordinators. Collectors were from local groups, including Active Retired Citizens, Headway Ireland (a community group for people with brain injuries), Doras Lumini (a group that supports asylum seekers and migrants that are new to Limerick), and other community and parish groups, which incorporated a diverse population of participants from different religious, socio-economic, age, and ethnic affiliations (Dillane & Langlois, 2015). Many participants were new to sound recording, and especially field recording that carries its own challenges of capturing sound in an uncontrolled environment (in contrast to the controlled space of a sound studio or sound booth). They were provided with digital recorders, high quality wind-jammers, and headphones along with being trained in how to use this equipment in ongoing dialogues and face-to-face workshops in order to ensure that recordings were of appropriate quality and to cultivate relations between facilitators and participants. In addition, workshops considered the ethical implications of field recording, which is absent or unacknowledged in other sound maps, including Radio Aporee. The collected material was collated and presented on an online sound map that utilises a Google Maps API.

From its inception, emphasis was placed on community involvement, or what the key members of the project Dillane and Langlois (2015) refer to as "critical citizenship." The aim of LimerickSoundscapes was not just to engage with the city through sound or to create a digital map of Limerick, but to treat the act of collecting sound and engaging with participants "as a means of mobilising civic engagement, in order to create a participatory and creative citizenship for the diverse array of people living in cities" (Dillane & Langlois, 2015, p. 136). The citizen in this context is defined as a dweller of a city who has a vested interest in that place (Dillane et al., 2015). Thus the project co-ordinators accentuate the importance of developing a methodology for teaching people to engage in field recording as perhaps even more vital to the project than the output of the sound map. Placing emphasis on engaging with and training collectors through community groups foster relations between participants that exceed the acts of recording and mapping sound while highlighting the place-particular qualities of the project.

One way the LimerickSoundscape differs from Radio Aporee is that the collaborative process of production is not limited to the final sound map presented on the Internet. However, as with Radio Aporee, LimerickSoundscapes remains reliant on its base map, provided through Google Maps, perpetuating some of the concerns related to use of these maps noted above. While there is an interactive layer of engagement introduced, the base map remains unchanged. This outcome of the project does not negate the emphasis that LimerickSoundscapes places on teaching its collectors how to record audio. Participants become sonically attuned to Limerick in distinctive ways, where the use of recording technology is placed in a wider discursive and practical framework, cultivating a pedagogy of listening through workshops and training, as opposed to relying on the technology to perform the task of listening. In order to take a pharmacological approach in analysis, though, requires acknowledging the limitations of the project in relation to its technological platforms, noting that the poisonous qualities of a technology are not in a zero-sum relation with the curative properties. Even through the use of dynamically collaborative methodologies and the development of a collective pedagogy of listening, certain un-noted restraints of digital technology are still present. In this context, the LimerickSoundscape alters how the city is presented cartographically, though the grammatization of the listener and the city through the final digital map interface remains unchallenged.

4.7 Locative Media

With the aforementioned sound maps, users are restricted to the stationary experience of engagement on a desktop computer in a different context from which the sounds were originally recorded. The possibilities for sound artists changed, however, with the advent of locative media. Locative media has become increasingly popular over the past decade as a means of integrating digital art into the physical environment, beyond the context of the gallery and the computer screen. The phrase "locative media" is attributed to Karlis Kalnins, who used it to differentiate digital art that is spatially localized from net-art, or art that can be experienced on a computer desktop, and was coined in 2002 during a workshop at the Latvian electronic art and media centre RIXC (Tuters & Varnelis, 2006). The rise of locative media is attributed to the increased ubiquity of GPS due to US President Bill Clinton's lifting of encryption on GPS signals in 2000 (Holmes, 2003) and the increased popular access of consumer products that utilise GPS technology, which initially included portable computer tablets and PDAs. The invention and dissemination of the Smartphone in 2007 further increased access to GPS (Frith, 2015), allowing both artists and audience members to more easily engage with GPS based projects through apps. According to Marc Tuters and Kazys Varnelis, its practitioners "have claimed an avant-garde position, insisting not only that their work is capable of creating a paradigmatic shift in the art world, but also that it can reconfigure our everyday life as well by renewing our sense of place in the world" (Tuters & Varnelis, 2006, p. 358). With utopian aspirations, locative media artists find ways to alter perceptions of geographies in situ.

US artist Teri Rueb has been creating locative media sound walks since 1999, utilizing GPS in order to play particular sounds and sonic compositions when a listener reaches a specific location. Examples of these works include *Trace* (1999), *itinerant* (2005), *Core Sample* (2007), and *Other Order* (2014). Her sound walks tend to emerge from "urban wilds," or natural zones in urban centres. She regularly incorporates the practice of walking into the manifestation of the work; a combination of audience mobility and sonic attunement referred to as sound walks. The sound walk has gained popularity as a particular form of sound art, which emphasises the movements of a listener in a specific geographic location. Generally, sound walks include an audio component that a person listens to through headphones while in situ. One notable quality of the sound walk is that along with taking advantage of the visual and sonic qualities of the city, there is also engagement with proprioception. The listener is immersed in the environment of the sound work, opening up potential for unpredictability and interactivity with the geographic site as experienced through the body.

Core Sample (2007) is a GPS enabled, locative media sound walk that Rueb created for Spectacle Island located in Boston Harbour. Now a park land, Spectacle Island has a rich history that includes being the site of a casino and hotels, a horse rendering plant, and city dumps. The current form of the island was shaped with excavated material from Boston's "Big Dig," which was a massive urban reconstruction process, lasting from 1982 to 2007, that involved moving the major highway traversing Boston, interstate 93, underground (LeBlanc, 2007). *Core Sample* provides a cultural topography of Spectacle Island's past and reclamation, as Rueb crafted a soundscape that blends field recordings taken on the island, interviews of former inhabitants, and electro-acoustic sounds and manipulations that shift as the listener changes elevation while walking in the environment. In addition, the listener uses headphones that are designed to allow in ambient noises as they navigate the place, blending the recorded, virtual soundscape with the natural environment.

A significant quality of locative media projects like *Core Sample* is that they rely on the listener being immersed in a particular setting in order to engage in the work. Thus, the listener becomes attuned to the acoustic environment while sound retains its fleeting and dynamic qualities. Voegelin describes how sound influences geographies by introducing "dynamic trajectories of individuals moving, being moved and remaining in place" (Voegelin, 2010, loc. 143). Through her sonic overlay, Rueb connects sound to particular locations of the environment, which shifts and emerges in relation to the listeners' movements. In this sense, there is some directing of the listeners' movements, but these connections of sound to place are not finite. Instead, the listeners' movements inform the unfolding of the work, creating invisible connections of the listener to the environment, as well as to each other.

As noted, when locative media gained momentum during the first decade of the twenty-first century, its practitioners celebrated the ability to extend the aesthetic experience of digital art beyond the computer desktop and gallery context. Despite the utopian aspirations of proponents of locative media, it is not without its critics. The early debates concerning the artistic forays into digital mapping have tended to focus on its reliance on GPS, a military technology. Some critics, including Andreas Broeckmann (Graham, 2004), Coco Fusco (2004), and Brian Holmes (2003), have questioned the political efficacy of these art practices. For instance, Broeckman emphasises how "'locative' in regards to locative media does not only relate to the user in the space using media to enrich experience of a place, but also the ability to locate or track down anyone wearing such a device" (as quoted in Graham, 2004, np). Thus, attention is drawn to dystopian potential of this technological reliance, focusing on GPS and the roots of this technology in the US military. By easily contributing data through the networks that enable locative media, users are facilitating the ability of institutional network surveillance.

In addition, while artists may have led early experimentation with locative media, this process was overtaken by commercial interests with the advent of the Smartphone in 2007 (McQuire, 2017). Through her practice, Rueb has taken advantage of different technological developments that have taken place since the beginning of the twenty-first century, this includes utilising GPS technology that makes users locatable. *Core Sample* is one sound walk that was created prior to the advent of Smartphones, along with *Trace* (1999) and *itinerant* (2005). These earlier projects were dependent on the listener being provided with the appropriate equipment to engage with the work. When *Trace* was first created for the Yoho National Park in British Columbia, it required listeners to wear a backpack with a computer containing a database of recordings that were enabled with a GPS receiver. The backpack was also fitted with headphones for the listener to use while hiking through the park (Rueb, n.d.). In 2012, Rueb developed an iPhone app that provides access to the project, no longer restricting listeners to specialised equipment, correlating with a broader trend of technology becoming obsolete and being replaced with commercial alternatives (Jethani & Leorke, 2013). The sound walk *Other Order* (2014) that was created for Harvard University's Arnold Arboretum, located in Jamaica Plain, Boston, MA, is native to the Smartphone app. The risk with using commercial technologies, according to Jethani and Leoke, is that the radical ideals of Locative Media, which is heavily influenced by Guy Debord and the mid-century revolutionary artist group the Situationist International (see Zeffiro, 2012), "become a ready-made, prepackaged resource for any number of artists and designers to appropriate and 'recuperate' through their work" (Jethani & Leorke, 2013, p. 497). As indicated, these resources are connected to private industrial structures that profit off of the collection of contributed data while homogenising and restricting the participatory potential of such projects as listeners are restricted by the limits of the database, geospatial data, and the technological affordances of the base map.

In addition to considering the implications of utilising locative media technologies that are potentially locative mediating, it is also important to acknowledge how Rueb plays with memory through the organisation of perception. In *Core Sample*, Rueb uses audio to shift through Spectacle Island's material, cultural, and social histories creating a topographic palimpsest that mimics the landscape of the island. Her sound walks are experienced through headphones, an individual form of sound delivery. In his empirical and theoretical study of personal stereos in the city, Michael Bull (2000) points out how these devices have the potential to empower subjects through the cultivation of an aesthetic construction of urban space using technology. Engaging with the often overlooked act of everyday mobility in the city, Bull considers how listening to personal stereos enable an mediation of the city through communication technology, as it enacts an ordering of the senses in relation to experience. Even though his study focuses on listening to music selected by the listener, his points regarding the interplay of listening to personal audio devices in relation to moving through cities is relevant to sound walks. However, instead of the listener selecting the soundtrack, as in Bull's studies, GPS enabled sound walks such as *Core Sample* are activated through the movements of the listener, though are composed spatially through the relation of sounds to GPS enabled specifications, making it more reactive than interactive. While each listening experience may vary, making it an individualised experience, ultimately the walk is directed by the artist. Also, as with the sound maps discussed above, user choice is restricted by the affordances of the database. In terms of sound walks, this does not just impact how users see or listen to a place, but also how they move through it, contributing to the grammatization of the listener in the city.

4.8 Electrical Walks: Transforming the Base Map

While Rueb's projects invite different ways of multisensory perception in relation to listening, these are still constrained through the technology she uses. Perhaps the most effective way to adopt technology to radically alter listening, therefore, is to not engage with it directly. German artist Christina Kubisch's *Electrical Walks* provides an alternative approach to artistic mapping as a means of challenging the navigation of urban space while also breaking from reliance on specific base maps. In these works, which began in 2004 in Cologne, Germany, she invites audience members to put on a specially designed set of headphones, equipped with a coil that translates electromagnetic fields into audible noises through induction, which is an analogue process. With this equipment and a paper map that indicates "hot spots" (suggested locations such as ATMs, security systems, tram lines, and other locations that emit strong signals) listeners move throughout the urban environment, allowing sound to function as a guide. The map provides only suggestions for listening, not informing what sounds are to appear in particular locations, as in the sound maps and locative media projects utilising GPS discussed above. To fully appreciate the aesthetic experience of the *Electrical Walks* involves being immersed in the cityscape, exploring the scene with curiosity and wonder. Attuned to a different perception of a place, participants engage in a distinctive manner sonically, visually, and proprioceptively with that particular locale. Their gestures and actions shift in order to more effectively listen to sounds that to passers-by are undetectable. For instance, participants of *Electrical Walks* bend their heads to listen to ATMs and ticket machines—they are adjusting their movements to hear objects that emit high levels of electromagnetic waves, with each movement changing the quality of sound being produced.

The listeners embody these acts of performative mapping as they negotiate the terrain in a novel way. To date, Kubisch has staged over 40 personal and public *Electrical Walks* in different parts of the world, including Germany, Ireland, France, Japan, Russia, Switzerland, Spain, and the United States. According to Seth Kim-Cohen:

"Kubisch and others have discussed this work in terms of revealing a hitherto hidden aspect of the city. The suggestion is that these sounds constitute a kind of secret about—or secret life of—the city" (Kim-Cohen, 2009, p. 110). In other words, in *Electrical Walks,* Kubisch does not direct the listeners where to go, but instead, provides them with specialised equipment and encourages them to roam. She appropriates what can be considered a by-product of many technological devices—electromagnetic radiation—that commonly goes unnoticed, transforming this energy into a new means of tuning into the world, inviting audience members to adjust sonically and proprioceptively to the surrounding environment. The listeners performatively carve new experiential geographies into each urban centre.

Moreover, the *Electrical Walks* invites new, multisensory behaviours of listening, which contrasts with the GPS-enabled projects discussed above. Even though Kubisch's work depends on mobile, solo listeners—a characteristic of today's auditory practices—she encourages them to modify their physical behaviours in a collective, public scenario as these participants complete the works through their embodied presence. It is only through the physicality of the listeners' presence that the *Electrical Walk* is made manifest, engaging in what Seth Kim-Cohen describes as "a slew of concerns having little or nothing to do with the sonic" (Kim-Cohen, 2009, p. 119). Instead, it is the play of sociality, the relations engendered between participants and their surrounding, multi-layered context, and the challenging of the implied movements within the city that constitute the significance of her work: not the sonic properties (though the resulting soundtracks are distinctive and notable compositions created through movement). Kubisch invites participants to engage with familiar environments in a distinctive way, thereby altering the prescribed Cartesian mapping of place. The listeners become producers of an improvised choreography, undermining norms of urban spatial etiquette that in turn create minute sites of resistance. As such, Christina Kubisch's *Electrical Walks* offer a distinctive spin on locative media that challenges particular grammars of place and listening of the sound map and sound walks. Listeners defy the implied performances of the city as ubiquitous electromagnetic fields become the soundtrack and guide of a new means of engaging with the urban environment in a fleeting experience.

LaBelle's definition of sonic agency is useful for considering Kubisch's *Electrical Walks*, particularly in the re-conceptualisation of public space. The headphones utilised by participants in an *Electrical Walk* work by transduction as the electromagnetic waves that exceed the human senses are sonified, thus making the act of listening the predominate means of engaging with place. Sound in this context leads the listener through curiosity, as opposed to the unfolding of narrative, as in Teri Rueb's *Core Sample*. Thus, the agency of the listener is made more apparent, with the capacity of the listener to affect place made manifest through the execution of embodied gestures that challenge the implied performances of that place. Sonic agency in this context functions as a means of engaging attention (LaBelle, 2017), inviting a mode of listening that encompasses new forms relating to place, things, and bodies (both human and non-human), contributing to and reconfiguring the sonic ecology of an environment by tapping into an energy that is not inherently sound-based.

Through the *Electrical Walks*, Kubisch evokes different modes of perception by sonifying what is commonly inaudible. In turn, there is a shift in the performative engagement with a space as bodies attune to the sounds of the environment. The sounds are not pre-recorded, but unfold through perception.[1] While an online-based sound map, like Radio Aporee, has a higher degree of accessibility than the experiences of Kubisch's electrical walks, it is important to not equate accessibility with an inherent potential for instigating changes to how place is experienced. As noted previously, the behaviours invited by users of the Radio Aporee sound map reinforce particular grammatization of the listener in relation to public space, where sounds become affiliated with a particular visualisation of space through the map, listening experiences are shared yet risk homogenisation, and participation is restricted to the affordances of the interface dedicated to extracting value from the acquisition of data. In short, Radio Aporee does not challenge the base map. While Kubisch's walks may immediately impact a smaller group of participants, the actions that are invited through listening allow for a subversion of how public spaces are utilised. These sonic experiences not only challenge how people listen in relation to public space, but also contribute to redefining the base map of a place as listening and walking are constitutive of place.

4.9 Conclusion

While the examples discussed above were created as art projects, what they invite in terms of listening behaviours extend beyond these initial contexts of presentation. Geographer Harriet Hawkins (2012) emphasises the significance of artistic practice in expanding understandings of a place, as art can alter orientation to site, defamiliarize the body's phenomenological experience, and draw attention to materialities that otherwise go unnoticed. Treating art as a "mode of critical exploration" (Hawkins, 2012, p. 53), as opposed to focusing on representational qualities, enables expanded modes of inquiry that exceed aesthetic appreciation. Focusing on acts of listening in the city, these projects offer new modes of acoustic engagement that enhances multisensory experiences of urban environments. Listening is treated as an active process where participants develop or alter judgments of sound through sensory reception, becoming attuned to the specific sonic qualities of a place, but the use of technology by artists may have unforeseen consequences. Taking a pharmacological approach, based on Stiegler's definition of the *pharmakon*, to artistic sound mapping projects in the city explores some of the unacknowledged implications of artists that use mapping and sound technologies. The acknowledgement of these implications is not meant to condemn these works, but to highlight the limitations of technological affordances, databases, and the base map in terms of user

[1] Kubisch has recorded the sounds emitted from the electric walks, though these function as documentation of the art work than the work itself.

participation, which can lead to the homogenization of listening behaviours through standardization resulting in the grammatization of the listener in the city. However, as the examination of Kubisch's *Electrical Walks* illustrates, sound can also invite new forms of agency through listening. These implications extend beyond the artistic works discussed to raise questions and insights of what it means to listen to a place and how the experience of listening can be mediated or dramatically altered.

References

Adams, M., Cox, T., Moore, G., Croxford, B., Refaee, M., & Sharples, S. (2006). Sustainable soundscapes: Noise policy and the urban experience. *Urban Studies, 43*(13), 2385–2398.

Atkinson, R. (2007). Ecology of sound: The sonic order of urban space. *Urban Studies, 44*(10), 1905–1917.

Atkinson, R. (2011). Ears have walls: Thoughts on the listening body in urban space. *Aether: The Journal of Media Geography, VII*, 12–26.

Blesser, B., & Salter, L.-R. (2009). *Spaces speak, are you listening?: Experiencing aural architecture*. Cambridge, MA: MIT Press.

Bull, M. (2000). *Sounding out the City: Personal stereos and the management of everyday life*. New York: Bloomsbury Academic.

Butler, T. (2006). A walk of art: The potential of the sound walk as practice in cultural geography. *Social & Cultural Geography, 7*(6), 889–908.

Cornell, L., & Halter, E. (Eds.). (2015). *Mass effect: Art and the internet in the twenty-first century*. Cambridge, MA and London: MIT Press.

Crowley, M. (2013). The artist and the amateur, from misery to invention. In C. Howells & G. Moore (Eds.), *Stiegler and technics* (pp. 119–134). Edinburgh: Edinburgh University Press.

Dalton, C. (2013). Sovereigns, spooks, and hackers: An early history of Google geo services and map mashups. *Cartographica, 48*(4), 261–274.

Deleuze, G. (1992). Postscript on the societies of control. *October, 59*(Winter), 3–7.

Dillane, A., & Langlois, T. (2015). Our sounds, our city: Urban soundscapes, critical citizenship and the "LimerickSoundscapes" project. *Journal of Urban Cultural Studies, 2*(1 & 2), 135–150.

Dillane, A., Langlois, T., Power, M. J., & Bhriain, O. N. (2015). Urban soundscapes and critical citizenship: Explorations in activating a 'sonic turn' in urban cultural studies. *Journal of Urban Cultural Studies, 2*(1–2), 89–105.

Frith, J. (2015). *Smartphones as locative media*. Cambridge and Malden, MA: Polity Press.

Fusco, C. (2004, December 16). Questioning the frame. *In These Times*. Retrieved February 23, 2018, from http://www.inthesetimes.com/site/main/article/1750/

Gallagher, M. (2016). Sound as affect: Difference, power and spatiality. *Emotion, Space and Society, 20*, 42–48.

Graham, B. (2004, May 20). Exhibiting locative media: (CRUMB discussion postings). *Metamute*. Retrieved March 9, 2018, from http://www.metamute.org/editorial/articles/exhibiting-locative-media-crumb-discussion-postings

Handel, S. (1993). *Listening: An introduction to the perception of auditory events*. Cambridge MA: MIT Press.

Hawkins, H. (2012). Geography and art. An expanded field: Site, the body and practice. *Progress in Human Geography, 37*(1), 52–71.

Hegarty, P. (2007). *Noise music: A history*. New York and London: Continuum.

Holmes, B. (2003). Drifting through the grid: Psychogeography and imperial infrastructure. *Springerin*. http://www.springerin.at/dyn/heft_text.php?textid=1523&lang=en

Jethani, S., & Leorke, D. (2013). Ideology, obsolescence and preservation in digital mapping and locative art. *International Communication Gazette, 75*(5–6), 484–501. https://doi.org/10.1177/1748048513491904.
Katz, M. (2010). *Capturing sound: how technology has changed music* (Revised). Berkeley, Los Angeles, and London: University of California Press.
Kim-Cohen, S. (2009). *In the blink of an ear: Toward a non-cochlear sonic art*. New York and London: Continuum.
LaBelle, B. (2015). *Background noise: Perspectives on sound art* (2nd ed.). New York and London: Bloomsbury.
LaBelle, B. (2017). *Sonic agency: Sound and emergent forms of resistance*. Cambridge, MA and London: MIT Press/Goldsmith.
LeBlanc, S. (2007, December 26). *On Dec. 31, it's official: Boston's big dig will be done*. Retrieved March 1, 2018, from http://www.washingtonpost.com/wp-dyn/content/article/2007/12/25/AR2007122500600.html
McQuire, S. (2017). *Geomedia: Networked cities and the future of public space*. Cambridge: Polity Press.
Noll, U. (2015, April 13). *Radio aporee: Maps*. http://aporee.org/maps/info/
Plantin, J.-C. (2018). Google maps as cartographic infrastructure: From participatory mapmaking to database maintenance. *International Journal of Communication, 12*, 489–506.
Plato. (1995). *Phaedrus* (A. Nehamas, Trans.). Indianapolis Indiana: Hackett Publishing.
Rueb, T. (n.d.). *TRACE*. Retrieved March 9, 2018, from http://terirueb.net/trace/
Sant, A. (2006). Reefing the basemap. *Intelligent Agent, 6*(2). Retrieved February 15, 2018, from http://www.intelligentagent.com/archive/Vol6_No2_interactive_city_sant.htm
Schafer, R. M. (1993). *The soundscape: Our sonic environment and the tuning of the world*. Rochester, VT: Destiny Books.
Stiegler, B. (2012). Relational ecology and the digital pharmakon. *Culture Machine, 13*, 1–19.
Stiegler, B. (2013). *What makes life worth living: On pharmacology* (D. Ross, Trans.). Malden, MA: Polity Press.
Stiegler, B. (2014). *Symbolic misery: The hyperindustrial epoch* (B. Norman, Trans.). Malden, MA: Polity Press.
Thompson, E. A. (2004). *The soundscape of modernity: Architectural acoustics and the culture of listening in America, 1900-1933*. Cambridge, MA: MIT Press.
Tuters, M., & Varnelis, K. (2006). Beyond locative media: Giving shape to the internet of things. *Leonardo, 39*(4), 357–363.
Voegelin, S. (2010). *Listening to noise and silence: Towards a philosophy of sound art*. London: Bloomsbury.
Wissmann, D. T. (2014). *Geographies of urban sound*. Farnham: Ashgate Publishing.
Wood, D. (2006). Map art. *Cartographic Perspectives, 53*, 5–14.
Zeffiro, A. (2012). A location of one's own: A genealogy of locative media. *Convergence, 18*(3), 249–266. https://doi.org/10.1177/1354856512441148.

El Putnam is Lecturer in Digital Media at National University Ireland Galway where she is director of the MA in Digital Media. She is an artist-philosopher writer working predominately in performance art and digital technologies. Her work focuses on borders and entanglements of gesture, particularly the interplay of the corporeal with the machinic. Her forthcoming book, *Strange Mothers: the Maternal, Digital Subjectivity, and the Aesthetics of Interruption*, is being published with Bloomsbury. She is member of the Mobius Artists Group (Boston) and holds a PhD from the Institute for Doctoral Studies in the Visual Arts.

Chapter 5
Theorizing Technologically Mediated Policing in Smart Cities: An Ethnographic Approach to Sensing Infrastructures in Security Practices

Vlad Niculescu-Dincă

Abstract Smart digital infrastructures predicated on myriads of sensors distributed in the environment are often rendered as key to contemporary urban security governance to detect risky or suspicious entities before or during a criminal event takes place. At the same time, they often involve surveillance of urban environments, and thus not only criminals but also large groups of people and entities unrelated to criminal phenomena can end up under close inspection.

This chapter makes its contribution on two levels. For one, it offers a theoretical framework to the research and conceptualization of the role of sensing infrastructures in urban security practices. It shows how insights from Philosophy of Technology and Science and Technology Studies can produce a nuanced understanding of the role of digital technologies in security practices, beyond standard conceptualizations of technology. Moreover, the chapter proposes a geological approach to enrich our repertoire of imagining and researching smart urban ecosystems.

Secondly, the chapter contributes to a higher level of transparency of these practices by presenting the results of ethnographic research performed in a set of police organizations that employ sensing infrastructures and algorithmic profiling in their practices. The chapter draws empirically on research performed in the Dutch police, both at municipal and national levels with some additional material gathered in a constabulary in England. In these organizations, resource allocation decisions are often predicated on automated number plate recognition technology that processes data from an array of smart cameras distributed in the environment. In these ways together, the chapter highlights a set of normative issues with implications for the effectiveness and legitimacy of urban security (surveillance) practices in smart environments.

V. Niculescu-Dincă (✉)
Institute of Security and Global Affairs, Faculty of Governance and Global Affairs, Leiden University, The Hague, The Netherlands
e-mail: v.niculescu-dinca@fgga.leidenuniv.nl

M. Nagenborg et al. (eds.), *Technology and the City*, Philosophy of Engineering and Technology 36, https://doi.org/10.1007/978-3-030-52313-8_5

Keywords Sensing · Infrastructures · Policing · Philosophy of technology · Urban ecosystems

5.1 Introduction

Contemporary urban ecosystems are increasingly pervaded by ubiquitous and networked sensors and devices communicating with each other and with other (online) services. Combining existing information infrastructures with a stream of initiatives and innovations these sensing infrastructures enable the gathering, communication and processing of data about the urban lifeworld. They support visions—referred to with umbrella terms such as Smart Cities and similar terms—that bring the promise of efficient, green, safe and secure cities (Townsend, 2013; Kitchin, 2014).

In the area of urban security governance, digital infrastructures have long been rendered as key enablers of efficient policing practices. For instance, intelligence-led policing styles promote not only specific tactics and ways of doing the business of policing but are also associated to a significant employment of overt and covert "technical means" (Tilley, 2008; Ratcliffe, 2008). Especially when combined with profiling and algorithms–ranging from rule-based algorithms to machine learning algorithms–digital infrastructures are relied upon in efficient resource allocation and decision-making processes. The role that digital technologies play in contemporary policing has made several authors link them to dramatic changes in the organization of the police (Chan, 2001; Harris, 2007; Byrne & Marx, 2011). At the same time, such solutions often imply and enable surveillance practices. Increasingly practiced by a plethora of public and private actors coming together in security networks, not only at national and international levels but also at local and regional levels (Whelan & Dupont, 2017), surveillance remains a key practice of modern approaches to policing (Leman-Langlois, 2012).

However, many scholars (Ericson & Haggerty, 1997; Maguire, 2000; Neyroud, 2008; Hildebrandt & Koops, 2010) have pointed out that not only criminals but also individuals and groups unrelated to criminal phenomena come under the close inspection of surveillance assemblages (Haggerty & Ericson, 2000) or are being subject to discriminatory decisions (Lyon, 2003; Wittkower, 2017). This is not to say that technology is a priori leading to these outcomes. It has been argued for instance that information technologies can play an important role in minimizing the discriminatory effects of surveillance (Kamiran, Karim, Verwer, & Goudriaan, 2012; Mancuhan & Clifton, 2014), while still enabling an effective policing (Schakel, Rienks, & Ruissen, 2013; Hellemons et al., 2013).

In this respect, empirical studies of the ways in which smart digital infrastructures are engaged in policing are viable ways to inform a higher level of transparency and human rights protection and, at the same time, not compromise the effectiveness of police work. That is, such kind of studies (Sanders, 2006; Van Ooijen & Nouwt, 2009) document technologically mediated practices in urban policing (after the fact and without operational information) and enable more clarity over the ways in which software-enabled technologies work in urban security practices.

However, there is still a scarcity of empirical studies of policing practices mediated by smart infrastructure technologies. Of course, there are at least two valid reasons for this gap in the literature. For one, police organizations require long and often difficult procedures for access to empirical research even if they are more transparent then ever (Manning, 2008). Although notable studies took an ethnographic approach to study technologies and surveillance in policing organizations (Norris & Armstrong, 1999; Manning, 2008), obtaining research access is often a difficult undertaking. Secondly, with a few exceptions with a few exceptions (Meijer & Thaens, 2018a, 2018b) the novelty of technologies themselves and the opacity of algorithms make it such that empirical studies of these technologies in practice do not abound.

Still, unlike many Smart City visions, some of the technologies are already here: radio frequency identification (RFID), automatic number plate recognition (ANPR), Bluetooth, ZigBee, and more. For instance, in policing literature, ANPR cameras are often regarded as one type of sensor employed by police to detect vehicles in traffic and monitor the road infrastructure from large control rooms that aggregate data flows from the environment. ANPR technology can network with other types of sensors to enable the police to also cover 'railways, waterways and cyberspace' (Schakel et al., 2013, p. 12). Many police organizations utilize ANPR in a variety of practices, from finding stolen vehicles and wanted criminals to profiling suspicious traffic behaviour. Therefore, empirically studying ANPR in these kinds of practices as, well as the ways in which the software code is designed, provides an opportunity to understand how urban security governance engages with smart city digital infrastructures.

Moreover, the chapter offers a vocabulary and approach to research and conceptualizes the complex role of sensing technologies in urban security practices. A vast majority of practitioners' guides, technical reports, policing managerial guidelines, promotional materials of technology producers, and a significant body of policing literature take a standard instrumentalist view on technologies as technical means. However, rendering technologies as neutral means tends to ignore their social, political and ethical charge. Remaining in this framing of technology, the conditions remain arbitrary for a more informed process of using and shaping smart infrastructures and their engagement in urban security governance.

The chapter brings insights from Philosophy of Technology and Science and Technology Studies to produce a more nuanced conceptualization of digital technologies in policing practices, beyond instrumentalist views. For one, it shows how an Actor-Network Theory inspired approach (Akrich, 1992; Akrich & Latour, 1992; Bowker & Star, 1999; Latour, 2005; Law, 2008; Mol, 2010) can be employed to 'dig up' and trace the ways in which suspicion and risk sediment in the layers of software code and get enacted in the broader materially-embedded policing practices. Engaging with the conceptual framework of mediation theory (Verbeek, 2011), the chapter reflects on how the practitioners' perceptions, experiences, decisions and actions are mediated by sensing infrastructures and algorithmically-generated entities. In these ways together, the chapter highlights a set of normative issues with implications for the effectiveness and legitimacy of urban security in smart environments.

The chapter is structured as follows. Section 5.2 gives a brief introduction to ANPR and documents policing practices that utilize this technology. Section 5.3 outlines an analytical framework to understand the role of technologies by building on research in philosophy of technology (PoT) and science and technology studies (STS). Section 5.4 shows the active mediation of sensing technologies–from the simplest, most common practice of detecting a priori known vehicles in traffic to the more sophisticated profiling techniques. Section 5.5 'digs' through the code of a set of profiles and renders the profile design process as a locus of reflection for ethical and effectiveness considerations. Section 5.6 concludes these investigations and outlines a set of considerations about the conditions that need to be met for these technologically enabled practices to deliver on their promises of enabling both effective and legitimate security practices.

5.2 Doing Fieldwork on ANPR in Policing

ANPR is a technology enabling the real-time identification of license plates of cars in traffic. Spread in large numbers across the environment or in mobile units that patrol particular areas, ANPR cameras sense the movements of vehicles, transmitting data in real-time to a back office facility. The technology dates from the late '70s and has slowly gained a reputation for an accurate technology with high reliability levels (Parker & Federl, 1996). Since its introduction, ANPR has been deployed in parking lots, entrances to areas, highways, roads, in marked and unmarked vehicles, and in any arrangement that needs to monitor road traffic and identify vehicles.

The most common way of engaging ANPR technology is by comparing number plates in real-time with reference lists. These reference lists of numbers are of particular interest to the organization running an ANPR system: road administrators, tax authorities or police organizations. For instance, monitoring the traffic is done to allow/block passage, to select for further investigation or to arrest. The choices of how many lists to define and how many numbers to include in a list are virtually unlimited. Lists may contain from one number plate up to tens of thousands. If the system detects a match in one of the lists, it returns a so-called 'hit'. The hit is accompanied by additional information on location, date, time, potential intelligence that may prompt further actions, and audible and visual alerts (coloured markers).

Another way to tackle criminal phenomena where the vehicles are not a priori known relies on designing real-time behavioural profiles based on crime models. These models are generalized abstractions of real world criminal phenomena (e.g. burglary, cargo theft, or drug trafficking) and capture various aspects such as locations, times, social networks or modus operandi (Schakel et al., 2013). Because it would be practically impossible to monitor locations all the time, let alone infer behavioural patterns, knowledge rules derived from these models are delegated to automated profiles that select suspicious behaviour based on sensor data and processing algorithms. In this way, the police can detect suspicious behaviour when those vehicles were not included in any police list. It is important to distinguish here between kinds of algorithms. One type of profile is the one generated with machine learning techniques, in which unexpected knowledge is produced, previously not hypothesized (Hildebrandt & Gurtwith, 2008). Another type of profiles are those that are directed, in which the algorithmic rules are previously derived from police knowledge, typically aggregated from different fields of the discipline, and then applied to the live traffic.

The chapter draws empirically on ethnographic research I performed in Dutch police organizations at local and national levels where, at that time, they engaged in the second type of profiling. That is, profiles were built through knowledge sharing between multiple branches of the police. To gain an additional vantage point, I've also studied the use of ANPR in a constabulary in England[1] which, at the time of the study (2012), had one of the densest networks of ANPR cameras in the UK, storing traffic data in persistent databases.

While both police services employed a similar ANPR system, purchased from the same vendor, they differ in the mode of engagement. Whereas the UK constabulary turned off part of its ANPR system in response to sustained criticism of its surveillance potential, the Dutch police started to explore alternative ways of engaging ANPR. That is, instead of storing all traffic data, they only select suspicious behaviour based on knowledge rules delegated to real-time algorithms that process ephemeral data streams, aiming to catch criminals red-handed (Schakel et al., 2013). With this way of engaging with sensing infrastructures, they aim at reducing the surveillance impact for most traffic participants, "protecting people's privacy by design" while simultaneously performing their policing tasks (Hellemons et al., 2013).

In both settings I performed semi-structured interviews with members of the ANPR teams. There were ten in-depth interviews in the Netherlands with multiple police officers involved in ANPR operation and with lead designers involved in building profiles. In England I had three in-depth interviews with the head of the ANPR team, control room operator and the senior analyst of the constabulary. In addition, I made ethnographic observations in these settings for more than 50 h taken together. These included participant observation sessions in control rooms to understand how the system enabled their operative decisions. Additionally, I participated in actions with road policing units to understand the implications of the ways in which officers worked with and talked about these technologies.

5.3 Theoretical Underpinnings

So how do policing practitioners typically talk about these technologies? How do they understand their role in daily policing work? What can philosophy of technology bring? This section first discusses some of the most common conceptions of

[1] To protect the confidentiality of the individual police officers the name of the organizations are anonymized in this study and the names of individual officers were made generic.

technology in policing and it analyses the assumptions, implications, and limitations of these conceptualizations of technology. It then outlines the theoretical framework of the chapter by bringing together insights from the areas of Science and Technologies Studies and Philosophy of Technology and showing how they can complement each other and offer a more nuanced and theoretically informed way of conceptualizing the role of digital technologies in policing practices.

5.3.1 Technology in Policing Work-Talk

Digital technologies have become constitutive of many contemporary professional practices and feature in their daily routines and ways of relating to the broader socio-technical infrastructures in which they feature. It is therefore no surprise that professionals develop vocabularies to come to grips with these entities that affect their practices so profoundly. The same is the case with policing practitioners. As the importance of technologies is unanimously recognized as a defining characteristic of contemporary policing, practitioners and scholars need to conceptualize their new and older technologies.

For many, technologies are 'technical means' (Tilley, 2008, p. 375), 'computer-based instruments' (Innes & Roberts, 2008, p. 250) or 'powerful tools' (Leipnik & Albert, 2003, p. 3). When technology is talked about on the beat, much of police officers' 'work-talk' involves a vocabulary that mainly covers the use, capabilities and powers of particular systems, equipment, machines and devices (Manning, 2008). They are "tools of communication like radios and call boxes, tools of investigation like cameras and fingerprint kits, and tools of surveillance like video and audio pickups. Increasingly, newer, more specific tools have been developed using emerging technologies. These innovations enhance the old tools and make some new tools possible" (Williams & Williams, 2008, p. 165).

This analysis is not meant to criticize police officers or the larger body of work of some of these authors but to point out the influence and limitations of particular conceptions of technology. In this conception, technologies are rendered as passive and obedient in the hands of their users. Tools do not determine human action. They are neither good nor bad, as people use them for their purposes; in this case the crime control purposes of policing practitioners. To illustrate this view we can find again a quote from the policing literature: "Just as technology can prevent or solve a crime, criminals can use technology to commit a crime. Computers can be used to hack into a company's computer system and alter information or transfer documents and falsify transactions" (Williams & Williams, 2008, p. 169). In this framing, technologies are value neutral. Technology does not entail consequences by itself but consequences are only stemming from choices made by the users (Sclove, 1995; Brey, 1998).

5.3.2 Theorizing the Role of Technologies in Policing Practices

So are technologies adequately understood as neutral tools in the hands of policing practitioners? This framing of technology may seem 'common sense' but it carries with it particular assumptions concerning the social role of technologies. To some extent it is understandable to want to think of technologies in policing as just tools in the hands of officers. In an environment pervaded with ever more sophisticated technologies, this conception places the responsibility for the outcomes in the hands of the practitioner. After all, nobody wants a police officer to wrongfully arrest someone or, when failing to catch a criminal, decline responsibility with the claim: 'The technology made me do it'. This conception of technology places responsibility for the outcomes in the hands of the users, however at a cost. Rendering technologies as neutral tools ignores the social, political and ethical dimensions of technologies as it silences their active role in inducing consequences. It lacks an account of the complex interactions between users, designers, organizations, institutions and legal regimes. Consequently, it provides a rather poor conceptual background for understanding the complex arrangements of smart urban environments and technologically-infused policing.

In the past decades, this view began to be contested by a body of work from an array of directions such as philosophy of technology, history of technology and Science and Technology Studies, to name but a few. Rather than assuming technologies as neutral or explicitly trying to get to a substance of modern technology, these studies analyse technology and science as thoroughly social activities. That is, they follow the technologists, scientists and/or users during their work and activities of producing and using knowledge and artefacts. They follow them as members of communities within which they interact with peers and get connected to broader cultural and societal processes, conflicts and debates. These studies delve into empirical investigations to understand how facts, knowledge, classifications, phenomena or artefacts are constructed or performed and what imprint these processes leave on social activities and their outcomes (Pinch & Bijker, 1987; MacKenzie & Wajcman, 1985; Bijker, 1992; Bowker & Star, 1999; Bijker, 2010; Law & Mol, 2001; Law, 2008).

In the STS tradition, Actor-Network Theory (Callon, 1987; Law, 1987; Latour, 1988; Akrich, 1992; Law & Mol, 2001; Law, 2009) is a framework, developed initially by Michel Callon, Bruno Latour and John Law, that emphasizes a symmetrical consideration of humans and of non-humans in analysing processes of construction (non-humans meaning anything else besides human beings, usually referring to technological artefacts but also, for instance, institutions, organizations, etc.).

This kind of analyses have been made in a wide variety of studies and, as a general feature, they treat symmetrically a diverse set of components that span materials, equipment, components, people, and institutions as simultaneously participating in the formation of *networks of relations*. For instance, Michel Callon (1987) uses the term '*engineer-sociologist*' to point out that technologies and visions of society are bounded to each other and neither would happen without the other. Similarly, John

Law (1987) uses the term '*heterogeneous engineering*' to argue that the work of technologists should not be separated as 'engineering' from the changing of society that is involved in introducing a new technology.

Similarly, for Latour (Latour, 1987, 1988, 1991, 2005), reality cannot be properly understood if humans and non-humans are analysed asymmetrically. Actor-Network Theory does not wish to prejudge the relative power or influence of any of the *actants*. The term is often used by Latour to emphasize even more the symmetric consideration of humans and non-humans as the term 'actor' usually suggests human agency. In this sense, he argues, what a thing is and does and also what a human being is and does, stems out of the *networks of relations and associations* that they have with other things and humans rather than from an essence that underlies them (Latour, 1988).

We may extend the argument to aggregates in a machine or to relations between machines and other entities such as standardization bodies, users or maintainers. For instance, a bicycle chain rotates the wheels in relation to a sprocket that matches and to a biker who pushes the pedals. A police officer assesses someone as suspect in relation to a risk profile that automatically generates a hit and in relation to an organization that assigns this task. If we want to understand the role of an automated profile in this process, we may want to make Latour's thought experiment and just imagine how much effort might be needed to obtain the same result without it: how many police officers in the field watching and quickly writing down number plates, how many meetings to discuss the findings, in the hope of finally agreeing on an identified pattern. In this sense, he argues, 'purely technical' artefacts should be understood as highly social and moral (Latour, 1988, p. 298).

5.3.3 *On Technological Mediation*

Therefore, we need a vocabulary that could help us talk and think about technologies in policing practices as actively doing something for and among police officers, organizational structures, legal provisions, inter-institutional arrangements and others without reducing them to neutral tools. We need a vocabulary that comes to grips with the active and nuanced role of technologies in influencing officers' practices, experiences, perceptions, their decisions and their actions.

5.3.3.1 Scripts and Mediation of Action

A contribution of ANT that conceptualizes the ways in which technologies relate to human action comes from Madeleine Akrich and Bruno Latour (Akrich, 1992; Akrich & Latour, 1992). They propose the notion of *script* and the related family of notions (prescription, inscription, subscription, etc.). In this vocabulary, rooted in ANT, rather than talking about technologies as tools, they *mediate* human action in the same manner in which theatre scripts influence the behaviour of actors: they are compelling enough to tell the actors what to do but do not determine how they will eventually perform.

In this understanding, technologies guide, inform, (firmly) suggest the user what to do/not to do with them (e.g. through user manuals, instructions, their shape but also their material properties and constraints of use). For example, 'two-hands' control mechanism require workers to use both hands to start a press in order to prevent work accidents. Weighted hotel keys are compelling guests to return them to reception if they don't wish to carry all day a bulky piece of metal. Speed bumps, or sleeping policeman as they are sometimes called, are compelling drivers to slow down in order not to wreck their cars or hit their heads, and in this way enforcing traffic norms (Latour, 1994).

At the same time, scripts imply scriptwriters. Designers translate their intentions and values, *delegating* them to the functions of artefacts. Technologies can be said to be more then neutral intermediaries but be involved in *translating programs of action*. This is the case when designers anticipate particular users with their behaviour, interests and motives in relation to the artefacts and they build in *preinscriptions*. For instance, designers of control room infrastructures in policing anticipate that officers need to be quickly alerted to certain situations from whatever state they were in, and they design bright and flashy visual alerts to wake up even the most bored officer. However, in the vocabulary of technological mediation, users can also give meanings to artefacts, thus also taking part in the inscription processes. For instance, officers can also choose to remain in the same state without even a casual glance at the alert.

Moreover, artefacts can be *prescriptive* without this originating in a designer's intention. For instance, visually impaired people were often unable to register for websites that implemented mechanisms for stopping 'robot registration' (website users need to introduce some random and distorted characters that are generated in a difficult to read image). In these cases, users are encouraged to behave in particular ways, as the scripts are guiding and altering their behaviour without the need of the designer's presence. Scripts are thus normative, transforming or reinforcing existing 'geographies of responsibilities' (Akrich, 1992), convincingly demonstrating the normative charge of technological artefacts.

While users may *subscribe* to the artefact's suggestions/constraints (e.g. workers may use both their hands to operate a 'two-hands' control, police officers may arrest those suggested by the profiling algorithms), it may also be the case that they define novel roles in relation to technologies, sometimes in opposition to the program of action (e.g. workers may still use only one hand if they really wish. For instance, by using a long stick that reaches both buttons of the control mechanism). Just like with a theatre script, an actor is able to appropriate it, change it and give it new meanings and interpretations. Even if this sometimes requires extra effort, the outcomes can be significantly different from what the designer intended. ANT does not a priori analyse technological artefacts as determining human behaviour and social relations.

In this light, the development of technologies appears as a heterogeneous process in which a whole set of actors bear parts of the responsibility for their effects. Rather than holding a heroic view of the user, fully responsible of the outcomes of technology, but also steering away from deterministic views in which 'we can't do anything about it', the notion of script conceptualizes both users and designers as actively inscribing their worldviews, interests and values in processes of technological development. Recognizing that technological designs and material artefacts are more than mere tools enables the study of interactions between users and technology with a vocabulary that allows more attention to their active mediation.

5.3.3.2 Mediating Perception and Experience

The vocabulary of technological mediation can be further expanded to help us understand more of the complex practices of police officers working with technologies in contemporary policing. Technologies mediate not only action but also human perception, practices and experiences of the world (Verbeek, 2005). What does it mean to 'smell something fishy' when reading indicators on a computer screen? What does it mean when officers say that 'something just didn't feel right' (Hess & Orthmann, 2010, p. 12)? How do police analysts perceive a rise/decrease in criminal phenomena from their computer desk by looking at numbers in a system? How do road policing units experience high-pitched sound alerts and flashy indicators of 'suspect vehicles'? How do mediated perceptions influence decision and action?

Peter-Paul Verbeek expands the vocabulary of technological mediation and shows how it can be applied to many other areas such as industrial design (Verbeek, 2005), engineering ethics (Verbeek, 2006, 2011) or interaction design (Verbeek, 2015). He builds on both ANT and the post-phenomenological approach of Don Ihde (1990). In the post-phenomenological approach technologies and human beings are understood as helping to constitute and shape each other into being. As Verbeek argues this point: "[Technologies] help shape how human beings can be present in the world and how the world can be present for human beings" (Verbeek, 2015, p. 29). For instance, they "help scientists to perceive the world. The reality of a star is profoundly mediated by telescopes, brain activity by MRI scanners, and the health condition of a fetus by ultrasound devices. Such mediations are not merely neutral "intermediaries": What a star, the brain, and an unborn child are for us cannot be understood without taking into account the mediating role of technologies in our perception and understanding of them" (Verbeek, 2015, p. 29).

Both Ihde and Verbeek steer away from approaching technologies in terms of essences or absolute foundations and from locating humans and technologies in two distinct spheres: one of the human subject, the other of the technological object. By understanding them as constituting each other, the approach aims to overcome some of the problems that classical phenomenology has been criticized for—seeking to describe 'reality itself' (Verbeek, 2005) or for "elevating a single person's self-ethnography to grandiose proportions" (Mol, 2010, p. 254). Rather, the approach of Ihde and Verbeek is called 'post' phenomenological as it aims to overcome these problems while still investigating the way in which actual technologies shape human access, experience and interpretation of reality. "Investigations of this type of mediation cannot possibly aim to return to 'the things themselves', but rather aim to clarify the structure of technological mediation and its hermeneutic implications" (Verbeek, 2008, p. 13). Issues of interpretation are central to this approach and its questions concern the mediating role of technologies from a hermeneutical perspective.

5.3.3.3 Types of Relations and Influences

Don Ihde identifies several ways in which humans interact with artefacts from this perspective and proposes a taxonomy to characterize these interactions. One type of relation in which artefacts mediate perception he calls *embodiment*. In this kind of relation technologies become a unity with the human being, so the artefacts is part of the experience. For instance, police officers look *through* binoculars, and not *at* binoculars, to follow someone or they perceive suspicious traffic activities on the streets *through* video surveillance systems.

The second relation in this taxonomy, Ihde calls *hermeneutic* and therefore implying the need for interpretation. In this relation of mediation, we experience the world with the explicit contribution of technology that calls our attention. For instance, we interpret the numbers on a thermometer to understand the temperature outside. A too high value means 'it is very hot'. In a similar manner, police agents read indicators on screens and they interpret their values, pitches or tones. They take some to mean, for instance, that 'the old boys are back in town' while others as being irrelevant or indicating 'business as usual'.

Another type of relation characterizes the interaction with the artefact itself. In this type of relation, that Ihde calls an *alterity* relation, human beings interact with technologies while the world is in the background of their interaction. This may be the case, for instance, when police officers operate a device or when police programmers code the criteria for a risk profile, while the criminal phenomenon still unfolds. In this case, Ihde argues, the world is in the background of the interaction with technologies, while attention is directed at the artefact.

In the *background* relation, technology shapes the relation to reality but without having an active role in the experience. In this type of relation technologies are the context for human experiences. In the policing context we can think of the sound of sirens while police officers approach and discuss with a suspect on the street, the humming noise of computers in a control room while officers analyse indicators on screens, or the warm air of the air-conditioning system of a police car. In these examples technologies shape officers' experience—for instance making them more relaxed, alert, stressed or irritated—but in the background of other activities.

Another type of relation that Verbeek (2011) identifies comes from technologies that merge with our environment. Smart environments are more than a background of human experience but have an *interactive* dimension. They form the context of human experience but can actively emerge from this context and call explicit attention. This kind of technology can actively detect vehicle number plates, recognize faces, sense movement and automatically give signals to police officers when detecting suspicious behaviour.

Finally, wearable technologies result in another kind of relation. Smart glasses, for instance, can be simultaneously embodied, just like normal glasses, but at the same time they can produce representations of remote events and indicators. This relation could be called *augmentation*, combining an embodiment relation and a hermeneutic relation. In a policing context we may see this kind of technology emerging with the rise of wearable technologies but we can already see them if we consider complex video surveillance systems. These kinds of systems allow officers to both look through their screens as they monitor the roads, while at the same time they display all kinds of indicators on the same screen (e.g. vehicle information such as speed, ownership or legal status, etc.) that require officer's interpretation.

Of course, the relations with artefacts can be simultaneously characterized by a multitude of relations from this taxonomy. For instance, we can often get lost in activities and embody the computer screen in our experience. Simultaneously, we can interpret various numbers on the screen (e.g. temperature in a weather application) and interpret it to mean "it's gonna be cold". Even more, if we really try, we can simultaneously pay attention to the humming noise of the device and realise how annoyed we are by it. All kinds of relations exist simultaneously and their separation becomes an analytic exercise. Still, it helps us to conceptualize the rich and diverse ways in which technologies mediate perceptions, experiences, decisions and actions.

Besides charting types of relations, the vocabulary of technological mediation is also mapping the kind of influence that technologies can have on human users. Tromp, Hekkert, and Verbeek (2011) identify the impact of technologies in terms of visibility and force, on a continuum between 'hidden' and 'apparent' and between 'weak' and 'strong'. On one end of the spectrum, strong and apparent influences can be called *coercive*. In the policing context we can think of access control mechanisms in information systems that require officers to have the proper credentials in order to access classified data (or make certain fields invisible). On the other side of the visibility spectrum we can find *persuasive* technologies. This kind of technology show their influence explicitly, without being overpowering on the user. In the policing context, persuasive technologies can be found in reward mechanisms that give police officers more credits/points when they achieve milestones. Without forcing the officers, persuasive mechanisms nudge them towards certain behaviours, patrol routes or sequences of action that can protect safety and other professional norms. Technologies can also have more soft and subtle influences, which work by *seducing* the users. The influence is hidden as people are generally not aware of the ways in which the technology shapes behaviour. For instance, police forces often place a coffee machine in a central place in the organization, at the junction of more departments, teams or units, to stimulate social interaction. A fourth type of influence in this taxonomy is both strong and hidden. Tromp et al. (2011) call it *decisive* or *implicative* because it exerts influence without this influence being noticed. In the policing context we can find cubicles that separate offices in a control room, for instance, effectively enforcing work division.

5.3.3.4 Technological Mediation and ANT

The approach to study technological mediation presented above differs from the mediation of action, as understood by Latour. Investigating human experience, this approach does not maintain the strict symmetry between humans and non-humans that ANT aims to keep. A post-phenomenological analysis of technological mediation takes the standpoint of the human being, experiencing, interpreting or perceiving, whereas in ANT the analyst can change the standpoint between humans and non-humans to chart the configurations and network of relations. Still, Verbeek and

Ihde identify similarities between the styles of investigating technological mediation. Both approaches are materially sensitive, recognizing the agency of artefacts, and both have abandoned the 'subject-object' dichotomy, conceptualizing humans and technologies as co-constituting each other.

Verbeek and Ihde argue for complementarity rather than competition between the approaches (Verbeek, 2005; Rosenberger & Verbeek, 2015, p. xv). "What postphenomenology contributes to actor-network theory is the situated perspective, the perspective 'from inside out', thanks to which part of the perceived associations and translations can be more closely analysed in terms of experience and action, existence and meaning [...]. Correspondingly, ANT contributes to postphenomenology a way to elucidate the networks of relations that allow entities to be present" (Verbeek, 2005, p. 168).

For the purposes of this chapter, the vocabulary of technological mediation becomes richer through this complementarity, offering a repertoire that can capture more nuances and dimensions of contemporary technologically mediated security practices. As Annemarie Mol suggests, "a contribution to ANT gently shifts the existing theoretical repertoire. And then, as the theoretical repertoire shifts, it becomes possible to describe further, different cases, and to articulate so far untold events (relations, phenomena, situations)" (Mol, 2010, p. 261). If we want to understand the complex work of technologies in contemporary policing, we need to understand the networks of relations between officers, artefacts, designers or policy makers. At the same time we need to understand what practitioners experience in their technologically-mediated work, how do they perceive criminal phenomena through a screen, or what does it mean to 'have a suspect' when they read indicators from an information system or hear a sound from an automated alert.

5.3.4 Investigating Digital Infrastructures

From this perspective, it should be clear by now that (digital) infrastructures are not merely applications of mathematics and physics but culturally contested artefacts, with social, ethical and political charge and in their turn are shaping, classifying and ordering the material and social realms alike. We can feel their force instantly if we would try, for instance, to ignore the sign posts at supermarkets indicating cash/pin only or on the streets indicating types and heights of allowed vehicles. Already implicit in these examples is that infrastructures incorporate systems of classifications, rules and procedures that often embody a highly normative charge (Bowker & Star, 1999, p. 4). In order to understand how these networked infrastructures are made and how they are often rendered invisible and taken-for-granted we need a type of 'archaeological approach' (Foucault, 1976) to 'dig up' the origins and consequences of these bureaucratic and technological infrastructures (Bowker & Star, 1999, p. 5).

But how to think of the kind of machine-learning algorithms that cannot be easily traced back to an algorithmic step or design decision? These kinds of algorithms, featuring in predictive policing and many more areas of urban ecosystems, are thus challenging our methodological assumptions. An archaeological approach, with its focus on human activities in the past, assumes and looks primarily for anthropogenic explanations. In this respect we could benefit from expanding and deepening the scope of investigations and bringing in the methodological and conceptual repertoire of *geology* to complement archaeology in our understanding of technological infrastructures. I have shown elsewhere how we can employ a *sedimentological* repertoire to understand phenomena with various degrees of dynamism and depth such as settling, debris, deposition, accumulation, sedimentation, explosion or volcanism in digital infrastructures (Niculescu-Dincă, 2018). Of course, human activity plays various roles in design processes. Still, we may need to be ready to account for a whole set of processes that may not be easily traced back to an initial human activity.

A geological mindset and vocabulary in the study of technological infrastructures gently shifts the theoretical repertoire and might prove a more adequate starting point in researching algorithms without assuming particular distributions of responsibility between the role of humans and non-humans in producing the outcomes. At the same time, geology retains the method of 'digging up strata' to understand the origins of our bureaucratic and technological infrastructures. The planetary scale of digital infrastructures and the far-reaching depth of these systems in our societies, with large software stacks running billions of lines of code and massive data flows running (through) our cities, warrants a new set of metaphors to help us come to grips with the spread and depth of these global phenomena. Moreover, bringing closer the repertoires of geology and technology is a timely step when we contemplate the recent efforts 'from the other direction' of the bridge, to expand the geological vocabulary with terms such as the 'Anthropocene' to account for human activities and the impact of our technological infrastructures on the earth's geology (Crutzen & Stoermer, 2000). With this conceptual expansion, even if not yet officially accepted, geology cannot be understood as a domain outside social, political and ethical influences thus warranting the conceptual undertaking to bring these disciplines closer.

A step in this direction it to expand our vocabulary with notions that can help us think about phenomena at this intersection. For instance, *sediment traps* is a rich notion that can be employed in a descriptive and methodological sense. On the one hand, sediment traps refers descriptively in geology to topographic depressions where sediments can accumulate. On the other hand, it refers in oceanography to scientific instruments that can be used to probe aquatic systems, usually oceans, to analyse particle samples. Of course, we need to test and adapt the notions of this interdisciplinary vocabulary.

In a descriptive sense, it can help us think about the ways in which software can accumulate sediments that thicken, become invisible and get taken for granted at runtime. For instance, this can be a design decision but also bias in a training data set that strengthens over time. While some of this bias can be traced back to the training data set, other outcomes cannot be accounted for by a design decision but they stem from accumulations in the layers of algorithmic code. In a methodological sense we can employ the notion of sediment traps to think about the ways in which we can probe and analyse the mesh of sensing infrastructures and the multitude of algorithms running in our cities. Once made available, it might be a gargantuan task to constantly review all algorithms. Therefore, we can think in terms of sediment traps to probe and analyse samples of data and algorithms and request further transparency upon finding 'explosive' or 'toxic' sediments.

5.4 Mediating Perception and Action

So, how can we understand the ways in which police officers engage with sensing infrastructures? How are particular ANPR systems and smart camera infrastructures *mediating* police perception and action in specific situations? This section discusses two vignettes. One is about the most common and widespread uses of the technology: checking traffic against predefined reference lists. The other is about the more recent uses of the technology in conjunction with profiles of suspicious behaviour. The first draws on empirical data I collected with a road policing unit during ethnographic research in a constabulary in England. The second draws on data from the control room of a police organization in The Netherlands. The analyses of both vignettes show the active mediation of perception and action of officers when they engage technology, sometimes weaker, sometimes stronger, more hidden or apparent.

In the constabulary where I performed this study the police made use of a large network of fixed ANPR cameras together with mobile cameras mounted on a fleet of marked and unmarked police vehicles. At the beginning of the shift, I was introduced to officers Jim and Morris, from the road policing department. They were preparing for their shift and they explained me the features of their vehicle and the capabilities of the ANPR system:

Officer Jim: "*Jump in I'll give you a quick tour. Ok, obviously we've got the police lights and everything [...] an air horn, like a truck horn if people don't see us [...] radio and everything else is standard. Then the screen: we have several different items on here. We have a mapping system, video capability as well, and ANPR. [...] The idea is to read number plates. It can read number plates like that* [officer snaps his fingers]. *It should give a 'hit' signal and who owns the car. So we can see who owns the car, see if it's got MOT on the car, which means if it's roadworthy and whatever year it has to be checked, the size of the engine, car type, etc.*"

After this short introduction we took several hours of driving, with the ANPR system turned on. One of the situations I experienced happened soon after. The situation provides an interesting starting point for an analysis that

(continued)

looks at the mediating role of technologies and their interaction with the officers in processes of enactment of suspicion. The officers decided to stop a car for investigation and they explained to me why they did so while they were pulling the car over:

Officer Jim: "*This Volvo has got four people in it and they just looked a bit...*".
Interviewer: "*So this wasn't a hit*" [triggered by an ANPR list]?
Officer Jim: "*No, not a hit, no.*"

Immediately after, Officer Morris intervened:

Officer Morris: "*Or actually it was, over here, 'no taxes' on the middle lane.* [Officer Morris points to the screen]".
Officer Morris [asking the back office for confirmation and potentially additional information]: "*What's the man's name and where was he from again please? Newcastle?*"
Radio in [from the back office]: *Owen Dumitru* [Romanian name].
Officer Morris: *We get a lot of problems with Romanians and Eastern Europeans travelling the country. This car is registered on a Romanian from Newcastle.*

The officers walk towards the stopped car and ask the driver to step out. After a few seconds, the front passenger steps out and opens the back trunk of the vehicle. One of the officers checks the interior carefully while the passenger awaits impatiently. After performing the search the officers consult for a moment. They turn to the nervously looking driver and decide to let him go, returning afterwards to the police vehicle.

Interviewer: "*So what was the problem?*"
Officer Jim: "*It's not a problem in the end.*"
Interviewer: "*So they had paid the taxes in the end?*"
Officer Jim: "*Yes, the tax database is often wrong. Tax is one of the unreliable ones.*"
Officer Morris [whispering for himself]: "*Every time*" [then louder]: "*It used to be a very good indicator for 'no insurance', but now it's not so good.*"

This situation helps to bring to the fore the mediating role of technologies in this police process. On the one hand, the quote shows that technology did not *coerce* the officers or rendered them into executants without responsibility. As we have seen in the dialogue, the officers initially stopped the car not because of the ANPR hit but because they assessed the passengers as 'looking' somehow suspicious. Unlike their colleagues in the control room, the officers in the car were able to look at the vehicle and assess both the screen information as well as the situation in the field.

On the other hand, the ANPR system subtly mediated officer behaviour. Taking the number plate as input, the ANPR system generated the name and nationality of the owners of vehicles on the road. Communication with the back office confirmed the information about the registered owner. Unlike with other vehicles in the 'tax list', they asked this driver to step out and open the trunk while they performed a close inspection of the vehicle. Being Romanian and traveling the country gave the officers extra incentives for verification which proved unjustified in the end. This situation is illustrative of what Norris and Armstrong argued elsewhere, namely that technologically mediated policing does not exclude target selection based on indices of race, age, appearance or demeanour (Norris & Armstrong, 1999). An identity attribute that would not be particularly sensitive in most contexts–nationality–reinforced in this case the officer's categorical suspicion and rendered the Romanian driver vulnerable to the officers, enacting him as suspicious.

Still, this situation brings to the fore the need to investigate further how the perception and experience of police officers in control rooms are mediated by these technologies. Being further away from the situation in the field, they are bound to react only on what the system displays and they can only check pictures of license plates as captured by ANPR systems. The following vignette explores this situation. It draws on data gathered in the control room of a Dutch police organization during a special action in which they were monitoring the traffic at particular locations and trying to see how they could translate the findings into profiles that could be automated on the ANPR network.

> It was a particularly busy day in the control room. Everyone was curious to see the results of monitoring the traffic at particular points in search for suspicious vehicles. From time to time a high pitched audible alert disturbed the room and bright colours flickered on the screen.
>
> The officer responsible for ANPR was particularly interested to see the results. He had the task to communicate the results to the team on the roads. I asked him about his expectations for the results of the profiling project and his response was:
>
> > I think that when it is fully developed, it is easier because I won't have to check it. It's just BAM, the hit comes. Ok people: check it, it's weird.

His reaction could be used to illustrate the way in which technologies mediate police work in control rooms. As the system runs the profiles on live traffic data, it automatically triggers an alert, actively emerging from the context of the officer's experience. Once it brings it to his attention he interprets the alert as something 'weird'. Of course, this does not necessarily need to lead to an arrest. Still, the quote shows that the officer in the control room expected these real-time profiles to deliver more than a mere indication but justifications for suspicion. His way of talking about it (i.e. the use of the interjection BAM) indicates the potential for the profile to strongly mediate his behaviour. He would have otherwise been inclined to perform an extra 'check' but (with the profile fully developed) he would delegate his trust to technology. In conjunction with the assertive visuals and strong audible alerts, technology tended to be trusted and profoundly mediate his perception of traffic participants.

However, being flagged on the screen as suspicious does not necessarily imply criminal behaviour or sufficient reasons for alertness. Technologies do not deliver a transparent and objective rendition of reality. Rather, they play an active role in enacting it. What it means to be a 'suspect' is partly what the screens in front of officers enact as such. Paradoxically, the better the profiles are (in delivering suggestions that turn out justified) the higher the risk to prescribe an overly confident suggestion for action. Therefore, an insight to take out from this analysis is that the less possibilities to understand the full situation (f.i. officers being far away from the field or only getting a few pieces of data to rely on) the less strong, coercive and confident the technology should be designed to perform suspicion. Rather than making similarly strong visual and audible alerts for every profile 'hit', the design of their intensity and assertiveness should correlate with the confidence and amount of information it relies on.

This analysis also implies the need to talk to not only officers in control rooms and in the field but also to designers and programmers, and analysing profiles. Profiles play an active role in the coordination between officers in action and police programmers and they have different meanings across these branches of the police. They are 'boundary objects' (Gerson & Star, 1986) and it is important to constantly investigate their design in order to prevent them "from becoming static and detached from practice" (Schakel et al., 2013, p. 6). This suggests that in addition to their role in mediating the perceptions and actions of officers we need to analyse profiles 'in the making' (Latour, 1987), 'excavating' the layers of their software code.

5.5 Profiles 'In the Making'

So how do designers and programmers talk about and engage with profiling and sensing technologies? How do they translate police knowledge in software code? What are their assumptions and some the potential implications that sediment in this process? This section draws on interviews with profile designers and observations of programmers at work in the Dutch police. It explores the ways in which designers relate to sensing technologies and writing the profile code and shows their main motivations and meanings associated with these technologies.

The following quote comes from an interview with a lead designer of profiles. He explains how ANPR is used in their project as a sensor, together with an array of other sensors, in order to build profiles of suspicious behaviour. These profiles would capture various aspects of criminal phenomena such as locations, times, social networks or modus operandi that can be detected by sensing technologies:

> We are looking for indicators. For instance, that theft is being planned or is being conducted right now. We know now from experience that about 90% of these indicators are only detectable by human beings. You see the conduct of a person, you see how they walk, what they wear. We can't always see that with sensing techniques. But 10% we can. [...] we try to find the things that we can manage digitally with ANPR cameras and other sensors like Bluetooth sniffers, GPS location detection and all kind of sensors.

As an example of successful behavioural profiling he explains the detection of the so-called 'canvas cutters'. These are thieves that steal from unattended trucks in parking lots by cutting their canvas in search of valuable goods. Their particular modus operandi is to hop from one parking lot to another–using a car that's not known to the police–in the search for unattended lorries. To arrive at this knowledge, several branches of the police, such as traffic police and the local police, came together with gas station owners, parking lot administrators and more, to share knowledge about this criminal phenomenon. As a result, they realised that one way to tackle this phenomenon is to detect this particular driving behaviour in the stream of vehicles: moving between multiple parking lots in a short amount of time. The police therefore, used the ANPR cameras of parking lots and designed a real-time profile that triggered an alert when a vehicle performed this behaviour.

The example showed how the police can successfully tackle a complex criminal phenomenon without the need of persistent databases. Instead of storing all traffic data and engage in data mining, the police can build a model of a criminal phenomenon by sharing knowledge between various branches of the police. In this case, criminal investigators, local neighbourhood officers, traffic police and administrators of parking lots got together and created a model of suspicious behaviour. After knowing the behaviour to look for, they translated it into measurable indicators and inscribed it into the code of a real-time profile.

Therefore, it is argued that this way of working "reduces the impact of privacy invasion" (Schakel et al., 2013, p. 7) while still catching criminals red-handed. The data of most traffic participants is not stored in police databases or quickly deleted after the profile runs its code. Any potential further investigation, whether unauthorised searches, curious officers of celebrity whereabouts and more, would not have the data to start with in the first place. At the same time, the quote suggests that this way of working contributes to inferring suspicion based on behaviour and not on identity attributes. Therefore, it contributes to lessening the risk of problematic discriminatory police actions.

Of course, behavioural profiling is often a highly dynamic process where knowledge is gained and lost, as criminals change behaviour and modus operandi, rendering the indicators obsolete and the profile in need of an update. In contrast with the high expectations of the officer in the control room (discussed in Sect. 5.3.4), the interviewed profile designer was well aware of this phenomenon and its potential risks.

> So it is not 100% safe. […] it's just an indication that something is going on. We see this guy driving up and down the parking places. That's what we really see. Whether there is really a car burglary in this case we don't know […] There are certain groups of vehicles that have the same conduct. We can white-list them (i.e. exclude the number plates of police vehicles from the profile), but a guy with a bladder problem….

The quote shows that the designer is aware of the potential false positives generated by the profiles which can prescribe an alerted attitude about someone who, for instance, needs to go often to the toilet. In other words, he knows the technology is not delivering certainty. Still, as we have seen in the previous section, his doubts and uncertainties tend to become black boxed at the operational level. The officer in the control room would have seen an assertive indicator that a potential theft is in the process and be inclined to send the nearest patrol to investigate the issue. Of course, we can only imagine from this point the surprise of 'the guy with a bladder problem' coming out of the toilet to be welcomed by two alerted police officers. And, of course, in hindsight this situation can be seen as funny or at least acceptable.

Still, while investigating the process of designing profiles, new issues of behavioural profiling came to the fore. Despite advertising a lack of identity attributes in the process of behavioural profiling, identity attributes can be inscribed as well when necessary. The following quote, from an interview with a police programmer in the project, shows the practice of translating police knowledge into profile code. The programmer was responsible for redesigning a profile that should give a hit whenever a set of known and problematic number plates were performing a certain behaviour: returning to a certain location by night. In other words, a profile hit would indicate to the police that 'the old boys are back in town'. The police programmer explained that he translated the specific police knowledge into the profile code by connecting to particular sensors and building a minimum threshold of 50 points, calculated from several parts:

> I made two indicators. Each one of them has 'a camera part' and 'a name and time part'. I'll explain. [Each vehicle] gets 20 points for passing by the cameras [near the location], additional 20 for passing it by night: then they have 40. When their name matches [compared to a list], they would get an additional 10 points [and trigger the alert].

Still, as he later mentioned, the 'name part' is problematic for both effectiveness and ethics:

> In my opinion the profile is not really strong because maybe they like that place so much they come there to go to the toilet or to buy some food, for pleasure, and not to steal some stuff. I would prefer a profile with some more behaviour elements, for example these movements [the programmer points on his screen to show a particular behaviour that could be suspicious]. I don't know if this one is accurate at the moment, but a few months ago it was accurate and at that moment this element was very valuable for us. At that moment we knew, we saw the boys and we saw them working.

The quotes above can be used to highlight three issues related to combining sensing technologies with persistent databases in an aggregated profile. For one, even if the profile is meant to assess behaviour it also requires an inquiry to the vehicle registration database. Therefore, even if behavioural elements are key, the designer inscribed indicators regarding identity attributes but he doubts their effectiveness.

Secondly, once connecting to persistent databases, knowledge-based behavioural profiling may lose its announced advantages with respect to privacy and non-discrimination. The translation of identity attributes into profile code can be fraught with normative charge. Of course, in this case, the profile looked for a list of already problematic vehicles but the practice raises a point of concern. When is it acceptable that vehicles are assessed for going to a particular location? Which identity attributes are part of a profile and which parameter values are assessed? Through connections to persistent databases and other sensor networks, profiles can aggregate identity related attributes such as name, age and nationality. Without careful assessment, this can result in the close surveillance of categories of people that are unrelated to criminal phenomena. It is clear that we cannot simply rely on 'hard-coding' human rights protection (Koops & Leenes, 2014). Therefore, because of the dynamic character of how suspicion is enacted in the midst of these relationships, we need approaches that offer the flexibility to keep neither law nor technology as mere means to specified ends (Hildebrandt & Koops, 2010; Hildebrandt, 2011). Although technical solutions can reduce the risks associated with surveillance, at the same time these practices may create new ones.

Thirdly, the relations between criminal behaviour, ubiquitous sensors and organizational factors are increasingly unpredictable and rapidly changing while software infrastructures tend to trap sediments of prejudice. As the programmer mentioned, he doubted the current effectiveness of the profile in delivering valuable actionable knowledge and implicitly acknowledged the above-mentioned risks for discriminatory actions. A recurring challenge for these issues remains the opacity of software (Pasquale, 2015; Kitchin, 2017) that increasingly enables decision-making in policing. Of course, to a certain extent opacity is constitutive and legitimate of policing practices. Criminals should not be able to easily anticipate their next move. At the same time, the software that enables police decisions is increasingly underlined by proprietary algorithms that can be seen as a competitive advantage. Thus, a case can be made prima facie not to make this kind of algorithms fully transparent to the public at large (but, at least presently, only to oversight bodies) (De Laat, 2018).

Even so, this line of argument contends that a higher level of transparency and a "spirited and open public discussion" about algorithms in general is still legitimate (De Laat, 2018). Especially when we contemplate how the perceptions, decisions and actions of policing practitioners are increasingly mediated in their daily routines by all kinds of software-enabled artefacts (e.g. automated alerts, risk profiles, rule-based suspicion algorithms, crime hotspots, etc.)–with implications for the effectiveness and legitimacy of urban security governance. An argument can be made that at least "over the years and the decades to come, they be made part of a public record available to us all" (Pasquale, 2015). The public needs to be able adequately conceptualise and relate to these entities that are profoundly affecting our cities and our societies in general. If we want an effective urban governance and if we aim for public acceptability of smart cities infrastructures, the public needs to be included in these debates.

However, from the examples discussed above we can see that adequately detecting criminal behaviour and simultaneously protecting human rights is not only a matter of algorithmic code but it depends on a whole set of socio-technical factors that need to be aligned. This way of working requires a concerted organizational, technological and imaginative effort. The following quotes from an interview with an officer some months later can be used to make this point:

> *[The project]* hasn't seen much progress other than one camera will be installed near *[location]* to see what kind of cars are passing at this specific point. So thinking in terms of what kind of movements we expect to see this is not taking place yet. And therefore the discussion of where to deploy other cameras or other sensors is being postponed.

> Availability of technology is not the problem but more the insight of what we can or must do with it seems to be the problem. Our imagination on how it can help us isn't growing as fast as I would like to see it. Changing old ways of working is hard. It'll have to take its time I guess.

His testimony illustrates both the promise and the difficulties that police officers face in engaging with sensing technologies in an effective and legitimate way. Behavioural profiling may offer better privacy protection and effective crime control, at least for a while, but *after* profiles become finished artefacts, ready to be inscribed in algorithms. Until then, the knowledge that can be used in building them can still be made by installing cameras 'to see what kind of cars are passing'. The process involves trial and error and the officer laments the slow progress. When 'the availability of technology is not the problem' and when perceiving the organizational arrangements as failing, monitoring seems to be the default option.

5.6 Conclusion

Knowledge-based behavioural profiling predicated on ephemeral sensor data streams holds the promise of effective policing actions while "reducing the impact of privacy invasion" and minimizing unfair discrimination (Schakel et al., 2013, p. 7). Still, this is not an automated discovery in which the police would find the relevant associations at a push of a button. This way of working requires certain conditions to deliver on its promises compared to storing and analysing data in bulk. These conditions imply that police organizations build profiles through knowledge sharing–and not thorough the use of persistent databases. It only works when many branches of the police do collaborate and continuously exchange knowledge rather than sharing data. In addition, they restrict themselves to behavioural profiling–and do not involve identity attributes–and the processing of data should be done locally (as much as possible) and not centralised. In short, sensing technologies and profiling algorithms are not flawlessly working surveillance machines that only select the perpetrators from the innocent, the bad from the good. Without concerted efforts to promote imagination in using sensing technologies in policing 'old habits' seem to die hard. Effective security governance while upholding our ideals of human rights in a world of smart cities *can* be achieved but with a heavy support for knowledge sharing between organizational, legal, policy and technological decision making, rather than just sharing data.

Still, engaging in this approach does not guarantee effectiveness and legitimacy. Police programmers are bound to continuously redesign profiles to catch criminals red-handed. This is done in order adapt to changes in criminal behaviour but also to prevent the erosion of the presumption of innocence when rising numbers of false positives get caught in their net. This process renders the profile design as a locus of analysis for both reasons of effective security governance as well as ethics. These evaluations depend not only on broad architectural choices to interconnect databases and sensor networks or not but on Boolean operators, algorithmic steps, and values given to parameters in suspicion and risk profiles. In other words, suspicion and the way it is enacted becomes a 'gateway' to the revocation of privacy. Preventing 'unnecessary infringement on the privacy of people' (Hoogewoning, 2006, p. 79) through technology design can also be read from the other direction: who *is* under suspicion can have their privacy rights legally lifted.

Therefore, an argument can be made that 'digging up' the layers of software code needs to be an ongoing endeavour just as technologies constantly and actively mediate their perceptions, actions and experiences of practitioners on a daily basis. By continuously tracing the enactment of suspicion in technologically mediated policing practices we can hope to counter more adequately some of the risks to police legitimacy while promoting an effective security governance within current and emerging forms of digitally infused urban environments. To be sure, the organizations where this study was conducted did have processes in place to evaluate profiles. Still, as this chapter suggests, the frequency of these evaluations did not keep up with the dynamics of police work and of criminal phenomena. At the same time, the normative charge of profiles was predicated on quickly changing parameter values and algorithmic steps.

Therefore, to promote an effective and legitimate security governance, requires the regular involvement of a broader set of disciplines with their sensibilities and theoretical insights in the evaluation of technologically mediated policing. This chapter showed how concepts and insights from Philosophy and Technology, Science and Technology Studies and Geology can be employed to 'dig up' and trace the ways in which suspicion is enacted in software code and in the broader materially-embedded policing practices. Moreover, it outlines and engages the conceptual framework of mediation theory to reflect on the ways in which the practitioners' perceptions, experiences and actions are mediated by sensing infrastructures and algorithmic profiles.

Acknowledgements The author wants to thank Tsjalling Swierstra, Irma van der Ploeg, and Jason Pridmore as well as the police staff, municipality officials and technology developers for their generous collaboration and support for this research.

References

Akrich, M. (1992). The de-scription of technical objects. In W. Bijker & J. Law (Eds.), *Shaping technology/building society: Studies in sociotechnical change* (pp. 205–224). Cambridge, MA: MIT Press.

Akrich, M., & Latour, B. (1992). A summary of a convenient vocabulary for the semiotics of human and nonhuman assemblies. In W. Bijker & J. Law (Eds.), *Shaping technology/building society: Studies in sociotechnical change* (pp. 259–264). Cambridge, MA: MIT Press.

Bijker, W. (1992). The social construction of fluorescent lighting, or how an artifact was invented in its diffusion stage. In W. Bijker & J. Law (Eds.), *Shaping technology/building society: Studies in sociotechnical change* (pp. 75–104). Cambridge, MA: MIT Press.
Bijker, W. (2010). How is technology made?—That is the question! *Cambridge Journal of Economics, 34*, 63–76.
Bowker, G. C., & Star, S. L. (1999). *Sorting things out classification and its consequences*. Cambridge, MA: MIT Press.
Brey, P. (1998). Artifacts as social agents. In H. Harbers (Ed.), *Inside the politics of technology. Agency and normativity in the coproduction of technology and society* (pp. 61–84). Amsterdam: Amsterdam University Press.
Byrne, J., & Marx, G. T. (2011). Technological innovations in crime prevention and policing. A review of the research on implementation and impact. In E. D. Pauw, P. Ponsaers, K. Van der Vijver, W. Bruggeman, & P. Deelman (Eds.), Technology-led policing (Vol. 2011/3, 20, pp. 17–40). Antwerpen: Maklu.
Callon, M. (1987). Society in the making: The study of technology as a tool for sociological analysis. In W. E. Bijker, T. P. Hughes, & T. J. Pinch (Eds.), *The social construction of technological systems* (pp. 83–103). Cambridge, MA: MIT Press.
Chan, J. B. L. (2001). The technology game: How information technology is transforming police practice. *Journal of Criminal Justice, 1*(2), 139–159.
Crutzen, P. J., & Stoermer, E. F. (2000). The 'Anthropocene'. *Global Change Newsletter, 41*, 17–18.
De Laat, P. (2018). Algorithmic decision-making based on machine learning from big data: Can transparency restore accountability? *Philosophy & Technology, 31*(4), 525–541. https://doi.org/10.1007/s13347-017-0293-z.
Ericson, R., & Haggerty, K. D. (1997). *Policing the risk society*. Toronto: University of Toronto Press.
Foucault, M. (1976). *The birth of the clinic: An archaeology of medical perception*. London: Tavistock.
Gerson, E., & Star, S. L. (1986). Analyzing due process in the workplace. *ACM Transactions on Information Systems (TOIS), 4*(3, Special issue: selected papers from the conference on office information systems), 257–270.
Haggerty, K. D., & Ericson, R. (2000). The surveillant assemblage. *British Journal of Sociology, 51*(4), 17.
Harris, C. J. (2007). Police and soft technology: How information technology contributes to police decision making. In J. Byrne & D. Rebovich (Eds.), *The new technology of crime, law and social control* (pp. 153–183). Monsey, NY: Criminal Justice Press.
Hellemons, A., Beek, P. V. D., Malenstein, J., Goor, A. A. T., Kuijten, C., & Schewe, W. (2013). *Final report on DEPET (Dissemination all over Europe of know-how of Privacy Enhancing Technologies). Prevention of and fight against crime 2010*. Brussels: European Commission. Directorate–General HOME. Directorate F – Security.
Hess, K. M., & Orthmann, C. H. (2010). *Criminal investigation* (9th ed.). Scarborough: Delmar Cenage Learning.
Hildebrandt, M. (2011). Legal protection by design: Objections and refutations. *Legisprudence, 2*(5), 223–248.
Hildebrandt, M., & Gurtwith, S. (Eds.). (2008). *Profiling the European citizen*. Berlin: Springer.
Hildebrandt, M., & Koops, B.-J. (2010). The challenges of ambient law and legal protection in the profiling era. *Modern Law Review, 73*, 428–460.
Hoogewoning, F. C. (Ed.). (2006). *The police in evolution - vision on policing*. The Hague: Project Group Vision on Policing. Dutch Police Institute.
Ihde, D. (1990). *Technology and the lifeworld*. Bloomington, IN: Indiana University Press.
Innes, M., & Roberts, C. (2008). Reassurance policing, community intelligence and the co-production of neighbourhood order. In T. Williamson (Ed.), *Handbook of knowledge based policing* (pp. 241–262). West Sussex: Wiley.

Kamiran, F., Karim, A., Verwer, S., & Goudriaan, H. (2012). Classifying socially sensitive data without discrimination: An analysis of a crime suspect dataset. In *Data mining workshops (ICDMW), IEEE 12th international conference, Brussels, 10–10 Dec. 2012* (pp. 370–377). Piscataway, NJ: IEEE. https://doi.org/10.1109/ICDMW.2012.117.

Kitchin, R. (2014). The real-time city? Big data and smart urbanism. *GeoJournal, 79*(1), 1–14.

Kitchin, R. (2017). Thinking critically about and researching algorithms. *Information, Communication & Society, 20*(1), 14–29. https://doi.org/10.1080/1369118X.2016.1154087.

Koops, B.-J., & Leenes, R. (2014). Privacy regulation cannot be hardcoded. A critical comment on the 'privacy by design' provision in data-protection law. *International Review of Law, Computers & Technology, 28*(2), 159–171.

Latour, B. (1987). *Science in action*. Cambridge: Harvard University Press.

Latour, B. (1988). Mixing humans and nonhumans together: The sociology of a door-closer. *Social Problems, 35*(3 Special Issue: The Sociology of Science and Technology), 298–310.

Latour, B. (1991). Technology is society made durable. In J. Law (Ed.), *A sociology of monsters: Essays on power, technology and domination* (pp. 103–131). London: Routledge.

Latour, B. (1994). On technological mediation: Philosophy, sociology, genealogy. *Common knowledge, 3*(2), 29–64.

Latour, B. (2005). *Reassembling the social: An introduction to actor-network-theory*. New York and Oxford: Oxford University Press.

Law, J. (1987). Technology and heterogeneous engineering: The case of Portuguese expansion. In W. E. Bijker, T. P. Hughes, & T. Pinch (Eds.), *The social construction of technological systems* (pp. 111–134). Cambridge MA: MIT Press.

Law, J. (2008). Actor-network theory and material semiotics. In B. S. Turner (Ed.), *The new blackwell companion to social theory* (3rd ed., pp. 141–158). Oxford: Blackwell.

Law, J. (2009). Seing like a survey. *Cultural Sociology, 3*(2), 239–256.

Law, J., & Mol, A. (2001). Situating technoscience: An inquiry into spatialities. *Environment and Planning D: Society and Space, 19*(5), 609–621.

Leipnik, M. R., & Albert, D. P. (Eds.). (2003). *GIS and law enforcement: Implementation issues and case studies*. London: Taylor & Francis.

Leman-Langlois, S. (2012). *Technocrime, policing, and surveillance* (Vol. 3). New York: Routledge.

Lyon, D. (Ed.). (2003). *Surveillance as social sorting (privacy, risk and digital discrimination)*. New York: Routledge.

MacKenzie, D., & Wajcman, J. (Eds.). (1985). *The social shaping of technology: How the refrigerator got its hum*. Philadelphia, PA: Open University Press.

Maguire, M. (2000). Policing by risks and targets: Some dimensions and implications of intelligence-led crime control. *Policing and Society, 9*(4), 315–336.

Mancuhan, K., & Clifton, C. (2014). Combating discrimination using Bayesian networks. *Artificial Intelligence and Law, 22*(2), 211–238. https://doi.org/10.1007/s10506-014-9156-4.

Manning, P. (2008). *The technology of policing: Crime mapping, information technology, and the rationality of crime control*. New York and London: New York University Press.

Meijer, A., & Thaens, M. (2018a). Quantified street: Smart governance of urban safety. *Information Polity, 23*(1), 29–41.

Meijer, A., & Thaens, M. (2018b). Urban technological innovation: Developing and testing a sociotechnical framework for studying smart city projects. *Urban Affairs Review, 54*(2), 363–387. https://doi.org/10.1177/1078087416670274.

Mol, A. (2010). Actor-network theory: Sensitive terms and enduring tensions. *Kölner Zeitschrift für Soziologie und Sozialpsychologie, 50*(1), 253–269.

Neyroud, P. (2008). Policing and ethics. In T. Newburn (Ed.), *The handbook of policing*. Cullompton: Willan publishing.

Niculescu-Dincă, V. (2018). Towards a sedimentology of information infrastructures: A geological approach for understanding the city. *Philosophy & Technology, 31*(3), 455–472. https://doi.org/10.1007/s13347-017-0298-7.

Norris, C., & Armstrong, G. (1999). *The maximum surveillance society*. Oxford: Berg Publishers.

Parker, J. R., & Federl, P. (1996). An approach to Licence plate recognition. *Computer Science Technical Reports, University of Calgary, Alberta Canada, 591*(11), 1–6.
Pasquale, F. (2015). *The black box society: The secret algorithms that control money and information*. Cambridge, MA: Harvard University Press.
Pinch, T., & Bijker, W. (1987). *The social construction of facts and artifacts: Or how the sociology of science and the sociology of technology might benefit each other*. Cambridge, MA: MIT Press.
Ratcliffe, J. H. (2008). *Intelligence-led policing*. Cullompton: Willan Publishing.
Rosenberger, R., & Verbeek, P.-P. (Eds.). (2015). *Postphenomenological investigations: Essays on human–technology relations (postphenomenology and the philosophy of technology)*. Lanham, MD: Lexington Books.
Sanders, C. (2006). Have you been identified? Hidden boundary work in emergency services classifications. *Information, Communication & Society, 9*(6), 714–736.
Schakel, J.-K., Rienks, R., & Ruissen, R. (2013). Knowledge-based policing: Augmenting reality with respect for privacy discrimination and privacy in the information society. In B. Custers, T. Calders, B. Schermer, & T. Zarsky (Eds.), *Discrimination and privacy in the information society. Data mining and profiling in large databases* (Studies in applied philosophy, epistemology and rational ethics) (Vol. 3, pp. 171–189). Heidelberg: Springer.
Sclove, R. (1995). *Democracy and technology*. New York: Guilford Press.
Tilley, N. (2008). Modern approaches to policing: Community, problem-oriented and intelligence-led. In T. Newburn (Ed.), *Handbook of policing* (pp. 373–403). Cullompton: Willan Publishing.
Townsend, A. M. (2013). *Smart cities: Big data, civic hackers, and the quest for a new utopia*. New York: W. W. Norton.
Tromp, N., Hekkert, P., & Verbeek, P.-P. (2011). Design for socially responsible behavior: A classification of influence based on intended user experience. *Design Issues, 27*(3), 3–19.
Van Ooijen, C., & Nouwt, S. (2009). Power and privacy: The use of LBS in Dutch public administration. In B. V. Loenen, J. W. J. Besemer, & J. A. Zevenbergen (Eds.), *SDI convergence. Research, emerging trends, and critical assessment*. Delft: NCG.
Verbeek, P.-P. (2005). *What things do: Philosophical reflections on technology, agency, and design*. University Park, PA: Penn State Press.
Verbeek, P.-P. (2006). Materializing morality. Design ethics and technological mediation. *Science, Technology, & Human Values, 31*(3), 361–380.
Verbeek, P.-P. (2008). Obstetric ultrasound and the technological mediation of morality: A postphenomenological analysis. *Human Studies, 31*, 11–26.
Verbeek, P.-P. (2011). *Moralizing technology: Understanding and designing the morality of things*. Chicago, IL: University of Chicago Press.
Verbeek, P.-P. (2015). COVER STORY beyond interaction: A short introduction to mediation theory. *Interactions, 22*(3), 26–31.
Whelan, C., & Dupont, B. (2017). Taking stock of networks across the security field: A review, typology and research agenda. *Policing and Society, 27*(6), 671–687. https://doi.org/10.1080/10439463.2017.1356297.
Williams, V. S., & Williams, B. O. (2008). Technology applications: Tools for law enforcement. In J. Ruiz & D. Hummer (Eds.), *Handbook of police administration* (Public administration and public policy) (pp. 165–173). New York: CRC Press.
Wittkower, D. E. (2017). Technology and discrimination. In J. C. Pitt & A. Shew (Eds.), *Spaces for the future: A companion to philosophy of technology*. New York: Routledge.

Vlad Niculescu-Dincă is Assistant Professor at Leiden University, Faculty of Governance and Global Affairs, Institute of Security and Global Affairs in The Hague. His research and teaching spans topics such as policing and technology, security networks, surveillance studies or urban policing. Dr. Vlad Niculescu-Dincă received his PhD in 2016 from Maastricht University for the thesis "Policing Matter(s). Towards a sedimentology of suspicion in technologically mediated surveillance." In 2017, he received the Early Career Award from the Society for Philosophy and Technology. In his research he examined various technologically mediated policing practices, drawing on ethnographic research which he performed at several police organizations.

Chapter 6
Now You See It (Now You Don't): Users, Maintainers and the Invisibility of Infrastructure

Mark Thomas Young

Abstract When infrastructural technology is functioning correctly, it is often considered to recede from view and become invisible. According to this perspective, visibility is restored in cases of breakdown and malfunction, which for this reason, are often understood to represent important epistemic opportunities for grasping previously hidden aspects of infrastructure. This article seeks to outline the limitations of the idea that infrastructural failure has a positive epistemic function by distinguishing between two fundamentally different ways in which the nature of technological function can be conceptualised; the first understands function as stemming primarily from the way a technology was designed and produced, the second as a process sustained by constant human activities of maintenance and repair. After illustrating how a number of recent studies of infrastructure build upon insights granted by the latter process perspective, this article aims to demonstrate how the experience of breakdown itself does little to facilitate a shift towards understanding infrastructure in this way. For this reason, this article ends with a suggestion for a shift in emphasis in studies of infrastructure, away from breakdown and towards a sustained examination of everyday practices of maintenance, repair and modification of infrastructural technologies.

Keywords Infrastructure · Phenomenology · Breakdown · Standpoint epistemology · Maintenance · Repair

6.1 Introduction

For the majority of the worlds inhabitants, the experience of life in the city is structured by systems of infrastructure. In the city, places where people live and work form nodes within integrated networks of water, electricity, waste, information and transport. Many of us, especially those living in the global north, can access these facilities

M. T. Young (✉)
Delft University of Technology, Delft, Netherlands
e-mail: Mark.Young@fof.uib.no

M. Nagenborg et al. (eds.), *Technology and the City*, Philosophy of Engineering and Technology 36, https://doi.org/10.1007/978-3-030-52313-8_6

from within our very homes at the push of a button or a twist of the tap. At the same time, homes and businesses in the city also represent sources of waste that are fed back into infrastructural networks along underground pipes and inner city roads, to be processed, recycled or disposed. The close associations in scholarly literature between infrastructure and "urbanization" reflect a general consensus that life in the city cannot be thought apart from the function of such networks and the roles we play within them. Yet inevitably however, living in a city also means that we will encounter cases where infrastructure fails to work as it should. Maybe we switch on the light one day and nothing happens due to a failure of the local electricity network, or maybe we are forced to take a detour on our drive to work due to road maintenance. More dramatic failures are encountered by those of us unfortunate enough to experience war or disaster, in which various infrastructural systems can collapse entirely.

According to a dominant line of thought in a variety of disciplinary approaches to the study of infrastructure, these different examples of infrastructural failure all involve a phenomenological shift in which a technical system that was previously invisible is suddenly brought to attention. As Susan Leigh Star suggests, 'the normally invisible quality of working infrastructure becomes visible when it breaks: the server is down, the bridge washes out, there is a power blackout' (Star, 1999). For this reason, it is common to ascribe a positive epistemic function to cases of breakdown; Stephen Graham for example, argues that 'studying moments when infrastructures cease to work as they normally do is perhaps the most powerful way of really penetrating and problematizing those very normalities of flow and circulation to an extent where they can be subjected to critical scrutiny' (Graham, 2010). Studies of infrastructure therefore often focus on calamities; events such as natural disasters or engineering failures, as moments where the nature of infrastructural technology is understood to be most visible.[1] Yet *which* aspects of infrastructure are rendered visible upon breakdown and the extent to which such experiences can actually facilitate a deeper understanding of the nature of infrastructural technologies are questions that have not yet received an adequate philosophical analysis.

The aim of this paper is to explore the limitations of the idea that infrastructural failure has a positive epistemic function by distinguishing between two fundamentally different ways in which the nature of technological function can be conceptualised; the first understands function as stemming primarily from the way a technology was designed and produced, the second as a process sustained by constant human activities of maintenance and repair. After illustrating how a number of recent studies of infrastructure build upon insights granted by the latter process perspective, this article aims to demonstrate how the experience of breakdown itself does little to facilitate a shift towards understanding infrastructure in this way. For this reason, this article ends with a suggestion for a shift in emphasis in studies of infrastructure, away from breakdown and towards a sustained examination of everyday practices of

[1] The idea that breakdown renders technology visible for analysis is by no means limited to studies of infrastructure. For a recent volume exploring the heuristic potential of material disruption for other kinds of technology, alongside the origins of this methodological strategy, see Strebel, Bovet, and Sormani (2019).

maintenance, repair and modification. Yet before the significance of such practices to our understanding of infrastructure can be outlined, it will be necessary to take a closer look at why infrastructure is so often regarded as invisible in the first place. For this reason, we will turn our attention in the next section to the different ways in which various schools of thought in sociology, history and philosophy have characterized the relationship between visibility and technology, in order to explore why infrastructure is so commonly considered a paradigm case of a technology that recedes from view.

6.2 Invisibility and Breakdown

In *Being and Time*, Martin Heidegger proposed a novel conception of technological experience. When a properly functioning technology is used to perform a certain task, Heidegger suggested, it recedes from view and disappears into the background. Because our awareness is focused on the goals we aim to achieve, in such situations a technology reveals itself to us as a source of efficiency stemming not only from the artifact itself, but also from the wider context of technologies and practices from which it derives its efficacy. However, when the same technology prevents the performance of a certain task; because it is broken or malfunctioning, it becomes visible for us as an object in its own right. Insofar as technologies can only appear in this way by obscuring the relations on which its usefulness had originally depended, Heidegger considered this mode of awareness to be a 'deficient mode of concern' (Heidegger, 1962, p. 103). Almost a century later, various features of Heidegger's phenomenological insight are reflected in a variety of different approaches to the study of technology. Perhaps the most similar account of technology in contemporary scholarship occurs in the field of science and technology studies, which employs the term "blackboxing" to capture 'the way scientific and technical work is made invisible by its own success' (Latour, 1999, p. 304). In *Pandora's Hope*, Bruno Latour explicates this idea with the example of an overhead projector in use at a lecture, which he characterizes as 'a silent and mute intermediary, taken for granted and completely determined by its function' (Latour, 1999, p. 183).

Yet while this perspective considers the normal function of a technology to render it invisible in a similar way to Heidegger's original conception, there is an important difference in that breakdown is ascribed a *positive* epistemic function; Latour's projector, for example, is suggested to reveal its components and their role in a wider network by failing to operate as it should;

> Now suppose the projector breaks down. The crisis reminds us of the projector's existence. As the repairmen swarm around it, adjusting this lens, tightening that bulb, we remember that the projector is made of several parts, each with its role and function and its relatively independent goals. Whereas a moment before the projector scarcely existed, now even its parts have an individual existence (Latour, 1999, p. 183)

Latour's account of the projector reflects one of the central presuppositions guiding sociological studies of science and technology; that important aspects of scientific and technological processes are opaque and therefore must be rendered visible before they can be subjected to critical analysis. The breakdown of social, scientific and technological processes is considered to play an important role in this process. The sociology of scientific knowledge (SSK), for example, contends that ordinarily, the social processes by which scientific consensus is formed are invisible to outsiders. However, periods of scientific controversy, when agreement amongst members of a particular scientific community break down, are understood to represent unique opportunities in which sociological factors influencing consensus formation remain visible for analysis. For this reason, studies in SSK often privilege periods of scientific controversy as important opportunities for empirical study (Martin & Richards, 1995).

The Social Construction of Technological Systems (SCOT) on the other hand, contend that the inherently social nature of the meaning and technical characteristics of particular technologies also become concealed from view as technologies develop. In particular, proponents of SCOT contend that during the development of a technology, artefacts become black-boxed through processes of closure and stabilization. Closure denotes the process by which the interpretive flexibility of the meaning of a particular artefact is lost when particular interpretations of an artefact become dominant at the expense of other competing interpretations. Stabilization, on the other hand, occurs through the increasing standardization of the material components and design of an artefact that occurs after closure. Together, these two processes work to obscure both the inherently ambiguous nature of technologies themselves, and the social processes of negotiation which result in their stabilization, thereby cultivating the illusion that certain paths of technological change are inevitable. In his seminal study of the development of the modern bicycle, Wiebe Bijke illustrates the tendency of alternative interpretations of the meaning of artifacts to become invisible over time by noting how;

> the process of closure is almost irreversible – almost, but not completely. It is now difficult to see the Ordinary bicycle as anything other than a very unsafe machine that is extremely difficult to ride. Trying to envisage the Ordinary as the artefact it was for its contemporaneous users seems to require not only the mental gymnastics of interpretive flexibility, but physical skills as well (Bijker, 1995, p. 87).

Like research in the SSK tradition, studies in SCOT aim to restore the sociocultural dimensions of the construction and interpretation of different technologies by focusing on controversies; periods in the development of a technology where competing interpretations of particular artefacts were subject to conflict and negotiation. Among proponents of SCOT, this process of rendering visible the historical contingency of different technologies is often described in terms of 'opening the black box' and reversing of the natural tendency of technological development to obscure the contingent nature of its origins.

Another perspective which accords a central place to invisibility in its account of technology, and one which often focuses specifically upon infrastructure, is the

Large Technical Systems (LTS) approach which emerged from historical studies of science and technology in the latter stages of the twentieth century. Unlike SCOT, research in the LTS tradition takes systems rather than artifacts as the proper object of analysis. Invisibility constitutes a central theme of research in the LTS tradition in two general ways; *Firstly*, because LTS studies aim to transcend the traditional focus upon artifacts, attention is directed towards the less visible, underlying systems, such as electricity networks, upon which the efficacy of artifacts, such as the light bulb, themselves depend. *Secondly*, like SCOT, studies in LTS often consider invisibility to characterize a specific historical phase in the development of a particular technology. As they mature, large technical systems are often considered to develop from a period of growth to a stabilization phase, in which they;

> function reliably and change only incrementally…(and) are largely "taken for granted" by those depending on their products and services. Even partial insight into their workings is constituted mainly around failures (Joerges, 1988).

This conception of large technical systems as invisible when functioning normally and rendered apparent in times of crisis, has become an increasingly common way to characterize people's experience of infrastructural technologies in a wide variety of disciplines. For example, one of the most well known studies of infrastructure, Susan Leigh Star's "The Ethnography of Infrastructure" (Star, 1999), argued that invisibility should be incorporated into the meaning of the term "infrastructure" itself, suggesting that for something to be considered a form of infrastructure, it *must* withdraw from our attention (Star, 1999). As Brian Larkin has recently noted, the tendency to consider infrastructure as invisible has been spectacularly influential, for;

> it is (now) commonplace, seemingly obligatory, for almost any study of infrastructure to repeat Star's (1999) assertion that infrastructures are "by definition invisible," taken for granted, and that they only "become visible on breakdown" (Larkin, 2013).

Yet invisibility is by no means a self evident notion. Despite the ubiquity of such assertions, the term itself is often employed to characterize infrastructural technologies in a variety of distinct senses. What follows provides a glimpse of the range of uses to which the notion of invisibility has been employed in the study of infrastructure;

Firstly, invisibility sometimes denotes the way certain forms of infrastructural systems in urban environments are often deliberately removed from view. As we walk inner city streets, for example, many of us are not aware of the sewerage pipes that carry waste underneath the ground simply because we do not see them. Yet as Kaika and Swyngedouw have argued, while today urban networks often become 'buried underground, invisible, rendered banal and relegated to an apparently marginal, subterranean world', this has not always been the case. Throughout early modernity, for example, infrastructural systems in Western urban environments were often celebrated as powerful symbols of progress and promise and for this reason located in highly visible spaces (Kaika & Swyngedouw, 2000). According to this perspective then, the invisibility which characterizes infrastructure in cities of the global north is an absence from view, and one that captures a particular phase in the historical development of infrastructural systems.

A second sense of invisibility pertains to knowledge surrounding infrastructures. As various theorists have argued, knowledge concerning the specific forms of social organization that emerge around infrastructural technologies and the particular configurations of activity that are permissible in relation to them, can only be acquired via participation in a community. Once attained, these forms of understanding develop into a naturalized familiarity which lend specific organizational arrangements and artifacts a 'taken for granted' character (Bowker & Star, 1999, p. 35). This kind of understanding that emerges around infrastructural technology can be conceived as a form of tacit knowledge insofar as it often cannot be consciously apprehended or explicitly stated;

> Belonging to a given culture means, in part, having fluency in its infrastructures. This is almost exactly like having fluency in a language: a pragmatic knowing-how, rather than an intellectual knowing-that, so that the bewildered questions of an outsider might strike one as not only hilarious but also unanswerable (Edwards, 2003)

Thirdly, Infrastructure is often suggested to be invisible because it functions as a backdrop to phenomena which themselves demand a larger share of our attention in daily activities. The widespread characterization of infrastructure as a substrate (Star, 1999) or background (Graham, 2010) upon which life in the city is played out invite comparisons to the stage of a theatre production; when the light dims and the play begins, our attention naturally shifts to the unfolding drama and away from the structure upon which the actors stand. In much the same way, for example, Brian Larkin suggests that;

> Infrastructures are matter that enable the movement of other matter…As things they are present to the senses, yet they are also displaced in the focus on the matter they move around. We often see computers not cables, light not electricity, taps and water but not pipes and sewers (Larkin, 2013)

By connecting invisibility to function, this sense of invisibility is also perhaps the closest to Heidegger's early phenomenological account of technology, which states that a tool withdraws (*zurückziehen*) when we purposefully employ it in practical contexts (Harman, 2002, p. 21).

Each of the above represent distinct ways in which theorists from different disciplines have understood infrastructure to be invisible. Despite their differences, what each sense has in common is that what is considered invisible is foregrounded, or made visible upon breakdown. In order to perform repairs on an inner city street, for example, it is often necessary to open the road; at which time, systems of pipes and cables that were previously hidden from view are suddenly rendered visible. Likewise, the bewilderment that someone from a distant culture experiences in trying to navigate bus, subway and road systems in our city of residence reminds us of the skills that are actually required for us to make our way to and from work each day. The third sense of invisibility also vanishes upon breakdown; during a blackout, for example, we can attend no longer to the information on our computer screens at the expense of considering the state of the local electricity network.

Yet despite the obvious popularity of conceptions of infrastructure as invisible when functioning correctly, the idea itself has not escaped criticism. A growing body of literature in philosophy and the social sciences has sought to explore the limitations of this perspective, leading some scholars, such as Brian Larkin to suggest that the idea that infrastructures are invisible;

> is a partial truth and, as a way of describing infrastructure as a whole, flatly untenable. Invisibility is certainly one aspect of infrastructure, but it is only one and at the extreme edge of a range of visibilities that move from unseen to grand spectacles and everything in between (Larkin, 2013).

One common strand of criticism draws attention to the way infrastructure remains visible through the effects it has on the character of our daily lives. For example, take the development and spread of artificial lighting made possible by the advent of electric power industries in the late nineteenth century. While it may indeed be common to take access to electric lighting for granted, the various ways in which many of us today experience life after dark are so closely coupled to the availability of this technology that it may be impossible to conceive of our experience apart from it. It is on the basis of similar concerns that philosopher of technology Don Ihde distinguishes the way in which a technology, such as eyeglasses, withdraws from our attention as a result of their embodied use, from the way in which automatic technologies such as refrigerators recede from view. The difference is that while eyeglasses *must* withdraw from our attention in order for us to see through them, automatic technologies recede from view insofar as they do not require constant interaction and can therefore also remain present in experience. For this reason, Ihde argues, we do not experience technologies such as electric lighting as transparent or opaque, but rather 'as a kind of "absence". The technology is, as it were, "to the side". Yet as a present absence, it nevertheless becomes part of the experienced field of the inhabitant, a piece of the immediate environment' (Ihde, 1990, p. 109). To illustrate this point, Ihde appeals to the example of;

> the electric home, (where) there is virtually a constant hum of one sort or the other, which is part of the technological texture. Ordinarily, this "white noise" may go unnoticed, although I am always reassured that it remains part of fringe awareness, as when guests visit my mountain home in Vermont. The inevitable comment is about the silence of the woods. At once, the absence of background hum becomes noticeable (Ihde, 1990, p. 109).

Another serious problem with the claim that infrastructure is invisible *by definition* however, concerns the way it generalizes the experience of inhabitants of cities in the global North while at the same time ignoring the diverse ways in which infrastructure is experienced in other geographical locations. For this reason, few theorists today affirm the invisibility of infrastructure without also qualifying this perspective by recognising how the;

> notion of infrastructure as an invisible, smooth-functioning background "works" only in the developed world. In the global South (for lack of a better term), norms for infrastructure can be considerably different. Electric power and telephone services routinely fail, often on a daily basis; highways may be clogged beyond utility or may not exist; computer networks operate (when they do) at a crawl (Edwards, 2003).

Examining a wider range of global perspectives has therefore revealed important limitations to the idea that it is the nature of infrastructure to withdraw from view. Yet while these insights are important, it is not my intention to pursue them further here. Rather I want to examine how considering different ways in which it is possible to conceive of the function of infrastructural technologies exposes a different set of concerns surrounding the common tendency to ascribe an epistemic function to infrastructural breakdown and malfunction. In other words, my goal will not be to challenge the traditional phenomenological conception of infrastructural technologies. For in spite of the problems outlined above, I think there remain valid senses in which infrastructure in certain parts of the world is experienced not only as withdrawing from attention when functioning correctly, but also as gaining visibility upon breakdown. At the same time however, most accounts which examine this transition have neglected to examine exactly *how* infrastructure is rendered visible in such situations and whether or not such a transition should be understood to serve a positive epistemic function. It is to this question that we turn in the next section.

6.3 Infrastructure as Form and Process

In this section we will explore two different ways in which it is possible to conceptualise how a technology functions. The first conception is what I call "technology as form" and results from considering the efficiency of technological artefacts to stem exclusively from the way they were produced; i.e. from the quality of their design, materials and construction. From this perspective, the agency of a technology is contained within the material components of the technology itself. Good examples of devices that are often understood in this way include those which are commonly considered to operate with minimal human intervention, such as robot vacuum cleaners or automated bread makers. However, understanding artefacts in this way often requires us to overlook various kinds of activities the user contributes to the function of these forms of technology. Consider the example of the mechanical bread maker, a device commonly considered to perform the task of baking on behalf of the user. As Harry Collins notes, this perception can only be sustained by overlooking the skills that are required in order to use the device, for 'the user of the mechanical bread maker has to know how to read and follow instructions, purchase ingredients, clean the machine and use it and repair it, or to get it repaired, if necessary' (Collins, 2010, p. 143). Similar considerations also apply for robot vacuum cleaners which require maintenance on a regular basis such as cleaning the filters, removing hair wrapped around the wheels and bearings or debris that collect in the inner parts of the device. Yet the status of such technologies as "labour saving devices" encourages us to overlook the significance of such activities when we consider how they function. The frequency with which these forms of activity are overlooked or downplayed can be understood to reflect not only common ways of considering particular artefacts, but also general tendencies in the history of technology to focus disproportionate attention on the production of technologies over their use and maintenance (Pursell, 1995).

The conception of technology as form therefore represents a powerful ideal which can structure not only how we think about technologies in use, but also how we understand cases of breakdown in which those same technologies cannot be used. For insofar as this perspective understands the agency of a technology to be bundled up within the device itself, when it does occur, breakdown and malfunction are often conceived in terms of a physical state of an artefact; when something comes loose and the device stops functioning as we expect. There are two main reasons why this perspective does not consider breakdown to grant insight into the nature of the technology itself; *firstly,* rather than reflecting an essential aspect of the nature of the technology itself, breakdown appears as an anomalous event stemming from a failure to attain ideals of permanence and stability in the design and construction of a particular technology. *Secondly,* because this conception is premised upon ideals of design as establishing capacities to operate with minimal human intervention, activities such as repair work or maintenance are not understood to play a central role in sustaining technological function and therefore hold little epistemic potential for helping us understand the nature of the technology itself.

While it may be most natural to identify this conception of technology in common attitudes towards devices like those mentioned above, similar assumptions are often visible also in the way larger technical systems, like infrastructures, are conceived. As Stephen Graham and Nigel Thrift note, for example, it is not uncommon for infrastructure to be regarded as permanent and stable technical systems, the efficiency of which derives from the physical design and production of their material components rather than the ongoing creative and improvisatory work of individuals;

> Cultures of of normalized and taken-for granted infrastructure use sustain widespread assumptions that urban 'infrastructure' is somehow a material and utterly fixed assemblage of hard technologies embedded stably in place, which is characterized by perfect order, completeness, immanence and internal homogeneity rather than leaky, partial and heterogeneous entities (Graham & Thrift, 2007).

In an influential paper, Kaika and Syngedouw argued that perceptions of infrastructure like this emerge as a result of a process of fetishization, in which the form of such systems as commodities are 'severed from (their) historical and geographical (hence social) process of production' (Kaika & Swyngedouw, 2000), a process which is intimately connected to the common tendency in the global North, to hide from view, not only the systems themselves, but also of the activities of maintenance and repair upon which they depend. Understanding infrastructure in this way, as an exclusively material achievement can also influence how cases of infrastructural breakdown are perceived. As Paul Edwards notes, the widespread tendency to refer to infrastructural breakdown as a "technological failure", functions to obscure the social dimensions of such events by reinforcing conceptions of infrastructure as a material system (Edwards, 2003). Understanding the apparent permanence and stability of infrastructural services to be a material achievement means that breakdown and malfunction can only appear as an anomaly; as a failure of design and construction. From this perspective, we have little reason to pay attention to the work performed by technicians in seeking to understand the nature of infrastructure, insofar as this perspective casts these forms of activity as addressing anomalous, rather than essential aspects of the nature of the technology itself.

Technology as form represents one amongst a variety of possible ways of conceptualising how a technology functions. In what follows, I want to contrast the foregoing standpoint with another which ascribes a more active role to users in sustaining the function of technologies. What I call "technology as process" is the view that the efficiency of a technology depends not only on the physical design and construction of a device, but also on practices of maintenance performed by users.[2] According to this perspective, maintenance is understood broadly, as any activity performed by users which contributes to the stable functioning of a technical artefact or system over time, a category which includes, but is not limited to, practices of cleaning, repair, modification or the replacement of parts. This conception of technology is most often applied to certain classes of objects, such as car engines which are naturally associated with practices of maintenance such as monitoring both oil and water levels, periodically replacing minor components and cleaning the contacts of the battery. For this reason, when purchasing a used car, potential buyers are typically interested not only in the quality of the production of the engine itself, but also in the way it has been previously used and maintained.

This way of understanding the function of a technology is best captured through analogy to the classic game of hoop rolling, in which players attempt to keep a hoop rolling upright for long periods of time by running alongside it while striking it with a stick. The rolling of the hoop provides an apt metaphor for this way of understanding technological function precisely because it is a process whose continuation relies not only on the physical form of the hoop, but also on constant improvised corrections made by the player. Rather than being bundled up within the device then, the agency of a technology understood in this way is distributed and emerges from the relation between a user and an artefact over time. For this reason, breakdown assumes a more ambiguous status in this context. Whether a device is broken depends here not only on the physical state of the artefact but also on the capabilities of the user. In this way, breakdown becomes relativized; while manufacturing flaws or problems arising through wear and tear may render a device inoperable in the hands of a novice, a skilful user can often find ways to work around such problems in order to maintain the function of the technology. Furthermore, because enduring function is premised upon the performance of maintenance activities by the user, many forms of breakdown are considered part of the normal life cycle of a technology rather than anomalous events. Understanding a technology as process therefore not only relativizes, but also *naturalizes* a wide range of different forms of breakdown.

Returning again to the question of infrastructure, we find that just as it can be experienced as form, so too can it be experienced in this way, as process. A central

[2] For a further elaboration on technology as process, see my Young (2021).

goal of much recent work on infrastructure in the social sciences is to illuminate exactly this perspective by revealing how the apparent stability of infrastructural systems emerges as a consequence of messy practices of support, maintenance and repair; that infrastructures are in fact 'always precarious achievements' (Graham, 2010). Denis and Pontille's recent study of Paris subway signs, for example, highlights how the efficiency of signs must be understood not only as a result of design, but also of the largely improvised practises of maintenance and repair which preserve the function of signs in the face of physical degradations such as mold, discoloration, graffiti, vandalism or theft (Denis & Pontille, 2014). In addition to emphasizing how signage systems are made possible through constant human activities of maintenance, repair and management, Denis and Pontille's study also illustrates how breakdown is often naturalized when a technology is approached from the process perspective;

> All infrastructural systems are prone to error and neglect and breakage and failure, whether as a result of erosion or decay or vandalism or even sabotage. Indeed, many such systems are premised on a certain degree of error or neglect or breakage or failure as a normal condition of their existence (Graham & Thrift, 2007)

Yet despite their importance, the crucial role that these activities play in sustaining infrastructural technologies is not always apparent. In the same way that we can overlook the significance of our own actions in sustaining the function of technologies such as automatic bread makers or robot vacuum cleaners, so too can we overlook the actions of others by interpreting maintenance and repair work as responses to anomalous occurrences rather than natural events in the life of a technology. Here we see how basic features of the process perspective are often obscured from view, and this marks another important sense in which a technology can be invisible. This invisibility is not only a result of failing to see the significance of maintenance, but sometimes of failing to see it at all. This occurs in part because, it is extremely common for maintenance and repair activities to be deliberately situated out of view. Maintenance of services such as subways and roads, for example, often occur 'after hours' when they are expected to cause least disruption. One of the central goals in the provision of infrastructural services is often to cultivate a sense of reliability and permanence that encourages repair work to be performed out of sight. Furthermore, the tendency for maintenance to recede from view is further exacerbated by the fact that these forms of work are often not evenly distributed throughout society. As Lee Vinsel and Andrew Russell note in their overview of the historical study of maintenance practices, 'maintenance work is (often) performed by people who do not hold positions of wealth or privilege in any given society. To focus on maintainers is therefore to look at the lives and labor of women, ethnic and racial minorities, migrants and laborers whose positions are contingent and vulnerable' (Russell & Vinsel, 2018). Similar observations are made by Graham and Thrift, who also note how 'it is often the poorest people in a society that are concentrated in maintenance and repair jobs' (Graham & Thrift, 2007). Like those aspects of infrastructure explored above then, a growing body of research points to ways in which the forms of activity required to keep such systems going also often become relegated to the margins of visibility.

6.4 The Perspectival Flexibility of Technology

Technology as form or process therefore represent two fundamentally different perspectives through which technologies can be viewed. When approached as form, technologies appear as permanent and stable material arrangements which derive their efficiency primarily from their design and construction. When viewed as processes on the other hand, the efficiency of technologies is understood to emerge, often precariously, at the interface between physical design and the constant and coordinated creative efforts of individuals distributed through time and space.

Its important to keep in mind that these different ways of conceptualising function have little to do with the *kinds* of technology that we consider. This is reflected simply by the fact that the same technology can appear *either* as an artefact whose capacities are given in its physical design, or as a process sustained by human activity. A concrete bridge provides us with a good example with which to illustrate this inherent flexibility in the way the function of technologies can be conceived. Partly due to the emphasis our culture places upon ideals of technological innovation, it is common to think of such large physical structures primarily as achievements of design and engineering. From this perspective, the standing of the bridge is understood to result from the way it was constructed; the conformity of its design with accepted principles of structural engineering and the extent to which its construction adhered closely to these requirements. This perception is reinforced by the kinds of stories that we often like to tell about failures of structural engineering, many of which revolve around catastrophes resulting from design flaws. Yet the emphasis our culture places upon design also makes it easy to overlook the fact that bridges are not just made, they are also maintained.

The maintenance of bridges can be understood to divide into two kinds; the first includes maintenance activities which are based on existing knowledge of predictable processes and thus can be often be scheduled in advance to prevent common forms of problems from occurring. These include forms of preventive maintenance such as painting and washing the bridge, lubricating bearings, fixing pavement cracks or cleaning and resealing expansion joints. The second are those based upon inspection or observation which respond to less predictable problems that emerge throughout the life of the structure and include some forms of preventative maintenance, such as snow and ice removal or wearing surface repairs, and all forms of corrective maintenance such as the repairing of corroded piers or structural steel, the replacement of bearings or the fixing of impact damage (Alampalli, 2014). Together, these forms of maintenance play a crucial role in enabling bridges to reach their life expectancy of 75–100 years (Alampalli, 2014). The importance of maintenance practices to the structural integrity of bridges signals how, like the rolling of the hoop, the standing of the bridge can also be understood to depend on constant, often improvised human activities which guide and support the structure through a constantly changing environment.

Technology as form or process therefore represent two fundamentally different perspectives through which the same technology can appear. Just as bridges can be apprehended in either way, so too can many other forms of technology. Consider smartphones as an example; while we don't normally consider such devices to depend on maintenance, their function can nonetheless be understood to depend on a host of activities performed after the production of the device, such as cleaning or charging the device, managing storage and software, not to mention the frequent upgrades required to maintain the integrity of the operating system in a digital environment which is constantly evolving. Whether we understand a particular technology as form or process, does not therefore depend on the *kind* of technology we consider, but rather on how we understand the significance of activities performed upon and with the device during use.

The fact that the same technology can be viewed either as form or process makes it possible not only to compare both modes of perception, but also to question how they are related to one another. The example of the bridge helps illustrate what is perhaps the most important feature of this relationship because it reveals the extent to which apprehending the technology as form is derivative of the deeper understanding gained through approaching it from the process perspective. There are two general ways in which an understanding of the bridge as form can be understood to be derivative; *Firstly*, understanding the structural integrity of the bridge to be exclusively a product of design and construction represents an impoverished conception because it overlooks much of the work that is actually required to keep bridges standing. Unlike the process perspective then, perceptions of technology as form are sustained through the negation of the significance of human labour to the ongoing life of a technology. *Secondly*, the perception of the bridge as form is derivative in the sense that it is itself made possible by the activities it ignores. For the apparent stability over time that this approach attributes to the production of a technology results not only from sound design, but also from continuous activities of maintenance and repair.

An increasingly common goal in studies of infrastructure in philosophy and the social sciences is to effect a shift in perspective, away from the more derivative understanding of infrastructural technologies as form and towards the deeper conception of infrastructure provided by the process perspective. However, as noted above, the key to achieving such a transition is almost invariably identified with the experience of breakdown. As Stephen Graham and Nigel Thrift note, for example, in cases of breakdown 'the sudden absence of infrastructural flow creates visibility…interruptions in infrastructural connections can be seen as a form of 'unblackboxing'' (Graham & Thrift, 2007). Similarly Steven Jackson draws attention to the 'world-disclosing properties of breakdown' by suggesting that it is 'precisely in moments of breakdown that we learn to see and engage our technologies in new and sometimes surprising ways' (Jackson, 2014). Yet whether or not breakdown should be ascribed such an epistemic function depends crucially on *how* infrastructure is rendered visible in such cases. While some studies have illustrated how breakdown makes visible our reliance upon infrastructural systems,[3] or previously hidden forms

[3] See for example, Graham and Simon (2001), p. 23 and Graham (2004), p. 247, n. 1.

of maintenance and repair work upon which it depends, it remains to be seen whether or not such experiences can actually yield insight into the nature of such technologies as objects of maintenance. In other words, can the failure of infrastructure reveal the nature of such technologies as processes? The final section of this paper seeks to outline different reasons why we might be inclined to answer this question in the negative.

6.5 Scrutinizing the Epistemic Potential of Breakdown

The first reason we might be sceptical about the epistemic function of breakdown is that according to the proposed framework, there appears to be nothing about the experience of breakdown or repair *in itself* which is incompatible with a conception of technology as form. As noted above, the diametrically opposed viewpoints of technology as form and process determine how breakdown and the work required to remedy it are experienced. When viewed as processes, activities of repair and maintenance appear as constitutive of the nature of infrastructure technologies. When viewed as form on the other hand, the same activities are understood as addressing the failure to attain an idealized permanence and stability in the design and construction of a technical system. In other words, while the experience of breakdown may create an opportunity to alter one's understanding of the function of a particular technology, it does nothing to *require* such a shift in perception.

Empirical studies which examine the relationship between users and maintainers of technology offer further reasons for thinking that experiences of breakdown often fail to result in a deeper understanding of technology. Ethnographic studies of repair (ESR) represent a branch of anthropology that explores the nature of work through the examination of detailed case studies of particular technological practices such as the maintenance of copy machines (Orr, 1996) or building systems (Henke, 2000). Because they provide a valuable resource for understanding the perspectives of maintainers on the nature of technology and practice, such studies yield important insights relevant to the question of the epistemic potential of breakdown.

A central premise of many studies in ESR is that repair work consists in fixing both things *and* people. Henke's study for example, demonstrates how repair work on heating and cooling systems in buildings often involves working with customers perceptions that a system is malfunctioning—a perception that can be 'fixed' by using the visible readout of instruments to demonstrate that the system is in fact functioning correctly (Henke, 2000). Orr's study of copy machine technicians draws similar conclusions in highlighting how the work of maintaining and repairing copy machines includes both fixing the actual machine itself and modifying the way in which customers use the machines through training, signs and modifications to the machines themselves. Such studies thus reinforce insights from the preceding section, in which it was suggested that for a technology encountered as process, breakdown represents not a physical state of a device, but rather a *relation* between a device and a user. Indeed, Orr utilizes just such a relational concept of breakdown to characterize the nature of copy machine repair work itself, noting that;

> the problems encountered by technicians are most fundamentally breakdowns of the interaction between customers and their machines, which may or may not include a malfunction or failure of some machine component (Henke, 2000).

Studies in ESR also provide support for the idea that maintainers inhabit fundamentally different perspectives to the user on the nature of technology itself. This is especially visible in Orr's study, which notes how users often misunderstand the nature of maintenance work itself, by failing to distinguish between work performed to clean or fix the machines or by neglecting the social dimensions of repair by understanding this practice exclusively in terms of fixing machines. Rather than describing the experience of breakdown as a point of possible convergence between the perspectives of the user and maintainer then, studies in ESR often emphasize the extent to which such differences remain irreconcilable throughout periods of malfunction. In doing so, such studies draw attention to cases in which the epistemic function of breakdown is limited, or even non existent, and therefore encourage us to question whether the same may apply for larger technical systems such as infrastructural services.

Insofar as the primary objective of research in ESR is often to examine the nature of work, case studies typically focus on everyday cases of repair rather than uncustomary repairs performed in the wake of catastrophic failures. This brings us to what is perhaps the central reason for questioning the epistemic function of breakdown. At least for the global north, disruptions to infrastructural systems represent the exception rather than the norm. If our goal is to gain insights into the nature of infrastructure then focusing our attention on anomalous events and the responses to them appears to overlook the most obvious, and direct source of information; the constant practices of maintenance and repair which sustain their function over time. A central theme in ESR literature is that breakdown and repair are not restricted to moments when infrastructure ceases to work as expected, but rather are both ubiquitous and necessary, awaiting discovery on the margins of visibility even when a system appears to be functioning normally. Henke, for example, seeks to challenge the idea that repair responds to abnormal events by arguing that;

> Repair is not at the margins of order, waiting to be deployed if something goes wrong. Instead it is a practice at the centre of social order: repair work makes workplaces normal (Henke, 2000)

Privileging cases of complete failure therefore appears to commit the same error as conceptions of technology as form in that it encourages us to overlook the many forms of breakdown upon which the normal function of infrastructural systems are premised; processes such as the gradual discoloration of subway system signage or the subtle corrosion of concrete bridge piers. For it is here, in the early hours of the morning, when expansion joints are cleaned and protective coatings are applied to bridge decks, that we find the forms of human activity necessary to understanding infrastructural technologies as processes.

It is perhaps ironic that repudiating the epistemic privileges traditionally accorded to breakdown represents an affirmation of important aspects of Heidegger's original phenomenological conception of technological experience. For despite widespread reference to Heidegger's work to illustrate the epistemic potential of breakdown in studies of infrastructure,[4] Heidegger did not himself consider the process by which technology is rendered visible through breakdown to possess a positive epistemic function. As noted above, Heidegger considered the way we experience a malfunctioning technology to be a 'deficient mode of concern' (Heidegger, 1962, p. 103). For Heidegger, things in the world reveal themselves to us originally both in relation to our practical goals and to the wider background of equipment and practices upon which their efficacy depends. Encountered in this way (*zuhanden*), entities are not visible to us as discrete objects but rather as aspects of an equipmental totality which enable us to perform specific tasks. As is well known, those same entities become visible when they, for various reasons, prevent us from achieving our goals. But visibility is not gained for free. As Heidegger himself notes, when we are truly unable to use something, it loses its character as ready-to-hand (Heidegger, 1962, p. 103) and is encountered instead as present-at-hand (*vorhanden*), a deficient mode of concern in which relations to our projects and the wider background of equipment are lost, allowing the entity in question to appear as a discrete object that possesses a specific set of properties. For Heidegger then, rendering something visible as an object for theoretical reflection functions to distort that which we seek to understand. As Graham Harman notes, 'it is the nature of tool-being (*Vorhendenheit*) to recede from every view. In the strict sense we can never know just *what* equipment is (because) tool-beings are encountered only once they have washed up dead on shore, no longer immersed in their withdrawn reality' (Harman, 2002, p. 4).

While I share Heidegger's scepticism surrounding the epistemic potential of breakdown, I also think we have good reasons today to resist equating visibility with distortion in this way. We need look no further than the social sciences for clear demonstrations of the capacity of detailed theoretical analysis to help us to gain a deeper understanding of aspects of everyday life which often reside in the background. Throughout the latter half of the twentieth century, sociologists and feminist theorists exploring the theme of invisible labour for example, produced a wealth of studies that succeeded in utilising ethnographic analysis to make visible activities which had previously been rendered invisible under then dominant modes of conceptualizing work. These studies, which often focused on gendered practices performed inside the home 'were crucial academically, enabling scholars to "see" the work and visualize workers in places previously invisible to conventional sociology' (Crain, Poster, & Cherry, 2016). If gaining a deeper understanding of technological systems depends, as Steven Jackson suggests, on a similar shift by which maintenance and repair are made 'empirically and conceptually familiar, even normal' (Jackson, 2014), a good place to begin will be by applying the same careful ethnographic methods of analysis to everyday practices of infrastructural maintenance.

[4] See for example, Graham and Thrift (2007) and Jackson (2014), p. 230.

6.6 Conclusion

While my goal in this article has been to question the extent to which experiences of breakdown can help us to understand the nature of infrastructural technologies, I do not mean to suggest that such experiences are void of epistemic potential altogether. For as we have seen, the experience of infrastructural failure can indeed render certain things visible. We may be surprised, for example, to find that the failure of electricity networks causes failures in other systems, such as water supplies, revealing connections among what we had previously considered were separate systems. Or the failure of local transportation networks can lead to a greater appreciation of the extent to which daily life depends on the consistent provision of such services. Yet such cases appear to provide more insight more into how we understand and engage infrastructure than yielding a deeper understanding of the ways in which the function of such systems are sustained over time. In particular, they fail to challenge what I have called perceptions of technology as form, which overlook how the stability of infrastructural systems is premised upon continuous practices of maintenance and repair. Instead, when viewed from this perspective, infrastructure appears as a material assemblage which attains efficiency as a technical system through the quality of both the design and construction of its physical components. As I have argued, such conceptions of the function of infrastructural technologies are fundamentally misguided and arise through the negation of specific forms of human activity. Electricity networks, for example, can only appear as autonomous and stable technical systems through the erasure of continuous activities of maintenance and repair work upon which they depend. This process of erasure can take different forms; it can occur as a result of positioning maintenance practices out of sight so that they go unnoticed by the majority of people who utilize infrastructural services. Alternatively, erasure can also occur when activities of maintenance and repair stand in plain view yet we fail to appreciate their full significance in sustaining the efficiency of such systems over time. Examples of such failures can include overlooking how the removal of ice from bridge decks contributes to the integrity of the structure itself over time, or misinterpreting cases of maintenance and repair as remedying failures of design or construction, rather than necessary elements upon which the efficiency of infrastructural systems depend.

Yet it is not only practices of maintenance and repair, but also the forms of breakdown towards which they are ultimately directed that can be experienced differently depending on how we conceive of the nature of technology itself. For when infrastructural breakdown does occur, it can be accommodated within the perspective of technology as form by explanations which attribute the cause of breakdown to technical failures and design flaws rather than deferred maintenance or the defective management of repair schedules. Because experiences of breakdown do not necessarily conflict with existing assumptions concerning the nature of certain technologies, they can serve to reinforce rather than replace fundamental misconceptions regarding the role of human agency in establishing the efficiency of infrastructural systems. For this reason, we should be wary of the special epistemic privileges that have traditionally been accorded to experiences of breakdown by researchers in the social sciences. For while the malfunction of infrastructural systems may indeed remind us of their presence, exactly *how* we should understand them to be rendered visible in such cases is not a question that should be overlooked. Throughout this paper I have sought to show how the interpretive flexibility of breakdown means that existing misconceptions regarding the nature of technologies can result in infrastructural technologies being made visible in a derivative way.

This study contends that if we want to understand more about the technologies we use then we might want to attend less to cases of breakdown in which such systems cannot be used, and focus more instead on the kinds of activities that occur while such technologies are *in use*. This will involve shifting the focus in historical studies of technology away from traditional narratives which privilege activities of design and construction and more towards the analysis of everyday practices of maintenance, repair and modification. While the recent turn towards use in historical studies of technology and the growing body of literature that has resulted[5] has prepared the ground for future studies of this kind, more work remains to be done. For if the black box of infrastructure can indeed be opened, it will be done so not by focusing on failure, but rather by turning an ethnographic eye towards everyday practices of maintenance and the conception of technology they imply.

References

Alampalli, S. (2014). Designing bridges for inspectability and maintainability. In M. Frangopol & Y. Tsompanakis (Eds.), *Maintenance and safety of aging infrastructure*. London: CRC Press.

Bijker, W. E. (1995). *Of bicycles, bakelites and bulbs: Towards a theory of sociotechnical change*. Cambridge: MIT Press.

Bowker, G. C., & Star, S. L. (1999). *Sorting things out: Classification and its consequences*. Cambridge: MIT Press.

Collins, H. M. (2010). *Tacit and explicit knowledge*. Chicago: University of Chicago Press.

Crain, M., Poster, W. R., & Cherry, M. (2016). Introduction: Conceptualising invisible labour. In M. Crain, R. Poster, & M. Cherry (Eds.), *Invisible labor: Hidden work in the contemporary world* (pp. 3–27). Oakland: University of California Press.

Denis, J., & Pontille, D. (2014). Maintenance work and the performativity of urban inscriptions: The case of Paris subway signs. *Environment and Planning D: Society and Space, 32*(3), 404–416.

Edgerton, D. (2006). *The shock of the old: Technology and global history since 1900*. London: Profile Books.

Edgerton, D. (2010). Innovation, technology or history: What is the historiography of technology about? *Technology and Culture, 51*(3), 680–697.

Edwards, P. N. (2003). Infrastructure and modernity: Force, time and social organization in the history of sociotechnical systems. In T. Misa, P. Brey, & A. Feenburg (Eds.), *Modernity and technology* (pp. 185–226). Cambridge: MIT Press.

Graham, S. (2004). Constructing premium network spaces: Reflections on infrastructure networks and contemporary urban developments. In R. Hanley (Ed.), *Moving people, goods and infor-*

[5] See for example, Pursell (1995), Edgerton (2006), Edgerton (2010), Oudshoorn and Pinch (2003), and Oldenziel (2001).

mation in the 21st century: The cutting edge infrastructures of networked cities (pp. 246–267). New York: Routledge.
Graham, S. (2010). When infrastructures fail. In S. Graham (Ed.), *Disrupted cities: When infrastructure fails* (pp. 1–40). New York: Routledge.
Graham, S., & Simon, M. (2001). *Splintering urbanism: Networked infrastructures, technological mobilities and the urban condition*. London: Routledge.
Graham, S., & Thrift, N. (2007). Out of order: Understanding repair and maintenance theory. *Culture and Society, 24*(3), 1–25.
Harman, G. (2002). *Tool being: Heidegger and the metaphysics of objects*. New York: Open Court.
Heidegger, M. (1962). *Being and time* (J. Mcquarrie & E. Robinson, Trans.). Oxford: Basil Blackwell.
Henke, C. (2000). The mechanics of workplace order: Towards a sociology of repair. *Berkeley Journal of Sociology, 44*, 55–81.
Ihde, D. (1990). *Technology and the lifeworld: From garden to earth*. Bloomington: Indiana University Press.
Jackson, S. J. (2014). Rethinking Repair. In T. Gillespie, P. Boczkowski, & K. Foot (Eds.), *Media technologies: Essays on communication, materiality and society* (pp. 221–240). Cambridge: MIT Press.
Joerges, B. (1988). Large technical systems: Concepts and issues. In R. Mayntz & T. Hughes (Eds.), *The development of large technical systems* (pp. 9–36). Frankfurt: Campus Verlag.
Kaika, M., & Swyngedouw, E. (2000). Fetishizing the modern city: The phantasmagoria of urban technological networks. *International Journal of Urban and Regional Research, 24*(1), 120–138.
Larkin, B. (2013). The politics and poetics of infrastructure. *Annual Review of Anthropology, 42*, 327–343.
Latour, B. (1999). *Pandora's hope: Essays on the reality of science studies*. Cambridge: Harvard University Press.
Martin, B., & Richards, E. (1995). Scientific knowledge, controversy and public decision making. In S. Jasanoff, E. Markle, J. Petersen, & T. Pinch (Eds.), *Handbook of science and technology studies* (Revised ed., pp. 506–526). Thousand Oaks: Sage Publications.
Oldenziel, R. (2001). Man the maker, woman the consumer: The consumption junction revisited. In A. Creager, E. Lunbeck, & L. Schiebinger (Eds.), *Feminism in twentieth century science, technology, and medicine* (pp. 128–148). Chicago: University of Chicago Press.
Orr, J. E. (1996). *Talking about machines: Ethnography of a modern job*. New York: ILR Press.
Oudshoorn, N., & Pinch, T. (Eds.). (2003). *How users matter: The co-construction of users and technology*. Cambridge: MIT Press.
Pursell, C. (1995). Seeing the invisible: New perspectives in the history of technology. *Icon, 1*, 9–15.
Russell, A., & Vinsel, L. (2018). After innovation, turn to maintenance. *Technology and Culture, 59*(1), 1–25.
Star, S. L. (1999). The ethnography of infrastructure. *The American Behavioral Scientist, 43*(3), 377–391.
Strebel, I., Bovet, A., & Sormani, P. (2019). *Repair work ethnographies: Revisiting breakdown, relocating materiality*. Singapore: Palgrave Macmillan.
Young, M. T. (2021). Maintenance. In D. Michelfelder & N. Doorn (Eds.), *The Routledge Handbook of the Philosophy of Engineering*.

Mark Thomas Young is a postdoctoral researcher in the Department for Technology, Policy and Management at the Delft University of Technology. His research covers two fields; the Philosophy of Technology, where he focuses on the use of automating technologies and the History and Philosophy of Science, where he explores instruments, craft practices and tacit knowledge in the early modern period.

Chapter 7
Structure and Background: The Influence of Infrastructures on Human Action

Marcel Müller

Abstract Infrastructures are widely considered integral to modern societies. As the built structures of modern cities, infrastructures provide the necessities for human life, economic growth and political stability. Infrastructures not only connect rural and urban areas with each other, they also link cities, nations and continents in vast and ever expanding networks of technologies, political decision making and regulatory requirements. The following chapter is an attempt to provide orientation regarding how infrastructures might be analyzed by philosophical means and how some of their features might be understood in relation to human action and experience. Such insight is especially beneficial for a philosophy of the city, since the concepts of city and infrastructure are so deeply connected that they can hardly be distinguished from one another.

Keywords Infrastructures · Socio-technical system · Postphenomenology · Praxis · Habit

7.1 Introduction

Exposure to changing weather conditions and the struggle to satisfy the elemental needs that arise from both body and mind are constants of the human condition. The ability to change our relationship with nature and adapt to its demands on a large scale is paramount for the establishment and adaptation of human societies. Technology has a major impact on this adaptive process. Tools and machines solidify how we interact through them with the objects of our actions. Contemporary human life is deeply mediated by various small and large technological artifacts that range from watches and smart phones to cars and public transport systems. These technologies deeply structure our daily routines. Especially in cities, the presence of ever larger technological systems and artifacts is noticeable. Drainpipes and power

M. Müller (✉)
Mainz, Germany
e-mail: mueller.phil@posteo.de

M. Nagenborg et al. (eds.), *Technology and the City*, Philosophy of Engineering and Technology 36, https://doi.org/10.1007/978-3-030-52313-8_7

cables stretch under widespread street networks and enable a constant flow of water, electricity, information and power within and across the borders of our settlements, nation states and even continents. Some of these systems and artifacts, such as streets, water supply and sanitation, date back to ancient civilizations. Compared to these age old constants of human settlements, electric power systems and the internet are quite new.

Basic facilities like drinking water and waste water disposal, roads, traffic and other facilities with a similar importance for human thriving are referred to as *infrastructures*. They are considered to be the lifeblood of modern societies (BMI, 2009). Modern urban infrastructures connect rural and urban areas and link cities, nations and continents in vast and ever expanding networks of technologies, political decision-making and regulatory requirements. Infrastructures pose a problem for philosophical inquiry. Compared to other technological artifacts that are analyzed by philosophy and philosophy of technology, infrastructures not only shape everyday urban life; they also provide its material background and structural framework.

In the following chapter, I explore the influence of infrastructures on human life and action by focusing on the dynamics between their concrete materiality on one hand and their role in practical relations and technological networks on the other. The chapter follows a three-part structure. Since my analysis requires a broad understanding of the term *infrastructure*, I will give a short introduction to infrastructure research in the first part. Infrastructures are the research object and working area of various disciplines, which leads to a wide array of different infrastructure-concepts and potential starting points for philosophical inquiry. The second part is concerned with infrastructures as material background of urban life. Here, I employ a postphenomenological approach towards infrastructural technologies and technological systems combined with more traditional thoughts of philosophy of technology about the role of technologies-in-use. This is done in order to analyze the direct and indirect influences of infrastructures on human action. In the third part, I focus on the more profound connection between human *praxis* and infrastructures as material background. This part is mostly concerned with Sartre's thoughts about technological artifacts and the way human society is organized.

7.2 A Brief Overview on Infrastructure Research

Even though the remains of ancient streets and water management systems give a hint of the crucial role they have for human civilizations through recorded history, the term, infrastructure' is a relatively recent invention. In the United States defense strategy, the term *infrastructure* was initially used across the development of defense strategies by the US Airforce shortly after Second World War. The principal purpose of the term was to identify susceptible ground facilities exposed to air raids. Later, the term was used for those structures whose disruption would have an impact on the safety and security of the United States, e.g. telecommunications, power systems, banking, transportation, water supply systems and political systems (Rinaldi,

Peerenboom, & Kelly, 2001). More recent definitions, for instance those of the Critical Infrastructure Assurance Office in the United States and the German Federal Ministry of the Interior, emphasize the interconnectedness of infrastructures and their crucial and thus *critical* role for a nation's well-being. These so-called critical infrastructures are all "organizational and physical structures and facilities of such vital importance to a nation's society and economy that their failure or degradation would result in sustained supply shortages, significant disruption of public safety and security, or other dramatic consequences" (BMI, 2009, p. 4). Infrastructures can be roughly divided into technical basic infrastructures and socio-economic services infrastructures. The former include singular technologies and complex technological systems and networks, such as power supply, ICT, transportation, roads and water supply; whereas the latter include public health, emergency and rescue services, disaster control, political institutions, finance and media (BMI, 2009).

Nowadays, the term 'infrastructure' applies to a wide range of different institutions and technological systems, whether they provide a social, political or technical service. Within disciplinary and interdisciplinary discourses, and depending on political agendas, the identification of key features of infrastructures may vary. For example, sociologists and STS scholars analyze the supposed invisibility of infrastructures (Edwards, 2003), the supposed politics of artifacts (Winner, 1980) and the social construction of large technological systems (Bijker, Hughes, & Pinch, 1987).

In addition, infrastructures provide the theoretical and practical framework for a vast amount of technology-oriented disciplines and sub-disciplines which are concerned with the intricacies of urban life. Engineers focus on the materiality of built structures and the varying standards and regulations that have to be adhered if one wishes to bring innovation to outdated systems. Some architects focus on the ways in which networks of wires and pipes can be hidden within building design. Others try to find creative ways of utilizing traditional knowledge regarding the interrelationships between infrastructures and society. Infrastructures are especially prominent in the field of urban planning. Since the very concept of *city* is closely linked with infrastructure's large-scale provision of the necessities of human life, their network-character is emphasized.

Another way of analyzing infrastructures is through certain key concepts of critical infrastructure research. These include criticality (i.e. relative importance of infrastructures), vulnerability, resilience in coping with disruptions, and preparatory and preventive measures for disruptions (Engels, 2018). These concepts are especially prominent in risk research. Here, the likeliness of potentially harmful events is assessed by focusing on critical vulnerabilities among others.

However, one core characteristic that constitutes the very concept of *infrastructure* can be identified. It is the fact that some institutions and technological systems provide specific services or knowledge that other sometimes more complex services, structures, institutions, individuals or the general public depend on. Modern governments, for instance, can barely function without the proper means of distributing information. These means, such as fiber cables and servers, are dependent on a vast network of electricity which is itself in need of some sort of governmental regulation in order to guarantee proper connectivity. In this example, the government as political infrastructure, the information infrastructure and the electricity infrastructure are heavily reliant on each other in order to guarantee individual and network functionality. Their individual status as infrastructures is derived from the role they play in service provision to the whole social system; which itself can be considered an important infrastructure of modern societies. Hospitals and electrical grids are not important *per se*. Their importance is relative to the range of hardly substitutable services they provide to society. At the same time, the technical basic and socio-economic services infrastructures are socio-technical systems themselves. Their internal functionality is the product of norm-guided, highly complex human-technology cooperation. This core characteristic of infrastructures is due to their twofold nature, which Larkin describes as follows: "Infrastructures are matter that enable the movement of other matter. Their peculiar ontology lies in the facts that they are things and also the relation between things" (Larkin, 2013, p. 329). Such a dynamic between rigid material elements and technical, social and political relations (amongst others) is what makes infrastructures so challenging for philosophical inquiry.

Since infrastructures consist of both technical elements and social relations, infrastructures are most commonly referred to as 'socio-technical systems'. Rooted in systems theory, the concept of *socio-technical systems* originated from the assumption that engineers need tools to take both technical and social aspects of technologies into account. In his "Allgemeine Technologie", Ropohl (2009) distinguishes what he calls 'action systems' (*Handlungssysteme*) and 'technical object systems' (*Sachsysteme*). Action systems are characterized according to their ability to act, which means that these systems have the capacity to both formulate goals and transform "a starting situation into a final situation" (Ropohl, 1999, p. 66). The internal structure of action systems consists of subsystems that are coupled to process inputs, outputs and system states. Individual human beings can be modeled as such systems since they can set goals for themselves and because they operate based on inputs from the outside world. This is a very basic and simplified way of conceptualizing action systems.

Technical object systems operate similarly, but with the exception that, unlike action systems, they cannot set goals for themselves. A car can be modeled as such a system. It consists of a complex arrangement of sensors and meters (*information subsystem*) as well as a motor, tires and a chassis that guarantees its structural integrity as a whole (*execution subsystem*). According to the principle of the division of labor, both action and technical object systems can perform their respective tasks as subsystems of a socio-technical system, i.e. "an action system that relies both on human and technical function carriers" (Ropohl, 1999, p. 66).

Following this theory, urban traffic infrastructures can be modeled as socio-technical systems. They consist of action systems, such as individual human beings and regulatory social and juridical institutions, as well as technical object systems, such as cars, roads, traffic lights, etc. Human action subsystems perform actions and set the goals of the larger system, for example, through their need for transportation. Technical object subsystems, such as cars for instance, perform some of the tasks of action systems in a more effective and efficient way, as these subsystems generally move faster with a slower deterioration of functionality. Since both subsystems are coupled within various feedback loops, they mutually process each other's outputs as inputs. In action systems this processing may trigger the awareness of a necessary traffic regulation through the reasonable implementation of traffic rules and laws. New needs arise with increased efficiency, such as wider roads and more connections to other cities.

Systems theory and the theory of socio-technical systems are used to grasp the complexity of large infrastructural systems, which makes it easier to highlight and analyze the complex intrasystem interactions between human beings and technologies. A large body of STS-literature addresses the problems and challenges of *large technological systems*; a concept that was introduced by Bijker et al. (1987). Monstadt (2007) addresses the socio-technicality of large infrastructure systems by pointing out network-like structures, large spatial networking and a high intensity of economic capital among others.

Conceptualizing infrastructures as socio-technical systems provides a shared vocabulary and joint concepts for interdisciplinary research. In addition, it allows retracing of the evolution of these large technological systems and illustrates their entanglement in political, scientific, technical and social processes. Furthermore, it illustrates the complexity of human-technology interactions and the interdependency of infrastructures. Fiber-optic and power cables may be providing the passive and material guidelines for information and electricity alongside other infrastructures, but without standardized rules and regulations as implemented by socio-political systems and without power generation, these infrastructures are non-functional. Their complexity and functionality is a direct product of their embeddedness in complex interdependent systems (Rinaldi et al., 2001). For philosophy in general and philosophy of technology in particular, the intricate relations engendered by the high-level interdependencies within infrastructure systems can be challenging. The twofold nature of infrastructures has to be constantly addressed in order to understand the full complexity of their interaction with society.

The somewhat challenging apparency of infrastructures, which ranges from transparent "artificial environment" (Edwards, 2003, p. 189) to opaque structures in case of emergency, catastrophe or disruption, is a direct outcome of their embeddedness within cities and settlements. In her ethnographic study on infrastructures, Star (1999) defines nine properties of infrastructures. Five of those properties, namely *embeddedness*, *transparency*, *learned as part of membership*, *embodiment of standards* and *becomes visible upon breakdown,* are directly related to infrastructures' apparency. Star points out that infrastructures can easily become obscured through everyday use, familiarization and standardization. For example, sewage and drinking water systems are embedded into the built environment of cities in such a way that these systems become almost indistinguishable from one another, even though they need to be physically separated for hygienic reasons and incorporate vastly different technologies and regulatory laws. Their status as embedded urban water infrastructure outlines their functionality as providing both the means for the distribution of fresh water and for the disposal of waste water. As such, the transparency of urban water infrastructures is due to the fact that they can be utilized without having to be "reinvented each time or assembled for each task" (Star, 1999, p. 381). In most developed countries, this functionality is not questioned because the convenient provision of water is the rule and not the exception. Since these urban water infrastructures usually do not cause a disturbance and work properly and reliably, they become transparent in such a way that their presence and service provision is not questioned anymore. However, this effect of infrastructural transparency only applies to a certain extent. In case of disruptive events such as blackouts or fires, the importance of infrastructures becomes apparent. This can be a problem in countries with usually reliable infrastructures, since the population's dependence on the reliable services of these infrastructures is much higher than it is in countries with less developed or less reliable infrastructures. During disruptions, missing infrastructural services can become very much visible. Speaking about transparency as an essential property of infrastructures is thus questionable (Larkin, 2013).

Edwards circumvents this problem. He does not consider transparency to be a property of infrastructures. Instead, he focuses on the mechanisms by which infrastructures become transparent or invisible in the first place. A rather technical aspect of this transparency is due to the fact that infrastructures are deliberately hidden. For instance, urban water infrastructures are usually walled in and buried alongside streets, tracks and other infrastructures for aesthetic, practical and precautionary reasons. These infrastructures are constructed to be unobtrusive for everyday users, while still allowing unhindered maintenance by technicians and engineers. Depending on whether infrastructures are handled by lay people or specialists, their apparency changes from background to foreground. Besides these technical aspects, habituation, habits and the learning of social norms seem to play a major role in infrastructure apparency. Since "[h]uman attention naturally focuses on what changes, rather than what remains constant" (Edwards, 2017, p. 329), the presence of infrastructures shifts out of everyday awareness. Habituation is "defined as a behavioral response decrement that results from repeated stimulation and that does not involve sensory adaptation/sensory fatigue or motor fatigue" (Rankin et al., 2009, p. 136). This applies to most acquired interactions with infrastructures, reaching from public transport and water utility to left-hand/right-hand driving and more. The comparison of both Star's and Edwards' approaches shows that infrastructure apparency is not purely a product of the fact that the complexity of infrastructures is deliberately hidden. The relative awareness of everyday users and their daily habits play a major role in how the presence of infrastructures is perceived.

7.3 Material Background: Life Within Infrastructures

Modern cities are almost fully comprised of small technologies and large technological systems. Hardly a *naturally* grown piece of land or tree can be found. Abandoned areas that have been left to their own devices are sometimes the only exception where nature might have taken its course. For humans, urban life takes place alongside technologies. Most activities of modern human beings depend on a strong interaction with technological artifacts, like alarm clocks, toothbrushes, clothes, smart phones and cars. We cook our food on the stove, we sleep in wooden beds and cover ourselves with rainproof jackets. Technologies are a reliable way to satisfy our needs.

Contemporary philosophy of technology is starting to focus on these urban technologies and technological systems. Although human beings do not directly interact with these systems in the same way as they do with smaller technologies, these large systems are nevertheless influential. In the following section, I explore the way infrastructures, as material background of urban life, influence the actions, perceptions and experiences of human beings by employing a postphenomenological approach towards urban infrastructures. I will follow up on this with further philosophical thoughts on infrastructure technologies.

One of the main claims of postphenomenology is that technologies and human beings mutually shape and constitute each other within the relations they form. However, unlike classical phenomenology, postphenomenology does so by taking individual technologies into account, instead of focusing on the essence of intentional objects as deduced by their phenomenal structure. The cornerstone of postphenomenological thought is the theory of technological mediation, which upholds the argumentative line that the lifeworld of human beings is mediated through technologies (Ihde, 1990). Postphenomenology provides several types of human-technology-world relations which may help to illustrate the role technological artifacts play in a person's world-experience. They are not so much prototypical types of relations but can be understood as heuristic means which uncover the different positions of technologies within this experience. Ihde (1990) proposes four human-technology-world relations; embodiment, hermeneutic, alterity and background. In all of these relations, technologies mediate between humans and world. Embodiment relations describe the relations in which technologies change the bodily perception of their users while becoming perceptually transparent themselves. Glasses and hearing aids are typical examples of this relation. The former refract light to correct for errors in their user's retina, while the latter render acoustic signals perceivable by their users. In doing so, these technologies alter their user's individual bodily sense of space since, for instance, the head has to be turned differently in order to properly see or hear. Although Ihde focuses more on perception instead of action he mentions that "to embody one's praxis *through* technologies is ultimately an *existential* relation with the world" (Ihde, 1990, p. 72). Accordingly, technologies not only expand or delimit the perception of human beings but affect the way these human beings are existentially directed towards their surrounding environment. This environment can be completely comprised of technologies. Ihde describes relations with semi-automated and automated technologies as background relations. He exemplifies this with toasters, washing machines and air conditioning as well as shelters. Other more exaggerated forms of the shelter example are nuclear submarines and space stations which seal off their inhabitants from an otherwise hostile surrounding. Ihde (1990) claims that human action is also mediated by technologies in the background. What differentiates background technologies from technologies in the foreground is the fact that background technologies do not constantly have to be interacted with in order to work and thus condition the context of human life. Most of the relations between human beings and infrastructures take place within the range of embodiment and background relations.

The material background of modern cities is comprised of concrete roads, buildings and power cables. Depending on region and culture, pedestrians, bicycles, cars or public transport systems are more prominent. The close interaction with infrastructure technologies is most apparent in the way we traverse through and out of our cities and settlements. How we act in cities and how we experience the urban framework depends on the technologies we use. Ihde gives the example of a sports car that mediates the road in a different way than bicycles or shoes do. Through finely calibrated shock absorbers every bump in the road can be felt accordingly and micro movements of the steering wheel can have tremendous consequences (Ihde, 1990). However, cars that are designed for urban usage are generally smaller and slower. Their design delimits possible options for action, which makes them more suitable for urban environments. It is easier to find parking spots and, due to their decreased speed, the likeliness of crashes is reduced. Some cars have a city drive mode that allows smoother steering under 30 kph. These technologies mediate the urban environment differently and thus provoke different modes of interaction. They change the perception of urban spaces. However, these interactions with technologies have to be acquired and do not come naturally. Across the range of relations with urban technologies, the relation with roads and the urban environment via bikes and cars is a more embodied one. Nevertheless, the relation depends on large urban technologies in the background, such as street lights, traffic barriers, concrete roads etc.

Compared to the embodiment of more individual traffic technologies, the embodying of public transport systems is less apparent but no less crucial for their utilization. The interaction with buses, trams and trains is less active. We enter the cabin, maybe find a place to sit, and the public transport system does the work for us. Still, we have to be cautious not to miss our stop. We have to embody the movement of the bus in a different way than we would embody the movement of self-driven bicycles or cars. This is especially challenging since we do not drive it ourselves, which can leave us unprepared for changes in direction or speed. Public transport systems are great examples of mixed human-technology-world relations, since part of the relation is naturalized by embodiment, whereas the technological artifact works without our interaction. The accelerated traversal of long distances with public transport systems has an immense influence on the perception of time and space in cities. Traveling from *Brandenburger Tor* to *Technische Universität Berlin* takes about 16 min if traversed by subway, but it takes about 45 min to walk the same distance.[1] Although this might sound rather obvious, the implications of traffic and transit technologies on everyday lives are far from small. By decreasing travel time, distance traveled is itself experienced to be less. In reverse, spatially

[1] Data obtained through https://www.google.de/maps (Mai 27th, 2019).

closer areas that can be hardly reached by common modes of transportation are experienced to be further away (Handel, 2017). For regions with a strong traffic infrastructure, consisting of various infrastructural technologies such as good road conditions and reliable public transport, this change in experience might not pose a problem. However, for individuals and social groups that do not have access to these means of transportation, it can cause a radical change in spatial perspective, as well as in self-awareness and external experience.

The aforementioned relations between humans and infrastructures illustrate their direct and indirect influence on the experience, perception and action of human beings. In his article "Do Artifacts Have Politics?", Winner (1980) illustrates the potential of large technological systems to generate effects that exceed their simple use-context. He points out various biases in the design and use of technologies—one of which is the focus on technologies-in-use only. Focusing on technologies as mere tools can obscure the larger effects these technologies have on urban life and society. According to Winner, the architect Robert Moses' low-hanging bridges were used to allow the traversal of cars in and out of Long Island. The downside of this was that the height of these bridges discouraged the presence of buses on parkways below them. Since mainly white people could afford cars and since black people, among other marginalized groups, could only afford public transit, they could hardly reach certain areas (Winner, 1980). Edwards (2017) illustrates a similar problem by analyzing the racial coding of public transport in pre- and post-Apartheid South Africa. Public buses and a railway line were used to bring otherwise segregated black laborers into white areas. Since these means of transport were less reliable, walking was the fastest way of commuting to work. The existential experience of racial inequality was, and still is, increased by the tension between the possible provision of public transport and the lack of its real-life instantiation (Edwards, 2017).

In the case of road construction in modern day Peru, Harvey and Knox (2012) identify at least three promises of emancipatory modernity which these newly built roads seem to instantiate. By connecting rural and urban areas, roads promise political freedom and economic prosperity. Since colonial times, such ideals have been far from the reality of marginalized groups (Harvey & Knox, 2012), not only in Peru. This way of thinking about infrastructures is usually situated in a horizon of technical systematicity, innovation, planning and construction. Infrastructures are associated as connectors between rural and urban areas and between nation states and even continents. Furthermore, they are also thought of as tools to potentially control and regulate societal processes. Infrastructures can thus be considered as having a powerful societal impact. They affect large groups of people far beyond their initial design.

The influence of infrastructure design becomes apparent in the research of *urban mobility cultures*. This concept is used to describe dominant modes of transport in connection to the social fabric of cities. Different clusters are distinguished that categorize cities according to their mobility culture. For instance, cycling cities are characterized by below average population size and settlement density, while generally being smaller than auto-oriented cities. In return, auto-oriented cities have more than average car-related businesses, combined with a generally negative *fun*-factor for cyclists and less than average road quality. Another cluster are transit cities. These are characterized by a generally lower household income in addition to a lower price for seasoned public transport tickets (Klinger, Kenworthy, & Lanzendorf, 2013). A similar concept is that of *urban fabrics*, which considers how the land use patterns change and react to the transport infrastructures and priorities of the city. This type of relation is illustrative of infrastructural systematicity, since it shows the complex relations between urban technologies and social processes. According to the transport priorities in the city, similar characteristics to that of urban mobility cultures can be derived. Walking cities, for example, have a high density, mixed land use and narrow streets; in contrast to transit cities, which generally exhibit medium density and a more diverse land use along tram routes. In their book "The End of Automobile Dependence", Newman and Kenworthy (2015) use the theory of *urban fabrics* to explain the relationship between what cities look like, what modes of transportation become dominant and their respective net energy utilized for transportation purposes, in order to highlight the vulnerabilities that arose through the existing automobile-based urban planning.

Infrastructures express the rationale of artificial enhancement of natural processes in order to satisfy human needs more effectively and efficiently. This rationale is most prominently discussed in Heidegger's "*Die Frage nach der Technik*" (2009). In this essay, the author compares ancient and modern technologies, especially in the way they change human's relation to nature. He claims that the rationale behind modern technologies reveals nature as a *standing reserve* (*Bestand*) for human utilization. Heidegger illustrates his thoughts by focusing on two infrastructures. The first infrastructure is an ancient windmill, which produces energy only if the wind blows. It is harmoniously embedded into nature and benefits from its forces. The second infrastructure is a modern power plant, which harvests the river's flow. Instead of being built on the river, the river itself is *built* (i.e. redirected) into the power plant. According to Heidegger, modern technologies challenge the forces of nature instead of working with them (Heidegger, 2009). This illustrates a tension between an ancient technology-concept, in which technologies are harmoniously situated in a more caring and preserving human-nature relation, and a modern one characterized by usurpation and utilization. Heidegger demonstrates the complexities of philosophic thought about technologies. He seems to assume a certain original human-world relation that is altered through the rise of modern technologies.

This alteration of human-world relations is seen in Graham's concept of cyborg cities, which mediate society, nature and culture. For instance, urban water infrastructures "radically shape urban biospheres whilst metabolizing waste matter from urban bodies" (Graham, 2010, p. 11). These infrastructures actively shape the manner in which nature is perceived from an urban point of view. In addition, the services these infrastructures provide become internalized as the prolonged arm of urban, and thus human, needs until both fuse together to become cybernetic organisms with a shared hybrid intentionality. Graham's thoughts shed new light on Heidegger's claims about modern technologies and the problems of their artifactuality. Infrastructures are constructed to satisfy human needs, but they do so on such a large scale that their undesired effects can become irreversible.

7.4 Infrastructures as Structuring Praxis

So far, the investigation of infrastructures as material background has shown the influence of omnipresent urban technologies and technological systems on human actions and experience. As discussed, traffic and transit technologies affect the way we commute and experience the urban environment, while the increased speed with which distant areas can be reached changes the individual perception of space and time. On a larger scale, infrastructures mediate our relation to nature, not only by providing an artificial environment, but also by potentially rendering nature a resource for large-scale utilization. Even though the investigation already illustrates both the direct and indirect influences of infrastructures on human action and experience, the connection between human life and the surrounding materiality can be even more profoundly analyzed. The following section presents an attempt to explore the more intricate practical relationship between human beings and their material environment by making use of Sartre's concepts of *praxis* and *hexis*.

In his "Critique of Dialectical Reason", Sartre (1978) identifies the relation between human beings and world as a transformative *praxis* which affects both human beings and material world. He derives his concept of *praxis from* the claim that human beings are neither self-sufficient when it comes to basic material needs (such as nutrition and shelter), nor when it comes to immaterial needs like knowledge, wisdom or societal organization. These needs all depend on material and social interaction in order to be satisfied. However, Sartre does not propose a materialistic determinism that considers material needs to be the causes for human action. For him, humans are as much beings of necessity as they are beings of freedom. This enables human beings to creatively employ strategies to more effectively satisfy their needs and to generate more complex levels of society. The most important aspect for Sartre is that human beings change their material conditions in such a way that the world itself becomes the passive residue of these needs and of the strategies which were employed to satisfy them. Through *praxis*, human beings leave a world of their end's material traces behind.

Sartre considers *praxis* to be a transformative act in which human beings materialize their ends in the world. He also considers technological artifacts to be such frozen and materialized *praxis*. Sartre claims that the *praxis* within technological artifacts stays active, which allows them to be used as tools. Usually, technological artifacts are designed to be more suitable for specific practical relations and less suitable for others. Even though technologies do not directly dictate how they are to be used, the suitability of certain tools is partially determined by how efficiently and effectively they can be used to satisfy human needs (Sartre, 1978).

Human beings materialize both their own needs and the strategies to satisfy those needs in the technological artifacts they use. These artifacts help to determine how specific needs are satisfied or interests are served. Depending on their suitability, these very artifacts influence further *praxis*. This process of materializing *praxis* within the materiality of the world can be observed in the implementation of infrastructure technologies. For instance, some roads started as paths that were literally trampled into nature. Through repeated utilization these paths become deeper and wider. Such paths are used to transport goods, people and information over longer distances. Since both transport and information are basic needs of more complex societies, more complex needs arise to secure their satisfaction. This leads to the fortification of paths and to the implementation of rules on how to use these newly built roads (Denecke, 2007); left-hand and right-hand traffic emerges. Similarly, rivers are used and redirected over time to allow a more effective fulfillment of human needs. Infrastructures are great examples for the co-shaping of human ends and technical means because they provide basic services. Water infrastructures are implemented to secure the procurement and disposal of water as a bare necessity of human life, usually in a reliable manner. Since most technologies utilize electricity, the same is true for power grids.

Alongside this continuous dialectical interaction between human beings and material conditions another important process takes place which binds together human beings, material conditions, technological artifacts and societal organization. Sartre believes that alongside every *praxis* a form of habituation takes place, through which successful strategies and utilizations of technical artifacts are passed on through time. Every success or failure leads to rules and norms which are again materialized in the way further *praxis* is actualized through technological artifacts. Sartre calls this form of habituation '*hexis*', Greek for disposition. In this *hexis,* not only strategies, but also norms and moral values are preserved. Sartre's concept of *hexis* has its roots in Aristotelian philosophy, where it is linked to the concept of *second nature*. For Aristotle, *hexis* is a "generic term for an entrenched psychic condition or state which develops through experience rather than congenitally" (Lockwood, 2013, p. 23). This can be connected to certain lines of thought in mobility studies, which analyze the cultivation and formation of habits regarding transportation infrastructures among others. Walker, Thomas, and Verplanken (2015) analyze how travel habits change after disruptive events. The authors show that it takes about four weeks for such habits to establish and that they slowly decay over roughly the same amount of time. Moreover, the authors suppose that the accessibility of infrastructures plays a determinant role in travel behavior and habituation after disruptive events. Schwanen, Banister, and Anable (2012) suggest that travel habits can be governed in order to promote less carbon-intensive means of transportation. For them, "a car habit is what makes it possible for car, driver, road, signs, passengers, traffic rules to merge into a seemingly effortless whole and what underpins the routine use of cars to access places" (Schwanen et al., 2012, p. 526). This again illustrates the potential of infrastructural relations to become stable equilibriums that reify order.

Similar to Sartre, Shove (2016) outlines three main aspects of how infrastructures and practical relations are connected. Firstly, infrastructures co-evolve together with individual practices. Depending on their design, infrastructures allow for the actualization of specific ends through them while making others impossible. Shove demonstrates this with street design. Starting from pathways for wanderers and horses, streets were continuously adapted to allow for their utilization through bicycles and early cars. Later, some roads are broadened and become highways that can only be used by fast cars and motorbikes. Needs, individual practices and built infrastructures are bound together in a push-pull of mutual adaptation. Secondly, infrastructures enable practices to circulate. The same push and pull not only leads to mutual adaptation, but also (thinking about the *hexis*) to new, innovative (and thus desired) forms of *praxis*. For instance, refrigerators are a phenomenon which came up in the last century. Not only did these freezing technologies change the way food storage works, they also changed the way people cook and live. In the case of large electrical grids and refrigerators, infrastructures directly influence normative orders and thus large areas of society. In the so called global north, changes like this have become the normal and modern way of living (Shove, 2016). Thirdly, infrastructures relate to changing complexes of practice. In accordance with Sartre's concept of human life as result of necessity and freedom, infrastructures do not dictate the way they have to be used. However, as built structures of society, they can be considered the materialized residue of those needs that they were built to satisfy. This presents both advantages and disadvantages for the continuation of individual lives as well as societies. On the one hand, infrastructures not only enable more complex forms of *praxis* to be actualized, they also secure the effective and efficient repetition of these practices. On the other hand, infrastructures may preserve past needs and strategies in such a way that individuals and societies are condemned to repeat whole sets of practices, together with normative orders, time and time again (Shove, 2016).

Considering infrastructures to be both products and producers of *praxis* and *hexis* allows us to grasp their fundamentally structuring role for human society. Human beings have needs and interests. Through an active engagement with their material environment, humans satisfy their needs and serve their interests. This changes the world into one of human needs or needfulness. Technical artifacts solidify specific strategies because they steer the actualization of *praxis* towards specific ends by virtue of being designed in certain ways. Finally, success and failure may determine which strategies are preserved and which are abandoned. This leads to a *hexis* (or disposition) through which various forms of *praxis*, the associated technological artifacts, specific strategies and norms are preserved. As large technological artifacts, infrastructures both form and preserve the very structure of societies.

7.5 Conclusion

Infrastructures not only influence individual action and societal organization by providing their material background, they also structure the manner in which human beings deal with their own needs in accordance to specific material arrangements. The contemporary globalized world is unthinkable without wide systems and networks of infrastructures which allow both control and freedom across national borders.

Infrastructures are both the passive material and the organizing structure of society; which has desired and undesired consequences. On the one hand, infrastructures store normative ordering patterns and provide orientation for the implementation of new technologies. Due to their rigidness, infrastructures store this order over time, which allows for the relative security of concomitant political systems. On the other hand, societies can easily become trapped by the way their infrastructure is materially arranged. Even though infrastructure technologies are implemented with specific ends in mind, these ends nevertheless are human ends. However, if infrastructures preserve a certain stage of historical development, the manner in which means and ends are connected at this stage can become obscured. This can lead to a repetition of practices without considering their initial purpose—a development that is called *path dependency* in infrastructure research. Sartre suggests that not only simple practical relations, but whole modes of production (and thus power relations) are preserved within the artifacts we use. In the case of infrastructures, the preservation is even less reversible.

With regard to philosophy of technology, the analysis shows two things. Firstly, it shows that infrastructures are still an underestimated research object in philosophy and philosophy of technology. This is partly due to a large body of literature from other scientific disciplines which deals with infrastructures in their own way, which makes it difficult to single out aspects of infrastructures that are promising for philosophical research. Nevertheless, this lack of research is also an opportunity for philosophy of technology. Even though I have tried to provide a wide spectrum of potential philosophical inquiry, other infrastructural aspects have been addressed only very briefly, such as regarding how infrastructures store power relations or how built structures can be used to include and exclude large parts of society. Another only briefly mentioned aspect is the apparency of infrastructures. On one hand, infrastructures are deliberately hidden by design. On the other hand, they become ‘invisible’ as a result of repeated interaction and habituation. Secondly, the analysis shows that a uniform philosophical approach towards infrastructures is still missing. This is partly due to infrastructures’ complex two-fold nature as things and as relations between things. However, this paper can be seen as an approach towards infrastructures from a praxeological point of view.

As human beings, the interaction with our built environment is an acquired one. We are thrown into a world of technological standards, old structures and innovations. We learn how to navigate through concrete jungles, how to use the things at hand and how to find meaning in the way we interact with other human beings and the material world. Infrastructures are the constant companions of our modern age; both in the way they provide us with opportunities and how they challenge us. The fact that such highly interdependent and complex technological structures can become second nature to us not only questions our conceptions of *first nature* but also the distinction between nature and culture in the first place.

References

Bijker, W. E., Hughes, T. P., & Pinch, T. (Eds.). (1987). *The social construction of large technological systems. New directions in the sociology and history of technology*. Cambridge: MIT Press.
Bundesministerium des Inneren. (2009). *National strategy for critical infrastructure protection*. Berlin: Author.
Denecke, D. (2007). Linienführung und Netzgestalt mittelalterlicher Verkehrswege. Eine raumstrukturelle Perspektive. In R. C. Schwinges (Ed.), *Straßen- und Verkehrswesen im Hohen und Späten Mittelalter* (pp. 49–70). Jan Thorbecke Verlag: Ostfildern.
Edwards, P. N. (2003). Infrastructure and modernity: Force, time, and social organization in the history of sociotechnical systems. In T. J. Misa, P. Brey & A. Feenberg (Eds.), *Modernity and technology* (pp. 185–225). Cambridge: MIT Press.
Edwards, P. N. (2017). The mechanics of invisibility: On habit and routine as elements of infrastructure. In I. Ruby & A. Ruby (Eds.), *Infrastructure space* (pp. 327–336). Berlin: Ruby Press.
Engels, J. I. (2018). *Key concepts for critical infrastructure research*. Wiesbaden: Springer VS.
Graham, S. (2010). When infrastructures fail. In S. Graham (Ed.), *Disrupted cities: When infrastructure fails* (pp. 1–26). New York: Routledge.
Handel, A. (2017). Distance matters: Mobilities and the politics of distance. *Mobilities, 13*, 473.
Harvey, P., & Knox, H. (2012). The enchantments of infrastructure. *Mobilities, 7*(4), 521–536.
Heidegger, M. (2009). Die Frage nach der Technik. In M. Heidegger (Ed.), *Vorträge und Aufsätze* (pp. 9–40). Stuttgart: Klett-Cotta.
Ihde, D. (1990). *Technology and the lifeworld. From garden to earth*. Bloomington: Indiana University Press.
Klinger, T., Kenworthy, J. R., & Lanzendorf, M. (2013). Dimensions of urban mobility cultures. A comparison of German cities. *Journal of Transport Geography, 31*, 18–29.
Larkin, B. (2013). The politics and poetics of infrastructure. *Annual Review of Anthropology, 42*(1), 327–343.
Lockwood, T. C. (2013). Habituation, habit, and character in Aristotle's Nicomachean Ethics. In T. Sparrow & A. Hutchinson (Eds.), *A history of habit. From Aristotle to Bourdieu* (pp. 19–36). Lanham: Lexington Books.
Monstadt, J. (2007). Großtechnische Systeme der Infrastrukturversorgung: Übergreifende Merkmale und räumlicher Wandel. In D. Gust (Ed.), *Wandel der Stromversorgung und räumliche Politik. Forschungs- und Sitzungsberichte der ARL 227* (pp. 7–34). Hannover: Verl. der ARL.
Newman, P., & Kenworthy, J. (2015). *The end of automobile dependence. How cities are moving beyond car-based planning*. Washington: Island Press.
Rankin, C. H., Abrams, T., Barry, R. J., Bhatnagar, S., Clayton, D. F., Colombo, J., et al. (2009). Habituation revisited: An updated and revised description of the behavioral characteristics of habituation. *Neurobiology of Learning and Memory, 92*, 135–138.
Rinaldi, S. M., Peerenboom, J. P., & Kelly, T. K. (2001). Identifying, understanding, and analyzing critical infrastructure independencies. *IEEE Control Systems Magazine, 21*(6), 11–25.
Ropohl, G. (1999). Philosophy of socio-technical systems. *Society for Philosophy and Technology Quarterly Electronic Journal, 4*(3), 186.
Ropohl, G. (2009). *Allgemeine Technologie: Eine Systemtheorie der Technik*. Karlsruhe: Universitätsverlag Karlsruhe.
Sartre, J. P. (1978). *Critique of dialectical reason I. Theory of practical ensembles*. London: NLB.
Schwanen, T., Banister, D., & Anable, J. (2012). Rethinking habits and their role in behaviour change: The case of low-carbon mobility. *Journal of Transport Geography, 24*, 522–533.
Shove, E. (2016). Infrastructures and practices: Networks beyond the city. In O. Coutard & J. Rutherford (Eds.), *Beyond the networked city: infrastructure reconfigurations and urban change in the North and South* (pp. 242–258). London: Routledge.

Star, S. (1999). The ethnography of infrastructure. *American Behavioral Scientist, 43*(3), 377–391.
Walker, I., Thomas, G. O., & Verplanken, B. (2015). Old habits die hard: Travel habit formation and decay during an office relocation. *Environment and Behavior, 47*, 1089–1106.
Winner, L. (1980). Do artifacts have politics? *Daedalus, 109*(1), 121–136.

Marcel Müller has a background in philosophy of technology with a focus on human-technology relations, existential, and dialectical philosophy. From October 2016 to September 2019, he was part of the interdisciplinary research training group KRITIS (GRK 2222 - Critical Infrastructures: Construction, Functional Crises and Protection in Cities) at TU Darmstadt, Germany. During this time, he started his PhD-thesis titled Needful Structures: The Dialectics of Practical Human-Technology Ensembles and their Role in the Formation of Urban Mobility, which he is currently finishing. In his thesis, he addresses the formation of urban infrastructures from a praxeological and philosophical perspective.

Chapter 8
Infrastructuring Bodies: Choreographies of Power in the Computational City

Jaana Parviainen and Seija Ridell

Abstract The aim of this chapter is to shed light on the power-related infrastructural dynamic that actualises in the interrelations of big data collection and the bodily movement of urbanites in contemporary cities. By drawing from Husserl's and Merleau-Ponty's phenomenologies of the body and combining them with recent theorisations on choreography, material media theory and critical technology studies, the authors address city dwellers' embodied relations with mobile devices and ambient technologies as integral to the micro-, meso- and macro-level (re)production of urban infrastructures. By way of discussing the technologically mediated kinaesthesia and movement trajectories of lived bodies, the chapter develops a novel conceptualisation of urban choreography for exploring the mechanisms through which dwelling-in-the-city today functions in a globally extensive cybernetic feedback loop with profit-motivated and surveillant big data operations.

Keywords Urban infrastructure · Lived bodies · Mediated bodily movement · Choreography · Big Data · Computational city · Phenomenology

8.1 Introduction

> We can't see how the street is immersed in a twitching, pulsing cloud of data. (…) /Such data emerges from the feet of three friends, grimly jogging past, whose Nike+ shoes track the frequency and duration of every step, comparing against pre-set targets for each individual runner. This is cross-referenced with playlist data emerging from their three iPods. Similar performance data is being captured in the engine control systems of a stationary BMW waiting at a traffic light, beaming information back to the BMW service centre associated with the car's owner. /The traffic light system itself is capturing and collating data

J. Parviainen (✉)
Faculty of Social Sciences, Tampere University, Tampere, Finland
e-mail: jaana.parviainen@tuni.fi

S. Ridell
Faculty of Information Technology and Communication Sciences, Tampere University, Tampere, Finland

M. Nagenborg et al. (eds.), *Technology and the City*, Philosophy of Engineering and Technology 36, https://doi.org/10.1007/978-3-030-52313-8_8

> about traffic and pedestrian flow, based on real-time patterns surrounding the light, and conveying the state of congestion in the neighbourhood to the traffic planning authority for that region, which alters the lights' behaviour accordingly. (Hill, 2008, para 1–3)

In this extract from an essay written in 2008, designer and urbanist Dan Hill illustrates how the urban environment, in the near future, would produce data on a vast scale, while the technical operations that afford this data collection would remain beyond the perception of ordinary city dwellers. Ten years later, the situation Hill imagined has increasingly become commonplace, as computation in diverse forms today literally envelops people in cities. In addition to incessant signal traffic pulsating wirelessly through the air, algorithmic technologies are embedded in urban physical structures, run the billboards, move along on the streets in vehicles or are carried around by people in their smartphones and other portable and wearable accessories. Meters and sensors are deployed on the streets, usually with little or no information about their actual purpose. The personal devices and their designs, which appear haptically attractive to urban consumers, often disguise the inconspicuous collection of data, busily performed by the gadgets and their numerous applications.

By envisioning how the future street would feel, Hill prompted a discussion about how ever more effective data generation and collection affect urban life. In more infrastructural terms, we can say that the technical systems enabling the production, collection and transmission of these data, as well as the quality of connections and their various levels of openness or privacy, all impact, in ways equally imperceptible and profound, how people dwell in the city, and how the urban environment, in turn, is furnished with new digitally embedded objects, products and services. Indeed, considering that cities have increasingly turned into all-encompassing computed environments, it makes sense to claim, as architecture and media studies scholar Mark Shepard (2009, 2012) does, that code-based technologies have become as important if not more important than buildings 'for organizing space, time, and the boundaries around the body in public space'.

What is striking in Hill's narrative is the crucial role people's bodies play in the processes of mass-scale data generation and collection in cities. Data are not collected merely, or even primarily, through autonomously observing external systems, but through the involvement of city dwellers' corporeality in diverse intentional and unintentional ways. In this chapter, we address the connections between moving bodies, portable and wearable technologies, and the pervasively computed urban ambience with the aim of grasping the power-related sociotechnical complexity that characterises the (re)production of infrastructures in twentyfirst century cities—infrastructure referring here to the fundamental sociotechnical structures that enable and guarantee the stability and continuity of urban life in the first place (see, e.g., Graham & Marvin, 2001). To do this, we develop around the notion of choreography a novel conceptualisation that directs attention to urbanites' mediated kinaesthesia and movement trajectories and, most particularly, sensitises us to their interconnections with large-scale computational systems.

8.2 Moving Bodies, Urban Mediation and Power

Movement and mobility are by now acknowledged to be crucial, in both research and politics, for understanding and facilitating the transformation of cities, including new lifestyles, sustainable transportation systems, creative architecture and innovative town planning.[1] To date, however, few theoretical attempts have been made to address how city dwellers' technologically mediated bodily movement and, more particularly, their embodied relations with algorithmic 'smart' technologies relate with increasingly digitalized and networked urban infrastructures. Our argument in this chapter is that the rhythms and paths of this movement, including diverse stopping points, are inherently infrastructural, and that this infrastructural dynamic can be captured fruitfully through a phenomenologically informed concept of choreography.

In approaching contemporary urban infrastructures from a choreographic perspective, we do not use the notion of choreography with reference to dancing and performing arts (e.g. Lepecki, 2006; Manning, 2009; Schiller & Rubidge, 2014), nor do we consider choreography in terms of inscription, as in studies inspired by actor–network theory (Loke & Kocaballi, 2016). Instead, we attach our understanding of choreography to how urbanites' media- and technology-related movements tend to form, through repetition, bodily habits, routines and practices, thereby contributing essentially to the constitution of urban sociotechnical structures (Ridell, 2019).

To begin to address the infrastructural dynamic that actualises in urbanites' mediated corporeality, a key reference point can be found in Merleau-Ponty's view of the lived body ('corps vivant'). In his *Phenomenology of Perception*, Merleau-Ponty (1962/1989) points out that bodily movements and gestures are the means through which any animate body explores its world and enacts intentions. These movements are an intermediary, since every action is performed within an intersubjective space (ibid., p. 98). Equally important, Merleau-Ponty conceives the lived body as both the lived centre of consciousness and one's own body as experienced.[2] In the distinction between the lived body and the physical body, the notion of the lived body captures the body in the embodied first-person perspective, whereas the physical body is an objectified, corporeal and material entity. Furthermore, following Husserl's phenomenology, Merleau-Ponty addresses the notion of motor intentionality, which concerns the lived body's intention towards objects, its directing itself towards goals, and its acting in a way that enables it to make sense of a collection of disparate bodily movements, unifying them into meaningful action (ibid., p. 137).

---

[1] See, e.g., European Commission's strategy on smart cities: *Cities of tomorrow. Challenges, visions, ways forward,* published in 2011. http://ec.europa.eu/regional_policy/sources/docgener/studies/pdf/citiesoftomorrow/citiesoftomorrow_final.pdf.

[2] More exactly, Merleau-Ponty states that '[c]onsciousness is being-towards-the-thing through the intermediary of the body' (Merleau-Ponty 1962/1989, pp. 138–139).

As for the notion of movement, we include in our discussion kinaesthesia, in the sense of both an individual's bodily movements and collectively formed movement patterns across the city. We conceive urban space, in turn, as increasingly software-dependent and data-intensive: as a computational environment that can be fruitfully described as a cybernetic system (see, e.g., Kitchin & Dodge, 2011; Hayles, 2009; Gabrys, 2016; Krivý, 2016).

To formulate the orientation of the chapter with respect to the contemporary urban condition, then, our focus is on how code-based algorithmic technologies engage city dwellers as corporeal beings in the (re)configuration of spatial power relations. From this starting point, it is not only that technologies as material components provide infrastructural support and control, but that mediated bodily practices with their entwinement in political and economic interests couple with and contribute to sociotechnical ensembles in multiple complex ways. In these constellations, urbanites' corporeality and, most particularly, the movements of the lived body become a constitutive aspect of infrastructural power.

More generally speaking, due to the overall digital mediation of both space and people's daily activities, the quantity of personal data generated daily in different spatial settings has reached an unprecedented scale in the past few years. Elmqvist (2011) and Haddadi, Mortier, McAuley, and Crowcroft (2013), in pointing to limitations in traditional human–computer interaction (HCI) research, suggest that it is time to pay attention to human–data interaction (HDI). The latter type of research is not only interested in relations between humans and large personal datasets but also takes into account that people often interact with technological structures that condition their activities in ways they do not necessarily recognise or understand, or which they would rather ignore. Unpacking how such infrastructures function poses specific challenges in the urban context. This is because the scale of these systems is much more extensive than that usually considered in studies focused narrowly on the level of human–computer interaction. In the present context, a pertinent question is how so-called 'big data' are generated and collected in urban environments, and how such data subsequently circulate back into these environments in diversely processed forms.

To grasp the infrastructural dynamic of data generation, collection and circulation in contemporary cities, it is important to consider the different levels of socio-technical activity and, above all, to pay direct attention to how these levels interrelate. Klauser and Albrechtstlund (2014) suggest that the micro and macro levels constitute the opposite ends of the big data spectrum, emphasising that these opposites are not separate from one another but connect through complex feedback systems. The choreographic approach we develop here also includes as an essential dimension the middle-range or meso level in considering city dwellers' everyday uses of smart devices, phones, wearables, cameras and sensors as bodily gestures that connect integrally to the macro-choreography of collecting big data for the purposes of governance and economic profit.

In what follows, we develop *urban choreography* as a concept that reaches across different technologically mediated sociospatial scales by reworking Husserl's notion of *kinaesthesia* relative to more recent theoretical discussions. As for the latter, we bring together the phenomenology of movement (Sheets-Johnstone, 1999), theorisations on choreography in the urban context (Parviainen, 2010; Robertson, Lycouris, & Johnson, 2007) and discussions on the mediated city (Frers & Meier, 2007; Mattern, 2015; McQuire, 2008; Parviainen, 2010; Parviainen, Tuuri, & Pirhonen, 2013; Ridell, 2010, 2013). In addition to kinaesthesia, valuable related concepts for defining urban choreography are *movement trajectory, kinaesthetic field* and *kinesphere*.

The chapter is structured as follows. In the next section, we proceed from the theoretical discussions on choreography to introduce our own conceptualisation of urban choreography; we also demonstrate how this notion can assist in analysing everyday movement trajectories and kinaesthesia as integral to the infrastructural dynamic of the computational city. In the following section, we direct particular attention to big data generation and collection in contemporary urban settings. We outline how urbanites' mediated bodily movements and their encounters with ambient technologies as part of their movement trajectories interconnect as micro- and meso-level choreographies with global big data developments, thereby contributing to urban macro-choreographies. We conclude with critical reflections on the recursive cybernetic logic by which the infrastructural macro level feeds back to lived urbanity, disciplining the embodied aspects of city life in novel ways and giving rise to new types of adaptive practices.

8.3 Conceiving Urban Choreography Phenomenologically

While in dance aesthetics choreography refers to dance composition, in recent discussions of interaction design and related disciplines, the notion of choreography seeks to capture the meaningful continuums of movement that humans, as individuals or as groups, experience during interaction with technology (Loke & Kocaballi, 2016). Typically, in computer science, human behaviour in urban settings has been approached at the level of physical bodies and their relations to other bodies and things, creating agent-based movement models to depict patterns of behaviour (Torrens, 2016). The discussion on urban choreography more specifically has raised interest in the movements of lived bodies (Lycouris, 2009; Schiller & Rubidge, 2014), and this discussion resonates with explorations in the fields of time geography (Crang, 2001; Hägerstrand, 1982; Pred, 3; see also Sullivan, 2017), urban anthropology and sociology, particularly as concerns the study of everyday pedestrian and crowd movement (see, e.g., Carter, 2009; Ingold, 2011; Sheller & Urry, 2006). The idea of choreography is also indirectly present in studies of urban polyrhythmicities at different sociospatial scales (Amin & Thrift, 2002; Edensor, 2010; Lefebvre, 2004; Smith & Hetherington, 2013).

In our consideration of urban choreography, we approach bodily movement in the computational city as a key element in a cybernetic system of thoroughly mediated interrelations. Choreography in this sense actualises as joint material performances by bodies, actions and environments rather than as simply bodies alone making movements to create choreography. There is a reasonable amount of theoretical discussion on choreography (e.g. Bench, 2014; Hirsch, 2014; Parviainen, 2010; Prytherch, 2015; Robertson, Lycouris, & Johnson, 2007) that can assist in developing a consistent choreographic take on the movement of data and bodies in the contemporary urban context. Broadly, the notion of choreography in our usage also contains an aspect of simultaneous multisitedness, as it refers to movements and activities in which the movements express relations and articulate meaningful interactions in and across lived and virtual spaces between various animate or inanimate agents. This means that no one choreographer can determine the overall constellation; rather, the connections between human and nonhuman agents create and sustain choreography in an ongoing manner. Treating human and nonhuman agents symmetrically, choreography arises from their reciprocal relations, actualising in movements, motions, patterns and rhythms.

This kind of conception of choreography has resonance with actor network theories (ANTs) (e.g. Latour, 1997; Law, 1992). In the ANT version developed by Latour (1997) and his colleagues, humans and objects become agents or *actants* influencing courses of action as well as being the results of them. An assemblage of people, objects and technologies are composed of heterogeneous elements that enter into mutual relations with one another. With regard to digital technologies, when action and materiality intertwine, human bodies do not remain independent from technologies. In such assemblages, embodied connections with technologies modify physical and lived bodies, forming new kinds of hybrid practices. One way to capture how the movement of human bodies and the motion of machines become synchronised in these practices would be to use geographer Nigel Thrift's (2004c) notion of *hybrid kinaesthesia*. Deborah Lupton (2013, p. 400), for one, borrows Peter Freund's (2004) term *technological habitus* to describe how bodies develop new habits and routines to blend in with the functions of technologies. Lupton suggests, however, that bodies do not intermesh smoothly or seamlessly with technology, but that there are disjunctions between bodies and objects (ibid.).

In Latour's (1997) articulation of ANT, the agency of actors and their assemblage is addressed without giving too much attention to the quality of action by the different actants in question; not showing interest in, for example, how affects and sensations of lived bodies may influence relations among actants. In contrast, the phenomenologically informed choreographic approach we are developing here is interested in the specific weights of actants' movement, i.e., how they make things happen and keep the action alive. To address the relational and affective dynamics of lived bodies and how their actions pull and push things towards a constellation, we draw inspiration from and develop further Husserl's idea of *kinaesthesia* (e.g. Husserl, 1997). This means that we sharpen the focus on embodied activities and transitions in an enmeshed network of heterogeneous elements, emphasising the reciprocal relations between hardware and software, bodies and environments in specific spatial configurations that can be partly planned and partly improvised. In contemporary cities, these dynamic interconnections form a socio- and body-technical membrane in which infrastructural power vibrates.

Husserl's (1997) particular relevance for our discussion, combined with Merleau-Ponty's view of the lived body, is that, for him, kinaesthesia is not interchangeable with the term *proprioception,* which denotes interoceptive information about limb position and how it contributes to motor programming. Instead, he refers to the notion of exteroceptive information of moving in space. In other words, kinaesthesia provides information about changes of location and motility as affective and about social interaction with other living and non-living beings. Hence, what is at stake is an ability to sense or feel movement or the sensation of movement. Such sensation is not necessarily associated with the body's own intentional movements; for example, motor vehicles or other living beings can cause movements in the body. This warrants an analytic distinction between two possible kinaesthetic situations: *self-motion* and *being-moved.* Self-motion refers to self-generated movements by which one intentionally moves one's body, and these generated movements physically, affectively or socially influence or resonate with other beings in proximity. At the same time, one can feel kinaesthetic sensations of being moved by the movements of other beings.[3] For instance, being-moved occurs when sitting on the bus and feeling how the bus transports one's body, sensing the vehicle's speedups, turns and brakes through the body but with very limited ability to influence the movements of the bus.

Kinaesthesia, then, is a central factor in addressing the interactions in and between lived bodies and their relations with objects and environments in the contemporary urban context. More generally, it is through kinaesthesia that individuals recognise differences and similarities in their own movement qualities and haptic sensations, as well as in the motions of the objects around them (Sheets-Johnstone, 1999). The movement of lifting a mobile phone reveals something about its weight. Rubbing one's fingers across the phone's surface reveals details about its texture, material and shape. Dropping the phone reveals something about its structure and resilience via the sound it makes when it hits the floor.

Furthermore, instead of focusing only on the body's internal kinaesthetic feelings, Husserl suggests that the kinaesthetic system should be analysed as a whole, i.e., as a *kinaesthetic field.* Motion appears to us as an integral part of relational systems, including visual and aural impressions of moving objects. Simplifying Husserl's own highly abstract formulation of the kinaesthetic field, we refer with this notion to the characteristic motions embedded in geographic locations and places and which vary according to the time of day and season (Parviainen, 2010). For instance, people gathering for a picnic in a city park on a sunny Sunday afternoon generate a kinaesthetic field that differs drastically from the kinaesthetic field co-created by the same people when driving in their cars to work on a highway on a rainy Monday morning. Such a formulation of the kinaesthetic field, in fact, comes

[3] In the contemporary urban context, this constellation has been rendered even more complex through the development of wireless technologies and, more precisely, through digital signal processing. While the materiality of the 'Herzian space' (Shepard, 2009) is not directly perceived by our senses, it affects the body's sensory orientation and relation to other bodies and things in urban space (Mackenzie, 2010; Ridell, 2019).

close to Husserl's (1989) concept of *Umwelt* and resonates with what Thrift later called the 'qualculative background' (see Thrift, 2004a). For Husserl, entities are 'meaningful' to us as a 'motivating' force similar to how we 'shake hands with one another in greeting' (Husserl, 1989, p. 192).

Proceeding from and developing further Husserl's notions enable us to formulate a conceptualisation of choreography that provides methodological tools to analyse more systemically how human and non-human movements in contemporary cities are tightly involved in kinaesthetic fields. In our previous studies (Parviainen, 2016; Parviainen, Tuuri, & Pirhonen, 2013), we developed a notion of choreography that specifies the movements of actants by dividing choreography analytically into micro, local and macro levels. To emphasise the power-related infrastructural aspect, we replace the notion of local level in this connection with *meso-level* choreography (for the analytic distinction between micro, meso and macro levels of infrastructure, see, e.g., Edwards, 2003). In a *micro-level* analysis, the focus can be, for instance, on the swiping movements of the index finger across a *screen*. More generally, these movements occur within one's kinesphere or internally. By *kinesphere* we refer to Rudolf Laban's definition of 'the sphere around the body whose periphery can be reached by easily extended limbs' (Laban, 1966, p. 10). Coming close to Edward T. Hall's (1966) notion of *personal space*, the notion of the kinesphere emphasises bodily movements, as well as their qualities, dynamism, tensions and efforts, as forming a lived space. The intersubjective spatiality that individuals live in and through their kinespheres plays a central role in *meso-level* choreographies through collectively shared kinaesthetic fields. At this level, choreography refers to bodies' socio-spatially attuned movements; for instance, when pedestrians navigate through the throng at a busy crossroad, or when passengers, sitting on a bus, use their phones to avoid eye contact with fellow passengers. While there is plenty of research on social intercourse from the perspective of situational conventions and 'grammars', particularly inspired by Erving Goffman's studies on behaviour in public places (Goffman, 1963, 1971; see e.g., Jensen, 2010), the corporeal dimension is not too often addressed (see Ridell, 2019). The *macro-level* choreography, in turn, refers to movements that connect individual bodies in their micro- and meso-level kinaesthesia with large-scale, even globally extensive, movement trajectories enabled by the physical infrastructures of transportation or the internet. Hence, passengers on a bus, hiding from those physically co-present behind their portable devices, may be reading text messages from friends on other continents or watching a video on YouTube that is being viewed simultaneously by hundreds of thousands of other people all over the globe.

To summarise, the phenomenologically informed approach to urban choreography we propose views singular bodily gestures and movements interrelated to digitally afforded objects and computationally mediated environments as integral parts of spatially extensive and sociotechnically complex trajectories and transitions. Understood this way, the choreographic analysis is not restricted to bodily movement in limited spaces and short time spans but instead takes into account the spatiotemporal multidimensionality involved when people respond to ambient technologies and use their portables and wearables in urban settings.

8.4 Choreographies of Data Generation in Urban Settings

Sociotechnical ensembles are now more complex and actualise on a greater scale than ever before. In western countries, a majority of people carry smartphones and other networked devices with them all the time; these devices have an inbuilt capacity to generate and collect data on how they are used. The phenomenon of 'personal data' has emerged only in recent years, representing an entirely new class of data. These data about one's locations, bodily movement trajectories and actions are produced by users themselves and are obtained automatically from their smartphones, activity trackers and other wearable devices, or are captured by sensors embedded in physical urban structures.

With regard to self-generated data, self-tracking applications, for instance, collect information about a variety of everyday activities, from status updates to discrete activity data that technology firms and platform owners can store, analyse and sell for profit to other companies. Smartphones, in turn, include multiple sophisticated technical functions and applications, such as cameras, navigation systems, proximity meters, heart rate sensors, accelerometers, gyroscopes, barometers and thermometers, that monitor individual bodies' movements and the state of one's physical body. Applications also assist users in generating data from their lived bodies that sensors currently cannot capture, such as data on food and drink consumption, aches, pains and even moods. Taken together, these data provide quantifiable, and thus profitable, information about urbanites' everyday movement trajectories as well as about their mediated bodily habits and routines in the city.

Technologies embedded in the urban environment and the collection of data about people's movements and daily activities in different locations often go unnoticed by city dwellers themselves. Moreover, urbanites mostly remain unaware of how the data gathered from them—for instance, by cameras with facial recognition systems—are stored, circulated and sold. In addition to commercial data collection, political and administrative interests shape the use of these algorithmic technologies in cities. We can take as an example current developments in China, the groundbreaker globally in the deployment of facial recognition systems and the integration of huge image databases of human faces for algorithmic processing. In several Chinese cities, facial data collected on streets and in banks, hotels, airports and even public toilets are used to verify people's identities.[4] The motivation for these

[4] Journalists from *South China Morning Post*, located in Hong Kong, and journalistic and other media in western countries have reported that a massive system is being established by the Chinese government to identify its citizens based on facial recognition technologies: see, e.g., 'China said to be testing facial recognition system to monitor Muslim-dominated Xinjiang region', http://www.scmp.com/news/china/society/article/2129473/china-testing-facial-recognition-system-monitor-muslim-dominated; 'China to build giant facial recognition database to identify any citizen within seconds' http://www.scmp.com/news/china/society/article/2115094/china-build-giant-facial-recognition-database-identify-any; 'China testing facial-recognition surveillance system in Xinjiang – report' https://www.theguardian.com/world/2018/jan/18/china-testing-facial-recognition-surveillance-system-in-xinjiang-report.

operations is to strengthen the political surveillance of citizens by both local administration and the central government. The far-reaching existential implication of the widespread deployment of facial recognition systems is that, as Gray (2003) points out, they will necessarily alter societal conceptions of privacy in public (urban) space as well as the dynamics of individual and group interactions.

The example of China demonstrates that the limits to how thoroughly politically motivated facial recognition may permeate urban space are flexible indeed. For example, in Ürümqi City, located in the Xinjiang region, street cameras can automatically classify pedestrians according to their gender, clothes and even hair length, and software allows people to be tracked from one surveillance camera to the next based on facial data alone. In 2018, China deployed a flock of 'bird drones', high-tech spy drones disguised to look like birds, to step up surveillance levels in the Muslim-majority Xinjiang region.[5] Chinese authorities also have plans to track, in the near future, urbanites' habitual movement routines and trajectories in order to spot 'problematic pedestrian behavior'. One of the directors at *Cloudwalk*, a Chongqing-based firm, provides a different example: 'If you know gambling takes place in a location, and someone goes there frequently, they become suspicious'.[6] When a suspicious person is identified, the system can examine her or his connections with other people by tracking the movement trajectories relative to their meeting points.

The notion of *movement trajectory* in urban settings refers to people's daily travels and patterns of movement in the city between home, work and social events. These trajectories often attach to and are supported (and sometimes disrupted) by ambient technologies and portable devices. In contemporary cities, movement trajectories form personal and collective choreographies, as people move regularly from home or place of work or study to having lunch or going shopping. Individual trajectories include different stopping points, with multiple networks of movement trajectories converging to form complex nodes of human–human, human–nonhuman and nonhuman–nonhuman interconnections (Gabrielli, Rinzivillo, Ronzano, & Villatoro, 2013). Another extract from Hill (2008) illustrates one such instance:

> As the bus departs, the new passengers on-board swipe their RFID-based integrated transport system ID cards, updating mass transit databases with every possible aspect that can be gleaned from this simple activity (time of day, location, frequency of use, favourite entry points etc.) The now-empty seat in the bus-stop registers that it is indeed now empty using simple sensors, and wirelessly logs this fact with a database monitoring the usage and state of street furniture in the neighbourhood. (Hill, 2008, para 12)

---

[5] See, for instance, 'China takes surveillance to new heights with flock of robotic doves, but do they come in peace?', https://www.scmp.com/news/china/society/article/2152027/china-takes-surveillance-new-heights-flock-robotic-doves-do-they; 'China is using robotic bird drones with cameras to monitor its citizens', https://bgr.com/2018/06/26/china-bird-drones-surveillance-robots/; 'China is testing creepy drones that look and fly like real birds to monitor citizens', https://www.businessinsider.com/china-is-testing-creepy-dove-drones-to-monitor-citizens-2018-6?r=US&IR=T&IR=T.

[6] See Simon Denyer's article 'Beijing bets on facial recognition in a big drive for total surveillance'

https://www.washingtonpost.com/news/world/wp/2018/01/07/feature/in-china-facial-recognition-is-sharp-end-of-a-drive-for-total-surveillance/?utm_term=.aff3000ae762.

Embedded algorithmic technologies, such as facial recognition systems, while often unnoticed by city dwellers, structure urban ambience unobtrusively, thereby contributing to the formation of what Thrift (2004a) called the urban qualculative background. With this notion, he refers to the abstractly material background of city life that is rarely questioned but merely assumed, i.e., 'the landscape which the body "naturally" adjusts to and which it regards as a normal part of its movement' (Thrift, 2004a, 584). Thrift's notion of hybrid kinaesthesia, in turn, captures how urbanites' bodily movements and gestures often take place in multiple spatial contexts simultaneously. Furthermore, Thrift (2004b) argued (over a decade ago) that we have entered a new phase of 'technological unconscious', one which is fundamentally shaping how corporeal activities are organised and coordinated in the city (see also Thrift & French, 2002). Unlike the Freudian or Lacanian notions of unconscious, this body-technical unconscious refers to the taken-for-granted processes that keep the corporeal body in the recursive loop of code-based technological infrastructure. What is especially noteworthy in this connection is that the repetitive uses of smart devices and complacency regarding what they afford automate urbanites' mediated bodily actions to a point of complete self-evidence.

The proliferation of networked portables is generally perceived as a positive development, because these gadgets provide people with easy access to information and social connections while on the move. At the same time, the amount of data collected, transmitted and stored about users through both these personal devices and via sensors and surveilling systems have grown exponentially. Increasingly in urban spaces, users can be tracked by nearby urbanites' devices, which can collect information about them. In all cases, personal data can be used to profile individuals as consumers and citizens based on their gender, age, social class, income, political opinions, habits, preferences, etc. with the aim of influencing, managing or controlling them. Considering this situation, surprisingly little is known about the repercussions of data collection in urban centres. In our view, the need to critically analyse the infrastructural dynamics of urban data generation and collection resonates with the broader need to grasp and make publicly visible the full implications of networked tracking for power relationships and social justice on the societal level (Gray, 2003; Lyon, 2001).

8.5 Bodies in the Feedback Loops of Computational Cities

To facilitate the generation and collection of personal data on a daily basis, digital gadgets are designed in ways that attach users in an affective loop of bodily habit formation via technologies. Nagamura (2015) suggests that the action of looking at a mobile phone display, for example, has become increasingly universal. There is an aspect of technologically induced compulsion at play, as users may look at their displays with rapid frequency and strong intent even when this action is nonessential. Research has also pointed to social functions of mobile phone display looking, as some users apparently rehearse this activity to avoid social encounters in public places (see, e.g., Baron & Campbell, 2012; Hampton & Gupta, 2008). In any case, be it intentional or unintentional, this simple kinaesthetic repetition can be fruitfully approached in terms of choreographic gestures on different infrastructural levels and at their intersections.

With regard to the micro-level choreography, the act of looking at a mobile phone display restricts the user's movement to turning the head and eyes towards the screen, while the fingers are needed to hold the phone, keep the display alive and access content. The choreography here consists of sequences of movement performed by individual bodies to check received messages or other content on the device. The length of checking sequences may vary depending on how often the user handles his or her gadget. The action becomes compulsive if the repetitive movements start to disturb essential daily activities (Roberts, Yaya, Honore, & Manolis, 2014). As Nagamura (2015) points out, the mobile phone is considered an insensitive and forcible medium because it keeps the user in an affective checking loop even without any external impulse. Importantly in the present context, there is a clear connection between compulsive behaviour and the macro-level choreographing of movement, as mobile devices and ambient urban technologies are designed to be immersive and to generate data imperceptibly.

To understand how city dwellers' bodily movements and gestures generate mass data comprising digital traces that technology companies collect, classify, refine and sell on a global scale, it is useful to take a look at the big data end of the urban feedback system. Big data is often defined as datasets whose size is beyond the capacity of ordinary database software tools to capture, store, manage and analyse (Manyika et al., 2011). In 2011, such a large amount of data was produced every two days; by 2013, the same amount was generated every 10 min (Ajana, 2015). Although scholars often place emphasis on sheer volume, it is worth recognising that big data is not merely about huge sets of bits. In fact, big data is, above all, about networked relational information (Ajana, 2015; Manovich, 2012). Size is certainly an important factor in that the more massive the volume of data, the better sorting algorithms become at spotting connections, creating and/or unlocking patterns, and rendering recognised relations in visible form—and, most importantly, *learning* from these operations. It is these latter aspects that make big data such a seductive field of investment for both commercial actors and political and administrative agencies.

Regarding the political economy of the feedback loop, the measurement of bodies by wearable devices, for example, is based on proprietary algorithms that track various aspects of their users' movements and actions. What is noteworthy is that these devices not only track and collect data on users but also function to steer their actions by, for example, suggesting that they exercise or pop into cafés or bars. When biosensors monitor how long someone has been sitting and working at a café, the apps installed on the device may prod him or her to walk around at specified intervals. Another example is that of a person taking a stroll along the street who receives automated 'fitness nudges' from his or her device stating, 'You're near a park. Why not go for a run?' In these cases, urbanites' bodies, while apparently in self-motion, are being-moved by their gadgets and the predatory applications installed on them. Put differently, embodied agency and individual decision-making are mediated by algorithms in a manner that renders bodily actions less intentional. Furthermore, as wearable devices continue to shape perceptions of the body and of bodily movement, people's individual and social activities may increasingly adapt to the algorithmic logic that runs their smart devices (Budish, 2015; Schüll, 2016), becoming integrated within urban algorhythmicities (cf. Miyazaki, 2013; also Coletta & Kitchin, 2017; but see Tironi & Valderrama, 2018). And the more seamless our experiences with wearables become, the more we may tend to rely on them to interpret the needs of our bodies. An extreme example of such a loop of embodied technology-dependency is the use of applications such as Apple's Screen Time app to reduce the time one spends using the applications installed on the device.

Companies and governments alike promote the idea that algorithmic processes and data-driven systems have been purged of human bias, error and interference, thereby ostensibly facilitating more neutral and objective decision-making (Ajana, 2015; Muller, 2004). Yet, computational classifications and the recommendations based on them are neither neutral nor objective; on the contrary, they are biased towards their own purposes (Burrell, 2016; Mackenzie, 2017). According to Dwork and Mulligan (2013), greater attention should be directed to the values embedded in big data classifications and to how their automation is transforming public and private life in diverse spatial settings (see also Gitelman, 2013).

From the perspective of technology manufacturers, it makes profitable sense to design wearable devices that are simultaneously immersive and pervasive in terms of data collection. Such devices are an integral part of a feedback loop that not only consists of the effective generation, collection, refinement and selling of data but also includes the transmission of the refined data back into urban settings as new products, services and smarter surveillance and control systems. In this inherently cybernetic circuit, an ever-increasing amount of data produced by human bodies is needed to hone algorithms to offer ever-better targeted suggestions for customers and citizens (Harwood, 2014).

What is particularly noteworthy in contemporary cities is that as the algorithms running wearables and surveillance cameras, for example, continue to advance, they become increasingly precise at recognising an individual's connections with other people, especially in physical spaces. Consequently, when technological systems can track relations with other potential actants around individual users and profile them as well, the feedback mechanism itself becomes significantly more complex than when it is based solely on a simple input–output logic. A logical next phase in this simultaneous broadening and deepening of the cybernetic cycle would be to socially rank individuals' movement trajectories relative to their associates' or nearby users' movement trajectories in urban environments. In the course of time, increasing accuracy in the relational recognition of individual movement and gesture patterns may enable the formation of practices in which algorithmically assessed relations with others are used to define the societal status and value of individuals. One may even propose, only mildly provocatively, that the current developments in China we mentioned above already demonstrate actual steps taken in this direction. In any case, following the cybernetic logic at play to the full promises (or threatens) to realise a sociotechnical tendency whereby the relationships, movement trajectories, daily routines and activities of individuals count as key signifiers of their infrastructurally co-conditioned social existence.

As we have stressed above, big data operations intertwine closely with urban physical infrastructures and city dwellers' embodied practices. Urbanites' bodies, in a few words, are not external to the hybrid constellations and cybernetic circuits to which their mediated movements and actions contribute (cf. Gabrys, 2016). Bodies themselves, intentionally and unwittingly, individually and jointly, are both shaped by and contribute to new adaptive habituations and practices. Apparently, it is possible to express embodied resistance to and even protect oneself from algorithmic tracking and surveillance through diverse bodily actions by, for example, avoiding particular urban hotspots, abandoning the use of smartphones or covering one's face with a mask or other form of camouflage when moving about in the city. Yet most urban technological systems are such that, in actual practice, urbanites cannot help but submit to their logic, which reaches beyond the local and the situational. To give a mundane example, customers paying for their groceries by flashing a smart card at the supermarket checkout or passengers swiping their travel cards on an integrated transport system on a bus are involved in a feedback loop that extends far beyond the micro gesture of paying for food or transportation or the meso gesture of attuning to co-present customers' bodily movements. With the constantly growing role of algorithmically processed big data in urban infrastructural systems, it remains to be seen what kinds of embodied practices and accompanying norms will evolve from the multiscale interplay of mediated choreographies.

8.6 Concluding Remarks

In this chapter, we have provided a phenomenologically informed take on the ways in which city dwellers' mediated bodily movements, their multispatial hybrid kinaesthesia, and their movement trajectories are integrally linked to contemporary urban infrastructural processes. Our proposal is that the conception of urban choreography as outlined above opens up a fruitful perspective for understanding how pervasively computed cities function increasingly as complex cybernetic systems. Furthermore, we suggest that the choreographic approach can pave the way for studying empirically how embodied adaptation to algorithmic technologies interrelates with digitalized urban structures, as well as how, among other things, the policies of city planning are affected and challenged in the process.

It seems, paradoxically, that the more individuals pursue making personal choices in urban settings, the more frantically they facilitate a world in which means and ends are dislocated. In other words, through their mediated bodily actions, urbanites contribute actively yet unintentionally to the acceleration of the circular feedback mechanisms that deepen technological saturation (Davison, 2004). In order to tackle the multiple ways in which city dwellers as corporeal beings are presently involved in power-related spatial production, we emphasise the importance of not restricting the choreographic analysis to the micro level. Nor is it enough to focus merely on the situational meso-level dimension of bodily movement. Unless we address these two levels as mutually connected and simultaneously interrelated with macro-choreographic forces, we will fail to uncover the broader political and economic implications of contemporary sociotechnical ensembles and their complex infrastructural dynamics.

To avoid reducing the discussion on human–technology relations in the urban context into approaches that overemphasise subjective experiences and unproblematic user intentionality (as is often the case in studies of human–computer interaction), the choreographic approach we have outlined addresses even the most personal movements and acts in the context of large-scale data-processing systems. These are driven by strong economic and political interests, and often jointly so, as in the case of the Smart City developments, for example. As we have illustrated, the couplings of lived bodies with physical urban spaces today are increasingly mediated by computation. This means that while smartphones and wearable trackers may be experienced as intimate technologies, they at the same time render corporeality into an external object to be (self-)surveilled, and the data generated in this way can be delivered to global technology companies for profit. Moreover, as we have suggested, various tools based on big data generation may easily be transduced into new mechanisms for steering and controlling both individual and social behaviour in urban space. Finally, due to affective attachment to personal devices, urbanites actively and willingly, yet in a deeply compulsive and unthinking vein, participate in the generation of big data, thereby assisting in the acceleration of cybernetic feedback mechanisms.

With regard to thinking about interaction design in the contemporary urban context, the choreographic approach underscores the importance of not concentrating exclusively on user–device relations or addressing some specific situational circumstances of use. The starting point here as well should be on how city dwellers' mediated bodily activities intertwine with macro-infrastructural processes across different spatial scales of bodily movement and movement trajectories, and how, ontologically speaking, corporeality in this process functions as a key element in the complex constitution of sociotechnical and institutional structures. The more practical implications of such an approach for interaction design would deserve a separate discussion.

To formulate the fundamental aspiration behind our sketching of the premises of the study of urban choreography in pervasively computed cities, a key aim concerns the politicisation of infrastructure (cf. Rossiter, 2016). One way to do such a choreographic analysis in practice would be to expose and scrutinise critically the concrete ways in which personal data promote an end to embodied agency as a personal and private issue in the sense of individuals' ownership of their lived bodies.

References

Ajana, B. (2015). Augmented borders: Big Data and the ethics of immigration control. *Journal of Information, Communication and Ethics in Society, 13*(1), 58–78. https://doi.org/10.1108/JICES-01-2014-0005

Amin, A., & Thrift, N. (2002). *Cities. Reimagining the urban*. Oxford: Blackwell Publishing.
Baron, N. S., & Campbell, E. M. (2012). Gender and mobile phones in cross-national context. *Language Sciences, 34*, 13–27. https://doi.org/10.1016/j.langsci.2011.06.018
Bench, H. (2014). Gestural choreographies: Embodied disciplines and digital media. In S. Gopinath & J. Stanyek (Eds.), *Oxford handbook of mobile music studies* (Vol. 2, pp. 238–256). Oxford: Oxford University Press.
Budish, R. (2015, October 20). What my hearing aid taught me about the future of wearables. *The Atlantic*. Retrieved from http://www.theatlantic.com/technology/archive/2015/02/what-my-hearing-aid-taught-me-about-the-future-of-wearables/385145/
Burrell, J. (2016). How the machine 'thinks': Understanding opacity in machine learning algorithms. *Big Data & Society*, 1–12. https://doi.org/10.1177/2053951715622512
Carter, P. (2009). *Dark writing: Geography, performance, design*. Honolulu, HI: University of Hawaii Press.
Coletta, C., & Kitchin, R. (2017). Algorhythmic governance. Regulating the heartbeat of a city using the Internet of Things. *Big Data & Society*, 1–16. https://doi.org/10.1177/2053951717742418
Crang, M. (2001). Rhythms of the city. Temporalised space and motion. In J. May & N. Thrift (Eds.), *Timespace: Geographies of temporality* (pp. 187–207). London & New York: Routledge.
Davison, A. (2004). Reinhabiting technology: Ends in means and the practice of place. *Technology in Society, 26*, 85–97. https://doi.org/10.1016/j.techsoc.2003.10.007
Dwork, C., & Mulligan, D. K. (2013, September 3). It's not privacy, and it's not fair. *Stanford Law Review* [online], 66, 35. Retrieved August 21, 2015, from: www.stanfordlawreview.org/online/privacy-and-big-data/its-not-privacy-and-its-not-fair
Edensor, T. (2010). Introduction: Thinking about rhythm and space. In T. Edensor (Ed.), *Geographies of rhythm* (pp. 1–18). Surey: Ashgate.
Edwards, P. N. (2003). Infrastructure and modernity. Force, time, and social organization in the history of sociotechnical systems. In T. Misa, P. Brey, & A. Feenberg (Eds.), *Modernity and technology* (pp. 185–225). Cambridge, MA: The MIT Press.
Elmqvist, N. (2011). Embodied human-data interaction. In A. N. Antle, P. Marshall, & E. Van Den Hoven (Eds.), *Proceedings of workshop embodied interaction: Theory and practice in HCI CHI 2011* (pp. 104–107). New York: ACM.
Frers, L., & Meier, L. (Eds.). (2007). *Encountering urban places: Visual and material performances in the city*. Hampshire: Ashgate.
Freund, P. (2004). Civilised bodies redux: Seams in the cyborg. Soc Theory Health, 2, 273–289. https://doi.org/10.1057/palgrave.sth.8700031
Gabrielli, L., Rinzivillo, S., Ronzano, F., & Villatoro, D. (2013). From tweets to semantic trajectories: Mining anomalous urban mobility patterns. In J. Nin & D. Villatoro (Eds.), *Citizen in sensor networks* (pp. 26–35). Heidelberg: Springer.
Gabrys, J. (2016). *Program Earth. Environmental sensing technology and the making of a computational planet*. Minneapolis, MN: University of Minnesota Press.
Gitelman, L. (Ed.). (2013). *"Raw data" is an oxymoron*. Cambridge, MA: The MIT Press.
Goffman, E. (1963). *Behavior in public places*. New York: The Free Press.
Goffman, E. (1971). *Relations in public*. New York: Basic Books.
Graham, S., & Marvin, S. (2001). *Splintering urbanism: Networked infrastructures, technological mobilities and the urban condition*. London: Routledge.
Gray, M. (2003). *Urban surveillance and panopticism: Will we recognize the facial recognition society?* Retrieved March 1, 2018, from: https://queens.scholarsportal.info/ojs-archive/index.php/surveillance-and-society/article/viewFile/3343/3305
Haddadi, H., Mortier, R., McAuley, D., & Crowcroft, J. (2013). *Human-data interaction* (Technical Report No. 837). Retrieved October 19, 2015, from The University of Cambridge Computer Laboratory website: http://www.cl.cam.ac.uk/techreports/
Hägerstrand, T. (1982). Diorama, path, and project. *Tijdschrift voor ekonomische em sociale geografie, 73*(6), 323–339. https://doi.org/10.1111/j.1467-9663.1982.tb01647
Hall, E. T. (1966). *The hidden dimension*. New York: Anchor Books.

Hampton, K., & Gupta, N. (2008). Community and social interaction in the wireless city: Wi-fi use in public and semi-public spaces. *New Media & Society, 10*(6), 831–850.
Harwood, K. (2014). *Algorithms: The next wearable tech frontier*. Retrieved October 19, 2015, from: http://www.wired.com/insights/2014/10/algorithms-wearable-tech-frontier/
Hayles, N. K. (2009). RFID: Human agency and meaning in information-intensive environments. *Theory, Culture & Society, 26*(2–3), 47–72.
Hill, D. (2008). *The street as a platform*. Retrieved March 1, 2018, from: http://www.cityofsound.com/blog/2008/02/the-street-as-p.html
Hirsch, A. B. (2014). *City choreographer: Lawrence Halprin in urban renewal America*. Minneapolis, MN: University of Minnesota Press.
Husserl, E. (1989). *Ideas pertaining to a pure phenomenology and to a phenomenological philosophy, second book: Studies in the phenomenology of constitution* (R. Rojcewicz & A. Schuwer, Trans.). Dordrecht: Kluwer Academic Publishers. (Original work Ideen zu einer reinen Phänomenologie und phänomenologischen Philosophie. Zweites Buch: Phänomenologische Untersuchungen zur Konstitution published in 1952).
Husserl, E. (1997). *Thing and space. Lectures 1907*. (R. Rojcewicz, Trans.). Dordrecht: Kluwer Academic Publishers. (Original work *Ding und Raum* published in 1973).
Ingold, T. (2011). *Being alive: Essays on movement, knowledge and description*. London: Routledge.
Jensen, O. B. (2010). Negotiation in motion: Unpacking a geography of mobility. *Space and Culture, 13*(4), 389–402.
Kitchin, R., & Dodge, M. (2011). *Code/Space: Software and everyday life*. Cambridge, MA: The MIT Press.
Klauser, F. R., & Albrechtslund, A. (2014). From self-tracking to smart urban infrastructures: Towards an interdisciplinary research agenda on Big Data. *Surveillance & Society, 12*(2), 273–286.
Krivý, M. (2016). Towards a critique of cybernetic urbanism: The smart city and the society of control. *Planning Theory, 17*(1), 8–30. https://doi.org/10.1177/1473095216645631
Laban, R. (1966). *Choreutics*. London: MacDonald and Evans.
Latour, B. (1997). *On actor–network theory: A few clarifications*. Retrieved August 21, 2015, from The Centre for Social Theory and Technology, Keele University, UK website: http://keele.ac.uk/depts/stt/stt/ant/latour.htm
Law, J. (1992). Notes on the theory of the actor network: Ordering, strategy and heterogeneity. *Systems Practice, 5*(4), 379–393.
Lefebvre, H. (2004). *Rhythmanalysis*. (S. Elden & G. Moore, Trans.). London: Bloomsbury. (Original work Éléments de rythmanalyse published in 1992).
Lepecki, A. (2006). *Exhausting dance: Performance and the politics of movement*. London: Routledge.
Loke, L., & Kocaballi, B. (2016). Choreographic inscriptions: A framework for exploring sociomaterial influences on qualities of movement for HCI. *Human Technology: An Interdisciplinary Journal on Humans in ICT Environments, 12*(1), 31–55.
Lupton, D. (2013). Quantifying the body: Monitoring and measuring health in the age of mHealth technologies. *Critical Public Health, 23*(4), 393–403. https://doi.org/10.1080/09581596.2013.794931
Lycouris, S. (2009). Choreographic environments: New technologies and movement-related artistic work. In J. Butterworth & L. Wildschut (Eds.), *Contemporary choreography: A critical reader* (pp. 346–361). London: Routledge.
Lyon, D. (2001). *Surveillance society: Monitoring everyday life*. Philadelphia, PA: Open University.
Mackenzie, A. (2010). *Wirelessness: Radical empiricism in network cultures*. Cambridge, MA: The MIT Press.
Mackenzie, A. (2017). *Machine Learners: Archaeology of a data practice*. Cambridge, MA: The MIT Press.
Manning, E. (2009). *Relationscapes: Movement, art, philosophy*. Cambridge, MA: The MIT Press.

Manovich, L. (2012). Trending: The promises and the challenges of big social data. In M. K. Gold (Ed.), *Debates in the digital humanities* (pp. 460–475). Minneapolis, MN: The University of Minnesota Press.

Manyika, J., Chui, M., Brown, B., Bughin, J., Dobbs, R., Roxburgh, C., et al. (2011). *Big Data: The next frontier for innovation, competition, and productivity*. Retrieved June 4, 2015, from the McKinsey Global Institute website: www.mckinsey.com/insights/business_technology/big_data_the_next_frontier_for_innovation

Mattern, S. (2015). *Deep mapping the media city*. Minneapolis, MN: The University of Minnesota Press.

McQuire, S. (2008). *The media city: Media, architecture and urban space*. London: Sage.

Merleau-Ponty, M. (1962/1989). *The phenomenology of perception*. (C. Smith, Trans.). London: Routledge. (Original work Phénoménologie de la perception published in 1945).

Miyazaki, S. (2013). Urban sounds unheard-of. A media archaeology of ubiquitous infospheres. *Continuum: Journal of Media & Cultural Studies, 27*(4), 514–522.

Muller, B. (2004). (Dis)qualified bodies: Securitization, citizenship and "identity management". *Citizenship Studies, 8*(3), 279–294. https://doi.org/10.1080/1362102042000257005

Nagamura, T. (2015). The action of looking at a mobile phone display as nonverbal behavior/communication: A theoretical perspective. *Computers in Human Behavior, 43*, 68–75. https://doi.org/10.1016/j.chb.2014.10.042

Parviainen, J. (2010). Choreographing resistances: Kinaesthetic intelligence and bodily knowledge as political tools in activist work. *Mobilities, 5*(3), 311–330.

Parviainen, J. (2016). Quantified bodies in the checking loop: Analyzing the choreographies of biomonitoring and generating Big Data. *Human Technology: An Interdisciplinary Journal on Humans in ICT Environments, 12*(1), 56–73.

Parviainen, J., Tuuri, K., & Pirhonen, A. (2013). Drifting down the technologization of life: Could choreography-based interaction design support us in engaging with the world and our embodied living? *Challenges, 4*(1), 103–115. https://doi.org/10.3390/challe4010103

Pred, A. (1981). Social reproduction and the time-geography of everyday life. *Geografiska Annaler, 63*(1), 5–22.

Prytherch, D. (2015). Rules of the road: Choreographing mobility in the everyday intersection. In J. Cidell & D. Prytherch (Eds.), *Transport, mobility, and the production of urban space, human geography* (pp. 45–63). London: Routledge.

Ridell, S. (2010). The cybercity as a medium: Public living and agency in the digitally shaped urban space. *IRIE: International Review of Information Ethics, 12*(3), 11–19.

Ridell, S. (2013). The city as a medium of media: Public life and agency at the intersections of the digitally shaped urban space. In S. Tosoni, M. Tarantino, & C. Giaccardi (Eds.), *Media and the city: Urbanism, technology and communication* (pp. 32–50). Newcastle: Cambridge Scholars Publishing.

Ridell, S. (2019). Mediated bodily routines as infrastructure in the algorhythmic city. *Media Theory, 3*(2), 27–62.

Roberts, J. A., Yaya, P., Honore, L., & Manolis, C. (2014). The invisible addiction: Cell-phone activities and addiction among male and female college students. *Journal of Behavioral Addiction, 3*(4), 254–265.

Robertson, A., Lycouris, S., & Johnson, J. (2007). An approach to the design of interactive environments, with reference to choreography, architecture, the science and the complex systems and 4D design. *International Journal of Performance Arts and Digital Media, 3*(2-3), 281–294.

Rossiter, N. (2016). *Software, infrastructure, labor: A media theory of logistical nightmares*. New York: Routledge.

Schiller, G., & Rubidge, S. (2014). Introduction. In G. Schiller & S. Rubidge (Eds.), *Choreographic dwellings: Practising place* (pp. 1–10). New York: Palgrave Macmillan.

Schüll, N. (2016). Data for life: Wearable technology and the design of self-care. *BioSocieties, 11*(3), 317–333.

Sheets-Johnstone, M. (1999). *The primacy of movement*. Amsterdam: John Benjamins.

Sheller, M., & Urry, J. (2006). *Mobile technologies of the city*. London: Routledge.
Shepard, M. (2009). Toward an architecture of Hertzian space. *ACADIA 09: reForm*, pp. 209–215.
Smith, R. J., & Hetherington, K. (2013). Urban rhythms: Mobilities, space and interactions in the contemporary city. In R. J. Smith & K. Hetherington (Eds.), *Urban rhythms: Mobilities, space and interaction in the contemporary city* (pp. 4–16). West Sussex: Wiley Blackwell.
Sullivan, R. (2017). *The Geography of the everyday: Toward an understanding of the given*. Athens, GA: University of Georgia Press.
Thrift, N. (2004a). Movement-space: The changing domain of thinking resulting from the development of new kinds of spatial awareness. *Economy and Society, 33*(4), 582–604.
Thrift, N. (2004b). Remembering the technological unconscious by foregrounding knowledges of position. *Environment and Planning D, 22*(1), 175–190.
Thrift, N. (2004c). Driving in the city. *Theory, Culture Society, 21*(4/5), 41–59.
Thrift, N., & French, S. (2002). The automatic production of space. *Transactions of the Institute of British Geographers, 27*(3), 309–325.
Tironi, M., & Valderrama, M. (2018). Unpacking a citizen self-tracking device: Smartness and idiocy in the accumulation of cycling mobility data. *Environment and Planning D, 36*(2), 294–312. https://doi.org/10.1177/0263775817744781
Torrens, P. M. (2016). Computational streetscapes. *Computation, 4*(3), 37. https://doi.org/10.3390/computation4030037

Jaana Parviainen is a philosopher and a senior researcher at Tampere University, the Faculty of Social Sciences. Her philosophical expertise covers social epistemology, the philosophy of technology, social ethics, and body studies. She has conducted and led several interdisciplinary research projects. In her ongoing research project, 'Struggling with Ignorance' (2018–2022), she considers the politics of artificial intelligence how predictive analytics implemented in decision-making systems can create prescriptive models and new regimes to manage unknown futures and insecurities in the society. Parviainen's research has been published in journals, such as, *Synthese*, *Mobilities* and *Techné: Research in Philosophy and Technology*.

Seija Ridell is Professor of Media Studies at the Faculty of Information Technology and Communication Sciences, Tampere University. Her research revolves around the conditions and forms of human agency in mediated environments, from journalistic media to social media platforms to the contemporary city. Her methodological expertise covers narratology, audience and user studies, and action research. Ridell's research has been published in journals, such as *Communications*, *International Journal of Communication* and *Media Theory*. She has co-edited an interdisciplinary book and several journal special issues on the topic of urban mediation. She is a vice-chair of ECREA's Media, Cities and Space.

Part II
Responsible Design of Urban Technologies

Chapter 9
Dockless App-Based Bicycle-Sharing Systems in China: Lessons from a Case of Emergent Technology

Rockwell F. Clancy and Aline Chevalier

Abstract Since cycling can contribute to sustainability, shared-bicycle schemes have been encouraged as a green technology. In Chinese cities, however, dockless app-based bicycle-sharing systems (DABS) have become a blight, resulting in tremendous waste. Ironically, this stems from the success of DABS—their rapid development and adoption. As an "emergent" technology, DABS in China consist in the confluence of existing technologies and extra-technological factors, situations different from the sum of their parts, where negative consequences are more difficult to identify and address. Additionally, DABS in China are a case of private, app-based technology providing a good traditionally regulated by municipal governments. For these reasons, it provides valuable lessons: The negative consequences associated with DABS in China result from a lack of policy formulation and implementation, failing to consider and address extra-technological dimensions. Private technology companies should not be left to ensure these goods, and municipalities must craft policies to ensure stakeholder interests.

Keywords Shared bicycles · China · Technological emergence · Policy

R. F. Clancy (✉)
University of Michigan-Shanghai Jiao Tong University Joint Institute,
Shanghai Jiao Tong University, Shanghai, China

Institute of Social Cognition and Decision-making, Shanghai Jiao Tong University,
Shanghai, China
e-mail: R.F.ClancyIII@tudelft.edu

A. Chevalier
Center for Urban Studies, Urban Cycling Institute, University of Amsterdam,
Amsterdam, The Netherlands

M. Nagenborg et al. (eds.), *Technology and the City*, Philosophy of Engineering and Technology 36, https://doi.org/10.1007/978-3-030-52313-8_9

9.1 Introduction

Since 2014, dockless app-based bicycle-sharing systems (DABS) have taken Chinese cities by storm. DABS companies have or plan to expand into urban markets in the US, Singapore, Japan, the UK, Italy, Malaysia, and Australia (Lin, 2017, Mar 24; Mobike launches, 2017, Aug 23; Pidd & Lavelle, 2017, Jun 29; Mobike Japan, 2017, Jun 22). According to recent figures by Mobike—the largest DABS in the world—since the concept was first introduced, the number of people using bicycles has roughly doubled (Mobike, 2017, Apr 12), and DABS have made Shanghai the world's largest bicycle-sharing city, with 280,000 shared bicycles (Wang, 2017, Jun 27).[1] Like the introduction of technology in general, DABS can address longstanding and important social problems, while at the same time resulting in additional negative consequences.

On the one hand, cycling plays an important role in urban sustainability, easing traffic congestion, alleviating air pollution, reducing CO_2 emissions, promoting energy efficiency, and enhancing public health and urban livability (Fraser & Lock, 2011; Gehl, 2010; Krizek, Handy, & Forsyth, 2009; Owen, Humpel, Leslie, Bauman, & Sallis, 2004; Pooley, 2013; Pucher, Dill, & Handy, 2010; Pucher, Buehler, & Seinen, 2011; Pucher & Dijkstra, 2003). As a result, bike-sharing schemes have been encouraged by municipal governments and are widely viewed as a green technology. On the other hand, in Chinese cities, DABS have become a veritable blight, resulting in tremendous waste and unfavorable perceptions of cycling and cyclists. Ironically, these negative consequences stem from the unparalleled success of DABS in China, the rapid production of bicycles by private bike-sharing companies and their use by the public—no other bike-sharing scheme in the world has been as successful. Both the successes of and negative effects associated with DABS in China are a result of their nature as an "emergent" technology, we argue, where the confluence of existing technologies and extra-technological factors spawn situations different from the sum of their parts, making it difficult to anticipate and address negative consequences effectively.

These same dynamics characterize concerns about the relation between private technology companies and public institutions outside of China, where technologies undercut the social goods they were originally meant to serve. Facebook, for instance, has faced criticisms for its role in Russian interference in the 2016 US Presidential election (Warzel, 2018, Feb 12; Vosoughi, Roy, & Aral, 2018; Michel, 2017, Oct 17), the dissemination of misinformation (Madrigal, 2017, Oct 2; Meyer, 2018, Mar 8), and polarization of the political sphere. Similarly, the operations of both Airbnb (Lieber, 2015, Aug 14) and Uber (Luegenbieh & Clancy, 2017, p. 95–96) have raised concerns regarding user safety, and concerns about the role of Airbnb in distorting real estate prices (Reid, 2018, Jan 31) and Uber in contributing to the elimination of public transportation (Morozov, 2016, Jan 31). Such consequences often result from failing to adequately consider and address the non-technical, cultural dimensions of technology, overlooking affected shareholders. These problems become more acute in urban areas, where more, different kinds of people live and interact.

[1] The exact number of DABS bicycles is unknown. Although Chinese cities have put limits on the number of DABS bicycles allowed, many speculate that these companies continue to illicitly introduce bicycles, such that it is difficult to determine exactly how many are in circulation for use at any one time. For this reason, figures reported throughout this chapter vary.

Examining DABS in China sheds light on these and similar cases, offering lessons about the ways private, technology companies can support goods traditionally provided and regulated by public municipalities, highlighting past shortcomings to avoid and future directions to take. This case is especially appropriate for examination since, first, DABS can be considered a green technology. Mobike, for instance, has always branded itself as such, with their slogan, "Save the world, ride Mobike."[2] In these cases, the risk exists of focusing on benefits to the exclusion of costs, especially when such costs are not foreseen or difficult to determine. As with other green technologies, such as electric cars and solar panels, the costs associated with DABS often fade from view. Overlooking negative costs associated with technology can result from innate and persistent psychological biases, especially "motivated blindness," which have been shown to occur in situations where intentions and anticipated results are good, as occurs with decisions related to green technologies (Bazerman & Tenbrunsel, 2013; Sachdeva, Jordan, & Mazar, 2015; Truelove, Carrico, Weber, Raimi, & Vandenbergh, 2014). Next, DABS are widely used among Chinese—the largest population in the world—and could be used by many others worldwide. Third, China currently has the highest number of urban areas in the world, and these are only projected to grow. Fourth, China is a non-Western country, and its development has been somewhat unique—insofar as its cultural background is non-Western, has embraced market reforms but has not substantially democratized—and countries developing today have and might increasingly look to China as a model of development (Luegenbieh & Clancy, 2017; Bell, 2015; Kurlantzick, 2013, Mar 21). Examining the non-Western cultural contexts surrounding DABS in China can illuminate and dispel common assumptions on which European and US policies might be based, but that do not necessarily hold in Asia. Finally, China is functionally a one-party state, where a tight relationship exists between the government and economy, such that government policies have the potential to directly and dramatically affect the economy and, therefore, the habits and preferences of citizens (Bell, 2015; McGregor, 2010).

Although both Chinese and foreign media have covered DABS in China (Wang 2017; Chinese discard, 2017, Jan 17; Horowitz, 2017, Aug 22; Moss, 2017, Apr 3; Rushe, 2017, Nov 5), discussions have tended to focus on their immediate and somewhat sensational effects (but see Spinney & Lin, 2017). This chapter focuses on the historical, cultural, technological, and political dimensions of DABS, to both diagnose problems and offer solutions. To do so, we begin by outlining China's history with bicycles and their cultural dimensions, which have contributed to both the adoption of and problems with DABS, which tend to be overlooked. Next, we describe the emergent nature of DABS—the technologies of which they are comprised, how this has facilitated the adoption of DABS, and the resulting benefits. Third, we highlight negative consequences associated with DABS, as well as their causes. Finally, we outline lessons that can be learned from DABS in China, for both DABS in China and the relation between private technology companies and public municipalities.

---

[2] In terms of their lifecycles – for example, their production and use in conjunction with a mobile-based app – DABS obviously depend on energy and, therefore, fossil fuels, which ultimately contribute to emissions. As with other green technologies, for example, solar panels and windmills, these costs often fall from view. We discuss such costs at greater length below.

9.2 A History of Cycling in China

Both the rapid adoption of DABS, and failure to develop adequate policies regulating their use, should be understood in terms of the history of cycling in China. Whereas North Americans or Europeans might be inclined to view the adoption of cycling as an uncontested good—such that individual citizens and municipal governments would promote their use—this has not been the case in Chinese cities. This reaction can be understood with reference to China's somewhat ambivalent relation to bicycles and, in this first section, we outline the history of this relation.

The ubiquity of bicycles in China might lead one to conclude the Chinese are culturally inclined to cycling, although this conclusion would be mistaken. Rather, socioeconomic and political conditions best explain China's development into the "bicycle nation of the 20th century" (Esfehani, 2004, p. 94).

The first bicycle imported into China arrived in Shanghai in 1868 (Xu, 2012). However, the use of bicycles in China was not initially widespread, since they were viewed as a foreign invention and their use put them in direct competition with the pulled rickshaw (Tao, 2012). Rickshaws provided work to a substantial segment of the lower class (Strand, 1993). Additionally, as a mode of transportation, rickshaws meshed well with existing social conventions, since they allowed those from the higher class to avoid the degrading physical effort of moving themselves.

This began to change in 1949, with the founding of the People's Republic of China (PRC), when a clampdown occurred on the use of rickshaws, due to their status as a symbol of working-class oppression. At this time, the Chinese Communist Party (CCP) encouraged the use of bicycles as an efficient and democratic means of transportation, where citizens using bicycles to commute were given additional benefits, which resulted in the rapid adoption of bicycles as a means of mass-transportation (Murray, 2011). Bicycles were identified as an important component of city planning, and PRC's first Five-Year Plan included a provision to grow the bicycle industry by 60%.

Since then, China's relationship with bicycles has been mixed: Bicycles have moved from being an ideological symbol of equality to a public nuisance and vestige of China's backwards past. Beginning in the early to mid 1990s, Chinese authorities encouraged the production and purchase of automobiles, since a variety of Chinese industries and companies have been central to the supply chains of automobile manufacturers. Cycling bore the brunt of this change. As with bicycles, cycling lanes in Chinese cities are ubiquitous, leading to the erroneous conclusion Chinese authorities prioritize cycling. However, most bicycle lanes date back to the 1990's and result from attempts to segregate bicycles from motor vehicles on the road, to the benefit of motorists (Zhang, Shaheen, & Chen, 2014). This same priority becomes clear in the evolution of national traffic regulations.

In 1995, for example, the central government released the Standard of Urban Road Traffic, which initiated a period of radical reduction in urban cycling (Zhang et al., 2014). Local governments have also adopted aggressive policies regarding the use of bicycles, sometimes going as far as establishing official targets by which to reduce the number of cyclists. In 1993, for instance, the Transport Master Plan of

Gurangzhou aimed at cutting the bicycle mode-share by a factor of almost three by 2010 (Ma, 2004). More recently, from 2000 to 2006, major Chinese cities, such as Shanghai, have banned the use of bicycles on central traffic arteries (Pan, 2011), despite conflicting trends—such as the 2002 White Paper of Shanghai Urban Transport Development, which promotes a more positive attitude towards and vision of cycling (Zhang et al., 2014).

In the last decade, several studies have called attention to deteriorating conditions for cyclists in Chinese cities—for example, encroachment on bicycle lanes by increasing amounts of motorized traffic, and the deterring effect this increased risk has created—in response to which few measures have been taken (Liu, Jia, & Cheng, 2012; Yang & Gakenheimer, 2007). Most recently, however, as Chinese cities have experienced problems related to motorized traffic, such as severe pollution and congestion, authorities are reassessing the use of bicycles.

9.3 DABS as an Emergent Technology

To explain the adoption of DABS in China—and lay the groundwork for understanding related problems—in this second section, we outline the technological characteristics and extra-technological factors that have contributed to the success of DABS in China. We begin by considering the difference between personal and shared bicycles—but not specifically DABS—going on to discuss the distinctively dockless and app-based features of DABS, as well as associated behaviors.

The adoption of new technology depends on not only technology but also the interplay between technology and extra-technological, social factors, such as user experience, historical, economic, and cultural factors (Basalla, 1988; Henrich, 2015; Srite & Karahanna, 2006). DABS have been more readily used and widely adopted in China than dock-based shared bicycles (DBSB) in either China or abroad. Chevalier et al. (2019), for example, found that, in Shanghai, 15% of DABS users are also car owners who do not own personal bicycles, a rate three times higher than DBSB users (Kumar & Taob, 2016; Yang, Chen, & Wang, 2015; Zhu, Pang, Wang, & Timmermans, 2013). This results from technological emergence, where the confluence of existing technologies and social practices contribute to situations qualitatively different from the sum of their parts, giving rise to novel behaviors and benefits that might not exist in conjunction with any one technology alone. As the rather awkward moniker makes clear, DABS are comprised of more basic technologies, specifically, bicycles, remote-operated locks, smart phones, and mobile-based applications.

In the first place, as opposed to owning and using a personal bicycle, both DBSB and DABS alleviate costs associated with monitoring and fixing a personal bicycle.[3] For instance, a survey conducted in 2015—the year DABS were first introduced in Shanghai—found that "fear of robbery" was the highest cited factor preventing

[3]Although this benefit could also result in harm, insofar as it undermines a sense of individual responsibility associated with ownership, further discussed below.

people from riding their bicycles in almost a quarter of respondents. Fear of maintaining one's own bicycle would be a hindrance to cycling, overcome with shared bicycle schemes (Chevalier & Xu, 2020).

In terms of the adoption of shared bicycles, however, the alleviation of personal costs seems to be negligible since, were it a sufficient condition for the adoption of shared bicycles, then DABS should fair no better than DBSB, but this has not been the case. Although 52% of the Shanghai population uses DABS, approximately this same amount also owns personal bicycles. This shows that, while DABS contribute to increased cycling, DABS do not necessarily contribute to the recruitment of new cyclists (Chevalier, et al., 2019; Chevalier & Xu, 2017). Hence, differences in user rates and frequency between DABS and DBSB seem to concern the dockless and app-based features of the latter.

Unlike DBSB, which have existed since the 1970s in cities like Copenhagen (Bycyklen), Paris (Vélib'), Berlin (Nextbike) and even Chinese cities, such as Hangzhou—which has the largest DBSB scheme worldwide—DABS can be picked up and dropped off anywhere. Since DABS do not have to be taken from and returned to specific docking-station locations, this would make them easier to use and, therefore, more attractive (Srite & Karahanna, 2006) than personal bicycles or DBSB, since one would not have to find a place near one's home or work for pick up and drop off, for instance. Additionally, since DABS are distributed throughout cities, people see them more often and see others using them, which could facilitate their adoption through two distinct but related psychological mechanisms. First, termed the "mere exposure" effect, familiarity through repetition has been shown to increase liking (Zajonc, 2001), such that the more that people see bicycles, the more likely they are to like and use them. Second, the transmission of beliefs and behaviors has been shown to occur through prestige bias and conformist transmission, where people tend to adopt beliefs and behaviors displayed by individuals who are perceived as prestigious (prestige bias), and by a majority or average of others (conformist transmission) (Henrich, 2015).[4] Finally, since DABS users transport bicycles from one point to another in the process of riding, dockless schemes could facilitate use—with DBSB, demand must be monitored and bicycles periodically transported among docking stations.

The integration of mobile-based apps into DABS is a second, technological factor differentiating DABS from DBSB. Although redistribution is still necessary with DABS, redistributing bicycles from low- to high-demand areas during particular times, the app-based nature of DABS in China makes this process easier. Through GPS tracking, DABS companies can monitor user behavior, identifying where individual bicycles are at any time, as well as patterns in use. This makes it easier for DABS companies to find, fix, and redistribute bicycles according to user demand, as

[4] Insofar as culture affects a variety of psychological processes and behaviors, the strength of these tendencies might not be uniform across different nations and cultures. Although a full consideration of this issue falls outside the scope of this chapter, with regard to DABS specifically and the adoption of technology in general, it would be a fruitful topic for further discussion and exploration, and below we consider historical and structural factors related to DABS in China.

Fig. 9.1 A touchpad to access a shared bicycle in Karlsruhe, Germany

well as for users to find DABS bicycles to ride. Additionally, the app-based nature of DABS makes it easier to pay than with DBSB. Although users still initially set up an account—linking it to a bank account, credit card, or other form of payment—once completed, accessing and using DABS in China consists in merely scanning a QR code, rather than a more complicated or tedious process—for example, entering a credit card or bank account number, making a phone call, receiving a pin code, etc. (Fig. 9.1).

This supposes, of course, that potential users are able and willing to use smart phones with app capacities, as well as DABS apps themselves, in conjunction with mobile payment, which might not be the case. Were it not the case that Chinese users have and are comfortable using smart phones and DABS apps, then it is unlikely DABS would be widely used and readily adopted. This has been a concern regarding the success of DABS in countries other than China, where these conditions might not be met.

9.4 Negative Consequences of DABS in China

Versus DBSB, DABS have unique features that have facilitated the success of these schemes. Ironically, these same features have resulted in negative effects and raise ethical issues. This section examines these effects, issues, and their technical causes in the case of DABS in China.

In August 2017, following a number of other regional authorities, the Shanghai Municipal Transportation Bureau implored DABS companies to stop inundating the

Fig. 9.2 DABS bicycles impeding pedestrian and bicycle traffic

city with bicycles and better organize the ones scattered throughout. Shanghai contained an estimated 1.5 million DABS bicycles at that time, approximately one for every 16 people (Horwitz, 2017b, Aug 22). By December 2017, it was estimated that close to 20 million bicycles had been introduced into Chinese cities through DABS (Li, 2017). This rapid deluge and subsequent disorder were caused by the business strategies of DABS companies, in conjunction with the nature of the bicycles themselves.

Like similar industries in China, such as ride-hailing apps, DABS companies have sought to capture as large a market share as quickly as possible, driving competitors out of business to obtain a monopoly (Horwitz, 2017a, Aug 3). For this reason, DABS companies have flooded Chinese cities with bicycles in excess of demand. Little effort has been devoted to either organizing or fixing bicycles, since the goal has been to simply saturate the market. Like DBSB, users do not own DABS bicycles, so have less motivation to take care of them than personal bicycles, resulting in damage. Unlike DBSB, DABS bicycles can be dropped off anywhere, such that bicycles can quickly accumulate and become damaged if not redistributed, collected, and/or fixed, impeding both pedestrian and bicycle traffic (Figs. 9.2 and 9.3).

The financial difficulties of Ofo, China's second largest DABS company, beginning in 2018, have made this problem worse—cheap, deteriorating yellow bicycles have been left to accumulate and become further damaged. Each Ofo bicycle costs less than $45 to produce, versus Mobikes, which are close to $300 (Lee, 2017, Sep 7). Citizens and municipal governments have incurred the costs of organizing, moving (Fig. 9.4), or destroying DABS bicycles, creating incredible waste in the process. Images of thousands of DABS bicycles for destruction have gone viral (Figs. 9.5 and 9.6), resulting in 300 thousand tons of waste metal from scrapping

Fig. 9.3 Damaged DABS bicycles piled up in a public space

Fig. 9.4 Shanghai city workers collect excess DABS bicycles

Fig. 9.5 Aerial view of a lot holding DABS bicycles in Hongkou district, Shanghai

Fig. 9.6 Ground view of a holding/disposal site for DABS bicycles in Shanghai

(Li, 2017). This incredible waste obviously calls into question characterizations of DABS as a green technology.

This disorder and waste likely precipitate more disorder and waste and, unsurprisingly, have precipitated negative views of cycling and cyclists among the public. Since perceptions and behaviors are shaped by repetition and the behaviors of others, as mentioned above, constantly seeing damaged and neglected—and others damaging and neglecting—DABS bicycles could result in descriptive or social norms (similar to "scripts") regarding the permissibility of this behavior (Bicchieri, 2016; Latour, 1991). A 2018 survey found that disorder and waste associated with DABS reinforce the historical view, discussed above, that cycling is a low-class,

Fig. 9.7 Sign forbidding DABS bicycles in a residential compound in Shanghai

outdated mode of public transportation, indicative of China's past (Chevalier, Charlemagne, & Xu, 2018). Many residential compounds in Chinese cities have posted signs forbidding DABS bicycles (Fig. 9.7) and—based on observations in Shanghai by the authors—these mainly appear at higher-end, more expensive compounds.

DABS also give rise to issues about user privacy, given the app-based nature of DABS and the relations of private Chinese companies to the government (Spinney & Lin, 2017; Gardiner, 2017, Nov 14; Mobike, ofo Share Bicycle Data, 2018, Jan 18). Some have speculated that DABS companies make or plan to make their revenue by selling user information and travel routes (Gardiner, 2017, Nov 14). User information and travel routes could be used by Chinese state organs to monitor not only Chinese citizens but also citizens of countries in which Chinese DABS companies operate. Of course, this concern would not be unique to China—Edward Snowden revealed US technology companies shared user information with US security and intelligence agencies (Greenwald & MacAskill, 2013, Jun 7). Technology firms in China, however, have tended to have closer relations with government bureaus, as these companies can act as mediums for collecting and disseminating huge amounts of information, to monitor and mold social opinions and discussions. DABS in China have received funding from and been aligned with much larger technology firms—MoBike with Tencent and Ofo with Alibaba—for example (Jing, 2017, Apr 27). Many involved in business in China have had either direct or indirect ties to the Chinese government and CCP, sometimes referred to as China Inc. (Fishman, 2006).[5]

This relationship has resulted in corruption and inefficiency, where projects might be approved and implemented to serve special interests rather than the broader population. Rather than allowing for, helping to craft, and complying with sensible municipal regulations regarding DABS, for instance, DABS companies could either meddle with or disregard such regulations—evidence of which is the way some DABS companies have disregarded calls to monitor and organize DABS,

[5] But see Hawes (2017, Mar 2) regarding the way and extent to which concerns regarding private companies acting as agents of the CCP are unfair and poorly founded.

as well as defying limits on the number of DABS allowed in cities. Similarly, in their foreign endeavors, Chinese DABS companies have tended to move into foreign markets with looser regulations of shared bicycles, for example, Italy rather than Germany.

9.5 Policies Surrounding DABS in China

Simply making available DABS bicycles will not result in an effective urban transportation system for cycling. The mere availability of bicycles in China has led to clutter, incredible waste, and negative perceptions of cycling and cyclists. These are a result of inefficient policies by both municipal governments and DABS companies. In this final section, we outline these shortcomings in, and make recommendations for the improvement of, policies surrounding DABS in China.

Inefficient public policies stem from the ambiguous legal status of DABS in China. DABS bicycles are no more regulated than personal bicycles, despite clear differences. For instance, they are branded products owned by companies rather than individuals. While DABS bicycles should not be parked on private property—implying they belong to the public realm—at present, few public policies exist regulating their use or stipulating in which public spaces they are allowed, or they remain quite general. There is no reason to believe the ambiguous regulatory status of DABS will be clarified in the near future.

In structured interviews with Shanghai urban planners, three of four participants answered "no" when asked whether DABS should be viewed as part of the city's public transportation network, and responded "no" when asked whether they anticipated any sort of partnership with DABS companies in the near future. These responses show urban planners view DABS as belonging to private companies and should not, therefore, be included in considerations of Transportation Planning.[6] This response might seem surprising, since private automobiles are generally accounted for in transportation planning, but urban planners do, indeed, think DABS should be ignored.[7] This view and resulting regulatory ambiguity likely stem from China's ambivalent historical relation to bicycles discussed above. As a result, responsibility for ensuring the safe and effective use of DABS has largely fallen to the companies themselves, in conjunction with users.

To do so, most DABS use credit systems to ensure the appropriate use of bicycles. In the case of MoBike, for instance, new users begin with a score of one-hundred points. This score increases by one point for every ride taken during which the user adheres to the terms and conditions, but decreases by fifteen to one-hundred points for every ride taken during which the user violates the terms and conditions,

[6] Interviews were conducted in March 2018, with four urban planners from the Shanghai Urban Planning and Design Research Institute.

[7] We are grateful to a reviewer for pointing this out.

where the amount of reduction depends on the nature of the infraction. This system is an instance of technological mediation, where technologies embody values, resulting in behaviors that reciprocally enforce values (Latour, 1991; Latour, 2002; Verbeek, 2005). Insofar as social monitoring has been shown to result in more pro-social behaviors (Bateson et al., 2015; Ernest-Jones, Nettle, & Bateson, 2011; Francey & Bergmüller, 2012), monitoring associated with DABS apps could motivate users to look after and take care of DABS, reinforcing a sense of responsibility potentially lost with shared versus personal bicycles, resulting in less clutter, damage, and waste.[8]

However, the priority given to corporate profits versus public goods are a major concern of this current credit system. For instance, sanctions for failing to lock up a bicycle after use are roughly equivalent to—and sometimes more severe—than sanctions for violations of parking and traffic rules: "[P]arking against the rules" results in a 20-point reduction, just slightly more than "forgetting to lock the bike, with the bike later found," which results in a 15-point reduction. However, "violating traffic rules, being stopped by traffic police, then ditching the bike to escape" results in only a one-hundred-point reduction, considerably less than "forgetting to lock the bike, causing its loss" or "installing private locks," that would prevent others from using it and cause a reduction in revenue for Mobike (Fig. 9.8), which results in an account being reset to zero and the user temporarily banned from using Mobike (Mobike Terms and Conditions).

These sanctioning policies are an example of private interests coming into conflict with public goods, an issue at the center of concerns regarding the role of technology and technology companies in deforming public institutions, mentioned

Fig. 9.8 DABS bicycle being kept in an apartment stairwell, preventing others from using it

[8]Although this could result in users refraining from using the app, since they do not want to be monitored and sanctioned, a survey in 2018 found that Chinese users were much less reluctant to share their personal data than foreign users. We are grateful to a reviewer for bringing this to our attention.

in the introduction. Insofar as fines associated with a loss of revenue are higher than those associated with ensuring safe and effective traffic regulations, DABS implicitly prioritize private, corporate interests at the expense of public goods. Insofar as the app promotes this prioritization, it could result in user behaviors that potentially impede the safe and effective use of bicycles in relation to other forms of transportation.

Better policies by both Chinese municipal governments and DABS companies are necessary to mitigate negative consequences resulting from DABS. In the first place, municipal governments simply need to consider DABS in city planning. Just because DABS bicycles are privately owned is a poor reason to exclude them from consideration, since DABS can contribute to or detract from public good. Public-private partnership models (PPP) have proven effective in integrating bike-sharing systems into transportation planning previously (Audikana, Ravalet, Baranger, & Kaufmann, 2017), which might be especially suited to China. The close relationship between government and business through the CCP has resulted in projects being quickly conceived and carried out. The benefits of authoritarian rule in a developing country such as China—where citizens might lack basic goods and services related to, for example, food, shelter, healthcare, and education—are often presented in contrast to shortcomings of democratic rule in other developing country, for example, India—where projects are often bogged down or left incomplete because of multiple layers of local, state, and federal government and associated bureaucracy (A bridge in Mumbai, 2012, Dec 22).

Based on the previous discussions, municipal governments should set and enforce clear guidelines for DABS companies, factoring them into transportation plans. At a minimal, these could consist in detailed, city-wide guidelines regarding where DABS bicycles are and are not allowed, how many are allowed in specific places at specific times—forcing DABS companies to organize or transport bicycles—and specifying minimal criteria for their design and manufacture, to minimize waste associated with their production and use. Municipal policies governing DABS companies should take priority over those governing users, since DABS companies are more influential and wealthy than individual users and, for this reason, have greater responsibilities. Since DABS companies reap financial benefits, they should also bear more of the costs associated with the (mis)use of DABS bicycles, rather than passing these off to municipalities or users. This should be reflected in the credit systems built into DABS apps, prioritizing public goods rather than financial benefits to DABS companies alone.

9.6 Conclusion

The promotion of cycling within cities can result in tremendous good, but this requires more than simply making bicycles available. DABS in China have been widely adopted as a result of the convergence of existing technologies, coupled with the cultural attitudes of users towards these technologies. However, DABS in China have also resulted in very bad effects, in part for the same reasons that have made them a success. This situation is characteristic of emergent technologies, where the confluence of technological and extra-technological factors give rise to unpredictable consequences. Technologies are increasingly emergent. To improve the integration of DABS into transportation planning, and private-public cooperation in general, the foregoing has examined DABS in China as a case of emergent technology. Critically, private technology companies should not be trusted to implement public goods on their own. Rather, public policies must be crafted and enforced to regulate this implementation and these companies, to maximize benefits and minimize harms.

References

A bridge in Mumbai: Halfway to paradise (2012, December 22). The Economist Retrieved March 15, 2018, from https://www.economist.com/christmas-specials/2012/12/22/halfway-to-paradise

Audikana, A., Ravalet, E., Baranger, V., & Kaufmann, V. (2017). Implementing bikesharing systems in small cities: Evidence from the Swiss experience. *Transportation Policy, 55*, 18–28.

Basalla, G. (1988). *The evolution of technology*. Cambridge: Cambridge University Press.

Bateson, M., Robinson, R., Abayomi-Cole, T., Greenlees, J., O'Connor, A., & Nettle, D. (2015). Watching eyes on potential litter can reduce littering: Evidence from two field experiments. *PeerJ, 3*, e1443. https://peerj.com/articles/1443/

Bazerman, M., & Tenbrunsel, A. (2013). *Blind spots: Why we fail to do what's right and what to do about it*. Princeton, NJ: Princeton University Press.

Bell, D. (2015). *The China model: Political meritocracy and the limits of democracy*. Princeton, NJ: Princeton University Press.

Bicchieri, C. (2016). *Norms in the wild: How to diagnose, measure, and change social norms*. New York: Cambridge University Press.

Chevalier, A., & Xu, L. (2017). On the applicability of a Western bikability index in the Chinese context. In *International Conference 2017 on Spatial Planning and Sustainable Development. Seoul, South Korea.*

Chevalier, A., & Xu, L. (2020). On the applicability of a Western bikeability index in the Chinese context. *International Review for Spatial Planning and Sustainable Development, 8*, 59–93. https://doi.org/10.14246/irspsd.8.1_59.

Chevalier, A., Charlemagne, M., & Xu, L. (2018). Improving Bicycle acceptance: On the impact of station-less shared-bikes and built environment. In *The Continuing City: People, planning, and the long haul to urban resurgence. 58th Annual ACSP Conference. Buffalo, New York, USA.*

Chevalier, A., Charlemagne, M., & Xu, L. (2019). Bicycle acceptance on campus: Influence of the built environment and shared bikes. *Transportation Research Part D: Transport and Environment, 76*, 211–235. https://doi.org/10.1016/j.trd.2019.09.01.

Chinese discard hundreds of cycles-for-hire in giant piles. (2017, January 17). *The Guardian*. Retrieved March 15, 2018, from https://www.theguardian.com/world/2017/jan/17/chinese-discard-hundreds-of-cycles-for-hire-in-giant-pile

Ernest-Jones, M., Nettle, D., & Bateson, M. (2011). Effects of eye images on everyday cooperative behavior: A field experiment. *Evolution and Human Behavior, 32*, 172–178.

Esfehani, A. (2004). The Bicycle's Long Way to China: The Appropriation of Cycling as a Foreign Cultural Technique. In *13th International Cycling History Conference.*

Fishman, T. (2006). *China, inc.: How the rise of the next superpower challenges America and the world*. New York: Scribner.

Francey, D., & Bergmüller, R. (2012). Images of eyes enhance investments in a real-life public good. *PLoS One, 7*, e37397.

Fraser, S., & Lock, K. (2011). Cycling for transport and public health: A systematic review of the effect of the environment on cycling. *European Journal of Public Health, 21*, 738–743.

Gardiner, S. (2017, November 14). How your personal information funds share bike schemes. In *The Sydney Morning Herald*. Retrieved March 15, 2018, from https://www.smh.com.au/business/consumer-affairs/how-your-personal-information-funds-share-bike-schemes-20171110-gzj05d.html

Gehl, J. (2010). *Cities for people*. Washington, DC: Island Press.

Greenwald, G. & MacAskill, E. (2013, June 7). NSA Prism program taps in to user data of Apple, Google and others. *The Guardian*. Retrieved March 15, 2018, from https://www.theguardian.com/world/2013/jun/06/us-tech-giants-nsa-data

Hawes, C. (2017, March 2). China's private companies are unjustly labeled as Communist Party plants. In *The Conversation*. Retrieved March 15, 2018, from https://theconversation.com/chinas-private-companies-are-unjustly-labeled-as-communist-party-plants-73834

Henrich, J. (2015). *The secret of our success: How culture is driving human evolution, domesticating our species, and making us smarter*. Princeton, NJ: Princeton University Press.

Horwitz, J. (2017a, August 3). One year after the Uber-Didi merger, it's only getting harder to hail a ride in China.. *Quartz Media*. Retrieved March 15, 2018, from https://qz.com/1045268/one-year-after-the-uber-didi-merger-its-only-getting-harder-to-hail-a-ride-in-china/

Horwitz, J. (2017b, August 22). Chinese cities are saying 'enough already' to bike-sharing services run rampant. In *Quartz Media*. Retrieved October 2, 2018, from https://qz.com/1058438/chinese-cities-saying-enough-already-to-chaos-generated-by-bike-sharing-services-like-ofo-and-mobike/

Jing, M. (2017, April 27). China's bike-sharing battle ratchets up a notch as Didi Chuxing enters the fray. In *South China Morning Post*. Retrieved March 15, 2018, from http://www.scmp.com/tech/start-ups/article/2091191/chinas-bike-sharing-battle-ratchets-notch-didi-chuxing-enters-fray

Krizek, K., Handy, S., & Forsyth, A. (2009). Explaining changes in walking and bicycling behavior: Challenges for transportation research. *Environment and Planning B, 36-4*, 725–740.

Kumar, K., & Taob, L. (2016). How accessible and convenient are the public bicycle sharing programs in China? Experiences from Suzhou city. *Habitat International, 53*, 188–194.

Kurlantzick, J. (2013, March 21). Why the 'China model' isn't going away: From Bangkok to Caracas, Beijing's style of authoritarian capitalism is gaining influence. In *The Atlantic*. Retrieved March 15, 2018, from http://www.theatlantic.com/china/archive/2013/03/why-the-china-model-isnt-going-away/274237/

Latour, B. (1991). Technology is society made durable. In J. Law (Ed.), *A sociology of monsters: Essays on power, technology and domination* (pp. 103–131). London: Routledge.

Latour, B. (2002). Morality and technology: The end of the means. *Theory, Culture & Society, 19*(5&6), 247–260.

Lee, E. (2017, September 7). Beijing joins list of cities saying no to new shared bikes. In *Technode*. Retrieved May 29, 2018, from https://technode.com/2017/09/07/beijing-joins-list-of-cities-saying-no-to-new-shared-bikes/

Li, D. (2017). Bike-sharing, how to bid farewell to "rubbish mountain"? Resour. Recycl. 34–37.

Lieber, R. (2015, August 14). Airbnb horror story points to need for precautions. *New York Times*. Retrieved March 14, 2018, from https://www.nytimes.com/2015/08/15/your-money/airbnb-horror-story-points-to-need-for-precautions.html?mtrref=www.google.com&gwh=9DC64BB39B2FE2C0BB93FFFEE47912CC&gwt=pay

Lin, L. (2017, March 24). China's bike-sharing startups are coming to U.S. cities. *The Wall Street Journal*. Retrieved March 15, 2018, from https://www.wsj.com/articles/chinas-bike-sharing-startups-are-coming-to-u-s-cities-1490287361

Liu, Z., Jia, X., & Cheng, W. (2012). Solving the last mile problem: Ensuring the success of public bicycle system in Beijing. *Procedia-Social and Behavioral Sciences, 43*, 73–78.

Luegenbieh, H., & Clancy, R. (2017). *Global engineering ethics*. New York: Elsevier.

Ma, X. (2004). Enlightenment on the variations in the structure of Guangzhou inhabitant mode share. *Urban Transport of China, 2*(2), 29–32.

Madrigal, A. (2017, October 2). Google and Facebook failed us. The Atlantic. Retrieved March 15, 2018

McGregor, R. (2010). *The party: The secret world of China's communist rulers*. New York: Harper Perennial.

Meyer, R. (2018, March 8). The grim conclusions of the largest-ever study of fake news. *The Atlantic*. Retrieved March 14, 2018, from https://www.theatlantic.com/technology/archive/2018/03/largest-study-ever-fake-news-mit-twitter/555104/?utm_source=atlfb

Michel, C. (2017, October 17). Commentary: The Russians pretended to be Texans – and the Texans believed them. *Washington Post*. Retrieved March 14, 2018, from https://www.washingtonpost.com/news/democracy-post/wp/2017/10/17/how-the-russians-pretended-to-be-texans-and-texans-believed-them/

Mobike (2017, April 12). Bicycle and urban development white paper 2017. *mobike.com*. Retrieved March 14, 2018, from http://mp.weixin.qq.com/s/MK8BafUZbklEvfl6xVR_uQ

Mobike, Ofo share bicycle data, reveal cycling heat map (2018, January 18). *Pandaily*. Retrieved March 15, 2018, from https://pandaily.com/mobike-ofo-share-bicycle-data-reveal-cycling-heat-map/

Mobike Japan in Fukuoka (2017, June 22). *mobike.com*. Retrieved March 14, 2018, from https://mobike.com/global/blog/post/mobike-japan

Mobike launches in Sapporo (2017, August 23). *Shanghai Daily*. Retrieved March 14, 2018, from https://www.shine.cn/archive/business/it/Mobike-launches-in-Japans-Sapporo/shdaily.shtml

Morozov, E. (2016, January 31). Cheap cab ride? You must have missed Uber's true cost. *The Guardian*. Retrieved March 12, 2018, from https://www.theguardian.com/commentisfree/2016/jan/31/cheap-cab-ride-uber-true-cost-google-wealth-taxation

Moss, T. (2017, April 3). Share bikes come to China, where some of them meet grisly fates. *The Wall Street Journal*. Retrieved March 15, 2018, from https://www.wsj.com/articles/braking-bad-thieves-and-pranksters-lay-waste-to-chinas-share-bikes-1491230284

Murray, H. (2011). *An historical and descriptive account of China: Its ancient and modern history, language, literature, religion, government, industry, manners and social state*. Edinburgh: Oliver and Boyd.

Owen, N., Humpel, N., Leslie, E., Bauman, A., & Sallis, J. (2004). Understanding environmental influences on walking: Review and research agenda. *American Journal of Preventive Medicine, 27*, 67–76.

Pan, H. (2011). Implementing sustainable urban travel policies in China. *International Transport Forum Discussion* Paper No. 2011-12.

Pidd, H. & Lavelle, D. (2017, June 29). Chinese bike-share scheme launches in rainy Manchester. *The Guardian*. Retrieved July 6, 2017, from https://www.theguardian.com/lifeandstyle/2017/jun/29/chinese-bike-share-scheme-mobike-launches-in-rainy-manchester

Pooley, C. (2013). *Promoting walking and cycling*. Cambridge: The Polity Press.

Pucher, J., & Dijkstra, L. (2003). Promoting safe walking and cycling to improve public health: Lessons from the Netherlands and Germany. *American Journal of Public Health, 93*, 1509–1516.

Pucher, J., Dill, J., & Handy, S. (2010). Infrastructure, programs, and policies to increase bicycling: An international review. *Preventive Medicine, 50*(Supplement), S106–S125.

Pucher, J., Buehler, R., & Seinen, M. (2011). Bicycling renaissance in North America? An update and re-appraisal of cycling trends and policies. *Transportation Research Part A: Policy and Practice, 45*, 451–475.

Reid, D. (2018, January 31). Airbnb is making rents in New York City spike as owners yank units off the market. *CNBC*. Retrieved March 14, 2018, from https://www.cnbc.com/2018/01/31/airbnbn-pushing-up-new-york-rent-study-claims.html

Rushe, D. (2017, November 5). Why we can't have nice things: Dockless bikes and the tragedy of the commons. *The Guardian*. Retrieved March 15, 2018, from https://www.theguardian.com/politics/2017/nov/05/why-we-cant-have-nice-things-dockless-bikes-and-the-tragedy-of-the-commons.

Sachdeva, S., Jordan, J., & Mazar, N. (2015). Green consumerism: Moral motivations to a sustainable future. *Current Opinion in Psychology, 6*, 60–65. https://doi.org/10.1016/j.copsyc.2015.03.029

Spinney, J., & Lin, W.-I. (2017). Are you being shared? Mobility, data and social relations in Shanghai's public bike sharing 2.0 sector. *Applied Mobilities*. https://doi.org/10.1080/23800127.2018.1437656

Srite, M., & Karahanna, E. (2006). The role of espoused national cultural values in technology acceptance. *MIS Quarterly, 30*(3), 679–704.

Strand, D. (1993). *Rickshaw Beijing: City people and politics in the 1920s*. Berkeley, CA: University of California Press.

Tao, X. (2012). 踏车而飞:自行车与中国骑车人(1868–1949 年) [A History of the Bicycle and Chinese Cyclists, 1868–1949]. *Cross-Currents: East Asian History and Culture Review E-Journal No. 3*. http://cross-currents.berkeley.edu/e-journal/issue-3

Truelove, H., Carrico, A., Weber, E., Raimi, K., & Vandenbergh, M. (2014). Positive and negative spillover of pro-environmental behavior: An integrative review and theoretical framework. *Global Environmental Change, 29*, 127–138. https://doi.org/10.1016/j.gloenvcha.2014.09.004

Verbeek, P. (2005). *What things do: Philosophical reflections on technology, agency, and design*. University Park: Pennsylvania State University Press.

Vosoughi, S., Roy, D., & Aral, S. (2018). The spread of true and false news online. *Science, 359*, 1146–1151. https://doi.org/10.1126/science.aap9559

Wang, S. (2017, June 27). Why bicycles are piling up in a Shanghai parking lot. *CNN*. Retrieved March 14, 2018, from https://edition.cnn.com/travel/article/china-shanghai-bikes/index.html

Warzel. (2018, February 12). Infocalypse now. *BuzzFeedNews*. Retrieved March 14, 2018, from https://www.buzzfeed.com/charliewarzel/the-terrifying-future-of-fake-news?utm_term=.gpmkYvBzx#.ilx9Grkq5

Xu, T. (2012). 踏车而飞:自行车与中国骑车人 (1868–1949年) [A History of the Bicycle and Chinese Cyclists, 1868–1949]. *Cross-Currents: East Asian History and Culture Review, E-Journal No. 3*. Retrieved March 18, 2019, from http://cross-currents.berkeley.edu/e-journal/issue-3.

Yang, J., & Gakenheimer, R. (2007). Assessing the transportation consequences of land use transformation in urban China. *Habitat International, 31*, 345–353.

Yang, J., Chen, J., & Wang, Z. (2015). Major issues for biking revival in urban China. *Habitat International, 47*, 176–182.

Zajonc, R. (2001). Mere exposure: A gateway to the subliminal. *Current Directions in Psychological Science., 10*(6), 224. https://doi.org/10.1111/1467-8721.00154

Zhang, H., Shaheen, S., & Chen, X. (2014). Bicycle evolution in China: From the 1900s to the present. *International Journal of Sustainable Transportation, 8*(5), 317–335.

Zhu, W., Pang, Y., Wang, D., & Timmermans, H. (2013). Travel behavior change after the introduction of public bicycle systems: Case study in Minhang district, Shanghai. In *92nd Annual Meeting of the Transportation Resej+1arch Board*.

Rockwell F. Clancy is a lecturer at Delft University of Technology. Before joining Delft, he was an Associate Teaching Professor at the University of Michigan-Shanghai Jiao Tong University Joint Institute, and a Research Fellow in the Institute of Social Cognition and Decision-making, Shanghai Jiao Tong University. His most recent book, Global Engineering Ethics, co-written with Heinz Luegenbiehl, was published in 2017 by Elsevier Press. Rockwell's work has appeared in Science and Engineering Ethics, New Directions in Children & Adolescent Psychology, and Philosophy and Literature. He completed his PhD at Purdue University, MA at the Katholieke Universiteit, Leuven, and BA at Fordham University. Further information can be found at www.rockwellfclancy.com.

Aline Chevalier is a trained architect who has practiced as a designer within four different countries before changing career paths to research in transportation planning and urban design. She is a permanent member of the Cycling Institute and a Research Fellow in the Centre for Urban Studies, University of Amsterdam. Her expertise in urban cycling is already established as she published in major international journals for the design, planning, and management of transportation systems.

Chapter 10
Universally Designed Urban Environments: "A Mindless Abuse of the Ideal of Equality" or a Matter of Social Justice?

Kevin Mintz

Abstract In "Justice and Nature," Thomas Nagel rejects the idea that social equality requires the universal design of urban environments to accommodate people with disabilities. Universal design is a movement in architecture and other arenas to minimize the need to provide individual accommodations for people with disabilities by designing environments that are accessible to a wide range of individuals. I advance that Nagel inappropriately categorizes universal design as a matter of humanitarianism or charity. I argue that for people with disabilities to receive equal access to a city, universal design must be a central consideration in urban policy and construction. I also suggest that assistive technologies, devices that intentionally or unintentionally help people with disabilities access the same opportunities that the able-bodied have, is critical to the realization of universal design. These technologies are critical to facilitating responsible design, and relational autonomy for people with disabilities.

Keywords Universal design · People with disabilities · Social justice · Shameful revelation · Equal access to advantage · Relational autonomy

10.1 Introduction

In his article "Justice and Nature," philosopher Thomas Nagel argues that disability is a matter of humanitarianism rather than justice. In the philosophical literature, humanitarianism is often equated with charity, whereas justice is often equated with the development of fair procedures that balance the interests of individual liberty with those of social equality. Humanitarianism, or charity, is often considered a

K. Mintz (✉)
Department of Clinical Bioethics, National Institutes of Health, Bethesda, MD, USA

M. Nagenborg et al. (eds.), *Technology and the City*, Philosophy of Engineering and Technology 36, https://doi.org/10.1007/978-3-030-52313-8_10

second best alternative to justice (Kymlicka, 2014, pp. 93–94). By way of clarifying his argument, Nagel asserts in a footnote:

It seems to me a mindless abuse of the ideal of equality, for example, that advocates for the handicapped have blocked the installation of free-standing pay toilets on Manhattan streets (of the kind that are common in Europe) unless they could all be large enough for wheelchair access (Nagel, 1997, p. 315).

With this statement, Nagel is rejecting the idea that social equality requires the universal design of urban environments to accommodate people with disabilities.[1] Universal design is a movement in architecture, technology, and other arenas to minimize the need to provide individual accommodations for people with disabilities by designing environments that are accessible to all. Although initially formulated in 1985 by Ronald Mace, universal design as a movement has matured in the years since the passage of the Americans with Disabilities Act (ADA) of 1990 (Hamraie, 2017, p. 5). Universal design is now a critical component of international disability rights discourse as well, underlying much of the substantive requirements of the United Nations Convention on the Rights of Persons with Disabilities (UNCRPD), an international human rights treaty aimed at promoting disability equality worldwide (UN General Assembly, 2007). The movement has also become the source of much scholarly debate in the field of disability studies, particularly with regard to whether it enhances or undermines the objectives of disability politics (Hamraie, 2017; Titchkosky, 2011).

In 1997, the Center for Universal Design at North Carolina State University released the following seven Principles of Universal Design, which have come to embody the central tenets of this design movement:

1. PRINCIPLE ONE: Equitable Use—The design is useful and marketable to people with diverse abilities.
2. PRINCIPLE TWO: Flexibility in Use—The design accommodates a wide range of individual preferences and abilities.

[1] I would like to clarify what I see as the relationship between justice and equality. Justice, as I refer to it in this chapter, is a set of minimal obligations that a society has towards its citizens. I use the term social justice in the same way. There are many different conceptions of justice in the philosophical literature, but when I refer to justice in this chapter I usually mean social justice as conceptualized by John Rawls and those who identify as Rawlsian liberals, including Nagel. Within Rawls' theory and related theories, equality usually refers to social equality with the two terms being used interchangeably. Nagel and I seem to agree that equality is a critical component of a just society and, therefore, it is reasonable to critique Nagel in terms of justice.Within this literature, in order for society to be just, it must distribute advantages in ways that strike an appropriate balance between liberty and equality. The precise meaning of social equality is highly debated, and in "Justice and Nature," Nagel argues that universal design promotes an inappropriate standard of social equality. As I discuss later, Rawls has been criticized for not taking into account the needs of people with disabilities. Thus, I do not necessarily endorse Rawls' framework in this chapter, but instead consider a broad conception of social justice that is indebted both to Rawls and his many critics that, despite their differences, share a fundamental commitment to balancing liberty and equality in a way that makes society more open and fair to all.

3. PRINCIPLE THREE: Simple and Intuitive Use—Use of the design is easy to understand, regardless of the user's experience, knowledge, language skills, or current concentration level.
4. PRINCIPLE FOUR: Perceptible Information—The design communicates necessary information effectively to the user, regardless of ambient conditions or the user's sensory abilities.
5. PRINCIPLE FIVE: Tolerance for Error—The design minimizes hazards and the adverse consequences of accidental or unintended actions.
6. PRINCIPLE SIX: Low Physical Effort—The design can be used efficiently and comfortably and with a minimum of fatigue.
7. PRINCIPLE SEVEN: Size and Space for Approach and Use—Appropriate size and space is provided for approach, reach, manipulation, and use regardless of user's body size, posture, or mobility (Connell et al., 1997).

To illustrate how these principles work in practice, the public toilets Nagel mentions would be consistent with these principles if they were all wide enough for wheelchair access and accessible to people with wider body sizes, those who use canes and other mobility aids besides wheelchairs. Signage for the toilets might also include braille for those who are blind or have low vision, automated voice options for those who have physical impairments, and pictures for those who do not speak English.

In this way, universal design has the potential to enhance access to urban facilities for all, regardless of disability status—a phenomenon that has been coined "the curb-cut effect" (Blackwell, 2017). Curb-cuts are ramps at the ends of sidewalks, usually intended for wheelchair users, meant to facilitate a pedestrian's safe transfer from sidewalks to streets. They not only provide physical access to streets for wheelchair users, but also make it easier for those pushing strollers to safely access streets. Accordingly, the curb-cut effect refers to the idea that accommodations made for people with disabilities can improve the lives of all. Because of its potential to reduce costs for disability accommodations and its potential to make design more responsible and inclusive, universal design has implications for nearly all aspects of urban planning, development, and governance. By responsible design, I mean design that takes into consideration legal and ethical requirements of equal access for people with disabilities and others. Relatedly, inclusive design means design that is aimed at increasing access for a wide variety of users of various abilities and social backgrounds. Responsible urban design, as I conceptualize it here, is a means of promoting social justice.

Although Nagel's assertion is 20 years old, it reflects an ongoing dilemma for urban governments which have to make difficult choices about the accessibility of both public and private facilities. My impetus for writing this response to Nagel is that I find, in my own day-to-day experiences as a person with cerebral palsy,[2] that

---

[2] In this chapter, I draw on many examples from my own life to motivate and substantiate my arguments. While this may be unusual to readers from other fields of philosophy, it is a rather common argumentative technique for philosophers of disability (Barnes, 2016; Kittay, 1998). I therefore follow the example of my colleagues in this subfield.

questions of facilities access in cities are still viewed as matters of goodwill and not justice. While many disability scholars with whom I discuss Nagel's argument view it as outdated or even "stupid," I believe his intuition about disability being a natural disadvantage still drives how many people view the challenges people with various disabilities face in accessing cities. Would anyone question, for example, that water fountains need to be built in a way that accommodate people of various heights, even though height is a "natural" characteristic of human diversity? I would hope not. In my view, Nagel's philosophical argument about disability reflects ableist assumptions that ubiquitously equate disability with "bad difference," meaning that it is always objectively better to be able-bodied rather than disabled (Barnes, 2016, pp. 54–77). Ableism is disability discrimination rooted in the misconception that people with impairments are inherently incapable of meeting the same social expectations as the able-bodied. As I argue in this paper, Nagel's thinking is problematic in that it can hinder urban design that fully takes into account the needs of those with disabilities.

This is particularly true in older urban spaces where too often I hear people excuse spaces that are inaccessible to me as a wheelchair user by saying a version of the following statement, "City X, or building Y is old. They simply were not thinking of you back then, but if we did anything now it would lose its charm." While this statement has validity in that older cities are harder to make accessible to those with disabilities, I often hear this statement in paternalistic tones that fail to think critically about how the situation came to be. These attitudes take for granted that disability is a natural deviation from some standard of normalcy that cannot be exacerbated by the choices of others. Examples of experiences like this are noted by many people with various kinds of disability (Barnes, 2016; Brownlee & Cureton, 2011). Even though Nagel's argument might be discredited among more informed audiences, his intuition and the implications it has for urban design and governance still warrant scrutiny.

Accordingly, this chapter explores three interrelated questions:

1. Should the universal design of urban environments be considered a matter of social justice?
2. How should philosophers conceptualize the role of assistive technologies in achieving the ideal of universal design?
3. How should philosophers and those involved in urban planning best contend with the limits of universal design?

In part two, I refute Nagel's claim that because disability is a "natural disadvantage," universal design does not fall within the appropriate scope of social equality, and is therefore not a matter of justice. I argue that the universal design of urban environments should be considered a matter of social justice, because when decision makers fail to design cities in ways that have disability in mind from the outset they risk denying people with disabilities access to the benefits of the city that would be afforded to them if they were not disabled. Such irresponsible design decisions reinforce differences between those with and without disabilities in ways that deny people with disabilities equal social respect in relation to others. As a matter of practicality, the decision not to construct urban environments in ways that are consistent with universal design could contribute to urban governments and private entities failing to meet their obligations under the ADA and other disability rights legislation worldwide. In this section, I also briefly argue for a standard of equal access akin to G.A. Cohen's equal access to advantage (Cohen, 1989), which I contend aligns with the spirit of most disability rights measures.

In part three, I discuss assistive technologies, devices that intentionally or unintentionally help people with disabilities access the same opportunities that the able-bodied have.[3] Here, I argue that these technologies are critical to the realization of universal design; however, while assistive technologies can facilitate progress in designing universally accessible environments, such technologies are only a means to an end. The effective use of assistive technologies in achieving the goals of universal design is contingent upon those involved in urban planning seeing their role as promoting a specific kind of relational autonomy between people with disabilities and their environment. In this way, the built environment becomes a resource for empowering people with disabilities to set the terms through which they relate to others in an urban community. In part four, I argue that in situations where universal design cannot help optimize urban accessibility, urban governments and private businesses still have obligations to retrofit existing city structures to enable access. I further suggest that in order to ensure the appropriate retrofitting of urban landscapes, localities and private businesses need to negotiate a formula of shared responsibility for costs. I conclude in part five by suggesting that future research should further scrutinize shared costs between governments and private businesses for implementation of universal design, as well as consider the question of how to advocate for universal design in developing countries where access to assistive technology might be difficult.

[3] It is important to note that I use the term assistive technologies more broadly than some other scholars. For example, Sven Ove Hansson categorizes assistive technologies as a subset of what he calls enabling technologies, those that "alleviate the impact of disease or disability" (Hansson, 2007, p. 258). Hansson defines assistive technology as "[technology that] makes it possible for the individual [with disabilities] to perform a task or activity despite an (uncompensated) disability or lack of function" (p. 258). While I appreciate the distinction Hansson is trying to make, I contend that any technology can potentially assist people with disabilities in accessing the same opportunities afforded to the able-bodied. I also find the term "enabling technologies" somewhat paternalistic and ableist in the sense that it implies that there are certain tasks that all people should be able to complete, regardless of disability. By contrast, my use of "assistive technologies" frames the use of devices in terms of how they can assist people with disabilities in taking advantage of the same opportunities that are afforded to those who are able-bodied.

10.2 Rebutting Nagel's Argument Against the Universal Design of Public Toilets in Manhattan

Nagel's motivation for rejecting the idea that universal accessibility is a matter of social justice stems from his overall argument in "Justice and Nature." Nagel expresses concern about the overexpansion of the scope of justice, particularly among luck egalitarians who generally hold that individuals are owed compensation for any inequality that results from "bad brute luck"—inequality that results from random or unforeseen circumstances outside of an individual's control (Cohen, 1989). He appeals to John Rawls' theory of justice as fairness to make this claim. Rawls' theory holds that political institutions, in an ideal liberal society, should follow two principles for the just distribution of primary goods[4]: (1) the equal basic liberties principle; (2a) the fair equality of opportunity principle; and (2b) the difference principle, which are given lexical priority in relation to each other (Rawls, 1999, pp. 37–39). The equal basic liberties principle is to be prioritized over the fair equality of opportunity principle, which in turn takes priority over the difference principle. Since justice as fairness is rooted in ideal theory, Rawls assumes citizens in the ideal society will not have any form of disability, a point that has been the source of much scholarship over the past four decades (Brownlee & Cureton, 2011; Kittay, 1998; Nussbaum, 2007; Richardson, 2006; Silvers & Francis, 2005; Wasserman, Asch, Blustein, & Putnam, 2013).[5] Rawls' omission of people with disabilities from considerations of justice establishes the conceptual space in which Nagel makes his argument against universally designed public toilets in Manhattan.

For Nagel, Rawls' ordering of the fair equality of opportunity principle over the difference principle gives absolute priority to purely "social" disadvantages for the purposes of justice (Nagel, 1997, p. 310). This is in juxtaposition to the purely "natural" disadvantages that Nagel views luck egalitarians as inappropriately placing within justice's scope. For Nagel, an inequality is appropriately treated as "natural," if it satisfies three conditions:

1. That there is a variable natural property of individuals that plays a significant causal role in the generation through social institutions of outcomes which differ substantially in value for those individuals,
2. That the institutions not aim to produce the differential results but have an independent and legitimate purpose,

[4] Primary goods are "what persons need in their status as free and equal citizens and as normal and fully cooperating members of a society over a complete life" (Rawls, 1999, p. xiii).

[5] As mentioned in footnote 1, Nagel places his argument in a Rawlsian framework and this is why I focus on Rawls here. The philosophers of disability cited above provide invaluable critiques of Rawls that have made important contributions to the egalitarian literature. At the same time, it is important to recognize that despite their individual disagreements with Rawls, each of these critics is also indebted to Rawls' legacy. Therefore, my choice to discuss Rawls here is also in recognition that any discussion of equality and justice must begin with some engagement with Rawls.

3. That to achieve that purpose without generating such differences would be significantly more difficult or costly (Nagel, 1997, p. 314).

On Nagel's terms, disability seems to cause an inequality where some public toilets would be accessible to all and others would not; the construction company acting on behalf of Manhattan did not make the decision to build two separate toilets to discriminate against wheelchair users—discrimination was merely a by-product of other legitimate concerns. Further, it would be difficult or costly for a construction company, on Nagel's account, to design a system of universally accessible toilets that could also have mechanisms in place to prevent misuse of the toilet facilities. While Nagel is correct that the needs of those with disabilities should be fairly considered against other public interests (as would the needs of any disadvantaged group), he incorrectly interprets the Public Toilet Example from multiple perspectives. First, he makes an inaccurate assumption about the excessive cost of universally accessible toilets. Second, he incorrectly claims that the universal accessibility of public toilets is an inequality resulting from nature, and thus a matter of humanitarianism, not justice.[6]

Nagel's motivation in claiming that disability is a matter of humanitarianism and not justice is rooted in the idea that if "natural" disadvantages, like disability, fall within the scope of justice, this will come at great costs to institutions that have to establish reasonable limits in the allocation of scarce resources. Accordingly, disability is an example that invokes what is now called the "bottomless pit" objection to Rawlsian distributive justice originally articulated by Arrow (1973). I grant that cost is a reasonable concern in the allocation of any public good or service; however, Nagel is incorrect to argue that universally designed public toilets would be more costly to Manhattan than a segregated system of public toilets, where one accessible unit would be provided for every non-accessible unit. A segregated system would be more costly to the city by nature of having to construct more toilets. Even if it were true that the city would spend more money on universally accessible units, the additional cost would likely be outweighed by benefits to users, both with and without disabilities. Parents who have small children in strollers, for example, would benefit from more spacious toilets just as much as wheelchair users would. This exemplifies the "curb-cut effect" discussed in the introduction and illustrates the value of universal design.

Moreover, Nagel inappropriately categorizes the inaccessibility of public toilets as a "natural" disadvantage. As philosophers such as Tom Shakespeare, Anita Silvers, Leslie Francis, and Lorella Terzi have noted, barriers of accessibility in the built environment are actually social disadvantages that arise from decisions not to design toilets, buildings, and other urban facilities with the needs of people with

[6] Specifically, Nagel asserts, "I do not think society has the same kind of responsibility, under justice, with respect to those inequalities that it has with respect to others that are socially caused. Straightforward humanitarian concern for the welfare of those afflicted will not be undermined by the fact that nature is responsible for their disadvantage, but the kinds of deontological judgments of justice that take precedence over the general welfare may be" (Nagel, 1997, p. 315).

disabilities in mind (Shakespeare, 2017; Silvers & Francis, 2005; Terzi, 2011). This thinking exemplifies the now widely acknowledged social model of disability, which interprets disability as resulting from the interaction of individual impairment with obstacles in built environments (Shapiro, 1993, p. 63). Nagel's interpretation of the Public Toilet Example is rooted, by contrast, in the medical model, as he seems to view disability as an individual pathology and experience that does not interact with the built environment.

To understand the merits of the social model, consider the following example. As a wheelchair user, I attend many classes that are held in rooms on different floors. Too frequently, there is only one elevator to transport me from floor to floor. Several times over the past 10 years, I have been unable to use elevators because they were being repaired, and there have often been no other ways to access various classrooms. In each situation, I had to miss class because it was too late to have them moved to an accessible location. It was not my inability to traverse stairs that prevented me from attending classes, but rather, I became "disabled" by a built environment that was temporarily inaccessible.

To consider this example from the perspective of someone who can use stairs, stairways are as much a part of the built environment as elevators are. If the stairways in a particular building ever became unusable, those who are able-bodied would become just as incapable of moving from floor to floor as I am when an elevator is not functioning. The inaccessibility of a public facility presents a physical barrier to my educational access. Here, my disability results from this physical man-made barrier, rather than my intrinsic physical limitations. This example speaks to the possibility that Nagel underestimates the extent to which the design of the built environment is the result of institutional decisions rather than solely the result of the natural lottery. The distinction between "natural" and "social" disadvantage is not finite as he contends.

As Jeremy Waldron has noted, the design of built environments, especially in cities, has historically been reflective of normative institutional judgments regarding physical and mental ability, race, gender, and other characteristics (Waldron, 1991). Often decisions regarding the segregation of public facilities have been rooted in arguments favoring eugenics and the "natural" superiority of certain types of urban residents over others. As David Serlin has documented in his provocative essay "Pissing on Pity: Disability, Gender, and the Public Toilet," public toilets have been important sites for the manifestation and rectification of social inequalities because of their uncomfortable placement in the conceptual space between public and private. Decisions over who has the right to perform the most intimate of bodily functions in public reflect broader values of social inclusion or exclusion (Serlin, 2010). Therefore, when Nagel's distinction is applied to the realm of urban justice in design, it takes on a potentially unjust dimension.

Universal design tries to avoid such historical legacies of injustice by encouraging designers to consider the widest variety of users possible. Deciding the precise scope of users for whom someone is designing is subjective and can potentially exclude individuals based on the designer's unfamiliarity with the needs of some populations. This speaks to the importance of both diversity and education in architectural and technological design and to the potential of urban governance for dictating specific standards of accessible design. Using the ADA as an example, the law's regulations dictate very explicit standards for the physical accessibility of public spaces. This guidance is intended to ensure that most people with physical disabilities who use assistive mobility devices will be able to physically access the goods and services that a city has to offer. As such, it becomes easier to use the social model to scrutinize Nagel's Public Toilet Example.

According to the *New York Times* articles that Nagel cites, disability advocates were protesting a bid by a major corporation, JCDecaux, which would have allowed for the construction of two separate toilets close to one another. In order to keep out the homeless and prevent the misuse of the wider toilets by sex workers and drug users, JCDecaux recommended that people with physical disabilities access the wider unit with a special keycard. The City of New York favored the position of disability activists and the primary debate was actually between these advocates and JCDecaux. Disability advocates were not preventing the installation of freestanding public toilets in Manhattan. They were, rather, trying to defeat a proposal that would not have included universally designed units. These advocates were encouraging the city's preference for proposals that did not by default include "separate but equal" toilets for people with physical disabilities (McKinley Jr., 1993). Other companies had submitted proposals for the project that designed all toilets to be universally accessible.

Recently, while in San Francisco, I saw one of these universally accessible public toilets constructed by JCDecaux. Unlike the toilets in JCDecaux's Manhattan proposal, these toilets are available to use free-of-charge and do not require a keycard to enter. Rather, patrons simply push a button, opening the door to the facility. I include photos of one of these toilets in Appendix A to aid readers in envisioning the type of facility in question. The fact that JCDecaux now constructs universally designed toilets illustrates how the framework of universal design has gained mainstream acceptance as a way of minimizing the need for special wheelchair accessible facilities.

However, the reality that JCDecaux's position has changed does not take away the theoretical question of whether the Public Toilet Example falls within the auspices of social justice. Even though the design of public toilets is a question of access to social advantage because it involves the construction of facilities in built urban environments, Nagel does highlight an important consideration in the Public Toilet Example. Design choices do not exist in a vacuum and, particularly in older cities, universally designed facilities may be infeasible for any number of reasons. I discuss how to think about such situations in Part IV. However, I do not think Nagel's Public Toilet Example is a case where universally designed facilities are not a feasible option. In particular, it is questionable whether JCDecaux had adequate justification for its two-toilet proposal because of the challenge of preventing the misuse of public toilets by sex workers or drug users. The *New York Times* articles that Nagel cites in relation to the example do not discuss whether JCDecaux fairly considered alternatives to the two-toilet system. However, in formulating my response to Nagel on this point, it seems reasonable to expect that all paying customers could have been given access to the same keycard that people with disabilities were expected to use for the purposes of accessing the same public toilet. This, once again, is an example of how universally designed procedures can minimize the need for separate facilities that reinforce differences between those with disabilities and the able-bodied.

Specifically, requiring wheelchair users to access the toilet using a keycard that no one else is required to use may serve as a symbol of their difference and result in what Jonathan Wolff has called "shameful revelation"—the phenomenon through which institutional procedures contribute to an individual recognizing they have a socially stigmatizing condition, likely contributing to a diminished sense of self-respect (Wolff, 1998, p. 97). This is not to say that developing unique procedures to accommodate the needs of people with disabilities will always be stigmatizing, but it likely would be in this case, where the key card system would be the result of JCDecaux's inability or unwillingness to provide universal access. Practices of segregation should always be scrutinized from the perspective of urban justice because they potentially reinforce the idea that only certain segments of the population should have access to the goods and services of a city. Segregation is not justifiable on the grounds that one construction company refuses to consider installing universally accessible public toilets.

In evaluating the permissibility of segregated public facilities from the perspective of justice, it is necessary to consider intent. There are certain instances—I am thinking particularly of special education for those in urban public schools—where segregation of those with certain kinds of disabilities might benefit the educational interests of the child, particularly if the child receives targeted support that cannot be fully provided in an inclusive setting. Legally, these decisions are highly scrutinized, and segregation is viewed as only acceptable as a last resort (Hehir, 2005). This logic does not hold true in Nagel's Public Toilet Example. JCDecaux was upset about potentially being left out of the bidding process because of its unwillingness to present a universally accessible toilet system. This is not sufficient justification for promoting segregation in public facilities. Further, JCDecaux's position exemplifies irresponsible design because it knowingly ignores the concerns of an urban constituency, namely people with physical disabilities, and the facilities access laws which guarantee their right to equality of access to public facilities. This does not mean that design should not consider how to minimize the negative consequences of a public facility being misused.

Consider the following parallel case. In 2011, after attending a film festival near Tufts University, I prepared to return home via the subway. The station was fully wheelchair accessible, but I could not use the elevator because someone was sleeping in it. The city of Somerville and the Massachusetts Bay Transportation Authority (MBTA) have interests in preventing those who are homeless from sleeping in subway station elevators because such use interferes with the operation of the elevators for their intended purpose. My experience raises the same issue of misuse as the concern about sex workers and drug users in Nagel's Public Toilet Example. These vulnerable populations should not be excluded from the use of a public facility for its intended purpose. At the same time, justice requires developing fair procedures that impartially allow everyone to benefit from the intended purpose of a public facility.

While most egalitarians would likely agree that these people are subject to systemic injustices, allowing misuse of public facilities does not provide long-term relief for the disadvantages that these populations confront. Cities should develop procedures that give these individuals protection from cruel treatment or exploitation while also ensuring that the subway station elevators and public restrooms are available for their intended purposes. In this situation, is the MBTA's obligation to provide universal access at a major subway station discharged solely on the grounds that people sometimes misuse? I would hope not.

What emerges from the analysis I have presented thus far is the idea that, contrary to Nagel's assertion, the universal design of urban facilities is a matter of responsible design. Further, because decisions related to design are often the result of institutional choices, they are a matter of social equality and justice. What this leaves open, however, is what metric of equality to use in the case of access to the benefits and services of cities. A full treatment of this question requires an entire chapter in and of itself, but I would like to suggest here that a good metric for evaluating questions of universal accessibility to urban life might lie in a refined version of G.A. Cohen's "equal access to advantage." This approach holds that people with disabilities should be given additional opportunities and resources to receive the same social benefits that are made available to those without disabilities (Cohen, 1989).

The general approach Cohen lays out was not originally formulated to think specifically about urban justice.[7] Further, Elizabeth Anderson has criticized the approach's general application for guiding the design of social welfare systems because of its potential to create a bottomless pit (see Arrow, 1973), whereby any subjective preference is a matter of social justice (Anderson, 1999). However, in the strict context of access to urban facilities and services, the approach seems to avoid the "bottomless pit of subjective preferences" objection in that urban design is distinct from the design of a robust social welfare system. In most cases, urban design involves shared responsibility between localized political institutions, private businesses, and corporations that shape the urban landscape. Often local governments do not assume the entire cost of designing urban environments that are accessible to all, regardless of ability. As a matter of law, the ADA, which has served as the model for most other major disability rights legislation, appeals to concepts of

[7] Cohen is specifically thinking about populations that are disadvantaged by circumstances outside of their control. This raises questions about the role of individual responsibility as it relates to state obligations to remedy social inequality. I bracket these questions here because I am not specifically endorsing the original formulation of Cohen's argument; rather, I am suggesting that his intuition about the appropriate scope of equality being access to advantage holds promise for questions of universal design and urban justice.

equal access, equal enjoyment of public accommodations, and equal benefit from goods and services that seem very much in line with what Cohen has in mind. Universal design is very much complementary to the standard of disability equality established by both equal access to advantage, as well as the ADA and other disability rights legislation.

So far, I have argued that Nagel incorrectly places universal design outside the scope of justice because he analyses disability from a strict medical model perspective. Because the design of public facilities can create inequalities of access, questions of wheelchair access to public facilities fall within the scope of justice. Specifically, the Public Toilet Example is an instance where a private company's choices could either exacerbate or alleviate the impact of a wheelchair user's disability on their capacity to benefit from a high-demand facility. Accordingly, constructing new facilities in ways that are consistent with universal design is both a form of responsible design and a matter of social justice in cities.

Putting universal design into practice seems to be contingent upon access to assistive technologies that broaden the range of accessibility features that a designer can build into an urban space. I consider this in the next section, where I argue that assistive technologies are an important means to the end of universally designed urban environments. I further argue that such technologies should be conceptualized in terms of relational autonomy, whereby the built environment becomes a tool to enable people with disabilities to set the terms through which they relate to others in urban communities.

10.3 Universal Design and Assistive Technologies as Means to the End of Relational Autonomy for People with Disabilities

By way of beginning to evaluate the relationship between people with disabilities, universal design, and assistive technologies, consider the following autobiographical reflection. My time as a college student was one of increased maturity and independence as it is for most students. However, as someone who has a physical disability, this was a time where I realized that in my case independence was synonymous with interdependence. I mean that my independence was contingent upon the use of assistive technologies and the help of other people who provided assistance when I needed it. Technologically, my independence was contingent upon access to a motorized wheelchair, the charger necessary to keep the battery well-charged in the Boston snow, power outlets throughout the city of Boston to plug the charger into my wheelchair, as well as the elevators, trains, and buses operated by the MBTA. I began to see my wheelchair and elevators as extensions of my legs, the wheelchair charger and power outlets as fuel for my legs, and the buses, trains, and subways as semi-automated chauffeurs. Boston became my home because of the advantages of an urban environment that, despite its long history, was mostly accessible to me.[8]

While not all people with disabilities view their relationship with medical and urban technologies in exactly these terms, many disability scholars have noted that maintaining independence while having a disability is largely contingent upon access to assistive technology (Anderberg, 2005; Barnes, 2016; Brownlee & Cureton, 2011). Universal design, on some accounts, has come to dominate disability discourse in urban settings only because assistive technology has progressed to the point where universal accessibility is an achievable ideal. The rise of computers and automation since the 1980s, for example, has made it possible for various forms of assistive technology to redefine the limits of urban accessibility.

However, from my perspective, assistive technology and universal design have a much more nuanced relationship than this line of thinking suggests. Assistive technology does not exist in a vacuum and neither does the normative ideal of universal design. The two work together to reinforce each other's purposes. Assistive technology, in many cases, enables universal design. In turn, designers who take universal design to heart can also create new assistive technology that makes the overarching ideal of universal accessibility more attainable.

The question remains, however, whether assistive technology is necessary to implement universal design. To answer this question, it is important to recognize that technology only comes into initial use because an individual designer sees a potential need for a particular innovation. This is not to say that the ways in which people use a technology cannot evolve over time from the designer's initial intent. However, without an initial motivation for developing a piece of technology, an innovator's design choices will likely lack purposefulness and not encapsulate a particular need that a user has. It is significant to keep in mind that assistive technology can be used for desirable or undesirable purposes, depending upon intent. The use of reproductive technology in urban hospitals, for example, can either be used to help those with fertility challenges conceive a child or to forcibly sterilize a person with an intellectual disability against the individual's will. Accordingly, assistive technology is only as helpful as the intent behind its use.

The unifying aim of both assistive technology and universal design, at least from the perspective of justice, is to aid in guaranteeing people with disabilities and others equal access to advantage. Underlying this conception of equal access is a conception of relational autonomy that recognizes how people with disabilities relate interdependently to those around them as well as features of the built environment

[8] Peter Anderberg has made a similar point, arguing that technology can be particularly empowering to people with disabilities if functionality is the central focus of design. He argues that, in order to assess whether technology is helping improve functionality for people with disabilities, one must not only look at the technology itself, but how the factors of attitude, control, and enabling shape users' experiences. Functionality, on his account, exists in the conceptual space between the technology and the user (Anderberg, 2005). In my case, the city of Boston made it easy for me to establish a relationship with subways, power outlets, and the like, such that the technology improved my functionality.

in order to go about their daily lives. Assistive technology and universal design should be understood as means that promote this end.

Relational autonomy is a concept rooted in feminist philosophy that challenges traditional liberal conceptions of self-sufficiency. Feminist critics of liberalism have argued that liberalism's emphasis on individualism and self-sufficiency often neglects the experiences of women and others who rely on networks of interdependence in, for example, providing care and performing other forms of domestic labor (Mackenzie & Stoljar, 2000; Nedelsky, 2013). Viewing autonomy relationally is meant to acknowledge that people are, in fact, interdependent on each other. In the disability context, this does not mean that a person with a disability becomes entirely self-sufficient nor does it mean that they are strictly viewed as recipients of humanitarian assistance, as Nagel postulates. Instead, relational autonomy, as I am using it here, refers to the idea that if people with disabilities have at their disposal a built environment that takes into account their unique limitations, they can be empowered to set the terms on which they rely on those around them.

Leslie P. Francis has argued that, in the instance of people with intellectual disabilities, assistive technology has the potential to change the relations between those with intellectual disabilities and non-disabled members of their community. With the presence of assistive technology, on Francis' account, people with intellectual disabilities can become relationally autonomous, because the technology can compensate for a person's intellectual challenges in particular contexts. To illustrate, individuals who have difficulty calculating sums by hand due to an intellectual impairment can potentially learn how to use a calculator (Francis, 2011). In urban contexts, similar logic could be applied to the operation of Automated Teller Machines (ATM), which can help people whose intellectual challenges make it difficult to manage their finances by making it easier to track balances and expenditures.

For the purposes of this chapter, what makes Francis' point particularly noteworthy is that all people could use assistive technology in this way, not only people with intellectual disabilities. Universal design is predicated on the logic of the "curb-cut effect" mentioned in the introduction, whereby individual disability accommodations have the potential to improve access for the general population as well. As an example, ATMs make banking more convenient for many people, not only people with intellectual disabilities.

There are some people with disabilities who will not be able to achieve the conception of relational autonomy I have in mind. I am thinking, in particular, of those whose conditions render them too severely physically, intellectually, or psychologically incapacitated to currently achieve relational autonomy.[9] For these individuals, it is important from the perspective of justice that technological innovation continues so that they might one day be able to achieve some standard of relational auton-

[9] Philosophically, it is interesting to note that these groups of individuals with disabilities are the most difficult to contend with in liberal contractualist accounts of justice because of their presumed inability to engage in reciprocal behaviors (Kittay, 1998; Silvers & Francis, 2005). However, from the perspective of justice in urban design that I am articulating here, these individuals do not present a considerable philosophical challenge.

omy. This is not to say there will not be some people with severe disabilities that will never be able to exercise relational autonomy. However, the fact that there will always be some people who have difficulty achieving relational autonomy should drive continued technological progress rather than hinder it. Therefore, the obligation to promote universal design and purposeful access to assistive technology is not mitigated by the severity of different kinds of disabilities.

More so than severity of disability, the cost of assistive technology is a stronger challenge to facilitating relational autonomy for people with disabilities in urban environments. As I see it, there are at least two layers of costs involved. First, individual medical devices can be expensive, and many people with disabilities do not have access to the kind of comprehensive health insurance that might alleviate such costs. Second, even though universal design can save cities money by eliminating the need to retrofit inaccessible structures, designing accessible urban infrastructures can still be quite costly. This is especially the case in cities contending with significant poverty, homelessness, and the like because these cities may have to prioritize promoting access against taking steps to alleviate economic inequality.

In thinking about how to contend with such difficult realities, it is useful to consider what Jonathan Wolff and Avner De-Shalit call the declustering of disadvantage (Wolff & De-Shalit, 2013, p. 103). On their account, disability, race, class, and other social characteristics can contribute to various forms of disadvantage, building upon each other, such that individuals experience clusters of disadvantage. For example, a person with a disability may be worse off by nature of their disability, but they can be even more worse off if the disability-related disadvantages are confounded by being poor or a member of a racial minority. To decluster disadvantage, Wolff and De-Shalit argue that societies have obligations to provide opportunities and resources that minimize the impact of various sources of marginalization on social opportunity. In the context of the present discussion, this might involve cities coming up with initiatives to both improve disability access and alleviate poverty simultaneously. A city could partner with a medical supply company to provide free or low-cost wheelchairs to those who cannot afford them. On the infrastructural level, a city could also collaborate with private businesses to come up with an understanding of shared costs in the construction of assistive technologies that potentially benefit the entire urban community. In this sense, contrary to Nagel's assumption, there are many cases where the cost of promoting disability access may not be in conflict with other social concerns.

Thus far, I have argued that universal design is consistent with the demands of both responsible urban design and social justice for people with disabilities. Assistive technology, while not absolutely necessary to the realization of universal design, makes implementation considerably easier and can redefine the limits of urban disability access. Accordingly, the ultimate goal of assistive technology and universal design should be to promote relational autonomy through equal opportunity for access to advantage. With all of this in mind, universal design should not be conceptualized as a panacea for all matters of urban disability access. In the next section, I consider the limits of universal design and the dilemmas that retrofitting for accessibility presents to a city.

10.4 The Limits of Universal Design and Dilemmas of Retrofitting for Accessibility

My discussion of universal design raises an important question. Going beyond new construction, to what degree does justice require the universal accessibility of all public facilities when retrofitting of old buildings is necessary? I have in mind two broad categories of urban facilities. The first category includes those structures that are considerably old and would cost a significant amount of money to retrofit, assuming retrofitting would even be plausible. The second category includes historical sites. In these cases, there could be a compelling reason to preserve the inaccessibility of a structure as a reminder of what the past was like. The distinction between these categories is not always finite, as historical structures are sometimes used in older cities, particularly in Europe, to provide housing, retail space, and the like. In these cases where universal design is infeasible, what obligations should cities have to promote disability access?

Generally speaking, cities have an obligation to retrofit insofar as retrofitting enables the delivery of a desired good or service to an individual with a disability. However, this obligation can be legitimately weighed against cost and the possibility that retrofitting would fundamentally alter the good or service being provided.[10] When retrofitting is evaluated to be inappropriate, a city or private corporation must make alternative arrangements to ensure that people with disabilities receive access to a comparable service, or in some cases, the same service in an accessible setting.

To better illustrate these considerations, I present the following five cases. The first returns us to a consideration of the MBTA. There are two subway stations in downtown Boston approximately one to two blocks away from each other. The first, Park Street (on the Red Line), has an elevator while the second, Boylston Street (on the Green Line), does not. The Park Street station is a modern facility on the MBTA's newest subway line. The Boylston Street station is much older than Park Street station is. Hypothetically, the fact that the Park Street station has an elevator and the Boylston Street station does not, would reinforce the same differences between those with physical disabilities and the able-bodied cited in my discussion of the Public Toilet Example.

In the second case, to consider another subway system, many stations that are part of the London Underground do not have elevators for wheelchairs. To be precise, one article I found in *The Guardian* suggests that only 24% of London Underground Stations are accessible and that until 1993, wheelchair users were explicitly prohibited from using the London Underground altogether due to fire safety concerns (Cathcart-Keays, 2014). Unlike the MBTA's system, the stations that do happen to have elevators are sporadically located around London, making it inconvenient for wheelchair users like me to efficiently use the Underground. Even

[10] I openly acknowledge that I am borrowing some of this language directly from the ADA, as I think the fundamental principles the law uses are helpful in determining reasonable limits for retrofitting urban facilities.

if retrofitting for universal accessibility were feasible, many of these stations have historical significance in helping the citizens of London survive the atrocities of World War I and World War II (Rabon, 2015). Considering the possibility of extensive costs and historical significance, there might be a different ethical standard for evaluating what constitutes equal access to the London Underground for people with disabilities than in the MBTA case.

The third and fourth cases involve two historic universities: Harvard and the London School of Economics (LSE). Both campuses have buildings that are not accessible to wheelchair users, due in part, no doubt, to the long histories of the campuses. I have participated in activities on both campuses, and often to accommodate me, personnel at both schools have moved the events to spaces that were accessible. In an ideal world, these two leading research universities would invest money into making all buildings universally accessible so that I would not have to have activities moved in order to participate. However, the realities are that these institutions, like urban governments, have to deal with limited financial resources. Harvard and LSE also have long histories that, in some instances, especially for Harvard, have shaped the architectural aesthetic of the surrounding urban area. Is it sufficient from the perspective of justice to have university policies to simply move events to wheelchair accessible locations rather than retrofitting all campus facilities?

The fifth example is Betsy Ross' house in the City of Philadelphia, Pennsylvania. Although Ross' historical significance as the creator of the first American flag is questionable, her house, dating back to the mid 1700s, nonetheless functions as an artifact of American colonial history (Historic Philadelphia, n.d.). It has many stairs and is, therefore, not fully accessible to wheelchair users. This case is somewhat different from the previous example in that, unlike subways and universities, the house's sole purpose is to help preserve and teach the colonial history of the United States. In this case, an individual skeptical of my overall argument in this chapter might point to the historical preservation of a site as a compelling interest that might override accessibility concerns. This leaves open the question of how cities can fairly make the determination of when historical preservation can be used as a legitimate rationale for not making a facility accessible to those with disabilities.

The two subway cases and the two university cases are more complex than the historical monument example. This is true for two reasons that are relevant to my argument. First, subway stations and universities are facilities that give people access to very different, but equally important social advantages. In the case of subways, access to fast and affordable public transportation makes it easier for people in cities to go about their day. People with disabilities, therefore, confront a considerable disadvantage if subways are not sufficiently accessible. Universities, by contrast, are in the position of giving people access to education and potential employment, whereby not being able to access university facilities might put an otherwise qualified person with a disability at a considerable disadvantage. This is not necessarily the case with historical sites like Betsy Ross' house, in that being able to experience historical sites does not preclude someone with a disability from receiving historical education in comparable ways.

Nonetheless, from my perspective, it is possible to derive an ethical standard of accessibility that applies in all five cases. In doing so, it is important to evaluate the specific advantages that a particular urban facility or service provides. Urban governments, private businesses, and other stakeholders can then weigh the benefits and costs associated with retrofitting the existing facility for service against providing comparable benefits in an alternate fashion. In the MBTA case, it seems acceptable to have an elevator for the Park Street Red Line stop but not the Boylston Street stop on the Green Line. This is because the distance between the two stops is negligible and, even in inclement weather, buses provide a comparable way of getting to Boylston Street for about the same cost as the Green Line.

The London Underground case, however, is deserving of further scrutiny. It should be said that London and surrounding areas do have alternatives to the Underground that are accessible to wheelchair users and others with disabilities. London's buses have ramps that provide wheelchair users with a way of boarding, as do all black taxis. What makes this case different from the MBTA is that the excessive inaccessibility of Underground stops necessitates that many people with disabilities rely on these alternative means of transport. Taking a bus or taxi from one part of London to another requires more time than using the London Underground. Additionally, the cost of a taxi is considerably higher than the cost of the Underground or the buses. Therefore, a person with a disability must be in a position to make a sufficient wage to live in Central London where it is easy to use the buses and taxis and/or plan out their days with enough time to accommodate the additional travel time.

The mostly inaccessible London Underground constitutes a considerable inequality of disability access. The history of some stations in the system further does not constitute a compelling argument for not at least deliberating how the city could make more stations accessible. This is because subways, like universities, provide services for which it is not critical to maintain the historic integrity of a built structure. Based on my argument in this chapter, various stakeholders in London and surrounding areas should engage in fair democratic deliberation to agree upon a plan for making more stations accessible over time. This process should include the perspectives of potential users with disabilities and include a range of viable options that promote access, while recognizing that retrofitting one of the world's oldest subway systems is likely to be very long and difficult.

The claim that the subway stations have historical significance should not be a factor in deciding to maintain the inaccessibility of the system because the purpose of the subway is to provide a means of transportation, not to remind people of the hardships Londoners faced in the two World Wars. Putting in an elevator in this case does not necessarily have to compete with preserving the historical legacy of the structure. Compromises will have to be made. In making difficult choices about retrofitting, it is important that all stakeholders commit to providing people with disabilities an equal opportunity to access the benefits of a good or service in a way that proactively minimizes the likelihood of them experiencing shameful revelation. In this way, the deliberation process can facilitate urban infrastructure that promotes relational autonomy for people with and without disabilities.

The cases of Harvard and LSE do not present the same degree of inequality that the London Underground case contains. Although some campus facilities remain critically inaccessible to those who use wheelchairs or other mobility devices, there are enough accessible locations on both campuses to make the procedure for moving events reasonably effective. That being said, such procedures should be relatively easy to navigate so that people with disabilities can request this accommodation in a non-burdensome way.

By the same token, a university like Harvard could not deny someone access to their facilities on the grounds that a student might be able to attend classes at a nearby institution with a more accessible campus. This is because a Harvard education is a particular service that cannot be provided, for example, on the Massachusetts Institute of Technology campus. The school could also not opt to provide exclusively online education to people with disabilities in order to avoid retrofitting when necessary, unless the institution decided to become an exclusively online campus for all. Based on 400 years of tradition, Harvard has evolved into a community where many of the educational benefits students receive have to do with the opportunity to interact with people on its historic campus. While some of Harvard's education can be accessed online, a considerable portion of the distinctive qualities of the school's program cannot be substituted by exclusively online education. A potential solution to the challenge of limited university resources might be to encourage Harvard, LSE, and similar institutions to make reasonable efforts to seek the support of state and private donors in gradually making these urban campuses more accessible to those with various disabilities. This should be a topic of further academic research and discussion within universities more generally.

This speaks to the notion of shared cost. Both government and non-government entities share obligations of cost with regard to urban design and accessibility. While I am not in a position here to present a finite formula of shared costs, it seems important to consider the size and economic position of the private entity when evaluating what proportion of accessibility costs stakeholders should incur. For example, the amount of financial responsibility the City of Cambridge might have to make a Harvard building accessible is likely to be considerably small given the substantial extent of Harvard's financial resources. That being said, there are some small businesses in Cambridge that might not be able to afford retrofitting like Harvard can.

In this case, the City of Cambridge might work with the business to arrive at a non-burdensome agreement of shared costs for all stakeholders. At the level of national policy, this argument might establish an ethical basis for expanding measures like the ADA to include a discretionary spending mechanism that allows more localized governments and private entities to receive grants to help with the design of accessible infrastructure. This and other policy innovations could make it considerably easier to realize the ideals of universal accessibility and relational autonomy for people with disabilities in a way that is sensitive to the realities of cost. It may be that in some cases stakeholders decide that retrofitting is simply too expensive, even with the shared cost. This is acceptable as long as people with disabilities are included in the decision making process and plans are put in place to revisit the issue in a reasonable period of time. I present these ideas as a tentative proposal. Future research in philosophy and urban policy should consider the details of this proposal more carefully.

To return to the case of historical sites like Betsy Ross' house, it is important to be sympathetic to the kind of skeptic I mentioned earlier in this section. Preserving history is an important value that a city has an interest in promoting. At the same time, it is important to keep in mind two considerations. First, it is possible to retrofit historical sites in ways that maintain the site's aesthetic integrity. I know from my own travels that the City of Athens, Greece has installed a lift to the top of the Acropolis that provides wheelchair access without changing the aesthetic beauty of the archeological structures or the natural landscape. The fact that the City of Athens is able to provide access to the Acropolis for people who cannot walk sets a rigorous standard for accessibility of historic sites. Like in the London Underground case, various city stakeholders should have fair democratic deliberations about how to retrofit historical sites in cost efficient ways that preserve the historical integrity of each site.

Second, in cases where retrofitting historical sites is truly infeasible, there are ways of providing people with disabilities comparable access to the educational value of the historic site. While visiting the Biltmore Estate in Asheville, North Carolina, I was unable to access the rooms in the mansion's basement. However, because of the availability of a DVD walkthrough of the basement and an audio guide that explained the important historical and architectural features of each room, I was given a comparable opportunity to learn about the rooms that were inaccessible to me as a wheelchair user (Biltmore, n.d.).

While this is not technically the same as being able to move from room to room, from the perspective of equal access to advantage, what seems to matter is that I was able to receive the educational benefit of learning about the rooms in a comparable and accessible manner. The purpose of accessible urban infrastructure is not to cure someone's disability, but to ensure that, to the greatest extent possible, people with disabilities are able to benefit from the advantages a city has to offer in ways that are comparable, if not equal to, the advantages available to those who do not have disabilities.[11] Universal design is an important means of promoting this ideal, but, as I hope to have demonstrated in this section, the retrofitting of older structures is as well.

10.5 Conclusion

This chapter has explored the ethical dilemmas that the framework of universal design presents for cities trying to accommodate people with disabilities. Using Nagel as a starting point, I have suggested that the design of accessible urban

[11] This allowance for a separate service as a reasonable accommodation should not be used to deny people with disabilities access to the historical sites of a city on the grounds that they can go to another city that is more accessible to learn about that city's history. The historical education provided by Boston and Philadelphia is substantially different. One cannot learn about Betsy Ross' house firsthand by visiting the Freedom Trail. Once again, the specialized service provided to a person with a disability must be substantially comparable to the service provided to non-disabled patrons.

facilities exemplifies the extent to which disability is a socially constructed human experience. Contrary to Nagel’s interpretation of the Public Toilet Example, the decision not to construct new urban facilities with the needs of people with disabilities in mind is a social inequality that falls within the scope of social justice broadly construed. In cases of new construction, following the principles of universal design should be considered a matter of responsible design. In providing access to the city, institutional actors should seek to provide people with disabilities opportunities to receive the same benefits from urban spaces that they would have access to if they did not have disabilities. In cases of retrofitting existing structures, following the principles of universal design might not be feasible or even desirable. When these situations arise, various stakeholders, including people with disabilities, should engage in open and constructive dialogues about what alternatives might be most appropriate. The ethical standard in these cases should be to provide people with disabilities access to comparable advantages afforded to those without disabilities in a way that makes every reasonable effort to avoid people with disabilities experiencing shameful revelation. These decisions are extremely difficult and need to be closely scrutinized on a case by case basis.

The relationship between assistive technology and the realization of the principles of universal design is philosophically nuanced. While assistive technology is not necessary for universal design to be implemented, access to such technology makes implementing universal design substantially easier. The positive relationship between assistive technology and increased accessibility exists for both individuals with disabilities and for the urban infrastructure around them. Wheelchairs and other assistive devices that serve as extensions of the body for those with disabilities provide opportunities for them to negotiate the built environment on an equal basis with others. This is contingent upon the built environment including curb-cuts, elevators, accessible pedestrian signals, and the like. These considerations reflect the reality that more wealthy and affluent people with disabilities living in well-resourced cities are more likely to have access to appropriate assistive technology and to the benefit from universal design.

In confronting the realities of scarce resources and the limits of universal design, urban governments and private entities should agree on specific formulas of shared cost. This normative ideal should serve as the basis for possible extensions to disability rights laws. Future research in philosophy and economics should consider the particulars of this formula. By taking a more proactive role in facilitating accessible design, governments at all levels can help make the ideal of universal access for people with disabilities more attainable. This should be done with the goal of promoting relational autonomy for people with disabilities, through which they are able to set the terms of their relationships to other members of urban communities.

The argument I have presented here is based on a critical assumption. Specifically, I have mainly discussed universal design in the context of societies that are committed to ideals of political liberalism and social equality. Yet, universal design and other mechanisms for promoting disability access are likely to be most needed in societies where commitments to these ideals are less firm or even non-existent. On the policy level, the maturity of the United Nations Convention on the Rights of

Persons with Disabilities (UNCRPD) places new obligations on developing and/or illiberal countries to promote human rights for populations of people with disabilities (UN General Assembly, 2007). The principles of universal design are enshrined in the UNCRPD. Yet, implementing them in cities across the globe likely requires financial resources that governments might be unable or unwilling to provide. Future work should aim to conceptualize the relationship between these cities, universal design, assistive technology, and global justice so that people of all abilities, nationalities, economic classes, and the like can have equal access to the goods and services that cities have to offer.

Acknowledgments This chapter represents the culmination of eight years of writing a critique of Nagel's argument against universal design being a matter of justice. I am grateful to my undergraduate mentors Eric Beerbohm, Katie Gallagher, Jennifer Page, and Emma Saunders-Hastings for their belief in this project from its infancy stages. I am also indebted to my graduate school mentors Joshua Cohen, Chandran Kukathas, Kristi Olson, Anne Phillips, Rob Reich, Debra Satz, and Laurie Shrage for their feedback and contributions to my thought process. Finally, I am indebted to my friends Cat Camacho, Mollie Gerver, Joanne Kalogeras, Nayeli Altamirano, Christopher Stucky, and Justin Wilck for their assistance over the years in typing the manuscript. I dedicate this chapter to the memory of Anita Silvers (1940–2019). Although I only met Professor Silvers briefly at a conference, her work on the social model of disability has profoundly shaped my argument here. I can only hope that future work in this area will remain true to her tremendous legacy.

Photos of Comparable Wheelchair Accessible Public Toilet

Accessible Public Toilet

References

Anderberg, P. (2005). Making both ends meet. *Disability Studies Quarterly, 25*(3). https://doi.org/10.18061/dsq.v25i3.585.
Anderson, E. (1999). What is the point of equality? *Ethics, 109*(2), 287–337.
Arrow, K. J. (1973). Some ordinalist-utilitarian notes on Rawls's theory of justice by John Rawls. *Journal of Philosophy, 70*(9), 245–263. https://doi.org/10.2307/2025006.
Barnes, E. (2016). *The minority body: A theory on disability*. Oxford: Oxford University Press.
Biltmore. (n.d.). *Accessibility*. Retrieved March 14, 2018, from http://www.biltmore.com/visit/visitor-information/estate-accessibility
Blackwell, A. G. (2017). The curb-cut effect. *Stanford Social Innovation Review, 2017*, 28–33. Retrieved March 14, 2018, from https://ssir.org/articles/entry/the_curb_cut_effect.
Brownlee, K., & Cureton, A. S. (2011). *Disability and disadvantage*. Oxford: Oxford University Press.
Cathcart-Keays, A. (2014). Disabled man takes on tube challenge. In *The Guardian*. 33. Retrieved March 14, 2018, from https://www.theguardian.com/cities/2014/oct/29/disabled-man-tube-challenge-london-underground-accessibility
Cohen, G. A. (1989). On the currency of egalitarian justice. *Ethics, 99*(4), 906–944. https://doi.org/10.1086/293126.
Connell, B. R., Jones, M., Mace, R., Mueller, J., Mullick, A., Ostroff, E., et al. (1997). *The principles of universal design*. Retrieved March 14, 2018, from https://projects.ncsu.edu/ncsu/design/cud/about_ud/udprinciplestext.htm
Francis, L. P. (2011). Understanding autonomy in light of intellectual disability. In K. Brownlee & A. Cureton (Eds.), *Disability and disadvantage* (pp. 200–215). Oxford: Oxford University Press.
Hamraie, A. (2017). *Building access: Universal design and the politics of disability*. Minneapolis: University of Minnesota Press.
Hansson, S. O. (2007). The ethics of enabling technology. *Cambridge Quarterly of Healthcare Ethics, 16*(03), 257–267. https://doi.org/10.1017/s0963180107070296.
Hehir, T. (2005). *New directions in special education: Eliminating ableism in policy and practice*. Cambridge, MA: Harvard Education Press.
Historic Philadelphia. (n.d.). *History of the betsy ross house*. Retrieved March 14, 2018, from http://historicphiladelphia.org/betsy-ross-house/history/
Kittay, E. F. (1998). *Love's labor: Essays on women, equality, and dependency*. London: Routledge.
Kymlicka, W. (2014). Altruism in philosophical and ethical traditions: Two views. In J. Phillips, B. Chapman, & D. Stevens (Eds.), *Between state and market: Essay on charities law and policy in Canada* (pp. 87–126). Montréal: McGill-Queens University Press.
Mackenzie, C., & Stoljar, N. (2000). *Relational autonomy: Feminist perspectives on autonomy, agency, and the social self*. New York: Oxford University Press.
McKinley, J. C., Jr. (1993). *Bids to build public toilets facing limits*. The New York Times, p. B00001. Retrieved March 14, 2018, from https://www.nytimes.com/1993/01/27/nyregion/bids-to-build-public-toilets-facing-limits.html
Nagel, T. (1997). Justice and nature. *Oxford Journal of Legal Studies, 17*(2), 303–322. https://doi.org/10.1093/ojls/17.2.303.
Nedelsky, J. (2013). *Laws relations: A relational theory of self, autonomy, and law*. New York: Oxford University Press.
Nussbaum, M. (2007). *Frontiers of justice: Disability, nationality, species membership*. Harvard, MA: Belknap Press.
Rabon, J. (2015), *London history: A look at the london underground during World War II*. Retrieved March 14, 2018, from https://londontopia.net/site-news/featured/london-history-look-london-underground-world-war-ii/
Rawls, J. (1999). *A theory of justice*. Oxford: Oxford University Press. Rev. ed.

Richardson, H. S. (2006). Rawlsian social-contract theory and the severely disabled. *The Journal of Ethics, 10*(4), 419–462. https://doi.org/10.1007/s10892-006-9000-5.

Serlin, D. (2010). Pissing without pity: Disability, gender, and the public toilet. In H. Molotch & L. Norén (Eds.), *Toilet: Public restrooms and the politics of sharing* (pp. 167–185). New York: New York University Press.

Shakespeare, T. (2017). The social model of disability. In L. J. Davis (Ed.), *The disability studies reader* (5th ed., pp. 195–203). New York: Routledge.

Shapiro, J. P. (1993). *No pity: People with disabilities forging a new civil rights movement*. New York: New York Times Books.

Silvers, A., & Francis, L. (2005). Justice through trust: Disability and the "outlier problem" in social contract theory. *Ethics, 116*(1), 40–76. https://doi.org/10.1086/454368.

Terzi, L. (2011). Vagaries of the natural lottery? Human diversity, disability, and justice: A capabilities approach. In K. Brownlee & A. Cureton (Eds.), *Disability and disadvantage* (pp. 86–111). Oxford: Oxford University Press.

Titchkosky, T. (2011). *The question of access: Disability, space, meaning*. Toronto: University of Toronto Press.

UN General Assembly. (2007). *Convention on the rights of persons with disabilities and optional protocol. 24 Jan 2007*. Retrieved March 14, 2018, from http://www.un.org/disabilities/documents/convention/convoptprot-e.pdf

Waldron, J. (1991). Homelessness. *UCLA Law Review, 39*(295), 295–324.

Wasserman, D., Asch, A., Blustein, J., & Putnam, D. (2013). Disability and justice. In *The Stanford Encyclopedia of Philosophy*. Retrieved March 14, 2018, from https://plato.stanford.edu/entries/disability-justice/

Wolff, J. (1998). Fairness, respect, and the egalitarian ethos. *Philosophy and Public Affairs, 27*(2), 97–122.

Wolff, J., & De-Shalit, A. (2013). *Disadvantage*. Oxford: Oxford University Press.

Kevin Mintz is a disability rights scholar and activist, and is currently a bioethics fellow at the NIH Clinical Center. He received his PhD in Political Science from Stanford University. The chapter herein was written while he was a graduate student. He also holds an AB in Government from Harvard College and an MSc in Political Theory from the London School of Economics. His work has appeared in a variety of academic publications including Disability & Society. He has also written opinion essays for the Orlando Sentinel and the Los Angeles Times on accessibility in the theme park industry.

Chapter 11
Towards a Darker Future? Designing Environmental Values into the Next Generation of Streetlights

Taylor Stone

Abstract This paper examines the ethical dimensions of a critical urban infrastructure: streetlights. The development and proliferation of nighttime lighting has been fundamental and formative for urban nights, and streetlights constitute the primary source of illumination. Recent developments to lighting technologies, namely LEDs and 'smart' systems, are spurring a new generation of streetlights, with retrofits being rapidly undertaken around the world. While they may offer substantial energy savings, their long-term environmental effects are still under debate. Concurrent to these technological developments, the adverse costs and impacts of nighttime lighting—known as light pollution—have emerged as an ecological, economic, and ethical issue. This confluence of technological innovations and moral evaluations creates new challenges, but also an opportunity to envision and enact new strategies. For this, *designing for darkness* is presented as a value-sensitive framework for responsible lighting strategies that strives to incorporate and foster both substantive environmental values and meaningful nighttime experiences into the next generation of streetlights. First steps are taken to explore how this framework can be operationalized, and three design concepts are put forward as a means to create darker urban nights.

Keywords Nighttime lighting · Streetlights · Light pollution · LEDs · Smart lighting · Environmental ethics · Responsible innovation · Philosophy of the city

---

T. Stone (✉)
Ethics and Philosophy of Technology Section, Delft University of Technology, Delft, The Netherlands
e-mail: T.W.Stone@tudelft.nl

M. Nagenborg et al. (eds.), *Technology and the City*, Philosophy of Engineering and Technology 36, https://doi.org/10.1007/978-3-030-52313-8_11

11.1 Introduction

Nighttime lighting is so ubiquitous, and such a commonplace component of urban spaces, that it readily fades into the backdrop of daily experience. The historian of technology David Nye (2010) remarked that nighttime illumination has become so normalized that it seems natural, and temporary blackouts now seem unnatural. But it is because of its omnipresence that it must be scrutinized, for artificial illumination is formative to our urban nighttime experiences and resultant moral judgments. Lighting shapes the spaces and defines the boundaries of urban nightscapes, effectively creating city nights on both a physical and symbolic level. Historians of city nights and lighting technologies have shown that illumination has been equally important for its practical applications as for its influence in shaping, and being shaped by, various societal values—for example, safety, policing, nightlife, and progress (e.g., Brox, 2014; Koslofsky, 2011; Nye, 1990; Schivelbusch, 1988; Schlör, 1998). Understood in this way, "Light is far more than "infrastructure" in a narrow sense of technological networks, but encompasses a range of developments including material and symbolic dimensions to state formation, the coevolutionary dynamics of different modes of governmentality, and the delineation of distinctive forms of public culture ranging from the mundane to the spectacular" (Gandy, 2017, p. 1096). In this paper I provide a forward-looking analysis of an additional dimension of increasing importance, namely the environmental values and human-environment relations influenced by artificial nighttime lighting. I will consider how these factors can inform the design of streetlights, and more generally, how we can re-think the design of our urban nightscapes via new technologies and evolving moral concerns.

We are on the precipice of two converging changes to outdoor lighting, which will profoundly change how we light cities at night. The various forms of electric lighting that have dominated the twentieth century are poised to be replaced by light-emitting diodes (LEDs), which has been described as a shift from electric to *electronic* lighting (Gandy, 2017). Concurrent to LEDs is the introduction of a variety of "smart lighting" applications and systems, adding another layer to the shift towards electronic outdoor lighting. The controllability and, most important, the efficiency of smart LED lighting has fostered the rapid development and uptake of this technology, with an increasing number of cities across the world undertaking lighting retrofit projects.

At the same time as these technological developments, a growing acknowledgment of the adverse effects of artificial nighttime lighting is causing a paradigmatic shift in evaluative judgments. Commonly referred to as *light pollution*, these negative impacts are far reaching: nighttime lighting has been shown to cost (and waste) billions of dollars and enormous amounts of energy, effect human health and well-being, disrupt ecosystems, and cut off experiences of the natural night sky (Stone, 2017). Growing recognition of light pollution has resulted in a shifted moral landscape, with over-abundant and poorly designed illumination being scrutinized for its inefficiency, as well as the resultant "disappearance of darkness" (e.g., Bogard,

2013). This has led to a re-evaluation of darkness, previously seen as foreboding and dangerous, now increasingly perceived as a threatened source of cultural and environmental value (Stone, 2018b). This complicates the adoption of smart LED lighting technologies, as current retrofit strategies—while championing lower costs and energy efficiency—are expected to exasperate many adverse effects of light pollution. Concerns have been raised regarding their impacts on ecosystems and human health, as well as the likelihood of further degrading dark skies and stellar visibility (e.g., Falchi et al., 2016). Furthermore, despite the promises of smart LED lighting, studies suggest they may ultimately lead to greater energy usage if not accompanied by new policies and strategies (e.g., Kyba, Hänel, & Hölker, 2014; Kyba et al., 2017).

This confluence of technological innovation and changing moral evaluations creates complex challenges for a ubiquitous but critical urban infrastructure. Yet, it also offers a rare opportunity to re-imagine how and why we light our nights, and to envision and enact new strategies. Taking into account the shifting terrain of technical and moral factors, this paper will articulate a morally engaged design strategy for the adoption of (smart) LED lighting. In particular, this paper will focus on how to incorporate environmental values into the next generation of streetlights. To do so, *designing for darkness* will be introduced as a framework that strives to foster a range of environmental values through the design of nighttime lighting. Though darkness is identified and presented as a design goal, nighttime illumination is not actively de-valued. Rather, the focus is on achieving a better balance of illumination and darkness, and letting dark skies—as a form of ecological restoration—back into our urban spaces. Through applying this framework, I will propose that the functionality of smart LED lighting can be exploited in creative and innovative ways that supersede a narrow focus on efficiency and achieve a broader range of environmental values.

It is important to recognize the momentous and rare opportunity created by the present convergence of technical and moral changes, as well as the stakes. In the longer history of public lighting, we can observe that major technological leaps happen at a somewhat gradual pace, roughly once every century or so. It seems that we are now in the early stages of the next leap, replacing the technologies that have lit the twentieth century and shaped contemporary debates. We can thus assume that the transition to smart LED lighting will come with a large degree of physical and temporal permanence, and may very well shape the next century of urban lighting—technically, politically, and morally. In the face of light pollution, as well as lighting's contribution to global problems such as climate change, we cannot let this opportunity pass us by. We are encountering a rare moment where we can re-imagine our urban nightscapes, and envision how twenty-first century nights could, and should, look. The values that have driven the proliferation of electric lighting must be met with scrutiny, and re-purposed to fit contemporary needs and goals.

Streetlights constitute the primary source of nighttime illumination, making them fundamental to any strategies addressing light pollution and incorporating the positive aspects of darkness into urban nightscapes.[1] Globally, there are estimated to be over 100 million streetlights in use. There is an estimated additional 55 million lights used for parking lots in OECD countries alone. Together, these sources make up over 90% of outdoor illumination (with traffic signals, billboards, and airports making up most of the remaining 10%). As could be expected, illumination outputs are not evenly distributed globally—the 35 nations of the OECD use 71% of the energy and produce 75% of the outdoor illumination (International Energy Agency, 2006). Streetlights are especially important for cities, being essential for urban nighttime activities and constituting a significant portion of electricity consumption and costs. In the United States, for example, outdoor lighting is estimated to cost about $10 billion annually (enough energy to power six million homes for a year), and can account for up to 60% of a city's public amenities electricity bill (Murthy, Han, Jiang, & Oliveira, 2015). Major cities typically have tens of thousands of streetlights, or more. For example: Washington, DC has over 70,000 publically owned streetlights (which are bound for LED retrofit in the coming years), Helsinki has approximately 85,000 streetlights, Milan's recent LED retrofit saw the installation of 100,000 new lamps, and New York City has approximately 262,000 streetlights. While such numbers serve as useful illustrations of the prevalence of streetlights, most important for this discussion is the recognition that streetlights constitute a massive and critical infrastructure. They are foundational to nighttime illumination, as well as the problems of light pollution.

The paper progresses as follows. The next section provides practical details to inform the later discussion, both introducing and describing the benefits and (potential) harms of LEDs and smart lighting systems. In Sect. 11.3, I articulate a move towards environmentally responsible lighting, using Responsible Innovation as a value-sensitive orientation to guide the development and design of nighttime lighting. *Designing for darkness* is then presented as a technology-specific form of Responsible (Urban) Innovation (Nagenborg, 2018), and defined via two interrelated components: the substantive environmental values to be fostered and preserved, and the striving for morally engaged nighttime experiences. Section 11.4 then explores how to operationalize a *designing for darkness* approach. First, it is translated into a tangible design goal aimed at addressing the issue of "shifting baseline syndrome." Next, three design concepts are put forward as actionable strategies to guide the re-introduction of darkness into urban nightscapes. Throughout, this paper relies on some (basic) lighting terminology—a quick reference guide for these various terms is included as an Appendix.

---

[1] Here, "streetlights" should be understood as synonymous with "outdoor lighting" or "outdoor stationary lighting." It refers to lighting used for roadways, parking lots, pedestrian walkways, etc., and typically lighting infrastructure that is publically owned and operated. Because of the level of analysis in this paper, there is no need to make a more fine-grained distinction between the functions and locations of streetlights.

11.2 LEDs: The Next Generation of Streetlights

Light-emitting diodes (LEDs) are a semiconductor light source, and a type of solid-state lighting (SSL). While they have been used in electronic devices for some time, their utilization for streetlights is a recent development. Early retrofits and prototypes began in the mid-2000s, and the trend has been quickly growing in the decade since. There is a high amount of enthusiasm for LEDs coming from both industry and municipalities, which typically site their improvements to efficiency, colour quality, and controllability as the rationale for adoption (see Table 11.1). Given the novelty of LED outdoor lighting, there has been a relatively quick uptake, with cities around the world rapidly adopting the technology. For example, a study done by the National Academic of Sciences (NAS, 2017) on SSL applications in the United States found that by 2015, LED lighting already accounted for 20% of area and roadway luminaires and 13.9% of parking lot lights. Overall, they found a total market penetration of about 18% for all exterior lighting applications. Further, they found that market penetration is increasing rapidly, and more than doubled (for all interior and exterior applications) between 2014 and 2015. A similar trend of rapid and increasing adoption is predicted for Europe (De Almeida, Santos, Paolo, & Quicheron, 2014). It would therefore be reasonable to assume that a large-scale conversion to LED outdoor lighting is no longer speculative or even prospective, but a dawning reality. All factors seem to point to the continued (and increasingly rapid) adoption of LED streetlights as their costs continue to go down and their efficacy continues to improve.

Discussions of potential improvements often use high-pressure sodium (HPS) lamps as a benchmark, as it is the most efficacious and affordable precursor. As of the International Energy Agency's (2006) report, about 62% of outdoor lighting was high or low-pressure sodium, and 30% mercury vapour.

Table 11.1 The beneficial characteristics of LEDs, adapted from De Almeida et al. (2014) and the National Academy of Sciences (NAS, 2017)

Lighting consideration	LED advantages
Efficacy	• High luminous efficacy (lumens/Watt), with rapid improvements in recent years; assuming that LEDs reach their projected efficacy of 200 lm/W by 2025, this can result in up to 40% energy savings for all lighting by 2030
Efficiency	• High efficiency (lower power consumption, lower operating voltage, reduced energy costs); can be up to 50% more efficient than HPS lamps • Longer lifespan; can last 3–6 times longer than HPS lamps
Environmental impacts	• No mercury; no UV or IV radiation
Controllability	• Easy dimming control compared to current technologies • Less light spillage, better directional light output
Lighting quality	• Vivid colour range and control • Cool white light (high CCT) allows for improved colour rendition

Alongside the conversion to LEDs, new streetlights are increasingly being fit with "smart" technologies, often with the stated goals of further improving the benefits described in Table 11.1. The label of "smart" for city systems and master plans has been met with scrutiny, for there is no clear consensus about the meaning or boundaries of the label, or its actual benefits (e.g., Kummitha & Crutzen, 2017; Sadowski & Pasquale, 2015). However, at the most general level, smart technologies can be understood as the use of ICT or Internet of Things (IoT) technologies for infrastructure and public services. We can therefore broadly understand "smart lighting" as the incorporation of ICT or IoT technologies to lighting—for example, adaptive controls, sensors, etc.—with the goal of improving performance (e.g., Juntunen et al., 2015; Murthy et al., 2015). For streetlights, this typically means incorporating sensors or cameras to light poles and fixtures towards the goals of increased (cost and energy) efficiency and safety (NAS, 2017). LEDs, given their controllability, are ideal for coupling with smart lighting systems, which has made for a high degree of complementarity between LEDs and smart lighting initiatives (Gandy, 2017). For example, Juntunen et al. (2015) prototyped a smart LED lighting system that was responsive to both natural lighting levels (e.g., increased brightness levels after a fresh snowfall, as well as during sunrise and sunset) and pedestrian activities. Through these innovations they were able to see significant power reductions, and further predict that smart LED lighting can offer up to 70% energy savings—far more than just switching lighting to LEDs.

Despite the potential benefits brought by (smart) LEDs, both scientific researchers and advocates for light pollution reduction have voiced concerns over the rapid proliferation of LED outdoor lighting. At the most general level, seeing LEDs as an environmentally sustainable infrastructure solution "is highly ambiguous because their introduction facilitates the development of more energy-efficient sources of light pollution" (Gandy, 2017, p. 1097). Gandy (2017) further asserts that LEDs, as a more efficient and cheaper source of lighting, could lead to a dramatic increase in light consumption. Kyba et al. (2014) also predict that an adoption of LEDs may not decrease energy consumption, unless linked with strategic and effective policies. A more recent study seems to support these warnings, finding that at a local scale there may be savings, but at a global (and often national) scale this is not the case—likely due to rebound effects and the installation of lights elsewhere (Kyba et al., 2017). There is also concern that LEDs will increase *skyglow*, the ambient atmospheric brightness caused by nighttime lighting, arguably the most conspicuous form of light pollution. It has been estimated that the replacement of current technologies with cool white LEDs could more than double skyglow levels in Europe (Falchi et al., 2016), further cutting off experiences of the night sky, especially in cities. Finally, there are concerns about the long-term health impacts caused by blue-rich white LEDs (American Medical Association, 2016), as well as the ecological impacts (Schoer & Hölker, 2017a). A proposed solution is to promote warmer colour temperatures, which may mitigate some of the harmful effects to humans as well as skyglow. However this decreases efficacy, and a recent study suggests this will not decrease the ecological impacts of LEDs (Pawson & Bader, 2014).

In sum, we can observe that LED outdoor lighting is being met with both optimism and scrutiny. As with any large-scale infrastructure, they carry a range of (potential) positive and negative effects. While the extent to which LEDs will exasperate negative environmental effects is still under debate, as is the exact costs of wasted light, these concerns nevertheless signal that there is a broad range of environmental issues at stake. And, the exact impacts will likely not be known with certainty until a greater number of LEDs have been installed and in operation for a longer period of time. Of course, by then the possibility of reversing any negative effects will be much more difficult. Regardless of these debates, it seems that barring some dramatic and unforeseen policy or market shift, their widespread adoption will continue over the coming years. We should therefore see this moment as an opportunity to anticipate downstream issues and explore alternative pathways that supersede a shortsighted focus on efficiency. For this, I propose that we strive to proactively incorporate *moral* values into the design and use of this new technology, without dismissing the potential benefits that LEDs (and smart lighting systems) can provide. By doing so, we can actively search for ways to reframe existing debates, take into account a full range of environmental concerns *before* the infrastructure's design and use is solidified, and articulate a value-sensitive framework for the adoption of this new technology.

11.3 Designing for Darkness: A Value-Sensitive Approach to Responsible Nighttime Lighting

How then to incorporate the instrumental benefits of new innovations, while also avoiding downstream detrimental impacts, in the development and design of new lighting technologies? To orient such an approach to urban nighttime lighting, I use Responsible Innovation as a starting point. At its most general level, this concept adds the qualifier of responsibility (via transparent, inclusive processes and the articulation of ethical end goals) to innovative practices (e.g., van den Hoven, 2013; Stilgoe, Owen, & Macnaghten, 2013). Nagenborg (2018), in the context of urban technologies, defines Responsible (Urban) Innovation via three key features. First, it requires that the ethical and societal implications of products and projects be considered. Second, it emphasizes a non-instrumental view of technologies, asserting that they are not value-neutral but instead value-laden. Third, innovations should strive for ways to include values in the design process that do not at the same time have a negative impact on other moral values. This problem of satisfying conflicting obligations or moral values is referred to as *moral overload* (van den Hoven, Lokhorst, & van de Poel, 2012). However, the very notion of Responsible Innovation—to innovate in a way so as to "expand the set of relevant feasible options regarding solving a set of moral problems" (van den Hoven, 2013, p. 82)—is conceptualized as a way to overcome such conflicts. Importantly, this necessitates a shift in how we conceptualize innovation—no longer just about technical improvements but also about ethics and moral values (van den Hoven, 2013). When applied to technologies embedded in urban contexts, this can take the form of either the responsible design and use of a technology in the urban environment, or more constructively exploring how new innovations can be developed to address specific urban challenges (Nagenborg, 2018).

As an example of an urban challenge specific to nighttime illumination, we can briefly look at the tensions between darkness and safety. In helping to avoid injury and find your way at night, lighting serves a necessary functional requirement for urban nightscapes. But beyond these basic practical functions, the effects of lighting are a debated issue. Recent studies examining the contentious relationship between crime and lighting—as well as the complex relationship between *feelings* of safety, *actual* safety, and fear of the dark—tend to show that while actual safety may not increase with more illumination, feelings of safety often do (e.g., Boomsma & Steg, 2012; Fotios, Unwin, & Farrall, 2015; Gaston, Gaston, Bennie, & Hopkins, 2015; Haans & de Kort, 2012; Marchant, 2004; Li et al., 2015; Pena-Garcia, Hurtado, & Aguilar-Luzon, 2015). Further, the associations between lighting and safety are deeply embedded connotations, with the goals of safety and security (as well as policing and surveillance) inexorably tied to the very foundations of modern public lighting efforts (e.g., Schivelbusch, 1988; Schlör, 1998). This relied on both a practical and symbolic interpretation of safety, as artificial illumination inherited a much longer history of lighting-as-good and darkness-as-evil cultural metaphors (Ekirch, 2005). We can thus appreciate that lighting-as-safe and darkness-as-dangerous associations are deeply embedded perceptions. However, neither are essential or static concepts, but influenced by our lighting practices. It would therefore appear that we are confronted with a situation of moral overload, and thus a challenge for responsible lighting strategies. While an in-depth analysis of the relationship between safety and darkness is outside the scope of this paper, Sect. 11.4 will put forward innovative approaches to incorporating darkness into urban nightscapes without necessarily compromising safety.

Taking Responsible (Urban) Innovation as an orientation for light strategies requires a re-framing of the goals driving innovations to nighttime lighting, moving beyond technical improvements as the primary goal for future lighting systems. The current focus on efficiency and cost-savings are certainly important goals for a municipality, however a value-sensitive approach emphasizes that other "non-functional" requirements also be incorporated into the development, design, and adoption of technologies and systems (van den Hoven, 2013). We must therefore begin with a value-level discussion, and *then* ask how these technical capabilities can be developed to serve desired goals. Only then, with goals clearly identified, can we judge the responsible usage of new lighting technologies, and how they may address the environmental impacts of nighttime lighting. For this, we need a framework that articulates morally relevant values as a starting point and uses them to better elucidate the "good" we can strive to achieve with smart systems and LED streetlights. Put otherwise, it forces us to ask, *exactly what we are designing for*?

A framework for responsible urban lighting must respond to the unique historical, social, and technological context of nighttime illumination. For conceptualizing the environmental values at stake, I propose the adoption of *darkness* as a key design feature to be sought after in future nighttime lighting strategies. To understand exactly what it can mean to *design for darkness*, and what such an approach would entail, I first present the substantive values associated with darkness. This leads into a longer discussion about how meaningful experiences can be wrought out of our urban nightscapes through a conscientious (re)introduction of dark skies. While focused on the positive features of darkness, this approach is not meant to de-value lighting or the various benefits nighttime illumination undoubtedly brings. Rather, it is about highlighting the many benefits that darker nights, combined with a restrained and conscientious use of lighting, can bring. Hence, it is ultimately about situating darkness as a design criterion and value-level consideration in nighttime lighting, and in the process achieving a better balance of lighting and darkness (Edensor, 2015, 2017). However, given the over-illuminated state (or at least poorly-designed illumination) of most of our cities, this requires an explicit focus on re-introducing darkness into our urban nightscapes. Also important to emphasize is that this approach is focused on elucidating and realizing environmental values. This is not to say that social or procedural values are less important, or should not be considered in future innovations to outdoor lighting, only that environmental impacts are the topic at issue here.

11.3.1 Valuing Darkness

A first step is to articulate what sorts of environmental goods should be pursued or protected in the policy-making and design of urban nighttime lighting. For this task, *darkness* can be conceptualized as a medium through which, or from which, a range of environmental values arise (Stone, 2018b). As such, darkness should be understood as an umbrella term encompassing various environmental concerns that are seen as negatively impacted (degraded, hindered, threatened, or otherwise) by light pollution, and nighttime lighting more generally. When arguing for a reduction in illumination levels for energy savings, or alternatively when striving to protect access to the starry night sky, there is the implication that darker nights—and hence *darkness* as an evaluative concept—contains or fosters positive value. To articulate the value of darkness, the commonly agreed upon effects of light pollution are re-framed as nine ways by which, or through which, value is derived from darkness (see Table 11.2; for an in-depth description of each value, as well as justification for their moral importance, see Stone (2018b, pp. 615–617)). Importantly, a key facet of focusing on darkness is that it allows for the assessment of *all* lighting and its quantitative and qualitative impact. Light pollution is a concept primarily concerned with identifying and mitigating the negative effects and thus undesirable uses of lighting. Preserving and fostering the valuable aspects of darkness incorporates these concerns, while also allowing for a more fundamental and holistic re-consideration of *all* urban lighting and its impact on environmental values.

Table 11.2 The nine values of darkness translated into prima facie obligations to consider in the design of nighttime lighting; adapted from Stone (2018b)

Value of darkness	Prima facie obligation derived from value
Efficiency	The responsible use of lighting where and when needed; money-saving
Sustainability	The responsible use of lighting where and when needed; energy-saving and preserving non-renewable resources
Ecological conservation	The protection and preservation of species and biodiversity; habitat conservation efforts
Healthiness	Promoting and fostering human health; physiological well-being
Happiness	Promoting and fostering happiness; emotional well-being
Connection to nature	Preserving a connection to the more-than-human world
Stellar visibility	Preserving conditions for access to the firmament
Heritage and tradition	Preserving the cultural heritage of the night sky for future generations
Wonder and beauty	Preserving the aesthetic appeal of the natural night sky

Towards operationalizing for the design and use of LED streetlights, the value of darkness can be further refined in two ways. First is a consideration of how darkness, as an environmental good, should be weighed in comparison with other values associated with both darkness and lighting. For incorporating darkness into this broader landscape of concerns, the nine values are further translated into *prima facie* moral obligations that future lighting strategies should—at the least—take into consideration (Table 11.2). Thus, each is presented as a first-order requirement in the design process, even if each obligation may not be achievable in every case. However, this creates an important shift in the burden of proof—to show why *not* to incorporate the obligation, and give justifiable cause for that choice (Stone, 2018b, pp. 622–623).

Second is a consideration of how to weigh and prioritize the nine identified values internal to darkness. Darkness has instrumental value, both from an anthropocentric and non-anthropocentric point of view. For example, it can be economically valuable, saving money and reducing energy usage, and can promote eco-tourism via "dark sky reserves." And, it is beneficial for many nocturnal species. Darkness also has intrinsic value, in the sense that there are reasons to preserve darkness for non-instrumental reasons. Certain phenomena and experiences, such as the starry night sky, are an inherent and inseparable quality of dark nights. In such cases, darkness is not a means to an end, but a facet of the end goal itself. These intrinsic qualities, I assert, should be the focus of a designing for darkness approach on two grounds. First, they are more robust, in the sense that they offer enduring, powerful experiences that are not contingent on external goals (e.g., cost-saving efforts). Second, achieving them will necessarily lead to instrumental benefits. Achieving darker skies offers the possibility of reconnecting urban spaces with their natural settings, but it will also lead to a reduction in energy usage, cost savings, and health and ecological benefits. To the contrary, installing energy-efficient lighting may achieve instrumental benefits while remaining neutral, or even having a negative effect, on access to the night sky. Thus, the cluster of values associated with dark skies (*ecological conservation*, *connection to nature, stellar visibility, heritage and tradition,* and *wonder and beauty*) is prioritized as the main goal for environmentally responsible urban lighting.

11.3.2 Experiencing Darkness

Darkness, as presented above, can be understood as an evaluative concept to define and categorize the environmental goods to be strived for in nighttime lighting. However, darkness is also a quality of lived experience—there is a physicality and spatiality to darkness, just as there is one to lighting. And, it is through such experiences that meaning ultimately arises. We must therefore understand darkness as both an evaluative concept *and* an experiential goal. This means that in addition to articulating darkness as something of value, we must also explore how darkness can be realized in urban nightscapes in ways that foster morally transformative, or at least morally engaged, experiences.

To properly situate the positive features of darkness into the context of lived urban spaces, it is important to understand the relationship between lighting and darkness, and how nighttime illumination shapes our perceptions of darkness. A basic presupposition of Responsible Innovation is that technologies are value-laden, and lighting is no exception. The many histories of nighttime illumination (e.g., Schivelbusch, 1988; Schlör, 1998) have done excellent jobs of highlighting the various values that have shaped, and been shaped by, nighttime lighting—a topic I summarize elsewhere (Stone, 2017). Here though, I will not discuss these various values further, but rather focus on the phenomenology of human-technology-environment interactions; or more specifically, the interaction between nighttime experiences, lighting technologies, and darkness. This requires that we appreciate the moralizing role lighting plays in shaping our perceptions and evaluations of darkness, and the capacity of lighting technologies to either foster or hinder those valuable aspects of darkness described above.

To understand the profound influence that technologies have on our morality, we can draw from recent work in *technological mediation* (Verbeek, 2011). Focused on human-technology relations, it interprets our actions and perceptions as "always closely interwoven with the material environment in which they play out" (Verbeek, 2011, p. 22). Technology is not something *out there* to be dealt with. Instead, our interactions and uses of technological artefacts play an active role in constituting our reality. Hence, we experience the world—and morality—*through* or *by way of* technological artefacts (Verbeek, 2011).[2] Urban infrastructures thus act as external

[2] The case of nighttime lighting arguably stretches the theory of *technological mediation*, focused primarily on individual artefacts. The moral implications of lighting can be traced back centuries and across multiple technological leaps (i.e., oil lamps, gaslight, and electric light). Furthermore, these moralizing effects are a product of the totality of lighting infrastructure, not individual lights. The paradigmatic examples of mediation show the moralizing effects of specific devices on

constraints, limiting our ability to perceive beyond them, both figuratively and literally, as well as shaping how we see ourselves in relation to the natural world (King, 2000).

As discussed in the introduction, streetlights dominate our nights. They are the most prominent source of illumination, both historically and in contemporary cities, effectively creating nighttime spaces (Major, 2017). Streetlights are an encompassing technology that not only mediate perceptions, but are foundational to our experiences and understanding of the city at night. They provide the backdrop and ambient lighting for all nighttime activities. In this sense, streetlights actively constitute our nights, both literally and figuratively. They are at the crux of the question of how darkness is perceived, and ultimately evaluated. Any attempt to re-introduce and re-imagine darkness must therefore begin with the lighting technologies themselves. To design for the value of darkness, we must focus on our lighting. Only by first appreciating that our cities are understood from *within* the illumination created by streetlights, can we move to re-design them in a conscientious way.

A designing for darkness approach must therefore acknowledge the profound mediating power of nighttime lighting, and strive for the incorporation of meaningful experiences of darkness *through* the design of streetlights. Given the hierarchy of substantive values identified above, such designs should strive to incorporate the intrinsic, future-oriented values of darkness into urban nightscapes. For hints at how such meaningful experiences could look, here I draw on the idea of a re-envisioned *nocturnal sublime* discussed in more detail elsewhere (Stone, 2018a). It is argued that focusing on incorporating the desirable aspects of dark nights—namely the starry night sky—into urban settings can engender aesthetic responses that are relevant to fostering an environmental ethic. The proposed re-envisioned nocturnal sublime is further situated as a form of holistic urban restoration (de-Shalit, 2003; Stone, 2018a). On a practical level, it helps to mitigate the negative impacts of light pollution, and satisfies the instrumental benefits of darkness (e.g., a reduction in energy usage). Yet it also attends to the moralizing effects of lighting infrastructure, in the process fostering the intrinsic value of dark skies. It allows for a glimpse beyond the lighting that now envelops our urban nightscapes, and opens the possibility of cities and their inhabitants (re)discovering an ecological and cosmological sense of place. Speaking about the future of "green" design generally, the architect James Wines (2005, p.18) stated, "The mission now in architecture, as in all human endeavour, is to recover those fragile threads of connectedness with nature that have

---

individual experiences, such as the obstetric ultrasound. Such devices, argues Verbeek (2011, p. 39), expand the "moral community" to include nonhuman artefacts. However, here I am less interested in any individual artefact (or the agency thereof), but rather the infrastructure in which they operate. What can be said about the larger system of nighttime lighting in which individual devices (i.e., a single streetlight on a lamppost) operate? The moralizing effects of nighttime lighting have a far greater temporal and spatial resonance than any single lamppost, or any one technological development. Thus, it would be untenable to analyze LED streetlights as a completely novel technology, or assert that they should be assessed as isolated artefacts. Nevertheless, the relationship between technologies and morality presented in mediation theory provides useful insights into the evolving morality of darkness, and the central role of lighting technologies.

been lost for most of the century." Experiencing the sublime night sky in cities can help to re-discover one of those "fragile threads" of connectedness with nature, and bring it back into our urban spaces. Thus, we can extend ideas of "greening" cities to also include "darkening" city nights as a form of environmentally restorative urban design.

11.4 Designing for Darkness: Practical Possibilities for (Smart) LED Streetlights

Designing for darkness provides a value-sensitive framework for urban nighttime lighting, articulating a range of substantive environmental values that can be fostered through the achievement of darker night skies. It also situates darkness as a form of urban (ecological) restoration, which can help to create meaningful experiences for city inhabitants. A necessary next step is to translate these abstract philosophical ideas in actionable design requirements, and explore how smart LED streetlights can be put towards these goals. Thus, here I take preliminary steps in operationalizing designing for darkness. To do so, I first identify a target issue of relevance to our evaluations and experiences of darkness, namely *shifting baseline syndrome*. This is followed by three possible designing for darkness scenarios, presented as a means to overcome shifting baselines and re-introduce darkness into our cities.

Important to note is that steps have already been taken towards formulating responsible urban lighting strategies. There exists a growing body of research investigating ways to curb light pollution, with a number of recommendations proposed in recent years (see Table 11.3). And, various countries and municipalities have

Table 11.3 Overview of proposed recommendations and policies for mitigating light pollution; adapted from Falchi, Cinzano, Elvidge, Keith, and Haim (2011), Gaston, Davies, Bennie, and Hopkins (2012), IDA (2016), IDA-IES (2011), and Schoer and Hölker (2017b)

Recommendation(s)	Suggested policies
Reduce skyglow, light trespass, and glare	• Properly positioned installations—for overhead lights, angled below 70° from the downward vertical • Fully shielded lighting luminaires to further direct light
Eliminate unnecessary lighting	• Switching off or dimming lights when not in use, to both reduce light pollution and save energy consumption and costs
Proper illumination levels	• Establish proper levels of illumination required for safety and visibility. Research suggests only 1–3 lx required for facial recognition and visibility (save for roadways with higher speeds); current streetlights have illumination levels of 10–60 lx
Proper colour spectra	• Avoid cold white light due to effects on ecosystems, human health, and skyglow; use "warm" or filtered LEDs with a CCT below 3000 K
Ecological sensitivity	• Maintain naturally unlit areas while promoting and enforcing "dark sky reserves" • Harmonize light levels with the needs of local flora and fauna

already introduced ordinances to mitigate light pollution, towards the goals of cost savings and energy efficiency, and in some cases dark sky protection (Kyba et al., 2014). These existing strategies typically take the form of technical recommendations to minimize specific causes and effects of light pollution. The concepts presented in Sect. 11.4.2, however, are meant to go further than mitigating light pollution, instead evaluating *all* nighttime lighting and actively re-imagining our future urban nightscapes, and the relationship between lighting and darkness therein. In this sense, the concepts presented here are more encompassing than, but ultimately complimentary to, these existing strategies for addressing light pollution. They are focused on broader visions of urban nightscapes within which various technical strategies and policies can be implemented.

11.4.1 Shifting Baseline Syndrome

A *designing for darkness* approach can take many forms, and respond to the values and experiences articulated in Sect. 11.3 in varying ways, and to varying degrees. However, it does require a specific orientation: that we assess new innovations to, and uses of, artificial lighting via both their quantifiable impacts *and* as a manifestation of social and environmental values. For translating darkness into an actionable design goal for LED streetlights, the abstract ideas of "connection to nature" and "mediation" require a tangible focal point for design interventions. For this, I will elaborate on a specific facet of urban nights that makes evident the mediating (and moralizing) capabilities of nighttime lighting, and can be directly addressed through new technological innovations: *shifting baseline syndrome*. It has been applied to issues of light pollution by Lyytimäki (2013), and is a term originating in ecological conservation research (Papworth, Rist, Coad, & Milner-Gulland, 2009). Lyytimäki (2013, p. e46) explains that, "Shifting baseline syndrome refers to the changing human perceptions of biological systems due to loss of experience of past conditions. Simply put, people may view the current situation as the typical or normal state, even when the ecosystem is considerably degraded compared to earlier states." Broadly, there are two types of shifting baseline syndrome: *generational* and *personal amnesia*. Generational amnesia occurs when previous experiences are not passed on to new generations, causing past conditions to be forgotten; personal amnesia occurs when individuals forget or "update" their own experiences, and come to believe that current conditions are the same as past ones, and thus normal (Lyytimäki, 2013; Papworth et al., 2009). In both cases, it causes a shift in perceived baselines, and thus an evaluative shift in accepted norms.

Lyytimäki (2013) posits that the loss of the night sky, and experiences of (especially urban) nights, cause of both types of amnesia. This assumption is largely supported by scientific evidence that investigates the brightness (and brightening) of our contemporary nights. It is estimated that 83% of the world's population, and over 99% in the USA and EU, live in areas considered to be above the "polluted" threshold, and in cities the artificial brightness can be several magnitudes greater. The Milky Way—arguably the key feature of the sublime starry sky—is no longer visible to 60% of Europeans and almost 80% of North Americans (Falchi et al., 2016). This trend is only increasing, with nighttime brightness estimated to be increasing 3–6% annually on a global scale (Hölker et al., 2010). And, as mentioned in Sect. 11.2, these trends are not expected to decrease with the adoption of white LEDs, unless new policies are enacted.

In applying shifting baseline syndrome to nighttime lighting, Lyytimäki is mainly concerned with addressing the effects on ecosystem services, typically known as ecological light pollution (Longcore & Rich, 2004), and how this effects value-level discussions about the need and use of artificial lighting. However, Lyytimäki's argument provides much broader and profound possibilities, offering a new perspective on urban darkness. Shifting baselines of darkness can be understood as the manifestation of lighting technology's mediating presence in our perceptions of nighttime environments. The repercussions of nighttime lighting on our baseline understanding of darkness can be considerable, as living in constant artificial brightness can lead to perceptions of darkness as unnatural or unsafe (Lyytimäki, 2013). Yet it also reveals that darkness—as both a normative and phenomenological concept—is not essential or static. It highlights the relative nature of everyday perceptions of "darkness" and calls into question evaluative categories such as "too dark." Such labels are not absolute, but shaped by our abilities to see (or not see) past our current conditions, and our tendencies to update our accepted and expected norms of illumination levels. As nights continue to get brighter, and we introduce new, more efficient technologies, we risk continuing to increase our *generational* and *personal* baselines. However, with the controllability offered in new lighting technologies, we can actively challenge this trend. Thus, shifting baselines are where a designing for darkness approach can focus: on seeking to reverse our baselines and re-introduce darker nights, while still maintaining sufficient levels of illumination for nighttime activities. This can allow for meaningful experiences to emerge, as well as the achievement of the intrinsic (and instrumental) values associated with darkness. And importantly, it can allow for the achievement of these goals without creating (or exacerbating) a situation of moral overload between darkness and safety.

11.4.2 Realizing Darkness

As a practical strategy for realizing darker nights—and specifically realizing the values and experiences elucidated in Sect. 11.3—we can thus explore possibilities that directly address our personal and generational shifting baselines. Here three design concepts for darkening cities are presented with these goals in mind. As with any urban planning strategy, a great deal more research into the feasibility and place-specific manifestations is required. However, here these concepts are presented as broadly applicable guiding visions that could be included in planning instruments such as a lighting master plan, as a means to speculate on the real-world potentials of designing for darkness. Each concept should, in principle, incorporate the instrumental benefits of smart systems and LEDs (energy savings, etc.), but also go much further in helping to re-position the relationship between lighting and darkness in urban nightscapes.

11.4.2.1 Concept 1: Incremental Darkening

A straightforward design solution is to quite simply turn down the lights. Current lighting practices typically give rise to street-level illumination of between 10 and 60 lx, which is several magnitudes greater than natural conditions. With no artificial light, a clear sky with a full moon provides an illumination of circa 0.1–0.3 lx, a clear moonless starry sky circa 0.001 lx, and an overcast night sky as low as circa 0.0001 lx (Gaston et al., 2012). Further, current lighting is often very inconsistent in the amount of brightness depending on distance from the light source—something that can be addressed via the uniform lighting distribution of LEDs (Gaston et al., 2012). Recent studies have found that only 1–3 lx is necessary for visibility and facial recognition, although roadways with higher speeds require higher levels (Schoer & Hölker, 2017b). This would suggest that the lighting levels on many of our streets and public spaces could be significantly reduced without necessarily compromising visibility or nighttime activities. This would have a range of economic and ecological benefits, as well as potentially avoiding downstream health issues. And, it would decrease skyglow and allow for increased stellar visibility.

For realizing an incremental darkening strategy, we can take inspiration from other domains of urban planning—for example, Copenhagen's initiative to reduce vehicle traffic and increase cycling. For several decades, Copenhagen has been gradually "removing driving lanes and parking places in a deliberate process to create better and safer conditions for bicycle traffic" (Gehl, 2010, p. 11). This has been combined with investments in cycling infrastructure, yielding positive results: bicycle traffic doubled between 1995 and 2005 (Gehl, 2010). A similar lighting strategy can be enacted that conscientiously slows, and potentially reverses, shifting baselines. A radical or abrupt change in lighting levels would likely cause concern, and may not be socially accepted. If shifting baselines create the perception that current levels are the norm, any drastic reduction could seem "too dark," and decrease feelings of safety. To avoid this, we can exploit the potentials of this new technology, and specifically its dimming capabilities, to gradually reduce brightness. Both illumination levels, as well as colour temperatures, could be uniformly reduced in incremental steps, in a way that is not overly impactful on daily experiences. Like Copenhagen, this could be done over several years, and could include regular feedback loops with public input as well as studies on crime and safety. Importantly, such a strategy should be planned and implemented citywide after smart LED retrofits, to ensure a uniformity of gradual darkening and colour changes. Otherwise, this may lead to increased contrast between adjacent spaces, which may have the reverse effect on perceptions of darkness, and perhaps cause (or reinforce) segregation between spaces or neighbourhoods. In sum, this concept positions the act of darkening cities as a long-term process implemented uniformly across an entire city or region, to reverse our shifting baselines and gradually re-introduce darkness into our lived spaces.

11.4.2.2 Concept 2: Environmentally Responsive Lighting

A central goal of designing for darkness is to reconnect cities, and their inhabitants, with an ecological and cosmological sense of place. In doing so, the act of darkening cities can become a form of holistic urban restoration, that lets the night sky back into urban experiences. The smart sensors and systems that offer increased controllability and real-time responsiveness can be developed towards this goal. Illumination can thus be responsive to changing natural conditions (moonlight, clouds, etc.), creating a lighting strategy closely attuned to seasonal cycles and weather.

These technological capabilities already exist, or are under development. Street lighting has traditionally been controlled by a clock, with modern systems often controlled by a photoelectric cell to detect daylight and adjust operating times accordingly. Still, such systems are limited because they operate as binary on/off systems. However, with LEDs, greater control is now possible (Juntunen et al., 2015). Tests have been done in which lighting levels respond to changing environmental conditions. For example, a study in Helsinki found that lighting levels could be significantly reduced when there was fresh snowfall, due to ambient brightness. The same study found that lighting could similarly respond to the brightening sky during sunrise (Juntunen et al., 2015). A similar approach has been suggested elsewhere as a strategy for mitigating light pollution, positing that future lighting systems could respond to changing lighting conditions during twilight hours (Schoer & Hölker, 2017b). Future systems could even incorporate more creative or informative dimensions that are unique to local geographies and social circumstances. For example, lighting could be developed that changes colours or brightness in the minutes before rainfall, or in response to wind speeds, as both a warning system and safety measure.

The key here is to develop sensors towards the explicit goal of illumination that is responsive to, or symbiotic with, natural brightness levels; to see lighting not as eliminating the night, or of pushing back darkness, but of responding to it. At the very least, this will save energy and costs. More optimistically, it can re-connect lighting infrastructure—and city inhabitants—to diurnal and seasonal patterns. In this sense it is less about "darkening" in a literal sense (as in Concepts 1 and 3), but rather attuning cities to their environment's natural rhythm. Instead of leaving cities awash in a static, uniform illumination from sunset to sunrise, cities respond to their nocturnal context and its idiosyncrasies.

11.4.2.3 Concept 3: Urban (Dark) Acupuncture

The previous two ideas are more gradual, aimed at incremental changes to ambient brightness, and should be implemented somewhat uniformly across a city or region. Another option is to strive for more radical or transformative experiences aimed at a direct and explicit confrontation with darkness, bringing artificial lighting out of the backdrop of daily life. For this, specific spaces—public squares, pedestrian streets, parks, etc.—could be explicitly and conspicuously left dark, perhaps not without any lights, but at near natural levels to contrast our current (perceived) norms. Through this, we can create moments of accentuated darkness throughout cities, to give inhabitants a glimpse of another possibility for urban nightscapes.

As inspiration for such an approach, we can draw from Jaime Lerner's idea of *urban acupuncture* (Lerner, 2014). This planning theory posits that a few well-placed "pricks" of a certain service, amenity, or structure can begin a process of urban renewal. The concept is most often applied to social concerns, such as safety, or of "restoring the cultural identity of a place or community" (Lerner, 2014, p. 9). This could mean, for example, strategically placing a new public amenity such as a library so as to stimulate visitors to the surrounding neighbourhood (and thus economic activity, educational services, safer streets, etc.). As another example, Lerner discusses nighttime lighting as a means to accentuate a space and create dynamic, interactive moments. Why not attempt to likewise accentuate a public square or bridge with darkness (perhaps accompanied by some explanatory or educational materials)? Alternatively, a darkened space can serve as a backdrop for a lighting installation, highlighting both the presence of darkness and the beauty of illumination (as well as their complementarity). The controllability of the brightness and directionality of LED lighting could be developed so as to make such moments possible. As an even more radical approach, we can explore how other new and emerging technologies, such as autonomous vehicles, can allow for novel experiences of darkness—an idea discussed in detail elsewhere (Stone, Santoni de Sio, & Vermaas, 2020).

Experiencing darkness—or alternatively, purposeful illumination within a darkened space—could allow moments where people glimpse past the lighting that now mediates our urban lives, and see a nighttime space oriented otherwise. In doing so, we could be forced to confront our (often unquestioned) expectations and perceptions of lighting, and possibly plant the seeds for a new version of darkness to permeate though our cities.

11.5 Conclusion

This paper has presented a forward-looking analysis of nighttime lighting, and in particular streetlights, a critical and formative urban technology. It focused on two interrelated issues: the rapid introduction of smart LED outdoor lighting, and the incorporation of environmental values into the design of public nighttime lighting. To do so, the challenges posed by the introduction of these new technological innovations were re-framed using Responsible Innovation, to argue for a shift from "smart" to responsible lighting. This requires a shift in thinking regarding new lighting strategies—away from efficiency as a primary goal and towards the proactive inclusion of a broader range of values into the development and design processes. *Designing for darkness* was then put forward as a value-sensitive, and technology-specific, framework for incorporating substantive environmental values and meaningful experiences of dark skies into city nights. Potential strategies for realizing *designing for darkness* were put forward by first identifying *shifting baseline syndrome* as a target concern, and then presenting three design concepts that utilize the features of smart LED lighting systems: incremental darkening, environmentally responsive lighting, and urban (dark) acupuncture.

Re-introducing darkness into cities is the primary goal of the approach defined here, as a means to both address the costs and impacts of light pollution, while also more fundamentally re-imagining urban nighttime lighting. However, this goal does not de-value or wholly dismiss the importance of nighttime illumination. Rather, it strives to foster an appreciation of the quantitative and symbolic goods that can be achieved through a better balance of light and dark. It is thus about learning to appreciate the value of darkness, and to experience it—not as dangerous or evil—but as a feature that can reconnect with us with natural rhythms and the starry night sky. For cities, this will not be an impenetrable pitch dark, but a better lighting-darkness balance. "The presence of urban darkness, like a phantasmogoric manifestation of "weeds," has been characterized as a symbol of disorder, neglect, or abandonment, but there is an intermediate spectrum of illumination that lies between a disorientating gloom and the incessant glare of late modernity" (Gandy, 2017, p. 1102). Through this process, the power and spectacle of artificial lighting can also be re-appreciated. Instead of being seen as a pollutant, the new era of *electronic* lighting can revitalize the positive symbolism of nighttime lighting, but with the added dimension of environmental values.

If we accept that designing for darkness is a justified goal, and that this is achievable through smart LED lighting, a future question is how this compares to, and is prioritized within, the larger arena of moral and political concerns shaping decision-making for lighting, and cities more generally. This will require a consideration of how this conceptual framework, as well as strategies for realization, affects safety, traffic, and nightlife within localized settings. It will have to be put into dialogue with concerns related to smart city initiatives, such as privacy and surveillance. And, the processes for enacting such visions will need to be done in a transparent and inclusive manner. However we should also be wary of too quickly relegating darkness solely into the category of an "environmental concern," or worse fall back onto old symbolic connotations, when confronted with the broader landscape of social and moral values. I would instead propose that a greater consideration of the societal benefits (or at least disruptive qualities) of darkness also be considered. For this, we can take inspiration from Nick Dunn's "Manifesto for the Nocturnal City" (Dunn, 2016, p. 96), in which he reflects on the potentials of nightwalking:

> Walking in cities at night, therefore, enables us to sense, connect and think with the city around us. We are able to give things our undivided attention, a welcome respite from the ongoing erosion and subdivision of our time and sense of belonging in the world. Deliberately moving out of the glare and stare of our commoditized and highly structured daily routines and into the rich shadows and patina of our cities at night may be one of the few truly beautiful and sublime practices available to us. Far from being dead hours, for the wakeful the night affords investigation and liberation.

In sum, it would seem that a darker future might indeed be on the horizon for our urban nightscapes.

Acknowledgements Earlier versions of this paper were presented at the Forum on Philosophy, Engineering and Technology at the University of Maryland in May 2018, and the Philosophy of the City Colloquium at the University of Twente in June 2018. In addition to thanking the audiences for their feedback, I wish to thank Pieter Vermaas and the anonymous reviewer for their written comments on earlier drafts of this paper.

Appendix: Lighting Terminology Quick Reference Guide

CCT Correlated colour temperature; a description of the colour of (white) light sources as perceived by humans, measured in Kelvins (K).

Warm = white light with a yellow or reddish tint (≤ 3000 K)
Neutral = warm white (3–4000 K)
Cool = white light with a blush tint (4500–6500 K)
Daylight = ~6500 K

HPS High-pressure sodium, a commonly used lamp type for outdoor lighting.

Light Pollution Any adverse impact caused by artificial light at night. It is typically sub-categorized into *skyglow* (light sent upward and scattered in the atmosphere, causing artificial ambient brightness and the orange-ish haze above cities), *glare* (excessive brightness that reduces visibility, such as bright floodlights at eye level), *light trespass* (unwanted or unintended light, such as a streetlight shining into your bedroom), and *clutter* (over-illuminated clusters of light sources, often found in downtown areas). The effects of light pollution are typically also sub-categorized into *energy usage*, *ecological impacts*, *(human) health*, *safety*, and it's obscuring the visibility of the *night sky*.

Lumens Measure of the quantity of light (luminous flux) emitted by a source.

(Luminous) Efficacy Measurement of how effective a light source is at converting energy into lumens of visible light; measured in lumens-per-watt (lm/W).

Lux (lx) The unit of light (illuminance) falling onto a surface; 1 lux = 1 lumen per square meter.

LED Light-emitting diodes, a semiconductor light source used in solid-state lighting (SSL) devices.

References

American Medical Association. (2016, June 14). *AMA adopts guidance to reduce harm from high intensity street lights*. Retrieved September 10, 2016, from https://www.ama-assn.org/ama-adopts-guidance-reduce-harm-high-intensity-street-lights

Bogard, P. (2013). *The end of night: Searching for natural darkness in an age of artificial light*. New York: Back Bay Books.

Boomsma, C., & Steg, L. (2012). Feeling safe in the dark: Examining the effect of entrapment, lighting levels, and gender on feelings of safety and lighting policy acceptability. *Environment and Behavior, 46*(2), 193–212.

Brox, J. (2014). Out of the dark: A brief history of artificial light in outdoor spaces. In J. Meier, U. Hasenöhrl, K. Krause, & M. Pottharst (Eds.), *Urban lighting, light pollution and society* (pp. 13–29). New York: Taylor & Francis.

De Almeida, A., Santos, B., Paolo, B., & Quicheron, M. (2014). Solid state lighting review - Potential and challenges in Europe. *Renewable and Sustainable Energy Reviews, 34*, 30–48.

de-Shalit, A. (2003). Philosophy gone urban: Reflections on urban restoration. *Journal of Social Philosophy, 34*(1), 6–27.

Dunn, N. (2016). *Dark matters: A manifesto for the Nocturnal City*. Winchester: Zero Books.

Edensor, T. (2015). The gloomy city: Rethinking the relationship between light and dark. *Urban Studies, 52*(3), 422–438.

Edensor, T. (2017). *From light to dark: Daylight, illumination, and gloom*. Minneapolis, MN: University of Minnesota Press.

Ekirch, R. A. (2005). *At day's close: Night in times past*. New York: W.W. Norton & Company.

Falchi, F., Cinzano, P., Duriscoe, D., Kyba, C. C., Elvidge, C. D., Baugh, K., et al. (2016). The new world atlas of artificial night sky brightness. *Science Advances, 2*(6), 1–25.

Falchi, F., Cinzano, P., Elvidge, C., Keith, D., & Haim, A. (2011). Limiting the impact of light pollution on human health, environment and stellar visibility. *Journal of Environmental Management, 92*, 2714–2722.

Fotios, S., Unwin, J., & Farrall, S. (2015). Road lighting and pedestrian reassurance after dark: A review. *Lighting Research & Technology, 47*, 449–469.

Gandy, M. (2017). Negative luminescence. *Annals of the American Association of Geographers, 107*(5), 1090–1107.

Gaston, K., Davies, T., Bennie, J., & Hopkins, J. (2012). Reducing the ecological consequences of night-time light pollution: Options and developments. *Journal of Applied Ecology, 49*, 1256–1266.

Gaston, K., Gaston, S., Bennie, J., & Hopkins, J. (2015). Benefits and costs of artificial nighttime lighting on the environment. *Environmental Reviews, 23*, 14–23.

Gehl, J. (2010). *Cities for people*. Washington, DC: Island Press.

Haans, A., & de Kort, Y. (2012). Light distribution in dynamic street lighting: Two experimental studies on its effects on perceived safety, prospect, concealment, and escape. *Journal of Environmental Psychology, 32*, 342–352.

Hölker, F., Moss, T., Griefahn, B., Kloas, W., Voigt, C., Henckel, D., et al. (2010). The dark side of light: A transdisciplinary research agenda for light pollution policy. *Ecology and Society, 15*(4), 13.

IDA. (2016, June 21). *The promise and challenges of LED lighting: A practical guide.* Retrieved May 2, 2017, from International Dark-Sky Association: http://www.darksky.org/wp-content/uploads/bsk-pdf-manager/IDA_LED_handout_48.pdf

IDA-IES. (2011). *Model lighting ordinance (MLO) with user's guide*. Tucson, AZ: International Dark-Sky Association.

International Energy Agency. (2006). *Light's labour's lost: Policies for energy-efficient lighting*. Paris: OECD/IEA.

Juntunen, E., Tetri, E., Tapaninen, O., Yrjana, S., Kondratyev, V., Sitomaniemi, A., et al. (2015). A smart LED luminaire for energy savings in pedestrian road lighting. *Lighting Research & Technology, 47*, 103–115.

King, R. (2000). Environmental ethics and the built environment. *Environmental Ethics, 22*, 115–131.

Koslofsky, C. (2011). *Evening's empire: A history of the night in early modern Europe*. Cambridge: University of Cambridge Press.

Kummitha, R. K., & Crutzen, N. (2017). How do we understand smart cities? An evolutionary perspective. *Cities, 67*, 43–52.

Kyba, C., Hänel, A., & Hölker, F. (2014). Redefining efficiency for outdoor lighting. *Energy & Environmental Science, 7*, 1806–1809.

Kyba, C. C., Kuester, T., Sanchez de Miguel, A., Baugh, K., Jechow, A., Hölker, F., et al. (2017). Artificially lit surface of Earth at night increasing in radiance and extent. *Science Advances, 3*(e1701528), 1–8.

Lerner, J. (2014). *Urban acupuncture: Celebrating pinpricks of change that enrich city life* (M. Margolis, P. Muello, & A. Daher, Trans.). Washington, DC: Island Press.

Li, Y., Ma, W., Kang, Q., Qiao, L., Tang, D., Qiu, J., et al. (2015). Night or darkness, which intensifies the feeling of fear? *International Journal of Psychophysiology, 97*, 46–57.

Longcore, T., & Rich, C. (2004). Ecological light pollution. *Frontiers in Ecology and the Environment, 2*(4), 191–198.

Lyytimäki, J. (2013). Nature's nocturnal services: Light pollution as a non-recognised challenge for ecosystem services research and management. *Ecosystem Services, 3*, e44–e48.

Major, M. (2017). The colour of the night - light and darkness in the city. In J. Best (Ed.), *Colour design: Theories and applications* (2nd ed., pp. 371–395). Amsterdam: Elsevier.

Marchant, P. (2004). A demonstration that the claim that brighter lighting reduces crime is unfounded. *The British Journal of Criminology, 44*(3), 441–447.

Murthy, A., Han, D., Jiang, D., & Oliveira, T. (2015). Lighting-enabled smart city applications and ecosystems based on the IoT. In *2015 IEEE 2nd world forum on internet of things* (pp. 757–763). Piscataway, NJ: IEEE.

Nagenborg, M. (2018). Urban robotics and responsible urban innovation. *Ethics and Information Technology*. https://doi.org/10.1007/s10676-018-9446-8.

NAS. (2017). *Assessment of solid-state lighting, phase two. Committee on Assessment of Solid-State Lighting; Board on Energy and Environmental Systems; Division on Engineering and Physical Sciences; National Academies of Sciences, Engineering, and Medicine*. Washington, DC: The National Academies Press.

Nye, D. E. (1990). *Electrifying America: Social meanings of a new technology, 1880–1940*. Cambridge, MA: The MIT Press.

Nye, D. E. (2010). *When the lights went out: A history of blackouts in America*. Cambridge, MA: The MIT Press.

Papworth, S., Rist, J., Coad, L., & Milner-Gulland, E. (2009). Evidence for shifting baseline syndrome in conservation. *Conservation Letters, 2*, 93–100.

Pawson, S., & Bader, M. (2014). LED lighting increases the ecological impact of light pollution irrespective of color temperature. *Ecological Applications, 24*(7), 1561–1568.

Pena-Garcia, A., Hurtado, A., & Aguilar-Luzon, M. (2015). Impact of public lighting on pedestrians' perception of safety and Well-being. *Safety Science, 78*, 142–148.
Sadowski, J., & Pasquale, F. (2015). The spectrum of control: A social theory of the smart city. *First Monday, 20*(7), 1–22.
Schivelbusch, W. (1988). *Disenchanted night: The industrialization of light in the nineteenth century* (A. Davis, Trans.). London: University of California Press.
Schlör, J. (1998). *Nights in the big city: Paris, Berlin, London 1840–1930.* (P. G. Imhof, & D. R. Roberts, Trans.) London: Reaktion Books.
Schoer, S., & Hölker, F. (2017a). Impact of lighting on flora and fauna. In R. Karlicek, C.-C. Sun, G. Zissis, & R. Ma (Eds.), *Handbook of advanced lighting technology* (pp. 957–989). Cham: Springer.
Schoer, S., & Hölker, F. (2017b). Light pollution reduction: Methods to reduce the environmental impact of artificial light at night. In R. Karlicek, C.-C. Sun, G. Zissis, & R. Ma (Eds.), *Handbook of advanced lighting technology* (pp. 991–1010). Cham: Springer.
Stilgoe, J., Owen, R., & Macnaghten, P. (2013). Developing a framework for responsible innovation. *Research Policy, 42*, 1568–1580.
Stone, T. (2017). Light pollution: A case study in framing an environmental problem. *Ethics, Policy & Environment, 20*(3), 279–293.
Stone, T. (2018a). Re-envisioning the nocturnal sublime: On the ethics and aesthetics of nighttime. *Topoi*. https://doi.org/10.1007/s11245-018-9562-4.
Stone, T. (2018b). The value of darkness: A moral framework for urban nighttime lighting. *Science and Engineering Ethics, 24*(2), 607–628.
Stone, T., Santoni de Sio, F., & Vermaas, P. (2020). Driving in the dark: Designing autonomous vehicles for reducing light pollution. *Science and Engineering Ethics, 26*(1), 387–403.
van den Hoven, J. (2013). Value sensitive design and responsible innovation. In R. Owen, J. Bessant, & M. Heintz (Eds.), *Responsible innovation: Managing the responsible emergence of science and innovation in society* (pp. 75–83). West Sussex: Wiley.
van den Hoven, J., Lokhorst, G.-J., & van de Poel, I. (2012). Engineering and the problem of moral overload. *Science and Engineering Ethics, 18*, 143–155.
Verbeek, P.-P. (2011). *Moralizing technology: Understanding and designing the morality of things*. Chicago, IL: University of Chicago Press.
Wines, J. (2005). The art of architecture in the age of ecology. In D. Brown, M. Fox, & M. R. Pelletier (Eds.), *Sustainable architecture white papers* (pp. 12–18). New York: Earth Pledge.

Taylor Stone is currently cross-appointed as a Lecturer and Postdoctoral Researcher in Ethics of Technology at Delft University of Technology. He received his PhD in Ethics of Technology from Delft University of Technology and the 4TU.Centre for Ethics and Technology, and was previously a Postdoctoral Researcher in TU Delft's Department of Industrial Design. His research explores the convergence of philosophy and design, with an emphasis on cities and urban infrastructure. Within this larger field, he specializes in the ethics and aesthetics of nighttime lighting.

Chapter 12
Shape Grammar Systems as a Technology for Flexible Design for Values in Cities: Giving Architectural Design to Inhabitants

Pieter E. Vermaas and Sara Eloy

Abstract In this chapter we consider computer tools for architectural design based on shape grammar design systems, and evaluate the advantages and disadvantage of handing over these tools to inhabitants for the design of their apartments. This evaluation is qualitative by considering the values of inhabitants, architects, and cities that are affected by this hand-over. Shape grammar design systems when applied to architecture enable computer tools to generate new designs and adjustments of existing designs of apartments on the basis of design rules. First, focussing on individual housing, it is argued in this chapter that the use of shape grammar design systems by inhabitants for designing their apartment realises the values of customisation and inhabitant autonomy without compromising design quality and structural safety. Second, focussing on cities, it is argued that a wide-spread use of these grammar design systems creates for cities a bottom-up inhabitant-led renewal of neighbourhoods and urban structures that maintains the social fabric in neighbourhoods and the architectural identity of cities. Values that are less realised by the hand-over are financial feasibility and ecological sustainability. Third, shifting to philosophy of technology, we present shape grammar design systems as a technology for design for values in architecture that allows for flexibility in the values designed for over the longer time periods characteristic to architecture. These systems are a technology that enables the incorporation of values of inhabitants, architects, and cities by the designs of apartments, and they are a technology that enables inhabitants to adjust their apartments through time to their developing needs and wishes.

P. E. Vermaas (✉)
Department of Philosophy, Delft University of Technology, Delft, The Netherlands
e-mail: p.e.vermaas@tudelft.nl

S. Eloy
Department of Architecture and Urbanism, Instituto Universitário de Lisboa (ISCTE-IUL), ISTAR-IUL, Lisbon, Portugal
e-mail: sara.eloy@iscte-iul.pt

M. Nagenborg et al. (eds.), *Technology and the City*, Philosophy of Engineering and Technology 36, https://doi.org/10.1007/978-3-030-52313-8_12

Keywords Shape grammar systems · Architectural design · Design for values · Customisation of apartments · Autonomy of inhabitants

12.1 Introduction

In this chapter we analyse how computer tools for architectural design, when handed over to inhabitants, can realise a number of values of these inhabitants, of architects, and of cities. The value central to this hand-over is customisation of housing, yet the impact goes beyond customisation and involves also inhabitants' values such as autonomy and the structural safety of their housing. A large-scale use of these computer tools can moreover realise values of cities such as the social fabric of neighbourhoods and city identity. Other values are in turn less realised or maybe even not realised, such as urban architectural innovation and the authority of architects. Computer tools for architectural design can thus have a profound impact on inhabitants, architects, and cities. Moreover, these computer tools can materialise the vision of design for values as currently explored in philosophy of technology. The analysis we present in this chapter is about these two promises of computer tools for architectural design. And we introduce shape grammar design systems as the technology that can create these computer tools.

Architects have already for decades researched possibilities for realising the value of customisation by creating diversified designs for housing that satisfy the individual needs and aspirations of inhabitants. In 1933 Le Corbusier presented his Plan Obus for Algiers which included a monumental spiralling building looking like a shell on which highways would be built on the roofs and in which houses would be constructed by the inhabitants, as in a medina. In the same period other modernist architects proposed housing solutions that were attempts to give more freedom to inhabitants. Mies van der Rohe, for instance, proposes in 1927 for the Weissenhof neighbourhood in Stuttgart a housing building that by using a long continuous glass façade enables an open and flexible use of the interior space not constrained by predefined walls. Both solutions enable inhabitants to design and redesign their apartment layouts according to their needs without having to obey to a strict common building layout. Later, during the 1960s architects reacted against the large-scale development of mass-produced homogeneous residential buildings. The architect H.P. Berlage, for instance, characterised the standardised uniform housing that comes with those buildings as an assault upon the inhabitants, upon their "personality, upon their freedom, upon their humanity; this kind of housing turns one into a herd-animal, a serf, a dependent" (Habraken, 1972, p. 2). Several attempts were made in architecture to promote customised housing diversity and the solutions presented included aesthetically appealing ones, such as the utopic cells units of Archigram and Habitat 67 by Moshe Safdie. Other attempts followed the Open Building approach by John Habraken (1972), the mobile architecture approach by Yona Friedman (Friedman, 1958; Tan, 2008), or the incremental housing approach promoted by Reinhard Goethert (Goethert, 2010) of which the Elemental solutions for social housing (Aravena & Iacobelli, 2016) are a contemporary successful example. As a more radical extreme we may find Friedensreich Hundertwasser (1958) with his *Manifesto Against Rationalism in Architecture*, in which he argues that the apartment-house inhabitant "must [..] be allowed to cut up the walls and make all kinds of changes, even if this disturbs the architectural harmony of a so-called masterwork, and he must be able to fill his room with mud or children's modelling clay".

After the failed attempts to provide quality mass housing in the 1960s Habraken concluded that a return to the involvement of users was mandatory and had to be accepted by architects. According to Habraken, the multiple inhabitants' initiatives to construct, improve, or change their original houses are the primitive expression of their energy and their wishes to illustrate their position in life and establish their ego (Habraken, 1972, p. 14). To that Henry Sanoff added that enabling inhabitants to participate in the design process of their environment assures its success and has benefits such as inhabitant empowerment, increasing social capital, and promoting a sense of community (Sanoff, 2008). For bringing the Open Building approach further, participatory design was taken on board so that inhabitants could actively contribute to the design of their houses. By enabling participatory design in housing it is possible to provide inhabitants with greater flexibility in choosing and changing their living environment according to their changing needs and therefore achieve housing solutions that bring more satisfaction for them.

The work by Habraken in the 1960s on the Open Building approach has influenced many architects to address diversity in design responding to the individuality of the users. From the 1960s onward pioneering projects were completed in countries like the Netherlands, the UK, and Japan (Bosma, van Hoogstraten, & Vos, 2000; Kendall & Teicher, 2000). Despite the innovation of the Open Building approach and the success of these projects, several of them have remained at proposal stages, and the ones built are few. The reality of housing construction shows that Open Building is hardly used in mass housing architecture.

With the advent of increasingly more sophisticated computer tools for design, the search for customisation of housing design through the active participation of inhabitants gained new momentum. Although most computer-based design systems are designed for architects, there are solutions which target the final users of the designed spaces. Tools like Sweet home,[1] Roomsketcher,[2] and IKEA design tool[3] are online available for inhabitants to help them plan their houses. Such tools enable inhabitants to design themselves and explore how to customise their houses to their needs and wishes. The designs these tools generate do not yet comply either with architectural good practices or with building regulation, and therefore somewhat mislead users. The next step in making these tools available is therefore including

[1] See http://www.sweethome3d.com/

[2] See http://planner.roomsketcher.com/

[3] See http://www.ikea.com/

more architectural knowledge such that inhabitants indeed can take part in the housing design. ModRule[4] may be taken as such a tool, although still in a research stage (Lo, Schnabel, & Aydin, 2016). And HouseMaker© is a vision developed by the architectural office MVRDV that all together hands over the design of houses to the future inhabitants.

This brief description of the search for customisation in architecture suggests that it are computer tools for designing that may eventually realise it. These technologies moreover will not only bring to cities housing that is meeting the individual needs and aspirations of inhabitants, but will also realise values such as inhabitant empowerment and sense of community. It is this possibility that we analyse in this chapter and we broaden it in two ways. First, we extend the analysis from the construction of new housing to also the renewal of existing housing. Housing in cities is indeed created through new construction but also by the renewal of large stocks of older housing buildings, and the vision of customisation applies equally to that renewal. Second, we focus in the analysis on a generic technology for creating computer tools for customised designing, namely shape grammar design systems. Shape grammar systems are a design technology that, when applied to architecture, enables to generate new designs and adjustments of existing designs on the basis of design rules implemented in computer software. And by being based on design rules shape grammar systems and the design tools that can be created with them, can incorporate architectural good practices and building regulation.

As with any architectural design tool, the regular perspective on shape grammar systems is to take them as tools for architects. Architects then can create and use shape grammar systems for generating different options to design tasks, for exploring these tasks, and for possibly selecting one of those options for further development. On this perspective the architects remain to keep the sole authority to design, and shape grammar systems are subordinated to that authority along all the stages of the design process. The alternative perspective is to take shape grammars as systems that can generate designs within the boundaries defined by the architect who created the design system but independently of his or her guidance. This alternative opens the possibility to offer these systems as stand-alone design systems by which inhabitants can themselves adapt their apartments. This chapter focuses on this alternative use of shape grammar design systems and explores the impact on inhabitants, architects, and the city as a whole when in this way architectural design is given to inhabitants.

After introducing shape grammar design systems in Sect. 12.2 we consider in Sect. 12.3 their impact on inhabitants in the case of the design of individual apartments. We articulate the advantages and disadvantages of handing over this design to inhabitants in the case of new housing and in the case of the renewal of existing housing. Then we move in Sect. 12.4 to the higher urban level of neighbourhoods and consider the advantages and disadvantages for cities of handing over housing design to inhabitants. We again focus on the construction of new neighbourhoods

[4] See http://modrule.com/session/login

and on the renewal of existing neighbourhoods. The evaluation of the impact of the use of shape grammar systems as stand-alone design systems for inhabitants is qualitative. We compare this use relative to the main extant alternatives, and focus in the comparisons on the values of inhabitant, architects, and cities.

By our analysis shape grammar design systems are a technology by which values are realised through the construction and renewal of housing and neighbourhoods in cities. Shape grammar design systems can therefore be identified as a technology that fits the *design for values* vision as currently developed in philosophy of technology. By this vision our values—not only practical and economical, but also ethical and societal ones—are to be realised by designing them into products, technologies, and services (Friedman, Kahn Jr., & Borning, 2006; Van den Hoven, Vermaas, & Van de Poel, 2015). Shape grammar design systems present, moreover, a case of a design for values technology that allows for flexibility in the values designed for over the longer time periods characteristic to architecture. As such they can deal with two concerns about the usefulness of the design for values vision in architecture, as voiced in, for instance, (Schrijver, 2015). These concerns, discussed in Sect. 12.5, are that architects typically design housing for longer time periods in which the values of inhabitants may change, and that inhabitants are inventive in countering any values that architects incorporate in housing.

In Sect. 12.6 we give our conclusion about the impact that shape grammar design systems can have on inhabitants, architects, and cities. They can allow for customisation and value flexibility in architecture since they enable inhabitants to adjust their apartments to their developing needs and wishes. And they can create for cities a bottom-up inhabitant-led renewal of neighbourhoods and urban structures that maintains the social fabric in neighbourhoods and the architectural identity of cities.

We employ two running cases in this chapter that we introduce in Sect. 12.2. The first is based on the vision of HouseMaker© by MVRDV; assuming that this vision can be implemented by shape grammar systems, we present HouseGrammar as the hypothetical design system by which inhabitants can create the floorplans for apartments in newly constructed building complexes. The second case is the Rabo-de-Bacalhau Transformation Grammar, which is a shape grammar system developed for the refurbishment of *Rabo-de-Bacalhau* apartments built in the mid twentieth century in Lisbon.

12.2 Shape Grammar Design Systems

A shape grammar design system is a generative design system that uses a vocabulary of shapes and shape rules that are applied to shapes to generate design alternatives that comply with a design language. A shape grammar is a design system that aims at providing alternative design solutions to a design problem instead of just one solution. A design is generated with a shape grammar by beginning with an initial shape and recursively applying shape rules. And by choosing different rules to apply the grammar generates different designs (Fig. 12.1).

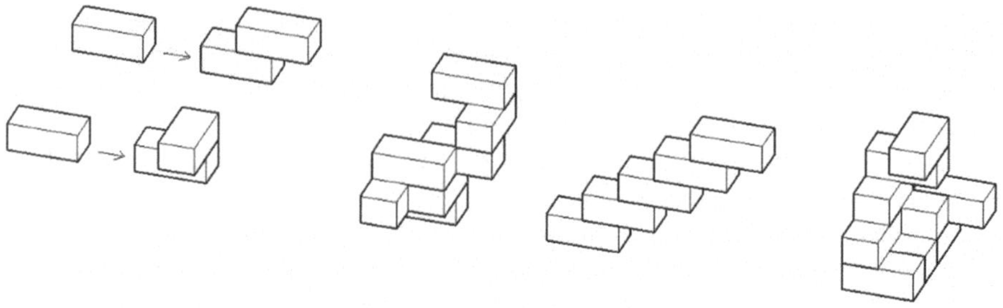

Fig. 12.1 Illustration of a shape grammar: the two shape rules (on the left) are used to generate the three designs (on the right) by applying one or the two rules in different orderings

Shape grammar systems were introduced in a seminal article by Stiny and Gips (1972), and since then these systems have been used in several fields such as painting, architecture, product design, and engineering. Shape grammar systems come in a number of flavours. First, *analytical grammars* are grammar systems that enable analysing and understanding existing design languages as shown in the work of Stiny and Mitchell (1978) on the villas by Andrea Palladio. With the Palladian grammar the original villas as well as new villas complying with Palladio's style can be generated. Since then, numerous analytical grammars have been developed for the work of several architects, such as the prairie houses of Frank Lloyd Wright (grammar by Koning and Eizenberg (1981)), the Malagueira housing complex of Álvaro Siza Vieira (grammar by Duarte (2005)), and domestic architecture of John Portman (grammar by Ligler and Economou (2015)). For delivering such a grammar a group of designs (a corpus) is studied and rules are inferred from them. Second, *original grammars*, illustrated by Stiny (1980) with Froebel's building gifts, enable to define new design languages from scratch. With such grammars designers may define their language of design and apply it to new design problems. The application of original grammars is less known but nevertheless used in architecture, especially when mass customised housing is envisioned, as in shelter housing (Benrós, Granadero, Duarte, & Knight, 2011), and new wood houses (Kwiecinski, Santos, Almeida, Taborda, & Eloy, 2016).[5] Third, *transformation grammars* are grammar systems that deal with transforming existing designs into new ones that comply with new design requirements. Examples of this type of grammars are the ones that transform existing house layouts into new ones.[6] A transformation grammar for housing includes shape rules and conditions that meet design requirements (e.g., interval of recommended areas, room dimensions, and recommended connections between rooms) thus incorporating architectural knowledge and expertise in the new designs.

For delivering architectural designs, grammars have to deal both with shape (geometry and syntax) and conditions (meaning and semantic) so that the social

[5] More examples are (Economou, 2000; Knight, 2000).

[6] See (Colakoglu, 2005; Eloy, 2012; Coimbra & Romão, 2013).

purpose of architecture is taken into account. According to Terry Knight (2000) the constraints of a design problem (e.g., site, economic, construction, or functional requirements) and the designer's style or design philosophy are added to the grammar and determine the design alternatives it creates. The rules for a shape grammar design system for a specific type of architectural task are derived from the design practices of architects, and in this sense capture their design knowledge and expertise. By using a shape grammar we can define a design system that generates house layout solutions for given design requirements. With shape grammar design systems design alternatives that respond to *a priori* requirements can be generated and users can choose between them. The outcome design solutions capture what is known in architecture from building codes to design languages. As shown in recent work by Sara Eloy, Pieter Vermaas, and Marina Andrade shape grammar design systems can generate solutions with a design quality comparable of that of designs created by professional architects (Eloy, Vermaas, & Andrade, 2017).

The regular perspective on shape grammar design systems is, as said in the introduction, to see them as design tools by which architects can generate different options to design tasks, can explore these tasks, and possibly select one of those options. And as tools for architects, their value lies, among others, in the possibility of creating large quantities of housing alternatives within a single design. For larger projects both for the private market and for social housing this possibility would enable diversity of apartments and thus social diversity and inclusiveness for inhabitants. In this chapter we explore the ramifications of the alternative perspective to also take shape grammars as stand-alone design systems by which inhabitants can themselves adapt their apartments, or future apartments, independently of guidance by architects. The advantages of this alternative seem also clear. Inhabitants could use the shape grammars to customise their apartments with, for adjusting them to their evolving wishes and needs, as adding new sanitary facilities, including a room for a child, creating an office for work, or adjusting access and layout in case of impairments. Our exploration of taking shape grammars as stand-alone design systems for inhabitants focusses qualitatively on the advantages and disadvantages for the inhabitants, architects, and cities, as compare to the main extant ways for designing the renewal of housing and apartments. We use two running cases of shape grammar systems—the Rabo-de-Bacalhau transformation grammar (Eloy, 2012) and the HouseMaker© vision presented by the architectural office of MVRDV (MVRDV and Axis.fm, 2012)—which we now introduce.

12.2.1 First Case: HouseGrammar

The known Dutch architectural office MVRDV is for long concerned with how architects can "handle the endless individual desires of prospective homebuyers" (MVRDV and Axis.fm, 2012). To respond to this need MVRDV proposes the development of a tool that helps homebuyers to design their own dream houses. The HouseMaker© (MVRDV and Axis.fm, 2012) is an attempt to develop a software

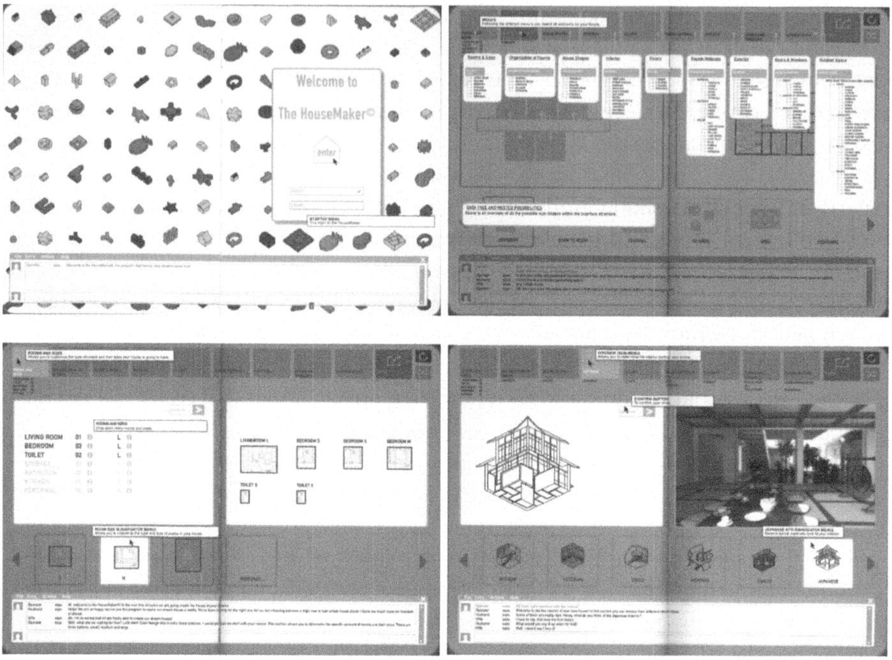

Fig. 12.2 Images of the HouseMaker© interface: opening menu, choosing the type of rooms, customising the rooms, choosing the interior style of the house (MVRDV & Axis.fm, 2012)

program that could address the many requirements of prospective homebuyers. The aim was to develop software that categorises all the elements of a house and offers a user-friendly interface with which inhabitants could customise their future houses regarding materials, shapes, rooms, sizes, layout, etc. (Fig. 12.2). HouseMaker© also allows the user to design functionality, spaciousness (e.g., small, medium, and large rooms), expressiveness (e.g., Modern, Victorian, Swiss, Chalet, Japanese), and luxuriousness, while simultaneously visualising the result and the costs involved. The HouseMaker© interface enables the inhabitants to customise their houses by selecting their preferences and visualise them directly as the house is virtually being generated. HouseMaker© was designed to be available online and includes a dialogue box where multiple users, say, e.g., all family members, can interact simultaneously with an operator or to a friend on social media.

It is our assumption that the HouseMaker© vision can be implemented with shape grammar technology. The styles the prospective inhabitants of a building complex favour (e.g., Modern, Victorian, etc.) can be captured in rules, including more contemporary styles the architects of the complex may propose. Moreover, safety and other regulation concerning escape route, physical accessibility to buildings, and technical equipment can be implemented. As mentioned before such a design system has advantages when used in large scale housing developments such as social housing and large private housing developments. As we will argue later the economic feasibility of using such a system requires that a large group of housing is to be designed independently of the economic or social status of their future inhabitants. We will use *HouseGrammar* for referring to this implementation, noting that this implementation also has at this point merely the status of a vision.

HouseGrammar could, for instance, be made available as an online tool such that inhabitants could easily access it for, first, inputting the needed data and, then, obtaining design solutions for their apartments. Such an online tool could be made available through, e.g., the website of housing corporations or other housing developers. As envisioned such an interface would not require any technology expertise from the inhabitant point of view and could be as easy to use as, e.g., the IKEA design tool. The availability of such a shape grammar design system to inhabitants would imply that they can design their apartments individually and according to their wishes while still part of a multifamily building.

12.2.2 Second Case: The Rabo-de-Bacalhau Transformation Grammar

The Rabo-de-Bacalhau (RdB) transformation grammar is a design system for *Rabo-de-Bacalhau* apartments, which are apartments built in Lisbon from the 1930s to the 1950s (Fig. 12.3). The development of the shape rules of this transformation grammar was based on two sorts of data: data from building codes, building construction techniques, house design recommendations, and refurbishment solutions done by current inhabitants of *Rabo-de-Bacalhau* apartments; and data obtained from refurbishment design proposals for these apartments as delivered by

Fig. 12.3 Existing Rabo-de-Bacalhau apartments, photos of the street, façade, collective stairs, and interior of apartments

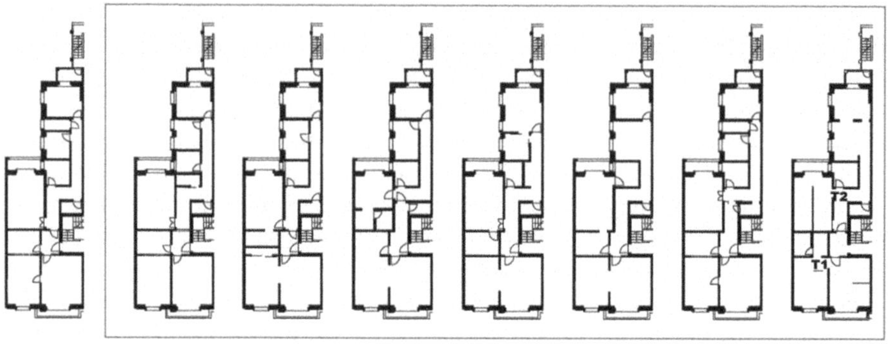

Fig. 12.4 Original floor plan of a Rabo-de-Bacalhau apartment (left) and solutions of refurbishment designs delivered for different households

several architects who were put 'into the mode of doing' as stated by Schön (1992). The system operates on an existing house layout using a set of functional requirements derived from the inhabitants' needs and wishes and using state of the art knowledge in architectural design. For meeting the functional requirements the system applies the shape rules in a given order to generate one or more design solutions that, at the end of the process, are presented to the inhabitants.[7] The RdB transformation grammar generates apartment layouts that are syntactically correct because the grammar follows the design language of *Rabo-de-Bacalhau* architecture, and that are semantically correct because the generated layouts meet the design requirements (Fig. 12.4).

The perspective on shape grammar systems that we here explore means making this shape grammar design system available to inhabitants and owners for enabling them to refurbish their *Rabo-de-Bacalhau* apartments themselves. As for HouseGrammar, the RdB transformation grammar could be made available as an online tool such that inhabitants could easily access it for, first, inputting the needed data and, then, obtaining design solutions for the refurbishment of their apartments. Such an online tool could be made available through, e.g., the website of the City Council of the city of Lisbon, or of any other city concerned, as a service to its citizens. A pipeline for the use of this online tool is sketched in Fig. 12.5.

The availability of such a shape grammar design system to inhabitants would imply that they can access expertise that inhabitants typically do not use when they refurbish their apartments themselves. Through the architectural knowledge included in the shape grammar design system, this availability would therefore contribute to a larger adoption of construction expertise in the individual refurbishment works leading to better quality solutions.

[7] A computer implementation of the RdB transformation grammar can be found in (Strobbe et al., 2016).

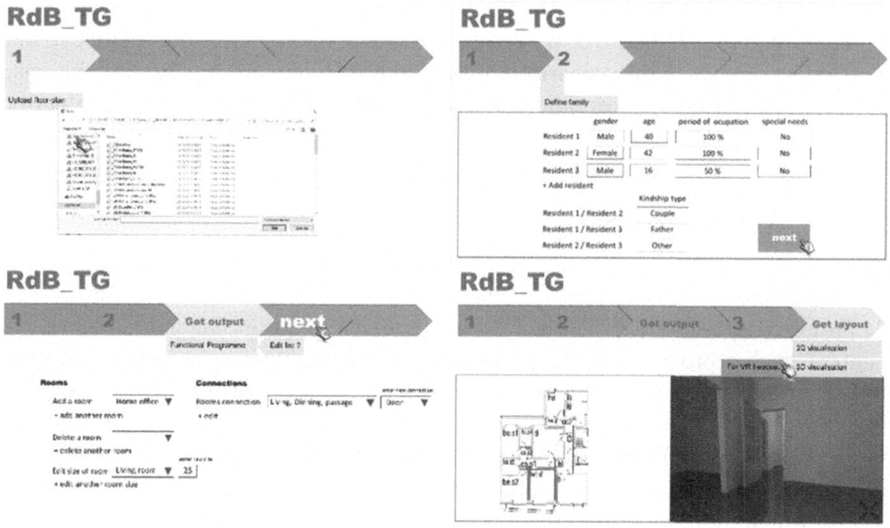

Fig. 12.5 Mock-up of the online system: the top screens and the left-hand bottom screen are for the input of data by the user, the right-hand bottom screen is the visualisation of the design solution with a floorplan and a 3D model that could be visualised with a VR headset

12.3 Inhabitants and Apartments

Consider now the possibility of handing over architectural design to inhabitants[8] in cities by making shape grammar systems available as stand-alone design systems. For new housing architects can deliver these grammar systems as part of their designs of the housing, say by HouseGrammar, such that the (future) inhabitants can determine the designs of their apartments according to their needs and wishes. For existing housing these grammar systems can be developed by architectural researchers, as the RdB transformation grammar was for Rabo-de-Bacalhau apartments, and inhabitants can then renew the design of their apartments in case their needs and wishes evolve.

We start our evaluation of this possibility by looking in this section at the advantages and disadvantages for inhabitants and architects in two scenarios: in the case of the construction of new housing buildings, and in the case of the renewal of existing housing. In the next section we broaden the evaluation by looking at the impact on cities when making stand-alone grammar design systems available to inhabitants.

[8] In this chapter we mean with inhabitants of an apartment the persons who use the apartment for living. In the case of an apartment that is designed to be constructed, these persons are more precisely characterised as the *future* inhabitants. Inhabitants of apartments may also be the owners of their apartments but owners of apartments – landlords, housing corporations, investors, cities – need not be their inhabitants. In this chapter inhabitants are our main focus, yet when it is relevant to our arguments, we introduce also owners.

Our evaluation of the impact of the use of shape grammar systems by inhabitants is qualitative and comparative. We compare for each scenario this use relative to the main extant options in the scenario. And we focus on the main values of inhabitants, architects, and cities that are realised and not realised by the options in each scenario.

12.3.1 New Apartments

Consider firstly the construction of a new housing building or complex with multiple projected apartments. The main extant approach to design the individual apartments in such a project is that the architects of the building or complex also give these apartment designs. In the extreme case all the apartments have the same standard design, yet regularly architects offer a series of different designs. The alternative to this extant approach is to not let architects design the apartments but to make a shape grammar design tool as HouseGrammar available to the prospective inhabitants by which they can design the apartments themselves. Hence in our evaluation of this alternative we distinguish two options for the design of new apartments:

Option 1. New construction design of individual apartments by architects.
Option 2. New construction design of individual apartments by inhabitants guided by shape grammar systems.

The central value in this chapter is the one of *customisation*. Other values relevant in the evaluation of the use of shape grammar design systems by inhabitants are *inhabitant autonomy*, *financial feasibility* and *architect authority*.[9] One can continue this list with more values relevant to inhabitants, cities, and architects, but for keeping the evaluation concise, we focus on only the main values that discriminate between the use of shape grammar design systems by inhabitants and the extant option 1. Hence, values as *design quality* and *structural safety* are of course important to the designs of new apartments, but assumed to be realised in both options.

[9]Autonomy is in this paper understood in the basic meaning of what in philosophy literature is called personal autonomy, i.e., the power of a person to act competently on the basis of his or her own needs and wishes. Making this meaning more precise, using the elaborate body of work in philosophy on this concept, is useful but lies outside the scope of this chapter. Yet we can note that in case of inhabitant autonomy the 'person' involved is regularly a social unit as a family, introducing questions about autonomy of groups. Furthermore, one can analyse the ability of inhabitants to determine their 'own' wishes and demands; for determining more functional needs such as shelter, heating and the number of sleeping places, this capacity may be taken as less problematic; whereas it may be criticised for more lifestyle related wishes as say the number of required bathrooms. The issue of the competence of inhabitants brings us to a second remark. We opt for considering the value of architect authority instead of architect autonomy because the hand-over of design tools to inhabitants affects primarily the way in which skills and knowledge of architects are employed; focusing of the autonomy of architects would frame the hand-over in a limiting way as a power struggle between inhabitants and architects.

Table 12.1 Inhabitant and architect values involved in new apartment construction

Inhabitant autonomy	The situation in which inhabitants have the power to decide about the apartment design, say about what their needs and wishes are, and about how these needs and wishes are accomplished in the new apartment
Customisation	The situation in which apartments are customised for inhabitants, say by offering different layouts for different family profiles or by architects customising the layout solution for their own individual apartment clients
Financial feasibility	The situation in which the costs of the construction or acquisition of apartment are kept low
Architect authority	The situation in which architects are the prime experts with the authority to design and re-design apartments, say since only architects have sufficient knowledge about apartment design

These latter two values will however surface as discriminative in the next subsection when we look at the renewal of existing apartments.

We characterise what we mean with these values by describing the situation, or state of affairs, in which they are realised (by this we admittedly bypass a more careful analysis of what values are). In Table 12.1 we do so for the values in our comparison of the two options for the design of new apartments.

When assessing the two options for new apartment design with respect to these four values, the following qualitative arguments can be given.

The value of *inhabitant autonomy* is clearly compromised in option 1. When architects design new housing buildings and their apartments, they typically do so with an anonymous final inhabitant in mind (when it is private development) or a group of standard inhabitants (say, when housing corporations lead the development). In both cases, the final layout solutions of individual apartments are typically not created with direct participation of the inhabitants who will live in the apartments. Hence, with option 1 inhabitant autonomy is not realised. When however the apartment designs are created by a shape grammar system, the final decision about individual apartment layouts is given to the inhabitants. Although these decisions will be constrained by the overall building plan as, e.g., position of the apartment entrance, position of the loadbearing structure, and acceptable positions of the wet areas (bathrooms and kitchens), a shape grammar system will empower inhabitants to have autonomy within the apartment design. Option 2 thus realises the value of *inhabitant autonomy*.

The value of *customisation* may be addressed in several ways in option 1. One way is that architects design buildings with one standard apartment, thus not realising the value of customisation. Or architects can include some diversity in their solutions, say four or six different types of apartments meeting different inhabitant profiles. In that latter case some customisation is realised in option 1. Yet, when inhabitants guided by a shape grammar system can themselves create individual design solutions, as in option 2, the apartments will fit their needs and wishes more fully. This option is thus fully realising the value of *customisation*.

The value of *financial feasibility* is guaranteed by option 1 since costs of the construction of the apartment designs can be kept lower by the scale advantages of

Table 12.2 Comparison of the options for new apartment construction; '+' means realised, '–' means not realised, '±' means partly realised, and '?' means uncertain whether realised

	Option 1: new apartment design by architects	Option 2: new apartment design by grammars
Inhabitant autonomy	–	+
Customisation	?	+
Financial feasibility	+	?
Architect authority	+	±

building numerous similar apartments. Option 2 adds in contrast the costs of creating the shape grammar. The diversity of solutions that can be adopted by inhabitants may make the apartments cheaper when inhabitants choose for simple designs, yet it may also lead to additional costs for inhabitants if more expensive solutions are chosen. Additionally, the scale effect is lost in option 2 and therefore the value of *financial feasibility* is not fully realised in that option.

When architects design the new apartments (option 1), their *architect authority* is clearly realised. For option 2 the assessment is less black-and-white. When the apartment designs are generated by shape grammar systems, architects are not directly involved *on site* in the actual projects, which may be interpreted as that the architects are excluded and their authority in fixing designs is shifted to the inhabitants. Yet, the shape grammars involved in option 2 are to be created with architectural knowledge by the architects as part of their design of the new housing buildings. Hence, an alternative perspective on this option is that architects still are having control over the shape grammars and relay by the grammars their expert knowledge to inhabitants. When this perspective is accepted, architects still exercise an indirect authority over the apartment design by the inhabitants.

When summarising these arguments, the following overall picture emerges (see Table 12.2). The possibility of letting inhabitants use a shape grammar system to design new apartments brings advantages to these inhabitants since it realises their autonomy in deciding about their apartments as well as enables the apartments to be better customised to their needs. Yet this solution ranks less well in financial feasibility.

12.3.2 Renewed Apartments

Let us now analyse the scenario of renewing existing apartments. We again do so by comparing the use of shape grammar design systems by inhabitants in this scenario with the main extant alternatives, and again we give the comparison in terms of the main values of inhabitants and architects.

The main way in which inhabitants can renew individual apartments is by refurbishing them as inhabitants. And the two main extant ways for inhabitants to refur-

Table 12.3 Inhabitant and architect values involved in apartment renewal

Design quality	The situation in which an apartment meets the overall architectural design standards ranging from construction, habitability, user's satisfaction, safety, aesthetic, energy, and economy aspects of the design[a]
Structural safety	The situation in which the apartment does not collapse when subjected to the normal loads, and possible specified abnormal loads as say due to earthquakes

[a]Design quality value overlaps typically with other values, as say the value of structural safety, and financial feasibility, enabling a broader vision on architecture that is common when assessing quality in general.

bish their apartments are under the direction of an architect and without such direction, as when inhabitants design themselves the refurbishment, possibly with the assistance of constructers or other non-architectural professionals. In our evaluation we therefore compare the use of shape grammar systems for refurbishment with these two extant ways, i.e., we compare the following three options:[10]

Option 1. Refurbishment design of individual apartments by inhabitants directed by architects.
Option 2. Refurbishment design of individual apartments by inhabitants directed by themselves.
Option 3. Refurbishment design of individual apartments by inhabitants guided by shape grammar systems.

The value of *customisation* is again central. Another value that is involved is *modernisation*: the situation that apartments can be adjusted to meet new habitability standards as, say, better thermal insulation. These two values are realised in all the three options, hence not discriminative. We leave them therefore out of the evaluation. The relevant values that we submit as discriminating between the three options are *inhabitant autonomy*, *design quality* and *structural safety* of the apartments, *financial feasibility*, and *architect authority*. In Table 12.3 we add characterisations of these values as far as they are not yet given in Table 12.1.

When assessing the three options for apartment refurbishment with respect to these seven values, the following qualitative argument can be given.

The value of *inhabitant autonomy* is obviously realised when inhabitants refurbish their houses themselves (option 2) since then it are the inhabitants who decide about, say, knocking out a wall, changing the old wooden window panes with modern ones, or installing a kitchen with modern appliances. Turner's vision on

[10]A fourth option that can be discerned emerges when the inhabitant of an apartment is not the owner of the apartment. In that case the actual owner can decide about the refurbishment of the apartment, and that can be done two ways: either in collaboration with the inhabitant or independently of the inhabitant. In the first case the three options described in the main text again cover the main ways in which the refurbishment is done by the collaborating owner and inhabitant. The only change in the argument is that one should read "owner plus inhabitant" whenever it refers to the inhabitant. In the second case the owner fixes the refurbishment design for the inhabitant, which infringes on the value of inhabitant autonomy and which probably does not realise the value of customisation.

autonomy states that if inhabitants are free to make their own contribution and make decisions, their well-being increases (Turner, 1976). Availability of the means for such interventions or constraints imposed by constructers may limit possibilities, yet it remains inhabitants who decide about the refurbishment designs. When the refurbishment is supervised by architects (option 1) this freedom may become compromised. Architects respond to inhabitants' needs and wishes, but rather than giving a blind response, often reframe the brief and give inhabitants an interpretation of their needs and wishes. Although this is what can be expected from architects and can be seen as their professional skills that lead to good design solutions, inhabitants may feel that their wishes are not well responded to. With the use of stand-alone grammar design systems, inhabitants can again feel empowered because it are they who operate the systems and thus can generate and explore multiple refurbishment design from which they can select the one they wish. Hence, with option 3 *inhabitant autonomy* is realised also.

The values of *design quality* and *structural safety* are in turn typically realised when architects supervise refurbishment (option 1). When inhabitants lead refurbishment (option 2), there is no guarantee that aspects related to design quality, as building style and building regulations, are observed. This situation may lead to incoherence between the original architectural design style and the new solutions implemented by inhabitants. The lack of expert knowledge by inhabitants may additional lead to the implementation of layout design solutions that prove not to respond well to the inhabitants' needs. Regarding building regulations, namely structural security, inhabitants do not have the knowledge to opt for design solutions that maintain and improve the physical conditions of the buildings. The use of shape grammar systems for generating refurbishment designs (option 3) does realise these two values of *design quality* and *structural safety* since grammar designs automatically comply with building codes and design languages. And as shown (Eloy et al., 2017) for the RdB transformation grammar it can be maintained that the design quality of such designs is comparable of that of designs created by professional architects.

The guarantee of the previous values is often conflicting with the guarantee of the *financial feasibility* for inhabitants (Turner, 1976). Costs associated to architects ranging from hiring them for designing to letting them give licensing of the design, may compromise the financial feasibility of refurbishment for some inhabitants. This may lead to favouring a do-it-yourself approach (option 2) which would avoid at least architectural honorariums for design work that come with option 1. It can be argued that shape grammars (option 3) integrate the best of the two other options, since it also avoids costs of architectural design yet offers designs that incorporate architectural knowledge. It may even be envisaged that costs of licensing can be avoided since cities could in principle automatically license the refurbishment designs generated by shape grammars, as these grammar designs comply automatically with architectural regulations. Yet, this argument about option 3 holds only if the costs of making a shape grammar are left out of the equation. These costs can be substantial, which rules out financial feasibility of creating a shape grammar system for apartments of which only one or a dozen exist. When there are however large

Table 12.4 Comparison of the options for apartment renewal; the sign '+' means realised, '–' means not realised, '±' means partly realised, and '?' means uncertain whether realised

	Option 1: refurbishment by architect	Option 2: refurbishment by inhabitant	Option 3: refurbishment by grammar
Inhabitant autonomy	?	+	+
Design quality	+	?	+
Safety	+	?	+
Financial feasibility	–	+	?
Architect authority	+	–	±

numbers of apartments, as is the case for *Rabo-de-Bacalhau* apartments, the argument still holds since then the costs of creating the shape grammar can be divided over that large number.

When architects supervise refurbishment (option 1), their *architect authority* is realised, while when they are excluded (option 2) it is violated. For option 3 the assessment is again less black-and-white. Architects are within this option not involved in the actual projects, yet shape grammars are to be created by architectural researchers hence are the products of architectural expert knowledge.

Summarising these arguments in Table 12.4, the following overall picture emerges. The possibility of handing over architectural design to inhabitants in cities by making shape grammar systems available as stand-alone design systems for refurbishment, seems to integrate for inhabitants the best of both worlds: inhabitants can renew their apartments with architectural design quality and structural safety, as when architects support them, yet now with autonomy and low prices, as when inhabitants refurbish their apartments themselves. This possibility changes however the role of architects from supervisors of individual projects to enablers of low-priced and qualitatively good refurbishment in the city.

12.4 Cities and Neighbourhoods

When shape grammar systems indeed would widely find their way to inhabitants for designing their new apartments and for refurbishing their existing ones, an aggregate impact on cities may be expected. As we will argue shape grammar systems then emerge as a technology that allows bottom-up inhabitants-led ways to renew neighbourhoods and urban structures. Architecture could stimulate this impact by standardly including HouseGrammar design tools for inhabitants when designing new housing buildings. And cities could stimulate it by commissioning architectural researchers to create shape grammars for apartment types that are common within them. The city of Lisbon, for instance, has in addition to the *Rabo-de-Bacalhau* other types of apartments, as the *Pombalino* and *Gaioleiro* apartments, and cities such as London, Paris, Amsterdam, etc., have their own specific types of apartments. Architectural researchers can create transformation grammars for all these types, as the RdB transformation grammar was for Lisbon's *Rabo-de-Bacalhau* apartments.

In this section we look at the advantages and disadvantages for cities as a whole in the case that apartments are continuously built and refurbished by means of grammar systems. For exploring these advantages and disadvantages we again qualitatively compare this option with the main extant ways by which this renewal takes place in cities.

12.4.1 New Neighbourhoods

At the city scale the extant way in which new neighbourhoods or parts thereof are designed is by a centralised effort in which an overall plan is delivered for several buildings to the involved client, say the city council, a housing corporation, or a private investor. The individual apartments in the new buildings are by this first option designed by the architects who design the buildings. This first option is a centralised task of designing new housing buildings. In this option we include all new building designs by architects, whatever option of diversity they include, ranging from all equal buildings and apartments' solutions to the inclusion of some diversity in apartments layout solutions and buildings. With shape grammar design systems available a second option emerges in which the apartments are designed by the future inhabitants. This second option may occur in two situations. The first is when the client sets the apartments for sale or hire immediately before its construction therefore giving time for future inhabitants to use the shape grammars system to customise their house and send it to the construction company that will build the entire building. The second situation is when the building promoter builds the building say with the basic infrastructures, as the Open Building approach advocates, and leaves the individual apartments to be built using the shape grammar system by the inhabitant and the construction company they hire. In our evaluation we therefore compare the following two options:

Option 1. New neighbourhood construction with design of individual apartments by architects.

Option 2. New neighbourhood construction with design of individual apartments by inhabitants guided by shape grammar systems.

The values we introduce for comparing these two options are *social cultural diversity*, *ecological sustainability*, and again *financial feasibility*. We submit all of them as discriminative and characterise the meaning by which we use them in the comparison in Table 12.5.

The value of *social cultural diversity* is realised when the new construction of buildings enables to create social and cultural diversity within the neighbourhoods. The worst situation in option 1 for realising such a value is the one where architects

Table 12.5 City and architect values involved in new neighbourhood construction

Social cultural diversity	The situation in which neighbourhoods in cities enable social and cultural diversity of inhabitants
Ecological sustainability	The situation in which the ecological impact of the construction of or in neighbourhoods is minimal
Financial feasibility	The situation in which the costs of the construction or renewal of neighbourhoods are kept low

design building solutions where all the apartments are equal, leading to socially and culturally homogeneous neighbourhoods. In this case *social cultural diversity* is not realised. A better situation in option 1 is that the architects include different types of apartment designs, in which this value can be realised. Option 2 brings however the possibility that inhabitants can create with shape grammars design systems different apartments' layouts responding to different needs, wishes, and cultural practices, thus realising the value of *social cultural diversity*.

The value of *ecological sustainability* can be fully realised by option 1 since within centralised projects construction methods can be adopted that are state of the art in reducing the ecological impact of the construction. For option 2 the same result might be achieved or not. In terms of ecological sustainability option 2 will differ from option 1 by bringing new layout solutions to the apartments and new material options for the finishing of the apartments. Although layout solutions may not have a worse ecological sustainability performance than option 1, new material options may include less ecological sustainable ones and therefore option 2 compromises the realisation of the *ecological sustainability* value.

The value of the *financial feasibility* of the construction of new neighbourhoods can be realised by option 1. This option brings the scale advantages of building large numbers of similar apartments, which regularly reduces the construction costs per apartment. These advantages are however lost when many different apartments are built in the neighbourhoods.[11] This effect can in principle already emerge in option 1 if architects include lots of diversity in their designs (Ruby & Ruby, 2013). In option 1 this diversity can be avoided, yet it is part and parcel of option 2. Hence option 2 will typically be the more expensive one, also because it involves the costs of creating shape grammar systems. One may argue that option 2 allows the possibility to transfer these extra costs from the clients to the inhabitants. Hence this second option may be financially feasible for cities, private investors, or housing corporations. Yet, overall option 2 realises the value of *financial feasibility* less than option 1.

[11] The emergence of Digital Fabrication and the Do It Yourself (DIY) vision presents the possibilities of fabricating individual customised product through decentralising production and of addressing local problems with local solutions (Gershenfeld, 2005). Supporters of the Digital Fabrication approach challenge the assumption that prices decrease only with an increase in the scale of production; they argue that prices become lower too by changes in production lines with generative-design-systems and ready to fabricate equipment.

Table 12.6 Comparison of the options for new neighbourhood construction; '+' means realised, '–' means not realised, '±' means partly realised, and '?' means uncertain whether realised

	Option 1: new construction by architects	Option 2: new construction with grammars
Social cultural diversity	?	+
Ecological sustainability	+	?
Financial feasibility	±	–

Table 12.6 summarises the arguments. Both options respond to the values identified for the urban scale and they differ on the *social cultural diversity* and on the *financial feasibility* values. While option 1 performs less good than option 2 regarding *social cultural diversity* since it enables to create less diversity, option 2 performs less good regarding *financial feasibility* than option 1 since the development of a shape grammar system is expensive. This comparison can lead to the conclusion that cities should choose for option 1 in cases that the financial value is of most importance but option 2 should be chosen if financial constraints are not trumping the quality of the social and cultural diversity of cities.

12.4.2 Renewed Neighbourhoods

At the city scale the main ways in which apartments and buildings are renewed can be divided in centralised operations in which an overall plan is delivered for one or several buildings, and decentralised operations in which every single apartment renewal is made independently of each other. These centralised and decentralised ways of renewal can be further differentiated by considering the type of renewal (refurbishment or renovation) and the agents who direct the design of the renewal (architects or inhabitants). Not all combinations are regular, as, say, centralised renovation directed by inhabitants. In our evaluation we choose four regular options and compare them with the option of decentralised refurbishment of neighbourhoods by inhabitants using shape grammar systems for refurbishing their individual apartments:[12]

[12] When the inhabitant of an apartment is not the owner of the apartment, a number of variations of the given options can be discerned. In that case the owner can decide to renovate or refurbish an apartment building. When an owner renovates an apartment building, one has option 2. When an owner refurbishing the building in a centralised way with the help of an architect, one has option 1. When the owner refurbishes the building in a decentralised way listening to the wishes of the inhabitants, one has options 3 and 4 depending on whether an architect is involved. A new option may be defined when an owner fixes the refurbishment designs for the inhabitants without the expert help of an architect. We take this case as in instance of option 1 in which the values of design quality and safety is less well realised because no architect is involved.

Option 1. Neighbourhood renewal by centralised refurbishment of housing stock directed by architects.
Option 2. Neighbourhood renewal by centralised renovation of housing stock directed by architects.
Option 3. Neighbourhood renewal by decentralised refurbishment of individual housing by inhabitants directed by architects.
Option 4. Neighbourhood renewal by decentralised refurbishment of individual housing by inhabitants directed by themselves.
Option 5. Neighbourhood renewal by decentralised refurbishment of individual housing by inhabitants guided by shape grammar systems.

The first and second options are centralised efforts at renewal and consist of apartment buildings or whole housing blocks being refurbished as a whole in one project, or of demolishing whole buildings and blocks in the existing city fabric for building new housing. These two options are regularly carried out under the supervision of architects. Options 3 and 4 emerge through the first two options for apartment refurbishment as considered in the previous section: the third option is that in a neighbourhood inhabitants refurbish housing as directed by architects, and the fourth option is that inhabitants do such refurbishment by themselves. Options 3 and 4 lead both to decentralised renewal, since one may assume that the refurbishment of apartments within say a neighbourhood is under both options not substantially coordinated. The fifth and final option is the one that shape grammar systems add: urban renewal by inhabitants adjusting their apartments and housing in a decentralised way.

The values we introduce for making the comparison are the ones of *city identity*, *social fabric*, *urban architectural innovation*, and again *ecological sustainability* and *financial feasibility*. We submit all of them as discriminative and characterise the meaning of the newly introduced ones in Table 12.7.

The value of *city identity* is realised when the renewal has the shape of a refurbishment led by architects (options 1 and 3). In such cases architects opt to maintain the original architectural style while introducing some improvements in buildings. On this level of action Habraken (1972, p. 39) says that "we have to make possible the creation of districts which may grow old without becoming obsolete, which can absorb the latest ideas and yet have a sense of history. Districts in which the population can live for generations, and which yet incorporate the potential for change."

Table 12.7 City and architect values involved in neighbourhood renewal

City identity	The situation in which the neighbourhoods in cities maintain to fit the original architectural style that originally characterised them, say the canal houses in Amsterdam
Social fabric	The situation in which the neighbourhoods in cities maintain the social relations between their inhabitants, say the social composition and economic ties of the inhabitants
Urban architectural innovation	The situation in which cities can renew their neighbourhoods with new architectural solutions and styles, say contemporary design buildings and built infrastructures

Since shape grammars incorporate rules that capture the style of apartments, option 5 also realises the value of city identity. However, when inhabitants lead the refurbishment of their individual apartments (option 4) there is no guarantee that city identity is preserved. In fact, a decentralised refurbishment by inhabitants involves private interventions in, e.g., façades and other design elements of housing. Examples of such interventions are the individual substitution of window frames leading to a heterogeneous façade and the total substitution of the interior wood work that deprive the original apartments from their original characteristics. These interventions may damage the architectural identity of the apartments and buildings leading to the loss of the original identity of the city. Also in option 2, a centralised renovation by architects, the value of city identity may be compromised since when substituting old buildings architects do not imitate the old buildings but create new buildings by contemporary standards and styles. This practice, although architecturally correct, leads to the loss of the architectural specificity of cities as happened with the *Brusselisation* phenomenon in Brussels during the 1960s and 1970s when certain areas of the old urban tissue were destroyed to give place to new modern buildings (Romańczyk, 2012).

The value of *social fabric* may not be realised when centralised refurbishment and renovation (options 1 and 2) is performed on larger scales and is involving mass displacement of inhabitants and mass introduction of new inhabitants who usually have higher economic incomes. Centralised renewal may then lead to gentrification by eroding the original communities living in neighbourhoods and excluding specifically the older and economically less powerful (Smith & LeFaivre, 1984), and therefore has negative impacts on cities through segregation and social and health inequities (Causa Justa, 2009). Only when these centralised efforts are done in a more dispersed way and on smaller scale, the social fabric can be maintained in neighbourhoods. Yet, when refurbishment is decentralised (options 3, 4, and 5), and done per individual apartments, possible replacements of people are kept local, meaning that the value of social fabric is kept by maintaining the existing diversity.

We can say that the *urban architectural innovation* value is realised by the central renovation of neighbourhoods as conducted by architects (option 2) since this option, by pulling down the existing buildings and building new ones, creates opportunities for adopting contemporary design strategies and styles. When renewal is done by architects through refurbishment of buildings, centralised or decentralised, the existing buildings and urban structure are on average kept as they were. Architects then usually opt for maintaining the existing building style and may only add new infrastructures, as heating and telecommunication. Hence, options 1 and 3 allow for limited modernisation or innovation, but realise the value of *urban architectural innovation* only partly. Decentralised refurbishment by inhabitants (option 4) lacks expert knowledge of modernisation and innovation and therefore realises this value even less. Shape grammars (option 5) have the same performance as options 1 and 3 since only part of the architectural innovation may be realised.

There is an emerging recognition that renovation (option 2) is less ecological sustainable as compared to refurbishment. According to Power "in all but the cases where *de facto* rebuilding would be required, refurbishment is both cheaper and less

Table 12.8 comparison of the options for neighbourhood renewal; '+' means realised, '–' means not realised, '±' means partly realised, and '?' means uncertain whether realised

	Option 1: central refurbishment by architects	Option 2: central renovation by architects	Option 3: decentral refurbishment by architects	Option 4: decentral refurbishment by inhabitants	Option 5: decentral refurbishment by grammars
City identity	+	–	+	±	+
Social fabric	±	±	+	+	+
Urban architectural innovation	±	+	±	–	±
Ecological sustainability	+	–	?	?	?
Financial feasibility	+	+	–	±	±

damaging to the local and wider environment than demolition or new build" (Power, 2010). When refurbishment is done in a centralised way by architects (option 1) the *ecological sustainability* value is realised since not only the reuse of the existing buildings is achieved but also the possibility to do the refurbishment in an efficient way by the scale advantages of large projects. Decentralised refurbishment, however it is done (by architects, not-architects, or grammars), seems less sustainable than centralised efforts at renewal since advantages of scale and close monitoring of how projects are carried out seems to be absent. Therefore the realisation of the *ecological sustainability* value by options 3, 4, and 5 is a doubt and may differ according to several criteria.

In centralised efforts to renew neighbourhoods all the costs are divided over a large set of apartments being refurbished or renovated and therefore the costs are smaller because of the possibilities of standardisation and of scale advantages. Options 1 and 2 may therefore be judged as *financially feasible*. In renewal by decentralised refurbishment with architects (option 3), these architects have to be hired per apartment, leading to substantial costs that do not emerge if the refurbishment is led by inhabitants (option 4). The use of shape grammars (option 5), when assessed on the neighbourhood level, again avoids those architectural costs thus again making refurbishment more financially feasible.

Table 12.8 summarises the arguments. The two options that best respond to the values identified for the urban scale are option 1—central refurbishment by architects, and option 5—decentralised refurbishment by grammars. The other options all do not realise one of the considered values. What makes options 1 and 5 different is the realisation of three values. First, while in central refurbishment the *social fabric* of the neighbourhoods may be compromised, in decentralised refurbishment with grammars this value is guaranteed since it enables the social diversity to be maintained. Second, *ecological sustainability* is realised by central refurbishment while there are no guarantees that it is fully realised by letting inhabitants use shape grammars. Third, *financial feasibility* of refurbishment when using shape grammars may be less as compared to centralised refurbishment by architects. Handing over architectural design to inhabitants in cities by making shape grammar systems available as stand-alone design system may thus be seen as a feasible possibility for cities for the renewal of neighbourhoods. Our qualitative comparison of this possibility does not demonstrate that it is a possibility that is better than the regular ways in which this renewal is occurring now. But it is a possibility that has distinct qualitative advantages over renovation and decentralised refurbishment, and that competes with centralised refurbishment of neighbourhoods. And when one prioritises the value of the *social fabric* of neighbourhoods and, drawing from the analysis of Sect. 12.3, the value of the *autonomy* of inhabitants, it is the possibility that should be preferred.

12.5 Flexible Design for Values

By our analysis shape grammar design systems are a technology by which values are realised through the construction and renewal of housing and neighbourhoods in cities. Shape grammar design systems can therefore be identified as a technology that fits the *design for values* vision as currently developed in philosophy of technology (e.g., Friedman et al., 2006; Van den Hoven et al., 2015). The application of technologies is, of course, regularly aimed at realising generally shared values such as human wellbeing, efficiency, and economic prosperity. And throughout the decades additional values were added to this mix by letting applications of technologies also realise values such as safety, usability, and, more recently, sustainability. It is this recognition that applications of technologies can be designed to realise the values we uphold that motivated the design for values vision to also include moral, societal and legal values into this design: if designers can design for safety and usability, can learn to also design for sustainability and say inclusivity, then it may also be possible to let them design applications of technologies that realise for values that are typically still seen as outside of the realm of design, such as democracy, equality, autonomy, and privacy. Shape grammar design systems made available to inhabitants are by our analysis cases of this vision, for they are tools by which values are realised as *inhabitant autonomy*, *social cultural diversity*, and *city identity*.

Shape grammar design systems present, moreover, a case of design for values that allows for flexibility in the values designed for over the longer time periods characteristic to architecture. As such they can deal with two concerns about the usefulness of the design for values vision in architecture. Lara Schrijver (2015) explored this usefulness and raised a number of concerns. She, for instance, points out that architects design buildings and urban structures for longer time periods and that during these periods the values that inhabitants, cities, or societies endorse may change. Say, new values may emerge, existing values may become outmoded, or existing values may change meaning, yet the buildings architects design stay erect. An architectural design that initially incorporated desired values successfully may therefore after a few decades stop to be successful design for values. A second concern by Schrijver is that inhabitants can be quite inventive in countering the values that architects design in their buildings. Successful incorporation of values in housing may therefore also quite quickly stop to be so. It can now be argued that our case demonstrates that in architecture design for values can at least partly be done in a useful way. Schrijver's first concern holds in part for our approach on shape grammar design tools. If, for instance, in the second half of the twenty-first century people do not care anymore about social relations within their neighbourhoods, say since they moved their lives to the virtual realm, design for *social fabric* by shape grammar design systems becomes spurious Yet the point does not hold for changing values of inhabitants about what is good living in their apartments. By realising customisation and autonomy shape grammar design systems enable inhabitants to change the layout of their apartments to their changing values whenever they desire. Returning for a last time to our running examples, *Rabo-de-Bacalhau* apartments were originally designed for families with an in-house maid. Changes in society and living made however that such maids are less common, leaving the layout of the apartments outdated. The RdB transformation grammar allows however inhabitants to change that layout, and adjust it to their current value and lifestyles. This possibility may also give an answer to Schrijver's second point. The values of customisation and inhabitant autonomy express that inhabitants should be enabled to counter the values designed into their housing. And if *Rabo-de-Bacalhau* apartments had immediately come with a HouseGrammar tool for refurbishing them, already their first-generation inhabitants who, e.g., did not value life with a maid, would have right away been empowered to adjust their apartments to their values. Stand-alone shape grammar design systems by realising the values of customisation and inhabitant autonomy can therefore also be taken as a technology that allows for flexible design for values in architecture.

12.6 Conclusions

In this chapter we considered computer tools for architectural design of apartments based on shape grammar systems and evaluated the possibility of handing over these tools to inhabitants. Our evaluation was qualitative by considering the values of inhabitants, architects, and cities that are realised and not realised when inhabitants will be designing their apartment layouts themselves. In architecture this possibility can be positioned in the development towards customisation of housing and towards giving inhabitants autonomy in the design of their housing.

Our qualitative evaluation has demonstrated that the use of grammar design systems by inhabitants for designing new apartments and redesigning existing apartments will indeed realise the values of customisation and inhabitant autonomy without compromising design quality and structural safety. This possibility however less realises the values of financial feasibility and ecological sustainability.

Stand-alone shape grammar design systems are thus a technology that enables architects to realise their visions of customisation and inhabitant autonomy.

We extended our evaluation from the level of individual apartments to the level of neighbourhoods for considering the effects of this possibility on cities as a whole. Our evaluation has shown that for new neighbourhood construction a widespread use of shape grammar design systems by future inhabitants is not unconditionally preferable over the usual practice that architects fix the layouts of the newly constructed apartments: the first option better realises the value of social cultural diversity while the second better realises the value of financial feasibility. For the renewal of existing neighbourhoods a widespread use of shape grammar design systems by inhabitants would however compete as the optimal choice with architect-led centralised refurbishment of neighbourhoods. The use of shape grammars by inhabitants would better maintain the social fabric in neighbourhoods, equally preserve the identity of cities, and equally allow for limited urban architectural innovation. But architect-led centralised refurbishment would better realise the values of financial feasibility and ecological sustainability.

If making shape grammar design systems available to inhabitants is the way to accomplish the architectural vision of customisation and autonomy, then the next issue on the Open Building agenda of Habraken is to find ways to limit the costs and ecological impact of inhabitant-led construction of apartments. One way to do so may be to include in shape grammar design systems features that make explicit those costs and impacts. Cities could enable customisation in a financial and ecological responsible way by commissioning architects to create such shape grammars for the refurbishment of apartment types that are common. And architects could generically create such grammars for the customisation of apartments in new housing, where financial feasibility and ecological sustainability could be achieved by limiting the range of materials and construction systems.

Our evaluation has also shown that the role of architects, for design contexts as the one described in this paper, will change from being authors of individual designs to providers of knowledge through the creation of design tools for inhabitants. Handing over shape grammar systems to inhabitants has the further consequence that it limits urban architectural innovation. Renovation of older neighbourhoods by replacing them with new architecture is then not anymore a natural option. By the efforts of inhabitants apartments in older neighbourhoods will be continuously modernised and renewed for meeting their current needs and wishes, creating a bottom-up inhabitants-led renewal of neighbourhoods and urban structures, and avoiding that such neighbourhoods deteriorate to a point that architecturally they better can be demolished.

Shape grammar design systems are a technology for design for values in architecture that allows for flexibility in the values designed for over the longer time periods characteristic to architecture. They are a technology that enables the incorporation of values of inhabitants, architects, and cities by the designs of apartments, and they are a technology that enables inhabitants to adjust their apartments through time to their developing needs and wishes.

References

Aravena, A., & Iacobelli, A. (2016). *Elemental: Incremental housing and participatory design manual*. Ostfildern: Hatje Cantz Verlag.

Benrós, D., Granadero, V., Duarte, J. P., & Knight, T. (2011). Automated design and delivery of relief housing: The case of post-earthquake Haiti. In *CAAD futures 2011: Designing together* (pp. 247–264). Liege: Eds De L'Univ De Liege.

Bosma, K., van Hoogstraten, D., & Vos, M. (2000). *Housing for the millions: John Habraken and the SAR (1960–2000)*. Rotterdam: NIA Publishers.

Causa Justa. (2009). *Development without displacement, development with diversity, Causa Justa*. http://abag.ca.gov/files/DevelopmentwithoutDisplacement.pdf

Coimbra, E., & Romão, L. (2013). The rehabilitation design process of the bourgeois house of Oporto: Shape grammar simplification. In *eCAADe 2013 computation and performance* (pp. 677–686). Delft: Delft University of Technology.

Colakoglu, B. (2005). Design by grammar: An interpretation and generation of vernacular hayat houses in contemporary context. *Environment and Planning B: Planning and Design, 32*(1), 141–149. https://doi.org/10.1068/b3096.

Duarte, J. P. (2005). Towards the mass customization of housing: The grammar Siza's houses at Malagueira. *Environment and Planning B: Planning and Design, 32*(3), 347–380. https://doi.org/10.1068/b31124.

Economou, A. (2000). Shape grammars in architectural design studio. In *2000 ACSA technology conference* (pp. 75–81). New York: ACSA.

Eloy, S. (2012). *A transformation grammar-based methodology for housing rehabilitation*. Lisbon: Universidade Técnica de Lisboa.

Eloy, S., Vermaas, P. E., & Andrade, M. (2017). The quality of designs by shape grammar systems and architects: A comparative test on refurbishing Lisbon's Rabo-de-Bacalhau apartments. *Journal of Architecture and Planning Research, 34*(4), 271–294.

Friedman, B., Kahn, P. H., Jr., & Borning, A. (2006). Value sensitive design and information systems. In P. Zhang & D. Galletta (Eds.), *Human-computer interaction in management information systems: Foundations* (pp. 348–372). M.E. Sharpe: Armonk, NY.

Friedman, Y. (1958). *Manifesto architecture mobile*. http://www.yonafriedman.nl/

Gershenfeld, N. (2005). *Fab: The coming revolution on your desktop – From personal computers to personal fabrication*. New York: Basic Books.

Goethert, R. (2010). Incremental housing: A proactive urban strategy. *Monday Developments, September*, 23–25.

Habraken, N. J. (1972). *Supports: An alternative to mass housing*. London: The Architectural Press.

Hundertwasser, F. (1958). *Mouldiness manifesto against rationalism in architecture*. http://www.hundertwasser.at/pdf/verschimmelungsmanifest_eng.pdf

Kendall, S., & Teicher, J. (2000). *Residential open building*. London: E & FN Spon.

Knight, T. (2000). *Shape Grammars in education and practice: History and prospects*. http://web.mit.edu/tknight/www/IJDC/

Koning, H., & Eizenberg, J. (1981). The language of the prairie: Frank Lloyd Wright's prairie houses. *Environment and Planning B: Planning and Design, 8*, 295–323.

Kwiecinski, K., Santos, F., Almeida, A. D., Taborda, B., & Eloy, S. (2016). Wood mass-customized housing a dual computer implementation design strategy. In *Complexity & Simplicity - proceedings of the 34th eCAADe conference* (pp. 349–358). Oulu: University of Oulu.

Ligler, H., & Economou, A. (2015). Entelechy I: Towards a formal specification of John Portman's domestic architecture. In B. Martens, G. Wurzer, T. Grasl, W. E. Lorenz, & R. Schaffranek (Eds.), *Real time - proceedings of the 33rd eCAADe conference* (pp. 445–452). Vienna: Vienna University of Technology.

Lo, T. T., Schnabel, M. A., & Aydin, S. (2016). Modrule: A gamified design communication platform Tian. In *2016 design communication European conference*. Brussels: EAAE.

MVRDV, & Axis.fm. (2012). Architecture in mind. The HouseMaker. In MVRDV & The Why Factory (Eds.), *The vertical village. Individual, informal, intense* (pp. 169–194). Rotterdam: NAi Publishers.

Power, A. (2010). Housing and sustainability: Demolition or refurbishment? In *Proceedings of the ICE - Urban design and planning* (pp. 205–216). London: ICE Publishing. https://doi.org/10.1680/udap.2010.163.4.205.

Romańczyk, K. M. (2012). Transforming Brussels into an international city – Reflections on 'Brusselization'. *Cities, 29*(2), 126–132. https://doi.org/10.1016/j.cities.2011.08.007.

Ruby, I., & Ruby, A. (Eds.). (2013). *MVRDV buildings*. Rotterdam: Nai010 Publishers.

Sanoff, H. (2008). Multiple views of participatory design. *International Journal of Architectural Research, 23*(1), 131–143. https://doi.org/10.15368/focus.2011v8n1.1.

Schön, D. (1992). Designing as reflective conversation with the materials of a design situation. *Research in Engineering Design, 3*(3), 131–147.

Schrijver, L. (2015). Design for values in architecture. In J. Van den Hoven, P. E. Vermaas, & I. van de Poel (Eds.), *Handbook of ethics, values and technological design* (pp. 589–611). Dordrecht: Springer.

Smith, N., & LeFaivre, M. (1984). A class analysis of gentrification. In J. J. Palen & B. London (Eds.), *Gentrification, displacement, and neighborhood revitalization* (pp. 43–64). Albany: State University of New York Press.

Stiny, G. (1980). Kindergarten grammars: Designing with Froebel's building gifts. *Environment and Planning B: Planning and Design, 7*, 409–462.

Stiny, G., & Gips, J. (1972). Shape grammars and the generative specification of painting and sculpture. In O. Petrocelli (Ed.), *The best computer papers of 1971*. River Edge: Auerbach.

Stiny, G., & Mitchell, W. J. (1978). The palladian grammar. *Environment and Planning B: Planning and Design, 5*(1), 5–18. https://doi.org/10.1068/b050005.

Strobbe, T., Eloy, S., Pauwels, P., Verstraeten, R., De Meyer, R., & Campenhout, J. V. (2016). A graph-theoretic implementation of the Rabo-de-Bacalhau transformation grammar. *Artificial Intelligence for Engineering Design, Analysis and Manufacturing, 30*(2), 138–158. https://doi.org/10.1017/S0890060416000032.

Tan, P. (2008). Yona Friedman and GEAM: "Mobile architecture" and the right of dwelling. In S. Van der Ley & M. Richter (Eds.), *Visionary architecture and urban design of the sixties reflected by contemporary artists*. Hatje Cantz: Ostfildern.

Turner, J. F. C. (1976). *Housing by people: Towards autonomy in building environments*. New York: Pantheon Books. https://doi.org/10.1017/CBO9781107415324.004.

Van den Hoven, J., Vermaas, P. E., & Van de Poel, I. (Eds.). (2015). *Handbook of ethics, values and technological design*. Dordrecht: Springer.

Pieter E. Vermaas is associate professor at the Philosophy Department of TU Delft, the Netherlands. One of his research lines concerns the structure of design processes and their applications in engineering, architecture and society at large. A second line focusses on the emerging quantum technologies and ways to explore their future impact on society. Vermaas is editor in chief of the book series "Philosophy of Engineering and Technology" and "Design Research Foundations", and President of the Society for Philosophy and Technology.

Sara Eloy Architect, PhD in Architecture. Eloy's main areas of research include Digital Technologies applied to Architecture, Shape Grammars, Virtual and Augmented Reality, Analysis of the building space namely considering space perception, and Housing Rehabilitation. Currently Eloy's research concerns the promotion of users' participation in the design processes enhanced by digital technologies. She is the director of the Information Sciences Technologies and Architecture Research Center (ISTAR-IUL) where she is a fellow researcher. She has participated and coordinated national and international research projects and published her work in several journals. Eloy is an expert evaluator for the European Commission since 2015.

Chapter 13
Authenticity and the 'Authentic City'

Ryan Mitchell Wittingslow

Abstract In this paper, I argue that the benefits that smart cities purport to provide cohere poorly with a number of our shared phenomenological intuitions about the relationships(s) between authentic experience and technologised society. While many of these intuitions are, strictly speaking, pseudo-problems, they deserve our attention. These issues will only grow more pressing as our 'dumb cities', already so opaque to experience, give way to hyper-technologised 'smart cities'. However, it is possible to design our way out of these pseudo-problems. Assuming we accept my argument that the distinction between authenticity and the device paradigm is premised upon a certain kind of category error, there is no categorical or definitional reason why it is not possible for urbanised, technologised spaces to feel authentic, whether by virtue of their aesthetic properties, or because they facilitate 'authentic' behaviour. Indeed, I argue that 'inauthenticity' is an aesthetic rather than an ontological category (much like 'ugliness', or 'boring-ness'), with feelings of inauthenticity serving as evidence of a basic failure of design. Redressing these failures of design requires that we adopt a novel approach to the design and use of technical objects. Consequently, in the concluding analysis of the chapter I outline how the feeling of authenticity can be invoked in the smart city and, consequently, how these failures of design can be avoided.

Keywords Borgmann · Authenticity · Urbanism · Postphenomenology · Smart cities

---

R. M. Wittingslow (✉)
University College Groningen, Rijksuniversiteit Groningen, Groningen, Netherlands
e-mail: r.m.wittingslow@rug.nl

M. Nagenborg et al. (eds.), *Technology and the City*, Philosophy of Engineering and Technology 36, https://doi.org/10.1007/978-3-030-52313-8_13

13.1 Introduction

On paper, 'smart cities' are an easy sell. Thanks to the transformative power of information and communication technologies (the much-vaunted 'internet of things'), smart cities purport to offer managers and bureaucrats a more harmonious and efficient means of reducing traffic, managing assets, and increasing public safety. Proponents hope that these cities will demonstrate the utility of measuring and mapping human behaviour in a manner that is both large-scale and fine-grained, rather than relying on the rough-hewn predictive models of yesteryear. However, I am dubious of these utopian sentiments. Indeed, I argue that the benefits that smart cities purport to provide cohere poorly with a number of compelling phenomenological intuitions about technologised urban spaces, in this case as expressed by Albert Borgmann.

While I think that Borgmann makes an excellent case, I make clear that many of these intuitions are, strictly speaking, pseudo-problems premised upon the assumption that "authenticity" possesses an ontological character. However, while these intuitions may be errors, due to their undeniable resonance, they also deserve our attention. The belief that technologised urban spaces somehow corrode our potential for authentic experience is one that carries enormous power; as Charles Taylor argues, authenticity is nothing less than the contemporary moral ideal (1991, p. 17). Consequently, it indelibly colours the relationships that we forge with our artefacts. These issues will only grow more pressing as our 'dumb cities', already so opaque to experience, give way to hyper-technologised 'smart cities'. Thankfully, it is possible to design our way out of these pseudo-problems.

Assuming we accept my argument that the distinction between authenticity and the device paradigm is premised upon the afore-mentioned error, there is no categorical or definitional reason why it is not possible for urbanised, technologised spaces to feel authentic, whether by virtue of their aesthetic properties, or because they facilitate 'authentic' behaviour. Indeed, I argue that 'authenticity' is an aesthetic rather than an ontological category (much like 'beauty', or 'sublimity'), and that feelings of inauthenticity are less ontological warnings than they are evidence of a basic failure of design. Redressing these failures of design requires that we adopt a novel approach to the design and use of technical objects—and particularly when these objects are the almost invisible, semi-autonomous constituents of the 'smart city'.

Consequently, in the concluding analysis of the chapter I develop Barry Allen's notions of 'technical coherence' and 'aesthetic presence', in conjunction with Don Ihde's post-phenomenological work on technical relations, to outline how the feeling of authenticity can be invoked in the smart city and, consequently, how these failures of design can be avoided.

13.2 Things and Devices

Borgmann has, throughout his career, given a frank assessment of the anomie that he believes exists at the heart of Western civilisation. Responsible for this regrettable state of affairs are our relationships to the technologies we develop and employ: an invisible system of devices that lurks like a vast, silent monster beneath the world of phenomenological experience. Befitting his Heideggerian inheritance, Borgmann differentiates these 'devices' from more benign 'things': a distinction broadly analogous to the one Heidegger poses between *Gegenstände* and *Dinge*. Under this rubric, 'things' are those entities whose functions, operations, and applications are transparent to intuition. It is obvious what authentic things are *for*, in that they "tend to include in their appearance the invisible presence of what is needed to fulfil their function" (Arnheim, 1969, p. 89): in being phenomenologically available to intuition and experience they, if you like, wear their thing-ness on their sleeves. We can identify five features common to Borgmann's things (1984):

1. Things are *difficult*, in that they require a certain amount of labour in order to function properly. Using a fireplace, for example, is only possible if we are willing to find, chop, stack, and burn wood.
2. Things *take time to use*. Heating our homes with a fireplace incurs a temporal toll in addition to the impost of labour.
3. Things are not *safe*, either in use or design. Using a thing requires skill, and incurs a certain amount of risk. An incompetent woodcutter could, for example, chop off his fingers or burn down her house.
4. Things are *localised*. When we use a fireplace, it only heats the room in which the fireplace is situated.
5. Things are *available to intuition*. The means of use and mechanisms of action of things are perceivable and understandable, and do not require specialised knowledge to appreciate.

As a consequence of exhibiting these five features, things are necessarily 'focal' ('focus' is Latin for 'hearth'); they serve as phenomenological loci or anchors for shared activity. For the fireplace was the focal point of everyday activity in the pre-industrial home: the place where cooking, heating and interpersonal exchanges took place, and the designated space for the Roman house gods. Borgmann attempts to reclaim this pre-industrial use of 'focus', arguing that focal points are those locations like the pre-device paradigm hearth: a central point around which human beings engage in practices that "center, orient and enlighten our lives" (Tijmes, 2001, p. 22). These practices are unquantifiable: although we can place an explicit dollar value on the amount of hamburgers sold at a fast food restaurant in any given period, we cannot do the same for a family meal "thoughtfully prepared and celebrated at home" (Borgmann, 1984, p. 56). These things and practices serve to provide context for our lived experiences, whether it be via physical labour, meaningful social interaction or experiencing the unmediated trappings of nature (1984, pp. 190–191). While things do not offer up their goods without difficulty, they facilitate an appreciation and understanding of the richly-textured fabric of our social lives and our places within the world. In a method analogous to the Heideggerian process of *aletheia*, or 'revealing' (Heidegger, 1977), things *disclose* the world, making it clear for us. This is manifest in the necessarily shared and phenomenologically full experience of encountering things as they function; we can easily imagine a family eating and talking in front of the fireplace after a long day of fruitful and fulfilling activity (1984, p. 42). As a consequence of these factors, there is something seemingly (and seemingly profoundly) *authentic* about Borgmann's things.

> Amidst the complication of conditions, of the *Bedingungen*, we must uncover the simplicity of things, of the *Dinge*. A jug, an earthen vessel from which we pour wine, is such a thing. It teaches us what it is to hold, to offer, to pour and to give. In its clay, it gathers for us the earth as it does in containing the wine that has grown from the soil. It gathers the sky whose rain and sun are present in the wine. It refreshes and animates us in our mortality. And in the libation it acknowledges and calls on the divinities (2003, pp. 294–295).

Meanwhile, devices are technological solutions that liberate us from the physical and temporal tolls levied by things; unlike things, they are easy, instantaneous, safe, and ubiquitous. Thanks to an aggregation of technical systems, we are no longer required to heat our homes by the chopping and burning of wood. Instead, we can instead adjust the temperature on our thermostats, and let our radiators do all the work. Devices also offer opportunities that things do not. These immensely complex aggregations of technical systems enable us to perform tasks that would be categorically impossible with only the aid of things. For example, whereas a fireplace serves as the 'thing-analogue' for a hot water radiator, there is no thing-analogue for a computed tomography (CT) scanner. There is no phenomenologically rich thing that allows us to so readily look inside objects (such as human beings) without intruding with pins and scalpels.

However, Borgmann argues, there is a catch. While devices are very good at both meeting basic needs or desires, as well as providing services that things cannot, the mechanisms by which they function are opaque to intuition and experience. Warmed by the hot water radiators in our homes, we are not cognisant of the byzantine network of plumbing that lies just beneath the bricks or plasterboard, let alone the dams or desalination plants from which the water is ultimately claimed from the natural world. Central heating, like all devices, is an example of the "conjunction of an easily available commodity and a sophisticated and impenetrable machinery" (Borgmann, 2000, p. 420). With the prevalence of hot water radiators and ducted heating, one need not even be tied to a specific location in order to find oneself warmed; the process has been abstracted to the point where *place* no longer relevantly bears upon the experience of being heated. In the case of hot a water radiator the focus is upon delivering heat, whilst the device itself *withdraws from view*.

As a consequence of this process of withdrawal, Borgmann argues that we are unable to interface with devices such as central heating systems with the same degree of richness that we would with a fireplace. Instead, our relationship with a thermostat is strictly instrumental, transactive, and context-free. Moreover, the tenor of this relationship is common across all devices: while devices in many respects make our lives better, they do not and cannot offer the phenomenological, authentic plenary offered by things, nor can they facilitate focal practices. Due to the preponderance of devices, Borgmann claims that we have lost the taste for things. Rather than fulsomely encountering authentic things and engaging in authentic practices, we have begun to treat the world, the things that constitute it, and other people as if *they too* are devices: objects only to be considered in terms of their utility (economic, labour, or otherwise), rather than in terms of any other values they may have. This instrumental attitude, in so many words, is the device paradigm.

Given the tenor of Borgmann’s criticism of the device paradigm, it is unsurprising that analyses of urban spaces feature in his later work, particularly *Crossing the Postmodern Divide.*[1] A city—even a traditional ‘dumb city’, as distinct from the smart city—is itself a kind of enormous device: a concatenation of nested apparatuses and densely intertwined civic, commercial, public, and private systems. However, Borgmann argues, while cities may be devices, they need not participate in the device paradigm assuming that they encourage authenticity and focality in civic, social, and artefactual practices. While the specific methods by which focality can be encouraged are obviously and necessarily sensitive to local conditions, focal urban spaces are united in offering what Andrew Light calls ‘thick’ experiences (Light 2005, p. 112). These experiences, while not always pleasant or safe, offer socially and phenomenologically rich interaction: the joyful buzz of a village square on market day; the democratic belligerence of a town hall meeting; the myriad of small interactions one undergoes when popping down to the post office to collect a parcel. While not necessarily *convenient*, these experiences help render urban spaces both parseable and navigable.

Unfortunately, however, urban design over the last century has failed to facilitate focal experiences thanks to design norms that privilege commercial activity at the expense of collective action, induce psychological and personal atomisation, and serve to collapse the ‘public’ into the ‘social’ (c.f.: Arendt, 1998, pp. 38–49). Borgmann writes: “Mixed-used skyscrapers and carefully designed urban malls are self-centered little universes that turn their backs on the real city. Thoughtful observers find these constructions controlled, contrived, artificial, and superficial” (1992, p. 130). Instead of the vitality and risk of the public square, we are insulated from one another via a spongiform mass of devices. Services and experiences—food, recreation, social interaction—that we once had to seek outside the bounds of our homes can now be piped in, painlessly and without effort. While the universe of consumption stitches together in a Frankenstein’s monster of overlapping services, our social universes—even in traditional dumb cities—become atomised and profoundly inauthentic. Or, as Borgmann writes:

> We live in self-imposed exile from communal conversation and action. The public square is naked. American politics has lost its soul. The republic has become procedural, and we have

[1] In this work Borgmann refers to cities as being either ‘real’ or ‘hyperreal’. However, I agree with Andrew Light’s view that the “real-hyperreal distinction does not do any work that the device paradigm does not; instead, the old distinction should be have applied more straightforwardly” (Light 2005, p. 128). Consequently, I take Borgmann’s urban hyperreal to be equivalent to the device paradigm.

> become unencumbered selves. Individualism has become cancerous. We live in an age of narcissism and pursue loneliness. (1992, p. 3).

Moreover, given Borgmann's clear pessimism about the effects that information and communication technologies (ICT) have upon our capacity for authentic experience, it seems that the smart city presents an even greater threat to focality and authentic experience: "information technology is currently the prominent and most influential version of the device paradigm" (1999, p. 352). In *Holding on to Reality* Borgmann argues that 'technological' information, as distinct from focal 'natural' or 'cultural' information, inherently functions within the device paradigm. Instead of being tied to world, society, and circumstance like natural or cultural information, technological information is a formal simulation of the world: information presented *as* reality, rather than as *part of* reality. Indeed, when our informational map of the world becomes so accurate and finely-grained as to be treated as if it were itself the territory, it will, Borgmann claims, 'overflow and suffocate reality' (1999, p. 213). Consequently, and assuming that we accept this story, Borgmann's claims have profound ramifications for the smart city as, it seems, it is categorically incompatible with focal practices—and thus, with authenticity.

'Authenticity' is no mere cavil. Indeed, for many of us in the early twenty-first century, 'authenticity' is nothing less than an ideal form of life; a perfect mode of being. Just as Heidegger's *Dinge* are objects busily engaged in the activity of being themselves, for a human being to act authentically she must do the same: she is required to act for herself in a universe comprised of things acting for themselves. However, unlike other things, the phenomenologically authentic is free to deny her own nature; it is always our choice to affirm or reject the demands made upon us by our being. Consequently, authenticity has a profoundly normative dimension: the authentic seeks the form of life that best nourishes her, allowing her to flourish. Moreover, it is not simply enough that it nourish her; rather, there is a distinct implication that there is a *single best means* of nourishment. The spectre of Borgmann's authenticity calls to us, beckons us to act in accordance with some kind of internally-held but objectively-the-case standard of wellbeing. To refuse is to act in bad faith; it is a means of denying the very thing that she should be. Consequently, as an ethic, Borgmann's authenticity is both eudaimonic and teleological. The 'authentic' person flourishes best when participating in the right kinds of practices; practices determined by the demands of the natural world: "Nothing so engages the fullness of human capabilities as a coherent and focused world of natural information" (Borgmann, 1999, p. 219). It only is when she successfully acts in harmony with those demands that the authentic person actually succeeds in achieving authenticity.

At this particular moment in the early twenty-first century, Borgmann's authenticity functions as an object of common desire; indeed, and as noted in the introduction, Charles Taylor argues that authenticity might just be the contemporary moral ideal: "many people feel *called* to do this [the authentic act], feel they ought to do this, feel their lives would be somehow wasted or unfulfilled if they didn't do it" (1991, p. 17). Moreover, because authenticity is an object of common desire, it has become imbued with an undeniable cultural and economic cachet. Certain goods are marketed on the basis that they can instrumentally precipitate an authentic life in the owners of those objects. The advertising of authenticity succeeds when it convinces punters that the products or services in question will *complete* them in some profound, ontological way. What is being sold is a picture of the world where everyone is perfect and everything has its right place; all of the moving parts work in such a way as to help us realise our apotheoses. It's a marvellous piece of legerdemain: the object of desire is entirely intangible. Being intangible, authenticity is not a fungible asset. It has no explicit economic value, nor can it be isolated or traded. However, despite having no explicit economic value, it is by no means *valueless*; to the contrary, authenticity is a concept that commands extraordinary social and cultural value. Indeed, phenomenological authenticity is an object of aspiration, and we admire those who we believe manage to achieve it—after all, to live an authentic life means that you're somehow living *properly*, having escaped the spiny thickets of civil society and late capitalism.

13.3 A Suspect Ontology

On this level of description, Borgmann's picture of authenticity looks unproblematic. However, cracks start to appear when we look closely at some of the tacit assumptions that underpin this model. As noted previously, there is something deeply teleological about Borgmann's authenticity. Our being—our *Dasein*, if you like—reaches for authenticity like a forest sapling grows towards the sun. While we are free to refuse the demands made of us by authenticity, we can only achieve the best possible life when we act in accordance with those pressures. This, in so many words, constitutes 'human nature': we are ontologically united by an intrinsic need for rustic simplicity, hard work, and fulsome community practices, rather than constant motion and the hyperactive manipulation of devices. This understanding of human nature paints a distinctly Heideggerian picture of the human that is fundamentally absent any kind of sophisticated technology: one premised upon an ontological distinction between 'natural' and 'unnatural' behaviours, with focal practices in the former category, and device-driven practices firmly in the latter.[2] We should be suspicious of these sorts of intuitions, and not only because any posed distinction between 'natural' and 'unnatural' reeks of a crude Aristotelianism that is "no longer connected with ontological claims", in the words of Jürgen Habermas (2003, p. 44). Indeed, even were that distinction to obtain, it fails to engage with the fact that, insofar as humans have an ontological character, it is one that is necessarily and categorically technologised, regardless of whether those technologies are 'things' or 'devices'.

Barry Allen, in *Knowledge and Civilization*, makes this point very clear. Although we might not think of ourselves as necessarily technologised creatures,

[2] Readers familiar with Heidegger will immediately see the parallels between Borgmann's implicature and Heidegger's ruminations on the hydroelectric plant (1977, p. 16).

he makes the case that technical culture comprises the evolutionary loam from which we have sprouted. From before the beginnings of genus *Homo*, humans have been crafting tools with other tools, creating a dense economy of complementary, mutually contingent artefacts that are united by a common grammar. By example, by smashing two rocks together I can make a sharper rock; with that sharper rock I can kill a deer; with that deer I can make clothes to keep me warm and delicious food to eat. Allen argues that our artefacts do not make sense unless embedded within a matrix of other, complementary tools: a needle cannot be used as a needle without a thread; a mobile phone cannot be used as a mobile phone without a satellite; a microwave dinner is going to taste even more like cardboard without a microwave. Moreover, this economy of artefacts is not only of interest in terms of material culture. Allen, citing Roy Rappaport, argues that this technological economy is the source of everything else that makes us human, too—including, potentially the development of language, ethical norms, aesthetic preferences, and whatever forms of life that we practice: "[we] are necessarily constituted by our technology, for our technology is that which facilitates that other necessary condition for humanhood: the ability to use and navigate symbolic references in a systematic and intelligible way" (Wittingslow, 2016, p. 134). In short, technology is a necessary precondition for culture.[3] Allen writes:

> [Technical] culture [...] is as old as the genus *Homo*, as old as language, or protolanguage, or ritual, and much older than our belated species, for whom it has become as indispensable as fresh water. The culture of artifacts is bred in our bones; we are *born* to culture; for us culture is nature (2005, pp. 218–219).

This picture also maps onto the archaeological record. According to estimates based upon the rate of phonemic diversity expressed in a number of test languages, full-blown language use in humans developed somewhere between 150,000 and 350,000 years ago (Perreault & Mathew, 2012, p. 5). Meanwhile, the very earliest Australopithecine stone tools—tools made of other tools, already fully embedded within an economy of artefacts—are dated much earlier, at approximately 2.6 million years ago (Plummer, 2004, p. 118). This indicates that technology has been with use two million years before the development of language, and 2.5 million years before the evolution of anatomically modern humans. This is what Allen means when he says we are 'born' to technology: the very first *Homo sapiens* arrived already technologised, already linguistic, and already fully acculturated, embedded in the dense economy of complementary, mutually contingent artefacts of our ancestors.

Allen's scholarship also has interesting parallels with that of Bernard Stiegler: a thinker much more closely aligned with the Heideggerian tradition to which Borgmann belongs. Stiegler, like Allen, is also interested in the mutually constituting relationship with humankind and our tools. He considers the birth of mankind as something that emerges from the process of interfacing with the material of our

[3] Readers interested in the details of this claim should consult (Allen, 2005, pp. 189–203) and (Rappaport, 1979).

first tools: a union of eye and hand and stone. In acting upon the world, he argues, we find our own mind speaking to us. This capacity obliquely reflects itself, "somber, buried, freeing itself from the shadows like a statue out of a block of marble" (1998, p. 141): where we find our own internal intentions marked upon the world, though necessarily limited by the external facts that the world imposes. The creation of the first human is a paradox, for we speak of an exteriorisation (the first stone tool) without that exteriorisation being exterior to *something*, without an intelligent and intelligible homunculus having taken up residence behind one's eyes. Instead, echoing Allen's basic phenomenological point, the interior is constituted by that first tool-making act: "The movement inherent in this process of exteriorization is paradoxical: [...] it is the tool, that is, *tekhnē*, that invents the human, not the human who invents the technical. Or again: the human invents himself in the technical by inventing the tool—by becoming exteriorized techno-logically" (1998, p. 141). For Stiegler, humankind is defined by the tools we wield, but the tool itself is rendered impossible without the presence of the human. Supposing that one has an ontology independent of the other is both incoherent and impossible.[4]

While Borgmann's desire to separate us from our dense technological milieu and subsequently relocate human nature to a kind of rustic simplicity is undeniably appealing, his view that we will somehow lose touch with the natural world due to over-reliance on devices is frankly incompatible with the observation that we can *never* return to nature, un-technologised and naïve. It is, in short, a pseudo-problem, premised on a significant category error. This is what I mean when I say that, insofar as we have an ontological character, human beings are necessarily and categorically technologised. Existing in the world of our artefacts, whether things or devices, is as authentic to our being as we can possibly get. As Don Ihde writes: "it might be possible for humans to live non-technologically as a kind of abstract possibility [...]. But there is no such empirical-historical human form of life because, long before our remembering, humans moved from all gardens to inherit the Earth" (1990, p. 13). However, just because authenticity is not a meaningful ontological category does not mean that authenticity isn't important. To the contrary, I believe that the question of authenticity is profoundly important, particularly insofar as authenticity discourse influences the form and function—and, consequently, our experience—of designed objects and places. It's just that Heideggerians, such as Borgmann, are mistaken about the kind of thing that authenticity is. For authenticity, I argue, is not an ontological category at all; rather, it denotes a class of related experiences that are themselves parasitic upon a certain class of postphenomenological relations. Authenticity is, in short, an *aesthetic* category.

While much of the ink-spilling that occurs within philosophical aesthetics concerns the nature of the beautiful, 'beauty' is by no means the only aesthetic category. Outside of beauty, 'sublimity' is perhaps the category that has attracted the most

[4] "[Whereas] animals are positively endowed with qualities, it is *tekhnē* that forms the lot of humans, and *tekhnē* is prosthetic; that is, it is entirely artifice. The qualities of animals make up a sort of nature, in any case a positive gift of the gods: a predestination. The gift made to humanity is not positive: it is there to compensate" (Stiegler, 1998, p. 193).

scholarship, thanks in no small part to the philosophical spat between Edmund Burke and Immanuel Kant in *Origin of Our Ideas of the Sublime and Beautiful* and *The Critique of Judgement*, respectively. Furthermore, there are plausibly a great many additional categories: Albert Chandler, by example, identifies three categories in addition to the two already mentioned: the 'tragic', the 'comic', and the 'interesting' (1921); while Sianne Ngai identifies 'interesting', 'zany', and 'cute' (2010, 2015). Although I have no real interest in providing a complete taxonomy of aesthetic categories (a task that strikes me as both thankless and doomed to failure), I do think that the notion of an aesthetic category is an important one: it is obviously the case that objects or scenarios elicit qualitatively different kinds of aesthetic experiences.

Beauty, for instance (at least when discussed in a Kantian register), refers to a kind of disinterested yet pleasurable appraisal of formal and semantic features; a "delight consistent with reason", in the words of John Dennis (1931, p. 381). The sublime, meanwhile, refers to experiences that exceed our capacity for reason: the agreeable sort of horror one feels when confronted with something "transcending every standard of sense" (Kant, 1969, p. 98). Less straightforward experiences, such as Matthew Kieran's 'grotesque'—referring to the "actual delight we feel in attending to repellent, ugly and incoherent artwork" (1997)—also potentially constitute categories, in that they too refer to qualitatively distinct aesthetic modes. Furthermore, rather than being essential or intrinsic to our being, aesthetic categories are fundamentally artefacts of acculturation, with the individually aesthetic instances that constitute those categories united only by family resemblance. This is why, for instance, we rightly recognise that the religious sublime of Burke and the Romantic sublime of Rainer Maria Rilke are qualitatively distinct, while at the same time expressing no qualms about describing them both as instances of sublimity. They are united only by a common sense that the human being is a small and incidental thing when considered against the power and size of the universe.[5]

Amidst this plethora of options, I claim that Borgmann's 'authentic' also functions as an aesthetic category, rather than—as he claims—an ontological quality. In this manner, focality is simultaneously denuded and reinvigorated: rather than being saddled with groundless Heideggerian intuitions about the teleology of *Dasein*, authenticity describes a family relation of aesthetic experiences, united by a common pleasure in hard work, community, and simple things. It, unlike the disinterested appreciation of beauty, or the giddy terror of sublimity, or the carnivalesque and ironic smirk of the grotesque, is a kind of aesthetic pleasure derived from fulsome, phenomenologically rich, personally and/or socially gratifying activities. In short, the aesthetic category of authenticity is comprised of the experiences of focal practices. Furthermore, we should not be dogmatic about what constitutes a focal practice and what does not. The truth of authenticity is in the feeling of it, in the profound sense of phenomenological engagement. Whereas Borgmann categorises

[5] "For beauty is nothing/but the beginning of terror, which we are still just able to endure, / and we are so awed because it serenely disdains/to annihilate us" (Rilke, 2014, p. 3).

certain activities as either focal or non-focal, we can be a little more generous and claim that focality can emerge in all manner of places, and in all kinds of unexpected ways. Even the mere act of drinking water can stimulate the sense of authentic engagement, assuming the correct circumstances:

> To drink water from a waxed paper cup on the highway and to drink it from a crystal goblet are different gestures. In the first case, you almost forget that you exist as you drink. In the second, […] you realize that you have in your hands an instrument that makes you reflect upon how you are living at that moment. (Ettore Sottsass, quoted in Grudin, 2010, p. 55).

Moreover, it is possible to design objects or events in such a manner as to induce experiences within the register of a given aesthetic category, even though these experiences are obviously underdetermined by the objects or events in question (one could fail to find a J. M. Turner painting sufficiently sublime, for example). So, were one seeking to make a sculpture that would induce experiences of beauty in viewers, they could do so by employing those methods that have previously succeeded in doing so: by guaranteeing elegance, or splendour, or charm, or delicacy. So it is with the case of authenticity. Consequently, it is here that the richness and usefulness of Borgmann's analysis really makes itself apparent. Whilst I profoundly disagree with his ontological claims, his sensitivity to the kinds of practices that human beings generally find existentially or spiritually fulfilling provides an excellent scaffold from which to examine the features common to experiences that we consider 'authentic'. To that end, we already know the five features that make things phenomenologically (or aesthetically) satisfying and, consequently, appropriate loci for focality: that is, they should be difficult, slow, unsafe, local, and intuitable. However, no longer constrained by Borgmann's ontological restrictions, we can loosen these requirements: now, they must only *seem* difficult, slow, unsafe, local, and intuitable. We must encounter them as if they offer up their gifts begrudgingly.[6]

Importantly, when authenticity is treated as an aesthetic category, there is no longer any categorical or ontological reason why it is impossible for urbanised, device-riddled spaces to feel authentic. Devices, inauthenticity, and non-focality are, for Borgmann, existentially threatening because they separate us from the fulsomeness of the given. However, in clarifying that authenticity is an aesthetic preference rather than an ontological need, the stakes are entirely different to those that Borgmann implies: instead, inauthenticity becomes a solvable *design* challenge, rather than an insoluble ontological challenge. While good design can immeasurably improve the lives of users, and failures of design can make life immeasurably worse, we can at least take comfort in the fact that bad design does not impinge upon or corrode being itself. Of course, producing designed goods that are robustly

[6]A good Heideggerian would certainly disagree with my assumptions here. Contra my claim that that "the authentic" denotes a class of experiences, said Heideggerian might instead insist that Borgmann's five features are accidental attributes of authentic *Dinge* rather than additively constituting authenticity (c.f. Heidegger, 2008). Unfortunately, given the profound differences between my ontological assumptions and those of our hypothetical Heideggerian, I fear that this disagreement is intractable.

and satisfyingly focal is still profoundly important if we are to feel as if we are living authentically; it's just that we should not consider failure ontologically or existentially threatening.

13.4 Designing for Focality

In order to improve the lives of users, focal goods need to succeed on two fronts. The first is the same dimension in which all designed objects need to succeed: what Barry Allen describes as 'technical coherence', or "why things work" (2008, p. 97). Technical coherence is a measure of the extent to which a designed object does the job for which it was designed. A paperclip, for example, is technically coherent when it successfully holds together loose leaves of paper; a sewer system is technically coherent when it seamlessly and invisibly transports waste to a treatment plant. Meanwhile, if an object fails to perform as designed, then something in the design has failed: it is instead technically incoherent. The second front upon which focal goods need to succeed is what Allen calls 'aesthetic presence': why things "affect us, appeal to us, fascinate and delight us, make us linger, remember, and revisit" (2008, p. 97). Not all designed objects have aesthetic presence: Allen uses the examples of electrical power grids, drain systems, and nanotechnology as objects without presence that nonetheless exhibit technical coherence. These coherent but non-aesthetic goods can be described as 'non-focal' because they make no attempt at phenomenological participation: Borgmann's devices fall into this category. Meanwhile, objects that do have aesthetic presence are not only employed instrumentally; they are also themselves the objects of our attention because they possess distinct points of interface, facilitate phenomenologically rich activity, and exist at a human scale. Borgmann's things fall into, but do not exhaust, this category.

Unfortunately, Allen does not further develop his distinction between technical coherence and aesthetic presence. Nonetheless we can, by appealing to the post-phenomenological scholarship of Don Ihde, enrich this taxonomy. Ihde, in *Technology and the Lifeworld* argues that there are four fundamental kinds of technical relations: that is, relations that hold between humans and our artefacts. These relations are 'embodiment relations', 'hermeneutic relations', 'alterity relations', and 'background relations' (1990):

1. 'Embodiment' describes mediated technical relations wherein artefacts are integrated into our perceptual schemas. Eyeglasses provide relations of this type: I do not perceive the glasses themselves, but rather I perceive *through* them; they withdraw from my perception.
2. 'Hermeneutic' describes mediated technical relations wherein artefacts do not withdraw from our attention, as in embodiment relations, but rather reveal an aspect of the world that would have been hitherto inaccessible. Thermographic cameras provide relations of this type.

3. 'Alterity' describes unmediated technical relations wherein artefacts are themselves the object of attention: as a "quasi-other" (1990, p. 98). Spinning tops, books, gaming consoles, and automatic teller machines provide relations of this type.
4. 'Background' describes unmediated technical relations wherein we do not directly encounter the artefact at all. Instead, these artefacts provide and shape "the context of our experience in a way that is not consciously experienced" (Verbeek, 2001, p. 132). Borgmann's devices and Allen's coherent but non-aesthetic goods provide relations of this type.

Amongst these four types of relation, any of the artefacts participating in the first three types of relation can have aesthetic presence, in that they are directly experienced. However—and significantly—not all of those relations can facilitate focality, for the simple reason that only one of these relationships has the right kind of relationship with the world. Embodiment relations, by example, describe relations where the artefact integrates into the body seamlessly and pre-theoretically: it is broadly analogous to the Heideggerian *Zuhandenheit*, or 'readiness-to-hand'. The 'I' and 'technology' collapse into a single entity, united by what Ihde calls the 'enigma position'; consequently, it restores or amplifies human agency, but reveals nothing about the world itself. Alterity relations are similarly unable to reveal the world, in that the artefact itself is the object of attention and interaction and the world is rendered both marginal and silent. In hermeneutic relations, however, it is technology and world that is united at the enigma position: our artefacts provide novel means of perceiving and engaging with the world. Consequently, in pursuing hermeneutic relations, the groundwork is placed for focal goods to provide an aesthetic sense of *aletheia*, of world-disclosure—and thus, an aesthetic sense of authenticity.

This is why, for instance, we perceive imperfectly-made artisanal objects as being more authentic than perfectly-machined mass-produced objects. Focal goods appear created within the register of what David Pye called "the workmanship of risk", as opposed to the "workmanship of certainty". Pye characterises the workmanship of certainty as being the kind of workmanship presenting in quantity production, with its "pure state in full automation." Meanwhile, unlike the "exactly predetermined" results of the workmanship of certainty, the workmanship of risk describes the process of *craftsmanship*, wherein "the quality […] depends on the judgement, dexterity and care that the maker exercises as he works" (1968, pp. 4–5). Furthermore, in bearing the marks of their production thanks to the workmanship of risk, focal goods also appear sensitive to the circumstances in which they were made—and indeed, seem to suggest the hermeneutic story of how they were made. Unlike the grim, industrial homogeneity of non-focal goods, focal goods interface with the realities of small scale production: the availability and specific qualities of local resources; prevailing aesthetic norms; receptivity to historical contingency and lived experience. Rather than pretending to meet universal human needs, they appear designed to meet individual, specific needs under local circumstances.

While I dispute the grounds on which Borgmann expresses his ontological concerns, I think his fundamental intuition is correct: first, that a great many human beings have an aesthetic preference for authenticity; second, that this aesthetic preference is both important and worth taking seriously; and third, that contemporary society is suffering a crisis of authenticity. Moreover, I think evidence for this preference being in place is remarkably easy to find. By example: according to the June 2015 issue of *Fortune*, in 2014 major packaged-food companies lost US$4 billion in market share, with reports indicating that consumers were actively preferring fresh, organic products over their packaged counterparts. This was not an isolated hiccup, either; the top 25 food and beverage companies in the US have haemorrhaged some US$18 billion dollars in lost sales since 2009 as a direct result of shifting consumer demands against perceived artificiality and towards perceived authenticity: "The search for authenticity has led organic food sales to more than triple over the past decade and increase 11% last year alone to $35.9 billion" (Kowitt, 2015). This is but one example of a pattern that is manifest in all corners of consumer culture. Our motivations are basically Borgmannian as well. By buying bamboo laptop cases, organic vegetables, convenience-store kombucha, by celebrating urban beekeeping and the reinvigoration of Savile Row, we feel as if we are better able to return to some kind of idealised natural state. As Elizabeth Outka argues in *Consuming Traditions*, the commercial appeal of the authentic lies in its ostensive resistance to commercialism and the device-riddled life that commercialism seems to entail (2008).

Perversely, this collective desire for authenticity has catalysed the development of an homogeneous 'authenticity aesthetic': a set of formal properties shared in one form or another by items that aspire to authenticity (c.f.: Chayka, 2016). Although I shan't exhaustively list these properties here, they're immediately identifiable: cafés and bars in brown varnished wood and lit with Edison bulbs; organic, free range produce prepared with artisanal flair; luxuriant hipster beards and the accompanying ecosystem of expensive men's barbers; raincoats constructed to look relevantly like military parkas from the mid-twentieth century. Although not present at all levels of consumption, at certain social and economic strata the aesthetic is unavoidable, irrespective of location. We can observe this harmonisation of taste from Milan to Maputo. Of course, the tragic part of this shift in consumer trends is that while these objects bear the superficial marks of authenticity, they do exactly nothing to facilitate experiences of focality, because they disclose nothing of the world. These goods are not merely non-focal, like electrical power grids or sewer systems. Instead, they are pseudo-focal: we perceive them as actively *inauthentic*, because they aspire to focality but fail to achieve it, in much the same way that objects that aspire to beauty but fail can collapse into ugliness or camp. In failing, these failures of design only exacerbate the aforementioned crisis of authenticity.

Given the vast concatenation of both non-focal systems and pseudo-focal goods in urban spaces, all cities—smart or dumb—struggle to address the anomie inculcated by Borgmann's device paradigm specifically, and by the insipid commercialism of late capitalism more generally. Indeed, under a certain reading, the entire history of twentieth and twenty-first century urban design is a series of attempts (of varying success) to address this anomie: the garden city, Le Corbusier's modernist nightmares, Speer's *Welthauptstadt Germania*, suburban sprawl, Jane Jacobs' *The Death and Life of Great American Cities*, New Urbanism, and/or sustainable development. While the methods differ wildly—the atomising effect of Frank Lloyd Wright's commuter cities and the integrated walkability of the *Charter of New Urbanism* could not be less similar—all of these movements are, in one way or another, engaged with the question of how to make post-industrial life more existentially satisfying. However, not all of these attempts were equally successful. Perhaps unsurprisingly, the movements that best address this anomie are those producing spaces that facilitate phenomenologically rich, communal activity, by introducing public squares, mixed-use transport systems, and shared green spaces (Montgomery, 2013). In so doing, the city appears opened up, authenticated, *disclosed* to its residents.[7] Echoing Henri Lefebvre, the residents' 'right to the city' is constantly and implicitly reaffirmed: hermeneutic relations made manifest in urban design:

> Here and elsewhere we assert that there is no urban reality without a center, without a gathering together of all that can be born in space and can be produced in it, without an encounter, actual or possible, of all 'objects' and 'subjects' (Lefebvre, 1996).

.All of this brings us back to the smart city. If cities already struggle to deal adequately with urban anomie due to the concatenation of non-focal systems and pseudo-focal design failures, then 'smart' systems do nothing to ameliorate these issues. If anything, and if Borgmann's intuitions are correct, they are liable to make these concerns *worse*. Nonetheless, attempts have been made to reconcile Borgmann's scholarship with the smart city. In Muki Haklay's essay, "Beyond Quantification", he proposes a program of smart urban renewal wherein citizens are directly responsible for data creation and gathering: what he calls 'participatory mapping'. Drawing inspiration from the work of citizen scientists, he imagines urban spaces wherein people adopt a custodial attitude to their own small corner of the world, using smart technologies to record, log, and improve their environs. He writes: "[the city is] a place for collective action and communal activities with good potential for developing new focal practices around data collection, processing and use. The degree to which users of existing technology are allowed to change the meaning of how it is used, or to apply it toward other unintended uses, is central to its potential to serve as a 'thing' and not only as a 'device'" (2018, p. 219).

However, I confess a degree of trepidation. While I certainly agree that smart cities could potentially provide new means of social and civic engagement for residents, I am entirely unconvinced that the means of engagement that he proposes qualify as sufficiently focal. Focal practices, after all, are oriented around the right kinds of hermeneutic relations: relations with artefacts that are (or seem to be) some combination of difficult, slow, unsafe, local, and intuitable. Focal goods need to be objects with which users can interface fulsomely and richly, engaging with a broad

[7] The implication here—that authenticity is not only desirable because we like it, but because it serves a public good in the sense of better facilitating citizen participation—is an idea that requires further unpacking. Unfortunately, I cannot do justice to in this chapter. However, I am very aware of this lacuna, and I intend to deal with it more substantively in a forthcoming paper.

phenomenological plenary in order to bring about a sense of *aletheia*. While participatory mapping offers urban residents a taste of agency, the ICT mechanisms by which this mapping is offered is still riddled with non-focal goods. Indeed, in order for the smart city to *be smart*, it must rely upon processes and mechanisms that are opaque, intensely complicated black boxes: processes and mechanisms categorically unavailable to both experience and intuition.

Nonetheless, while Haklay's proposal is necessarily constrained by taking Borgmann's ontological picture seriously, it carries the kernel of a solution. Given that urban residents already struggle to engage authentically with dumb cities thanks to the infrastructure and systems they have inherited, Haklay's proposal for smart cities looks like a means by which urban spaces can be returned to their residents. However, designers and planners in this space need to consider the actual *means* of engagement with respect to Borgmann's intuitions but without respect to his ontology: the points of interface wherein residents of the smart city can interact with, influence, and feed back into municipal systems. This means that not only must residents be able to feel the effects of their participatory mapping in a relatively immediate and non-attenuated way, but also that the process by which the mapping occurs is itself sufficiently focal: phenomenologically rich, communal, and seemingly intuitable.

This will also help preserve the *desirable* features of urban devices, while still rendering them more transparent to the body politic. It is obviously true that not all designed objects should be capable of hosting authentic experience. Indeed, in the case of large urban systems like plumbing or electricity this is almost certainly undesirable: while it might be *authentic* to contest with a wet rag and a pit toilet, the safety and ease of use offered by sewerage infrastructure seems clearly worth the odd touch of anomie. Indeed, this sort of non-focal but convenient infrastructure is among the features that make cities attractive places to live. However, I do think it might be desirable to *open up* device systems—and particularly devices in smart cities—by providing focal points of interface with the public. The structural integrity of a given polity is premised, at least partially, in the collective faith that systems of governance are sufficiently robust. In prising open these non-focal systems for examination, analysis, and interaction, the robustness of those systems can be clearly and obviously demonstrated.

As a design problem, the difficulties cannot be overstated: making an immensely complex, hidden system appear agreeable and open is hard enough without also having to guarantee that residents be able to interface robustly and hermeneutically with that immensely complex, hidden system. Nonetheless, in order for the smart city to succeed it *requires* these focal points of interface; it is an aesthetic imperative to facilitate world-disclosure. Anything less will only exacerbate the anomie that Borgmann articulates.

13.5 Concluding Remarks

While I do not and cannot endorse Borgmann's entire critical apparatus, I believe that the intuitions that he articulates are both real and powerful. Moreover, if the administrative promise of the smart city is to see fruition, we cannot afford to ignore these intuitions. Instead of adding yet more non-focal systems to our complex urban machinery, we must strive for means that make these systems more transparent, world-disclosing, and phenomenally enriching. These systems must permit urban spaces that facilitate the interaction of differences, with an eye to increasing both the democratic potential of urban spaces and the privileging of users over commercial activity. We have already mentioned Henri Lefebvre's 'right to the city': the radical proposal that is, in the words of David Harvey:

> [...] far more than the individual liberty to access urban resources: it is a right to change ourselves by changing the city. It is, moreover, a common rather than an individual right since this transformation inevitably depends upon the exercise of a collective power to reshape the processes of urbanization. The freedom to make and remake our cities and ourselves is, I want to argue, one of the most precious yet most neglected of our human rights (2008, p. 23).

In light of these sentiments, it is my hope that designers of smart systems think seriously about the appropriate intersection between the smart city and the right to the city: in other words, what might be described as the 'right to the smart city'. For the utopian promise of the smart city to pay social dividends, automated systems must facilitate, even welcome, the plurality of both users and uses, whether serious, dull, administrative, libertine, or ludic. While we will not suffer ontological damage without the potential for robust, hermeneutic interaction, it is only in providing the possibility for focal interaction that the smart city can become aesthetically satisfying: from the smart city to the authentic city.

References

Allen, B. (2005). *Knowledge and civilization*. Boulder, CO: Westview Press.

Allen, B. (2008). *Artifice and design: Art and technology in human experience*. Ithaca, NY: Cornell University Press.

Arendt, H. (1998). *The human condition*. Chicago: Chicago University Press.

Arnheim, R. (1969). *Visual thinking*. Berkeley, CA: University of California Press.

Borgmann, A. (1984). *Technology and the character of contemporary life: A philosophical inquiry*. Chicago: Chicago University Press.

Borgmann, A. (1992). *Crossing the postmodern divide*. Chicago: Chicago University Press.

Borgmann, A. (1999). *Holding on to reality: the nature of information at the turn of the millennium*. Chicago: Chicago University Press.

Borgmann, A. (2000). The moral complexion of consumption. *Journal of Consumer Research, 26*(4), 418–422.

Borgmann, A. (2003). Focal things and practices. In R. C. Scharff & V. Dusek (Eds.), *Philosophy of technology: The technological condition* (pp. 293–315). Hoboken, NJ: Blackwell.

Chandler, A. R. (1921). The aesthetic categories. *The Monist, 31*(3), 409–419.

Chayka, K. (2016). Welcome to Airspace: How Silicon Valley helps spread the same sterile aesthetic across the world. *The Verge*.
Dennis, J. (1931). *The critical work of John Dennis*. Baltimore: Johns Hopkins University Press.
Grudin, R. (2010). *Design and truth*. New Haven, CT: Yale University Press.
Habermas, J. (2003). *The future of human nature*. Cambridge: Polity Press.
Haklay, M. (2018). Beyond quantification: A role for citizen science and community science in a smart city. In R. Kitchin, T. Lauriault, & G. McArdle (Eds.), *Data and the city* (pp. 213–224). London: Routledge.
Harvey, D. (2008). The right to the city. *New Left Review, 53*, 23–40.
Heidegger, M. (1977). The Question Concerning Technology. In W. Lovitt (Ed.), *The question concerning technology and other essays* (pp. 3–35). New York: Harper Torchbooks.
Heidegger, M. (2008). On the origin of the work of art. In D. F. Krell (Ed.), *Basic writings* (pp. 142–212). New York: HarperCollins.
Ihde, D. (1990). *Technology and the lifeworld*. Bloomington, IN: Indiana University Press.
Kant, I. (1969). *The critique of judgement*. Oxford: Clarendon Press.
Kieran, M. (1997). Aesthetic value: Beauty, ugliness and incoherence. *Philosophy, 72*(281), 383–399.
Kowitt, B. (2015). Inside the trillion-dollar war on big food. *Fortune*.
Lefebvre, H. (1996). Space and politics. In E. Kofman & E. Lebas (Eds.), *Writings on cities* (pp. 185–202). Hoboken, NJ: Blackwell.
Light, A. (2005). Wim Wenders's everyday aesthetics. In A. Light & J. Smith (Eds.), *The aesthetics of everyday life* (pp. 109–131). New York: Columbia University Press.
Montgomery, C. (2013). *Happy city: Transforming our lives through urban design*. New York: Farrar, Straus and Giroux.
Ngai, S. (2010). Our aesthetic categories. *PMLA, 125*(4), 948–958.
Ngai, S. (2015). *Our aesthetic categories: Zany, cute, interesting*. Cambridge, MA: Harvard University Press.
Outka, E. (2008). *Consuming traditions: Modernity, modernism, and the commodified authentic*. Oxford: Oxford University Press.
Perreault, C., & Sarah Mathew, S. (2012). Dating the origin of language using phonemic diversity. *PLoS One, 7*(4), e35289.
Plummer, T. (2004). Flaked stones and old bones: Biological and cultural evolution at the dawn of technology. In *Yearbook of Physical Anthropology* (Vol. 47, pp. 118–164). Hoboken, NJ: Wiley.
Pye, D. (1968). *The nature and art of workmanship*. Cambridge: Cambridge University Press.
Rappaport, R. (1979). *Ecology, meaning, and religion*. Berkeley, CA: North Atlantic Books.
Rilke, R. M. (2014). *The Duino Elegies & the sonnets to Orpheus*. New York: Knopf Doubleday Publishing Group.
Stiegler, B. (1998). *Technics and time, 1: The fault of Epimetheus*. Palo Alto, CA: Stanford University Press.
Taylor, C. (1991). *The ethics of authenticity*. Cambridge, MA: Harvard University Press.
Tijmes, P. (2001). Albert Borgmann: Technology and the character of everyday life. In H. Achterhuis (Ed.), *American philosophy of technology: The empirical turn* (pp. 11–37). Bloomington, IN: Indiana University Press.
Verbeek, P.-P. (2001). Don Ihde: The technological lifeworld. In H. Achterhuis (Ed.), *American philosophy of technology: The empirical turn* (pp. 119–146). Bloomington, IN: Indiana University Press.
Wittingslow, R. (2016). Bloody-minded metaphysics: Barry Allen vs. the world. *Contemporary Pragmatism, 13*(2), 129–142.

Ryan Mitchell Wittingslow is an assistant professor of philosophy of art & culture at the University of Groningen, and received his PhD from the University of Sydney in 2014. Most of his research sits at the intersection of aesthetics, philosophy of technology, and philosophy of design. He also has devastating opinions about art.

Chapter 14
Applying Biomimicry to Cities: The Forest as Model for Urban Planning and Design

Henry Dicks, Jean-Luc Bertrand-Krajewski, Christophe Ménézo, Yvan Rahbé, Jean Philippe Pierron, and Claire Harpet

Abstract The idea of applying biomimicry to cities is attracting increasing attention as a way of achieving sustainability. Undoubtedly the most frequently evoked natural model in this context is the forest, though it has not yet been investigated with any great scientific rigour. To overcome this lacuna, we provide: first, a justification of the model of the forest via what we call the arguments from "fittingness", "scale", and "complexity"; second, an exploration of various key innovations made possible by this model in the fields of urban planning, urban water systems, urban energy and transport systems, and urban food and nutrient systems.

Keywords Biomimetic architecture · Biomimetic urbanism · Eco-cities · Nature-based cities · Sustainable cities

14.1 Introduction

Biomimicry may be defined as the art or science of imitating Nature. Applied initially to relatively small-scale innovations, often under the names of biomimetics and bioinspiration, in recent years the rising belief that it is through imitating Nature

H. Dicks (✉)
University Jean Moulin Lyon 3, Lyon, France

J.-L. Bertrand-Krajewski
Lyon University, Lyon, France

C. Ménézo
University Savoie Mont Blanc, Chambery, France

Y. Rahbé
AgroParisTech, Paris, France

J. P. Pierron
University of Burgundy, Dijon, France

C. Harpet
University Jean Moulin Lyon 3, Lyon, France

M. Nagenborg et al. (eds.), *Technology and the City*, Philosophy of Engineering and Technology 36, https://doi.org/10.1007/978-3-030-52313-8_14

that we will be able to achieve the elusive goal of sustainability (Benyus, 1997) has led to increased interest in applying biomimicry at much larger scales. Permaculture (Hervé-Gruyer & Hervé-Gruyer, 2014) and certain variations of agro-ecology (Jackson, 2011) have advanced the idea that our agricultural systems should imitate Nature. Industrial ecology has sought to take the working of natural ecosystems as models for industrial systems (Erkman, 2004). And the emerging field of biomimetic urbanism has sought to take Nature as a model for cities. It seems fair to say, however, that this third major large-scale application of biomimicry is at present the least developed. Broadly speaking, there would appear to be two related reasons for this. The first is that cities are extremely complex, bringing together a wide variety of different technological objects and systems, and involving almost every major sector of technological innovation, including energy, water, the built environment, ICT, waste, food production, and transport. The second reason is that the way humans participate in cities is much more complex and introduces far more important issues relating to the human and social sciences than is the case in other artificial macro-systems, such as agricultural or industrial systems. Cities are not just places where technology and labour combine to produce things, but also economic, social, cultural, and political centres, a state of affairs that poses a major challenge to standard approaches to biomimicry operating only at the interface between the life sciences and the technical disciplines (CESE, 2015).

Given the complexity of cities, it is likely to be beyond the scope of a text of this length to address in a meaningful way the myriad issues raised by the application of biomimicry to the city. Our goal will thus be the more modest one of exploring—principally at the level of technologies and technological systems—a natural model that has occurred again and again in the recent literature on the subject: the forest. This model would appear to have been introduced first by Braungart and McDonough (2002, p. 139), who call on us to "imagine a building like a tree, a city like a forest", but without examining this idea in a scientific or systematic way. Similarly, the contemporary biomimetic architects, Schuiten (2010) and Callebaut (2015), both take up the idea of the forest as model for the city but without subjecting it to rigorous analysis. Callebaut's vision for the Paris of 2050, for example, is underpinned by his belief that we should "[t]ransform our cities into ecosystems, our districts into forests and our buildings into inhabited trees" (Callebaut, 2015, back cover, our translation), but when it comes to applying this model, his proposals are limited to stand-alone buildings and so can hardly be said to address the question of what it would mean for an entire city to be modelled on a forest. Other examples of architects attempting to apply the model of the forest to the city are similarly partial and unsystematic. Faced with the problem of chronic air pollution in China, Boeri (2015) has proposed the creation of what he describes as "forest cities". In practice, however, the model of the forest serves here only as a model for tall buildings covered in vegetation in order to purify the air and regulate internal temperatures. Other ways in which the forest may serve as a model for urban design are not addressed.

If, in keeping with our definition of biomimicry as the "art or science" of imitating Nature, contemporary architects have typically tended towards the "artistic" pole of biomimetic urbanism, many academics have developed more "scientific" approaches. Newman and Jennings (2008), for example, think cities should imitate various emergent or systemic properties characteristic of natural ecosystems, most notably their status as healthy, zero-waste, self-regulating, resilient, self-renewing, and flexible. Further, they also suggest that exploring an analogy between ecological succession and urban development may help us better implement the transition towards cities characterized by such characteristics of mature ecosystems as high levels of nutrient cycling and more efficient energy use. Another notable example in this field is the work of Pedersen Zari (2015), who has argued that the systematic study and quantification of the ecosystem services provided by the pre-development ecosystem may provide not just a *model*, but also a *standard* or *measure* capable of helping us set targets for the reproduction of these same ecosystem services in future bio-inspired cities.

If all these approaches have the advantage of adopting scientific methodologies and focusing on the city as a whole, rather than on individual buildings or isolated problems, it is also true that they do not explicitly address the model of the forest. So, while in Pedersen Zari's case study of the city of Wellington in New Zealand, the pre-development ecosystem was in fact a forest (low-land broadleaf podocarp), her concern is primarily with how "behavioral change" and "existing technologies" may help us reproduce the ecosystem services provided by this forest and not with how the various elements and relations within it—the organisms, abiotic elements, and the relations between them—might help us develop appropriate bio-inspired technologies and articulate them into a new urban form. The result is relative indifference to the means by which the pre-development ecosystem services are reproduced, such that, to take just one example, it matters little whether energy is produced by wind turbines or solar cells inspired by tree leaves, provided the basic ecosystem service—energy production—is reproduced to the required degree.

In view of all this, it would seem that there is something of a lacuna in current thinking about biomimetic urbanism regarding the systematic, scientific exploration of the model of the forest. The remainder of this article will be dedicated to filling in this lacuna. In the first part, we will concentrate on *justifying* the model of the forest. The idea of "buildings like trees, cities like forests" clearly holds a strong poetic appeal to many biomimetic architects and urbanists, but are there rational grounds for privileging this model, such that it could potentially become the standard model for the design of all or almost all future cities, or is the forest ultimately just one model amongst others? With a view to answering this question, we will consider three potential justifications for the idea that forests possess a privileged status in biomimetic urban design: (a) the argument from *fittingness*, the key idea of which is that in a large proportion of cases the forest would have been the native or pre-development ecosystem destroyed to make room for the city; (b) the argument from *scale*, which holds that, in contrast to many other natural systems, there is a relatively close correspondence in scale between forests and cities; (c) the argument from *complexity*, which maintains that, as generally the most complex of terrestrial ecosystems, forests hold the most significant potential for bio-inspired innovation. In the second part of the article, we will turn instead to *exploring* the model of the forest. This exploration will be divided into four parts, each focusing on a specific sector of biomimetic urbanism: urban planning, urban water systems, urban energy and transport systems, and urban food and nutrient systems.

14.2 Why the Forest?

14.2.1 The Argument from "Fittingness"

Much of the literature on biomimicry and biomimetic urbanism emphasizes the importance of "fitting in" (Benyus, 1997; Braungart & McDonough, 2002). Benyus, for example, argues that whereas the mentality underlying the first industrial revolution impelled us to "learn *about* nature so that we might circumvent or control her", the mentality underlying the second industrial revolution—the "biomimicry revolution", as she calls it—involves coming to "learn *from* nature, so that we might fit in, at last and for good, on the Earth from which we sprang" (Benyus, 1997, p. 9). Nature, as Benyus (1997, epigraph) points out, does not just tell us what "works" and what "lasts", but also what is "appropriate", that is to say, what allows us—including our artefacts and our systems—to fit in in a given context. In the case of artificial macro-systems, this is typically taken to mean that we should look to native ecosystems for guidance as to how to do things in the same location. The American agro-ecologist, Wes Jackson (2011), for example, argues that it is the pre-development ecosystem—in his case the prairie—that should teach us how to farm. And Pedersen Zari (2015), as we have already seen, thinks the pre-development ecosystem should act as both model and measure for urban design.

With this focus on the pre-development ecosystem in mind, it is interesting to consider George Perkins Marsh's claim that "[t]here is reason to believe that the surface of the habitable earth, in all climates and regions which have been the abodes of dense and civilized populations, was, with few exceptions, already covered with a forest growth when it first became the home of man" (Marsh, 1864, p. 114). While forests currently represent only around 31% of the earth's current land surface, in pre-historic times this figure would have stood at 45% (FAO, 2012), and there can be little doubt that forests have been disproportionately chosen by humans as places to build cities (subsequent to deforestation). One single forest biome—the temperate forest—is home to many of the great cities of North America, notably those of the East Coast, to the majority of important European cities, to the cities of northeastern Asia, including the whole of Japan, North and South Korea, and Eastern China, as well as to many major cities in Australia and New Zealand. When one further factors in cities located within the tropical and sub-tropical forest biomes of South America, West Africa, India, South East Asia, and Northern Australia, as well as those found in the boreal forests of Scandinavia, Canada and Russia, it is clear that the forest would indeed have been the pre-development ecosystem for a very large proportion of the world's existing cities. Further, the very principle that brings us to the model of the forest—the idea that Nature teaches us how to fit in in a specific place—also points towards imitating not just the features common to forests in general, but also to specific forests in specific locations.

There is, however, an obvious objection that could be put forward here. Not only are there many existing cities not located in forest biomes, but one may also wonder what is to prevent us from taking non-forest biomes as models for new bio-inspired cities. There are two important strategies one might use to respond to this objection. The first consists in exploring the idea that many existing cities located in other terrestrial biomes may either prove unsustainable in the long-term or would struggle to be viable were the native ecosystem to be taken as model. As a general rule, the two key parameters that enable forest growth are precipitation and temperature (Ghazoul, 2015). When precipitation is too low, either absolutely or relative to temperature, grassland or desert will generally result. Given that cities, like forests, require significant quantities of usable water per unit of land, the limited local precipitation characteristic of grassland and desert biomes often constitutes a significant challenge for cities, hence their frequent recourse to the exploitation of fossil aquifers or to transfers between rivers or between river basins. Both of these solutions, however, are problematic in this context. Exploiting aquifers more quickly than they are replenished—a particular problem in arid areas such as America's Great Plains (Ogallala aquifer)—is unsustainable, and large-scale artificial transfers clearly involve "circumventing or controlling" Nature, as opposed to following or imitating it, hence philosophical critiques of this practice as "unnatural" (Rolston, 1995). Of course, it could also be objected here that biomimicry provides a perfect response to both these problems. Rather than drawing on fossil aquifers or diverting water from distant rivers, cities located in grassland or desert biomes could mimic the water-saving techniques characteristic of their native ecosystems, while at the same drawing on the parameters of these native ecosystems to set limits for sustainable water consumption. There can be little doubt that this would make cities located in these biomes more sustainable, but even then it is possible that they would still suffer shortfalls of locally available water that would put an intolerable strain on the basic functioning of the city. As for cities located in tundra biomes, the principal long-term obstacles they face to sustainability are sourcing sufficient renewable energy, particularly in winter when sunlight is much reduced, and growing enough food locally, given the obstacles to long-distance food transfers posed by a post-carbon future.

The second strategy one might use to respond to the objection that other ecosystems could also furnish important models for biomimetic cities is to note that cities typically require much greater quantities of water than is available locally in the two important biomes of grasslands and deserts, and that, were these greater quantities of water available sustainably, this surplus water may allow them to take forests as models, thus benefitting from other advantages of this model, most notably their appropriate scale and tendency to exhibit greater complexity (see below). This is particularly relevant in the case of the many cities located close to rivers in grassland or desert biomes, which could potentially use water from local rivers to realize cities based on forests. Indeed, one could even argue that exploiting local rivers in this way is simply an extension of what already happens in Nature when riparian forests form along the banks of rivers in otherwise arid areas, and in this sense the practice may be regarded as biomimetic, in contradistinction to engineering projects that redirect water flows between rivers and basins in accordance with the modern logic criticized by Benyus of "circumventing and controlling" Nature.

14.2.2 The Argument from Scale

It could perhaps be thought that we have proceeded too quickly in assuming that it is some sort of ecosystem that should constitute the basic model for the city. After all, why not use other models present in Nature? While we cannot compare the model of the forest to every other possible natural model, it will nevertheless prove instructive to consider two models frequently evoked in the literature on urban design: biological organisms and the nests of social insects. As we will see, however, both of these alternative models face significant difficulties arising from disparities in scale.

One important feature of the organismic model that should give us pause for thought is that it has for centuries already served as an important model—or at least as an important analogy or structuring metaphor—for the conception of the city. Many architects of the early Italian Renaissance, such as Giorgio Martini, Filarete, and Alberti took the human body as model for the city (Choay, 1974). Similarly, from the nineteenth century onwards, the model of the biological organism was also very common. Cerda, for example, understood the city as an "urban organism", with the urbanist's role being that of its "anatomist" and "doctor" (Choay, 1996). Likewise, in *Urbanisme*, Le Corbusier (1925) constantly describes the city in biological and physiological terms, and, in keeping with this, sees the role of the urbanist as that of either "doctor" or "surgeon". Even today, this model is still very much present, whether in highly scientific approaches to cities (Batty, 2013) or in the context of smart cities (Picon, 2013).

One thing that proponents and theorists of the "urban organism" have typically overlooked, however, is the question of scale. For the architects of the early Italian Renaissance, the principal geometrical feature of the human body they sought to reproduce was its proportions, which, in keeping with the notion of Vitruvian man, were regarded as "perfect" (Wittkower, 1962). But taking the human or animal organism as model implies that the city will take on the ecological function of a giant consumer, incapable of carrying out analogous ecological functions to those of producers and decomposers. An unfortunate consequence will be huge and often unsustainable demands placed on external ecosystems both in terms of resource provision and waste absorption. Much the same may be said about contemporary attempts to theorize the smart city as an "urban organism" (Picon, 2013). While these have typically attempted to address the problem of unsustainability via the notion that information and communication technologies—conceived via the model of the sensory apparatus and nervous system—could make the city more efficient, thanks to the detection of leakages, rapid or pre-emptive reaction to external perturbations, and so on, they do not consider the possibility that the model of the organism may, in large part for reasons of scale, be fundamentally problematic no matter how efficient it may become.

A similar issue arises regarding the nests of social insects. Lovelock (2014), for example, thinks that the nests of termites provide a good overall model for the city, in part because they could allow us to solve the problem of excessive urban temperatures resulting from global warming. More specifically, he thinks it may be possible to build entire cities largely closed off from the external environment, as buildings generally are, and thereafter regulate their internal temperature using bio-inspired passive cooling techniques pioneered by termites. Again, however, one obvious danger with this model is that cities are orders of magnitude larger than insects' nests and that, like living organisms, these nests also assume the ecological role of consumers and are thus heavily dependent on their external environment both for the provision of resources and the absorption of wastes.

If the above arguments do indeed suggest that the forest may be a more appropriate model for the city than either biological organisms or the nests of social insects, it may nevertheless be objected that there is no reason why *one single model* should be taken as fundamental. Indeed, it could even be argued that privileging the model of the forest amounts to a sort of fundamentalism and that a more pragmatic and pluralist approach would have the advantage of enabling the articulation of multiple models for the city, without any one of them being accorded priority. The principal weakness with this objection is its assumption that highly diverse features of natural systems operating at radically different scales may be abstracted from their natural context and thereafter blended, modified, and manipulated at will. One important issue this assumption overlooks is the way that within living systems the presence of certain elements or properties will often strongly favour or disadvantage others. For example, it may well turn out that taking trees as models for energy generation at building-level will also strongly favour the adoption of a comparable mechanism to that found in the forest for temperature regulation, namely evapotranspiration; conversely, it may prove very difficult to combine tree-inspired energy generation systems with a gigantic city-wide air-conditioning system that requires the city to be largely sealed off from the external environment, as is the case in termites nests. To say this does not entail commitment to a strong holism, according to which specific parts of living systems cannot be taken as models and thereafter translated into other contexts, for that would by definition render all biomimicry impossible. Seeing the forest as the *fundamental* model for the city, understood as that model that gives the biomimetic city its fundamental characteristics, but without determining its every last detail, would still allow us to draw on biological models not found in forest ecosystems. To take a simple example, sea sponges have aroused the interest of biomimicry researchers thanks to the advanced fibre optics they possess (Kulchin et al., 2008). But the simple fact they are found in the sea is no reason not to import this technology into urban telecommunications networks. Indeed, the "appropriateness" of this natural technology has little to do with the specifically marine environment in which it first developed and more to do with its advantageous mechanical parameters and low-temperature production process. This in turn reminds us of the importance of not interpreting "appropriateness" or "fittingness" in too narrow or dogmatic a way, as if the sole relevant criterion were whether the model to be imitated already exists in the local ecosystem.

14.2.3 The Argument from Complexity

A third argument for taking forests as models for cities, and one which we have already alluded to in our earlier comparison of the earth's major biomes, concerns the fact that forests are by and large the most complex land-based ecosystems. This is of course a broad generalization. Barthlott, Erdelen, and Rafiqpoor (2014) note that biodiversity varies in large part as a function of what they call "geodiversity", understood as diversity of abiotic factors, such as climate, geomorphology, and geology. Biodiversity is interpreted here in a rather narrow way as species diversity, which is clearly not the same thing as complexity, but it does nevertheless suggest that some non-forest ecosystems may exhibit greater complexity than some forest ecosystems. This proviso aside, the basic reason for the tendency of forests to exhibit greater complexity is the availability of significant quantities of water, minerals, and organic compounds, which, coupled with appropriate temperatures, generally allow forests to contain more diverse elements and more diverse relationships between these elements than other land-based ecosystems. These elements and relations in turn make possible such important system-wide properties as relatively high rates of nutrient recycling, effective self-regulation, and resilience to perturbations. Further, it is also true that those forests that benefit from the most favourable climatic conditions, namely tropical rainforests, are not only generally the most diverse of forest ecosystems, but also the most stable and resilient (Ghazoul, 2015).

The high complexity often characteristic of forests constitutes a significant advantage for biomimetic urbanism, for it allows them to offer a greater variety of natural models to imitate, whether at the level of individual elements, relationships, or systemic properties. Further, given the great complexity of cities, it is also debatable whether ecosystems working at relatively low levels of complexity offer viable models in this context. Grasslands may well offer important models for biomimetic agricultural systems (Jackson, 2011), but, as we have already seen, the limited water resources and the reduced complexity that result may pose a significant obstacle to their viability as models for cities. Deserts may contain clusters of cacti, which could perhaps be taken as models for small villages that emulate the water-saving and energy generating capacities of these succulent plants, but it is doubtful whether they exhibit sufficient complexity to act as a model for an entire city. So, while it is beyond doubt that the fauna and flora of non-forest ecosystems constitute interesting models for biomimetic design, whether they can function as models for entire cities remains an open question.

14.3 Exploring the Model of the Forest

14.3.1 City Planning as "Biome-Mimicry"

In the previous sections, we referred to both forest *ecosystems* and forest *biomes*. Biomes differ from ecosystems in two respects: they are large-scale entities, whereas ecosystems may be defined at varying scales; and whereas ecosystems are defined by interactions between biotic and abiotic components, biomes are defined by their principal form of vegetation. This is not to say, however, that within a given biome various ecosystems not defined by their principal vegetation cannot also be present at smaller scales. The temperate forest biome, for example, is defined by its principal form of vegetation, though any given portion of this biome may also contain such ecosystems as rivers, lakes, wetlands, treeless floodplains, and steep slopes or cliffs. In what follows, we will draw on the distinction between ecosystems and biomes to propose a vision of city planning as the imitation of the native biome, or "biome-mimicry", an approach that has thus far been applied mainly to regional planning (Biomimicry Guild and HOK, 2013).

The traditional way that environmentally conscious urban planners have taken the different ecosystems of the native biome into account, as opposed to just smothering them by clear-cutting forests, burying rivers, draining wetlands, building on flood plains, and so, was through the method of "ecological determinism" advanced and practised most notably by McHarg (2006). Ecological determinism consists in mapping the various natural features—geological, hydrological, biological, ecological…—of the local site, categorizing them in terms of their relative ecological value as established via considerations of productivity, biodiversity, role in risk mitigation, and so on, and then privileging the development of those sites whose ecological value is lowest.

An important aspect of McHarg's ecological determinism is that, as Corner (2006) has noted, it continues to suppose a dichotomy between nature and the city. In doing so, it also remains within the "dualist" mind-set characteristic of what environmental philosopher, Mathews (2011, p. 364), calls "the traditional project of environmentalism", the basic focus of which is Nature preservation and conservation. Where Mathews thinks biomimicry differs from this dualist mind-set is in its belief that humans are not necessarily a destructive ecological force, since, through following Nature's basic rules and principles, such as running on solar energy, recycling nutrients, or using energy efficiently, we may come to participate constructively in the life of the biosphere, a change in perspective that Mathews thinks represents a "turning point in Western thinking" (Mathews, 2011, p. 368). Applied to the city, this would mean that any natural ecosystem within the forest biome could at least in principle be the subject of biomimetic development or, in those cases where it had already been smothered by traditional urban development, regenerated in such a way that it could participate effectively in a city working in analogous ways to a forest ecosystem. As an example of the former, undeveloped wetland areas could be taken as sources of inspiration for new wastewater treatment systems, whether through adapting existing wetlands to that purpose, building specialized sewage treatment plants based on wetland ecosystems (Todd & Todd, 1993), or by some combination of the two. An example of the latter would be the daylighting and regeneration of urban streams and rivers such that they could henceforth carry out various important ecological services—e.g., flood protection and water purification—important to the functioning of "cities like forests".

In keeping with the status of the forest as the defining ecosystem of forest biomes, it is also true, however, that it is this ecosystem that would constitute the principal model for the built urban environment. One important planning strategy this model suggests is increasing density through verticality, whether through upward or downward growth, both of which could draw inspiration from a key feature of tree growth: modularity, as seen in the growth of trunks, leaves, branches, and roots. This does not imply, however, that skyscrapers would become the norm. Within forests, upward growth is subject to limiting factors and the same would also be true of biomimetic cities; cities composed primarily of skyscrapers, for example, would no doubt struggle to generate sufficient solar energy, for increasing volumes would lead energy demand to increase disproportionately with respect to supply. As for downward growth, it is interesting to note that the celebrated French architect, Dominique Perrault, often frames his key concept for the densification of cities—the "groundscape"—in biomimetic terms (Degioanni, 2015), a topic to which we will return when we come to examine various sector-specific biomimetic innovations.

14.3.2 Urban Water Systems

In most forest ecosystems, the predominant source of water is precipitation and only to a minor extent streams, rivers and groundwater. This model implies the generalization of decentralized rainwater harvesting, whether for non-potable uses (irrigation and gardening, laundry, industry, fire-fighting reserves), to reduce surface runoff and flooding, to refill aquifers (Han & Mun, 2011), or even—especially in places with water scarcity or uneven precipitation regimes—for drinking water production. The transition to decentralized rainwater harvesting would require improved water quality and thus also a reduction of anthropogenic pollutant emissions and the development of appropriate rainwater treatment technologies. The forest biome offers appropriate models for both of these, for it not only runs on non-toxic biological nutrients, but also filters water in ways we might profitably imitate, including physical filtration, phytodepuration, and mycofiltration (Stammets, 2005). In addition, decentralized rainwater harvesting would significantly increase resilience in the face of technological failures, terrorist or military attacks, natural disasters, and other disruptions.

Another important feature of the forest biome is the use of rainwater for cleaning purposes. The leaves of many plants, including most famously the lotus, have hydrophobic surfaces which enable rainwater droplets to collect and remove impurities. This property has already been imitated in the form of self-cleaning paints and surfaces (Forbes, 2005), but its natural deployment in leaves suggests a particularly important application: self-cleaning solar panels, which could prevent efficiency reductions of up to 30% caused by surface impurities (Crawford, 2012). Further, forest rainwater that has previously been used for cleaning purposes will later reach the ground where it can be re-used. Drawing on this model, the local recycling and reuse of greywater—generally used initially for cleaning purposes (washing buildings and vehicles, showers, laundry…)—could be developed for, amongst others uses, irrigation or gardening (Li, Wichmann, & Otterpohl, 2009).

The model of the forest could also underlie a paradigm shift in stormwater management. In existing densely urbanised environments, more than 50% of precipitation is transformed into surface runoff and evacuated by means of artificial systems (drains, sewers) into surface waters or aquifers, approximately 15% is infiltrated into the soil through the remaining pervious areas, and the last 30–35% either evaporates directly or evapotranspires. In forest ecosystems, by contrast, surface runoff is typically only around 10%, infiltration 50% and evapotranspiration 40% or above (Barraud et al., 2009). Taking the forest ecosystem as model would thus imply a significant reduction in the imperviousness of urban ground surfaces, by means of porous pavements and roads, infiltration ponds and swales, raingardens, and biofilters. Further, drawing on the model of the tree, architects could design buildings in such a way that they mimic tree roots in facilitating the infiltration of water into the soil, thus reducing the need to dedicate large surface areas to artificial infiltration basins. Increasing infiltration in these ways would offer many advantages, in particular groundwater recharge, contribution to river base flow, and, with adequate storage similar to water retention in forest soils, the generation of additional water resources for use in dry periods, as advocated by proponents of the "sponge city" (Tu & Tian, 2015). As for evapotranspiration, it could be imitated via a diversity of techniques, including vegetated roofs and walls (which also reduce runoff), growing creepers up buildings, raingardens, natural and constructed wetlands, and tree planting (Dover, 2015). The principal advantages of these techniques include a reduction of the urban heat island effect, aesthetic and psychological benefits for humans, and an increase in urban biodiversity.

Wastewater treatment is another sector the forest biome allows us to rethink in a more ecological manner. The model of the forest biome implies both the greater use of decentralized wastewater treatment and the development of treatment techniques that approximate more closely to the complex ecological treatment of wastes characteristic of natural wetlands. Regarding decentralized treatment, dry toilets, faeces composting, and separate urine treatment systems could not only reduce water consumption and the pollution of surface waters (especially in combined sewer systems), but also make possible efficient treatment of pharmaceutical wastes and nutrient recovery for re-use as fertilizers. As for dedicated wastewater treatment systems, it would be possible further to develop and generalize the use of wetland-inspired water-treatment plants, which not only facilitate the recovery of nutrients (biogas, nitrogen, phosphorous) but also reduce energy use for oxygenation (Todd & Todd, 1993). Further, as noted above, the restoration of natural wetlands or the construction of artificial wetlands could also play an important role in wastewater treatment, though the viability of this method would likely depend on the adoption of other biomimetic practices, notably a reduction of the quantity (via less wasteful consumption and greywater reuse) and an improvement of the quality (via avoidance or treatment of anthropogenic pollutants) of the wastewater entering these ecosystems (Kadlec et al., 2000).

14.3.3 Urban Energy and Transport Systems

The principal energy source of forests is sunlight. Taking trees as models for buildings not only suggests the possibility of fitting each individual building with an array of solar panels, but could potentially also lead to developments in biomimetic solar technologies, including water-splitting technologies that reproduce many of the functional properties of leaves (Chong, Colón, Ziesack, Silver, & Nocera, 2016; Reece et al., 2011). Similarly, the study of leaf arrangements on stems, phyllotaxis, may help us arrange these solar technologies in optimal ways, and the imitation of heliotropism could help maximise solar exposition. The French biomimicry specialist, Gauthier Chapelle (2015), even goes so far as to imagine translucent solar panels positioned at different levels and capable of capturing different wavelengths of light, as is the case in forests. Beyond plant photosynthesis, various other living organisms could offer useful models either for the improvement or development of other renewable energy sources (Harman, 2013) or for temperature regulation (Gruber, 2011).

A well-known limitation of solar energy is its intermittency. In forests, this issue is resolved via a number of adaptations, including seasonal changes (lower growth rates, leaf loss) and energy storage in the form of sugars, lipids, or other types of biomass. Regarding the former, it is quite possible that in strongly seasonal climates urban energy use would need to decline significantly in winter months, though drawing on geothermal energy—as some animals do for warmth—could partially offset this problem. Regarding the latter, while energy storage is a widely recognized necessity for the transition to renewables, the important question it raises is whether it would be possible to store energy in simple chemical form—hydrogen and hydrogen compounds being the obvious examples—rather than in conventional battery technologies. Indeed, were water-splitting artificial photosynthesis technologies to become viable at a large scale, then this could potentially signal a move away from the current focus on electricity storage and towards a hydrogen-based energy storage system closer to that found in forest ecosystems.

Another important feature of forests is the rhizosphere, which could provide a response to the problem of uneven energy access and demand. In forests, trees exchange energy for nutrients with an extensive network of mycelium, a process which also makes excess energy available to other trees. The net result is the transfer of energy from trees that have an excess of energy to those that, for a variety of reasons—especially differences in species, age and height—are energy deficient (Simard et al., 1997). This system may provide a model for a distributed underground energy network enabling buildings that are at any given moment "energy positive" to transfer their excess energy to those that are "energy negative". Further, the information flows that manage these transfers and exchanges in what has been referred to as "nature's internet" (Stammets, 2005) and the "underground information superhighway" (Bais, Park, Weir, Callaway, & Vivanco, 2004) could also provide a natural model for the convergence of distributed energy and communication technologies advocated by Rifkin (2011).

A further way in which energy use in forests could provide a model for cities concerns the transport sector. The lightweight bone structures and morphological adaptations of animals have long been used as a source of inspiration for vehicle design, often with a view to improving energy efficiency (Harman, 2013). As far as the model of the forest is concerned, however, perhaps the most important source of inspiration is the interactions between producers (especially trees) and consumers (insects and animals). Just as the insects and animals of the forest obtain their energy from the trees and other plants, so the vehicles of the city could obtain their energy directly from the buildings, as has also been advocated by Rifkin (2011). As for urban vehicle parks, it would again be possible to draw on the model of the forest to design multiple tree-shaped structures—posts on which an artificial canopy of solar technologies would be mounted—, which could not only recharge electric or hydrogen-powered vehicles, but also provide shade, just as the trees in the forest do for animals and other organisms. Lastly, nature could even provide models for innovative urban transport systems. Schuiten (2010), for example, has drawn on the model of pine processionary caterpillars to propose an automated transport system that can work in both "private" (separate) and "public" (processionary) ways, thus reducing both fuel consumption and congestion.

14.3.4 Urban Food and Nutrient Systems

The idea of "buildings like trees, cities like forests" was put forward in Braungart and McDonough's *Cradle to Cradle* (Braungart & McDonough, 2002), a book better known for the idea that there are two basic types of nutrients—biological (biodegradable) and technical (non-biodegradable), both of which may be infinitely recycled. In combination with the idea of "cities like forests", this implies a transition towards much greater recycling of both types of nutrients within the urban environment.

Despommier (2011, p. 3) has recently applied this principle to urban agriculture, arguing that in order to transform cities into the "functional urban equivalent of a natural ecosystem", we must both produce and recycle the food consumed in the city within the city itself, hence his advocacy of vertical farming. On our view, this argument overlooks two key issues. The first concerns the fact that, while biomimetic cities would come to resemble forests in certain important respects, they would nevertheless remain cities, a state of affairs which raises the important question of *what a city is*. Without advancing a full response to this question, it is important to note that ever since their initial development in Natufian and Neolithic cultures, cities have depended on surrounding agricultural systems capable of generating high quantities of calories in the form of cereals, meat, and dairy products (Mithen, 2004), in which case it seems plausible to say that dependency on extra-urban agricultural production has always been part of what makes a city a city. The second concerns the fact that, unlike forests, cities contain a very high concentration of one species, *homo sapiens* (and their domestic companions). Deploying sufficient numbers of vertical farms to feed all the human residents of the city would amount to a radical departure from the workings of the forest, for a forest could never produce enough food to feed current urban populations of humans. Further, it also seems likely that the attempt to increase urban food production to anything like 100% of local consumption would in many cases lead to problematic competition with other uses of available space. In view of all this, it is very hard to see how the calories currently provided by cereals and dairy animals—both of which inhabit non-forest biomes (grasslands)—could be produced in "cities like forests" using nothing other than the renewable resources directly available in the urban environment, a state of affairs that raises the important question of the models of both the prairie (Jackson, 2011) and the forest in biomimetic agriculture (Hervé-Gruyer & Hervé-Gruyer, 2014). To be sceptical of the idea that all food production could potentially take place in the urban environment does not imply, however, that urban farming is to be excluded. On the contrary, the forest offers many natural models one could draw on in this context, including fruit-producing trees, the production of herbs, vegetables and berry-producing plants in clearings, the cultivation of fungi by certain species of ants, honey production from bees, and raising forest-dwelling scavenging animals like pigs and chickens.

As far as the technical nutrients currently dominant in the manufacturing and construction sectors are concerned, while it is true that more or less closed-loop recycling of these nutrients implies a move away from the linear flows of the "extractive economy" (Benyus, 1997) and the imitation of the high-levels of nutrient cycling characteristic of forest ecosystems, it is also true that biomimicry has been critical of the "heat, beat, and treat" techniques used in the production and recycling of technical nutrients (Benyus, 1997). With this in mind, the generalization of the model of the forest would seem to point towards a reduction of technical nutrients and the transition to a bio-based, solar economy, as advocated by Scheer (2002). Of particular interest in the context of architecture and urbanism would be the generalization of infinitely recyclable, biological construction materials, the most obvious one—given the model of the forest—being wood. As for manufacturing, it is interesting to note that 3D printers and other decentralized manufacturing techniques, which could potentially be designed to run on biological nutrients (Kennedy, Fecheyr-Lippens, Hsiung, Niewiarowski, & Kolodziej, 2015), could enable us to approximate to the hyper-local production and recycling systems characteristic of forests, thus reducing ecologically disruptive and energy-intensive flows of materials at national, continental, and global scales.

14.4 Conclusion

This chapter had two principal objectives. The first was to justify the idea that the forest is at least the most important model, and, more controversially, perhaps even the only viable model, for biomimetic cities. If the three arguments put forward—from fittingness, scale, and complexity—were no doubt insufficient to prove the stronger of these two positions, such that it clearly remains an open question as to whether tundra, grassland, or desert biomes may function as models for cities in corresponding locations, they do nevertheless strongly suggest that the forest is likely to prove the most important model for biomimetic urbanism. Moreover, even if not presented as such, the exploration of this model with respect to urban planning and design carried out in the second part of the article is perhaps the model's most important justification, for it is through exploring the concrete possibilities it opens up that one becomes aware of the awe-inspiring potential of forests as models for sustainable cities.

It is also true, however, that our exploration of the model of the forest was highly selective, both in that it focused on certain features of the forest at the expense of others and in the sense that it adopted a sectorial approach, when other approaches would also no doubt have been possible. This in turn suggests that much further research is needed into this model, including: a more detailed study of a wider range of key sectors of forest-based urban planning and design, as well as explicit reflexion on their articulation and integration; analysis of the pertinence of different types of forest for biomimetic urban design; consideration of how the temporality of forest ecosystems, including such issues as seasonality, succession, evolution, resilience, and adaptation to climate change may serve as models for urban design; exploration of the spatiality of forest ecosystems qua models for urban design, an obvious starting point for which would be the common division of the forest into different levels (the rhizosphere, the forest floor, the understory, the canopy, and the emergent level), but which could also extend to a consideration of the embeddedness of forests at regional, continental and global scales; and, perhaps most difficult of all, critical reflection on the relation between forest-based cities and the human and social sciences, whether in the form of such relatively practical considerations as the institutional obstacles to realizing biomimetic cities, or deeper, anthropological and philosophical questions about what taking forests as models for cities would mean for our understanding of what it means to be human and of our overall place in Nature.

References

Bais, H. P., Park, S.-W., Weir, T. L., Callaway, R. M., & Vivanco, J. M. (2004). How plants communicate using the underground information superhighway. *Trends in Plant Science, 9*(1), 26–32.

Barraud, S., De Becdelièvre, L., Clozel, B., Gaboriau, H., Seron, A., Come, J.-M., et al. (2009). *L'infiltration en questions - Recommandations pour la faisabilité la conception et la gestion des ouvrages d'infiltration des eaux pluviales en milieu urbain*. Villeurbanne: Rapport ANR.

Barthlott, W., Erdelen, W., & Rafiqpoor, M. D. (2014). Biodiversity and Technical Innovation: Bionics. In D. Lanzerath & M. Friele (Eds.), *Concepts and values in biodiversity* (pp. 11–55). London: Routledge.

Batty, M. (2013). *The new science of cities*. Cambridge, MA: MIT.

Benyus, J. (1997). *Biomimicry: innovation inspired by nature*. New York: Harper Perennial.

Biomimicry Guild and HOK. (2013). *Genius of biome*. Retrieved April 25, 2017, from: http://issuu.com/hoknetwork/docs/geniusofbiome?e=3095950/2547879

Boeri, S. (2015). *Forest city*. Retrieved April 25, 2017, from https://www.stefanoboeriarchitetti.net/en/portfolios/forest-city/

Braungart, M., & McDonough, W. (2002). *Cradle to cradle: re-making the way we make things*. London: Vintage.

Callebaut, V. (2015). *Paris 2050: Les cités fertiles faces aux enjeux du XX siècle*. Paris: Michel Lafon.

CESE. (2015). *Le Biomimétisme: s'inspirer de la nature pour innover durablement, les avis du Conseil Economique, Social et Environnemental*. Retrieved April 25, 2017, from http://www.lecese.fr/sites/default/files/pdf/Avis/2015/2015_23_biomimetisme.pdf

Chapelle, G. (2015). *Le vivant comme modèle*. Paris: Albin Michel.

Choay, F. (1974). La ville et le domaine bâti comme corps dans les textes des architectes-théoriciens de la première renaissance italienne. *Nouvelle Revue de Psychanalyse, 9*, 239–251.

Choay, F. (1996). *La règle et le modèle: sur la théorie de l'architecture et de l'urbanisme*. Paris: Editions du Seuil.

Chong, L., Colón, B., Ziesack, M., Silver, P., & Nocera, D. (2016). Water splitting–biosynthetic system with CO_2 reduction efficiencies exceeding photosynthesis. *Science, 352*(6290), 1210–1213.

Corner, J. (2006). Terra fluxus. In C. Waldheim (Ed.), *The landscape urbanism reader* (pp. 23–33). New York: Princeton Architectural Press.

Crawford, M. (2012). *Self-cleaning solar panels maximize efficiency*. Retrieved April 20, 2017, from https://www.asme.org/engineering-topics/articles/energy/self-cleaning-solar-panels-maximize-efficiency

Degioanni, J.-F. (2015). Visite de l'exposition 'Groundscape, chroniques et fictions' avec Dominique Perrault. *Le Moniteur*, http://www.lemoniteur.fr/article/visite-de-l-exposition-groundscape-chroniques-et-fictions-avec-dominique-perrault-27609713

Despommier, D. (2011). *The vertical farm: Feeding the world in the 21st century*. New York: Picador.

Dover, J. W. (2015). *Green infrastructure: Incorporating plants and enhancing biodiversity in buildings and urban environments*. Oxford: Routledge.

Erkman, S. (2004). *Vers une Ecologie Industrielle: Comment mettre en pratique le développement durable dans une société hyper-industrielle*. Paris: Charles Leopold Meyer.

FAO. (2012). *The state of the world's forests*. Retrieved April 24, 2017, from http://www.fao.org/docrep/016/i3010e/i3010e.pdf

Forbes, P. (2005). *The Gecko's foot: Bio-inspiration – engineered from nature*. London: Fourth Estate.

Ghazoul, J. (2015). *Forests: a very short introduction*. Oxford: OUP.

Gruber, P. (2011). *Biomimetics in architecture: architecture of life and buildings*. New York: Springer.

Han, M. Y., & Mun, J. S. (2011). Operational data of the Star City rainwater harvesting system and its role as a climate change adaptation and a social influence. *Water Science and Technology, 63*(12), 2796–2801.

Harman, J. (2013). *The Sharks Paintbrush: Biomimicry and how nature is inspiring innovation*. Ashland: White Cloud Press.

Hervé-Gruyer, C., & Hervé-Gruyer, P. (2014). *Permaculture: guérir la terre, nourrir les hommes*. Arles: Actes Sud.

Jackson, W. (2011). *Nature as measure: The selected essays of Wes Jackson*. Berkeley: Counterpoint.

Kadlec, R., Knight, R., Vymazal, J., Brix, H., Cooper, P., & Haberl, R. (2000). *Constructed wetlands for pollution control*. London: IWA.

Kennedy, E., Fecheyr-Lippens, D., Hsiung, B.-K., Niewiarowski, P. H., & Kolodziej, M. (2015). Biomimicry: A path to sustainable innovation. *Design Issues, 31*(3), 66–73.

Kulchin, Y. N., Bukin, O. A., Voznesenskii, S. S., Galkina, A. N., Gnedenkov, S. V., Drozdov, A. L., et al. (2008). Optical fibres based on natural biological minerals—sea sponge spicules. *Quantum Electronics, 38*(1), 51–55.

Le Corbusier. (1925). *Urbanisme*. Paris: Flammarion.

Li, F., Wichmann, K., & Otterpohl, R. (2009). Evaluation of appropriate technologies for grey water treatments and reuses. *Water Science and Technology, 59*(2), 249–260.

Lovelock, J. (2014). *A rough ride to the future*. London: Allen Lane.

Marsh, G. P. (1864). *Man and nature or: Physical geography as modified by human action*. Seattle: University of Washington Press.

Mathews, F. (2011). Towards a deeper philosophy of biomimicry. *Organization & Environment, 24*(4), 364–387.

McHarg, I. (2006). Ecological determinism. In F. R. Steiner (Ed.), *The essential Ian McHarg: Writings on design and nature* (pp. 30–46). Washington: Island Press.

Mithen, S. (2004). *After the ice: A global human history 20,000 to 5000 BC*. London: Phoenix.

Newman, P., & Jennings, I. (2008). *Cities as sustainable ecosystems: Principles and practices*. Washington: Island Press.

Pedersen Zari, M. (2015). Ecosystem services analysis: Mimicking ecosystem services for regenerative urban design. *International Journal of Sustainable Built Environment, 4*(1), 145–157.

Picon, A. (2013). *Smart cities: théorie et critique d'un idéal auto-réalisateur* (p. 120). Paris: Collection Actualités.

Reece, S. Y., Hamel, J. A., Sung, K., Jarvi, T. D., Esswein, Z. A. J., Pijpers, J. J. H., et al. (2011). Wireless solar water splitting using silicon-based semiconductors and earth-abundant catalysts. *Science, 334*(6056), 645.

Rifkin, J. (2011). *The third industrial revolution: How lateral power is transforming energy, the economy, and the world*. New York: Palgrave Macmillan.

Rolston, H. (1995). Using water naturally. *Illahee: Journal for the Northwest Environment, 11*, 94–98.

Scheer, H. (2002). *The Solar Economy*. London: Earthscan.

Schuiten, L. (2010). *Vers une Cité Végétale*. Wavre: Madraga.

Simard, S. W., Perry, D. A., Jones, M. D., Myrold, D., Durall, D. M., & Molina, R. (1997). Net transfer of carbon between ectomycorrhizal tree species in the field. *Nature, 388*, 579–582.

Stammets, P. (2005). *Mycelium running*. New York: Crown Publishing.

Todd, N. J., & Todd, J. (1993). *From ecocities to living machines: Principles of ecological design*. Berkeley: North Atlantic Books.

Tu, X., & Tian, T. (2015). Six questions towards a sponge city – report on power of public policy: Sponge city and the trend of landscape architecture. *Landscape Architecture Frontiers/Views and Criticisms, 3*(2), 22–39.

Wittkower, R. (1962). *Architectural principles in the age of humanism*. London: Alec Tiranti.

Henry Dicks is an environmental philosopher and philosopher of technology. Having completed a post-doc on an interdisciplinary project on biomimetic cities at the University of Lyon, he is currently working as a visiting research fellow at the University of Leeds and as a lecturer in environmental ethics at Lyon 3 University and Shanghai University. His ongoing research project into the philosophy of biomimicry has given rise to publications in Philosophy and Technology, Environmental Ethics, Journal of Agricultural and Environmental Ethics, Ethics & the Environment, Architecture Philosophy, Environmental Philosophy, as well as a special issue on the subject in Environmental Values. He is currently writing a book entitled The Biomimicry Revolution: Foundations of a New Philosophy.

Jean-Luc Bertrand-Krajewski is Full Professor in urban hydrology at INSA, Lyon University, France and director of the DEEP laboratory (Laboratory on Wastes, Water, Environment and Pollution) since 2014. He obtained his doctoral degree in 1992 from Strasbourg University. His research topics are related to processes in urban sewer and stormwater systems, and integrated sustainable urban water management, with numerous national and international collaborations. He was member of the IWA (International Water Association) Strategic Council from 2010 to 2018. He is (co-) author of more than 220 papers in peer reviewed journals and conferences. More information can be found on his website at jlbkpro.free.fr.

Christophe Ménézo is Deputy Head of the Solar Academy and scientific director of the CITEE chair at University Savoie Mont Blanc. He is Director of the CNRS Research Federation on Solar Energy and in charge of the Green/smart building topic in the CNRS/NTU Network SINERGIE. He hosted the Chair "Habitats and Energy Innovations", INSA/EDF from 2011 to 2015 and was Associate Professor at University Lyon1 from 1999 to 2008. His research is carried out at the National Institute of Solar Energy (INES), LOCIE lab and focuses on solar buildings, design of building envelopes through bioinspired approaches and solar potential for cities.

Yvan Rahbé graduated in Agronomy at INA-PG (AgroParisTech, Biochemistry 1982) and took a PhD in Insect biochemistry (1984). After a move into tropical entomology (leaf-cutting ants, INRA Guadeloupe), he joined INRA Lyon in 1986, to build a group on aphid-plant physiology. INRA Research director in 2000, head of research unit BF2I in 2006 (UMR0203 INRA/INSA-Lyon), he moved to a CNRS Lab in microbiology in 2016 (UMR5240 MAP), to start a project on aphid-bacterial interactions and vector biology, and on the biomimetics of insect cuticles. He is nurturing a multidisciplinary group on biomimetics (BiG, bioinspiration group) at INSA de Lyon, a leading Poly-Tech University in France.

Jean Philippe Pierron is a full professor at the University of Burgundy, France, specializing in the philosophy of life, medicine, and care. A former dean of the University Jean Moulin Lyon 3, he is the director of the research chair "values of care" at Lyon 3 and co-director of the Franco-Mexican national research project, Val-Uses (Values and uses of sediments: sustainability issues in the valley of Usumacinta at the Mexico-Guatemala border (2018-2020). Director of the professional Masters, "Ethics, Ecology, and Sustainable Development", taught at Lyon 3 and Shanghai University, his publications include: La poétique de l'eau : Pour une autre écologie (2018) and Prendre soin de la nature et des humains: Médecine, travail et écologie (2019).

Claire Harpet is an anthropologist specialized in the interactions between humans and the environment, working as research engineer at University Jean Moulin Lyon 3. Her research focusses on discourses and practices related to nature in inter-tropical regions (Madagascar, Mayotte, Mexico) and in urban areas (metropolitan France). In this latter context, she worked on the transformations, representations and uses of water in the heart of 21st century cities and coordinated, with Jean-Philippe Pierron (director of the Chair, "Rationalities, Uses, and Imaginaries of Water"), several books including Ecologie politique de l'eau: Rationalités, usages et imaginaires, Ed. Hermann, 2017.

Part III
Urban Futures and "Smart" Cities

Chapter 15
From Liberalism to Experimentation: Reconstructing the Dimensions of Public Space

Udo Pesch

Abstract This paper describes how particular normative dimensions have become embedded in public space. It identifies four periods of urban history, starting with that of the *liberal city* which introduces a heterogeneous set of liberal and democratic values. The *sanitary city* of the nineteenth century came to depict public space as a system of flows that needed to managed. The twentieth century *rational city* endorsed functionality and efficiency as key values, as manifested in modernist architecture and neo-liberal policies. Recently the *experimental city* emerged, in which digital technologies give rise to 'smart' systems, but also to new forms of civic engagement. It is argued that the experimental city needs to accommodate the normative dimensions of public space developed in preceding historical periods, which necessitates that conflicts between these dimensions need to subjected to democratic trade-off processes.

Keywords Public space · Urban history · Public values · Urban politics

15.1 Introduction

Current attention for the role of public space in urban environments plays out against the background of neo-liberal and technocratic appropriations of public space. The privatization and homogenization of spatial contexts may threaten democratic qualities of public space, such as the presence of normative plurality and general accessibility (Barnett, 2007; Low & Iveson, 2016; Springer, 2011). These threats are especially urgent in combination with the ascent of the so-called 'smart city', which is based on the surveillance and control of technical and social infrastructures.

U. Pesch (✉)
Faculty of Technology, Policy and Management, Delft University of Technology, Delft, Netherlands
e-mail: U.Pesch@tudelft.nl

M. Nagenborg et al. (eds.), *Technology and the City*, Philosophy of Engineering and Technology 36, https://doi.org/10.1007/978-3-030-52313-8_15

Claims about public space often involve a number of theoretical and political assumptions. First of these is that public space tends to be seen as unambiguously separated from a heterogeneous set of private spaces. Second, people that enter public space are assumed to be *hermeneutically independent*: they bestow their own meanings and categorizations upon their lifeworld. In contrast, the third assumption is that neoliberalism and technocracy are forces that are external to this hermeneutic independence by imposing meanings and categorizations that are non-negotiable (Kohn, 2013; Low, 2006; Mitchell, 1995; Shaw & Graham, 2017). Fourth, technology is seen as something that exerts power or reproduces existing power asymmetries, while also being a homogenizing power in itself. Acknowledging their character as crude generalizations, these assumptions will be qualified in this paper by sketching out the historical development of public space in order to see how certain sets of values have become embedded in public space as normative dimensions. By showing how the conditions for democracy have coevolved with public space itself, this paper aims to contribute to a further understanding and uptake of the democratization of public space within in the context of contemporary challenges.

The paper will present four critical periods, in which new substantiations of public space become manifest that create new societal norms, values and relations. The first period is that of the *liberal city*, which is socially constituted by the omnipresence of strangers in daily life, and which was answered by the physical differentiation of a public from a private space. In turn, this setup of spheres was ideologically reconstructed in the dogmas of political and economic liberalism as well as in the development of a democratic doctrine. The nineteenth century gave birth to the *sanitary city*, in which governmental authorities became the designated manager of public space, while public space itself became a metabolic system that had to be arranged so as to allow flows of matter and people. The third period is that of the *rational city* in which modernist architecture transformed public space into a sphere of vacancy. Also the rise of organizations and the introduction of neo-liberal politics contributed to a public space that is vacant, denying any intrinsic qualities of publicness. Recently we have entered the age of the *experimental city*, in which new forms of relations between actors and institutions are explored, also because of the introduction of new technologies. At the same time, existing conceptualizations of public space continue to guide these explorations. As such, the experimental city should be further developed while taking full account of historically evolved normative demands.

It needs to be admitted that this account is historical only in a revisionist way. The concepts and principles, which are used to describe the four periods, have in some cases only been introduced later on. It should also be noted that the geographical scope of these periods varies over time. While the account of the liberal city has a clear euro-centric focus, the sanitary and rational city can be associated with developments in both Europe and North America. Urban initiatives connected to the experimental city may be found all around the globe. The purport of the paper above all is to explore the normative connotations of public space as have been developed over time. It aims to identify the *values* that are introduced in these different periods that are, in turn, shaped by discourse, institutions and technology. The four periods of urban history have provided us with normative dimensions of public space that co-exist, build upon each other, or conflict in a myriad of ways. In the light of urban challenges posed by new technological developments, it is highly important that these normative dimensions are reflected upon in order to allow for a trade-off between values that is both socially responsive and ethically acceptable.

15.2 The Liberal City

According to the famous lines of Jane Jacobs (1992a, p. 30): "Great cities are not like towns only larger; they are not like suburbs only denser. They differ from towns and suburbs in basic ways, and one of these is that cities are, by definition, full of strangers." This makes the essential challenge for any city how to deal effectively with the omnipresence of people that do not know each other. In most social structures, the confrontation with strangers is a recipe for violence and exclusion. In the liberal city, these threats have been resolved by a separation into different spheres of life, which entice different norms for interaction (Ariès, 1962; Silver, 1997; Valentine, 2008). It is here that we find the defining criterion for a social infrastructure that allows for a peaceful life while one is surrounded by strangers.

Most elementary, the separated social spheres can be equated with street life on the one hand and domestic life on the other. On the city streets, people maintain Goffman's disposition of civic inattentiveness: "one gives to another enough visual notice to demonstrate that one appreciates that the other is present (and one admits openly to having seen him), while at the next moment withdrawing one's attention from him so as to express that he does not constitute a target of special curiosity or design" (Goffman, 1963, p. 84). In the safe environment of one's house, more intimate forms of interaction may take place, and personal convictions may be pursued.

The liberal city has evolved as a context in which diversity and pluralism can prosper. Its historical seeds can be found in the Dutch cities of the seventeenth century in which economic prosperity, religious heterogeneity, and the presence of strangers gave rise to a new kind of urban context based on tolerance and the separation of the lifeworld into different spaces. Cities like Amsterdam could thrive because of their location as a node in the major marine networks in Europe. This brought along trade, money and many strangers from all over the continent (Braudel, 1982; Nijman, 1999; Shorto, 2013).

Characteristic of Dutch cities was the absence of nobility and ostentation in the urban architecture (Schama, 1988). Even the biggest houses were relatively small, modest and inconspicuous. Streets and homes are clearly and unequivocally separated by the presence of a wall, and not much more than that. The houses on the Amsterdam canals have no front garden, no zone which acts as a transferal from life outside to life inside. Outside on the street we will find the public space in which we are all equals, indifferent and restrained; the streets figure as a theater stage upon which we play a role. The inside of the house is where we lose our masks and where we can be the ones that we want to be. The home is the sphere of inequality, where we may reserve a special place for the ones we love, and where we may legitimately keep the door shut for others. It is in the private space where your wealth can be displayed and an intimate identity can be developed.

Critical for the development of the public sphere appears to be the emergence of and tolerance for semi-clandestine churches, that were used by different factions of the patchwork that characterizes the Dutch religious landscape. These churches were not recognizable from the street as houses of worship, only on the inside they figured as such. Admittedly, being secretive about religious practices was a necessity in the beginning as persecutions were still omnipresent. Later, however, these churches became tolerated, as long as they were distinctively private spaces (Kaplan, 2002). This ambiguity of the categories of publicness and privateness will be shown in this paper as one of their key attributes, which creates a highly adaptive social order that is capable of incorporating interests, values, and concerns that emerge in society.

15.2.1 Politics Without Politics

The separation of the lifeworld into a private and a public space can be seen as the starting point for the ideology of political liberalism. In the private sphere, an individual pursues what she finds worthy of pursuing. The public sphere is that of the state, which exists to protect the autonomy of the private sphere. With that, the public sphere is fully subservient to the private sphere, and should be arranged in such a way that individuals do no harm to each other (Macpherson, 1962; Minogue, 1963; Ryan, 1983).

The contingency of these dematerialized public and private spheres is legitimized by building on Hobbes's invention of the state of nature. The political state is derived not from the empirical state of things, but from an idealized reconstruction of society from the starting point of an individual agent with her own drives and convictions. Ironically, there is no state in the natural state, just people. Politics are a necessary evil that evolves from the fact that individuals are confronted in their daily lives with other people, who are similarly autonomous, and similarly subject to personal motivations and passions. This autonomy should be warranted in a liberal state, the physical, moral and psychological integrity of others should be left intact. In other words, one should not induce harm upon another, as has been described by John Stuart Mill (1985) who saw the private sphere as the sphere in which one's life and conduct only affected oneself, or if it affected others it was based on free and voluntary choice. In this sphere of action you, as an individual, are free to frame the plan of your own life to suit your character. The legitimacy of the public sphere of the state only evolves from the need to prevent individuals doing harm upon each other, other than that the state should retreat from intervening in the courses of people's lives. By the introduction of impartial rules, the strict adherence to the legal constitution, and by the separation of the public and private space, the biggest harm is to be prevented: the exertion of state authority in personal autonomy.

15.2.2 Seeking Prosperity: A Space for Trade

In our common discourse, the market has become an ideal-typical entity. As Michel Callon (1998) observes, we know a lot about the 'market', but not about markets. The 'market' presents a way to understand transactions between actors that exchange money, goods, and/or services. This theoretical understanding distracts the focus from the physical origins of the market as a location where goods are traded, where people come to sell and buy products.

In most European cities, market places—open spaces designed for trade—figured as the central squares. Products could be sold or acquired, giving rise to the laws of the market, in which the pursuit of individual gain could lead to collective well-being (Polanyi, 2001; Sedlacek, 2011). The concentration of people also facilitated the laws of specialization. A seemingly endless reservoir of workers could help to create the impression of laws that were based on the availability of infinite resources. As such, liberal economics is also liberal in the sense that it pertains to the influx of strangers. The expansion of cities has always been based on the supply of young people, predominantly men, that were looking for prospects for a better life (Sennett, 1993). Failed crops and uneven heritage arrangements drove them from the countryside towards the city.

In all, cities came to figure as the social contexts that inspired modern economic theory (Dumont, 1977). Cities presented economic actors as individuals without further feudal constraints, which came to constitute a system made up of atomic parts that pursued the same goals, namely economic well-being. Traditionally, the egoism and materialism that motivated these actions were found to be immoral, only acceptable out of sheer necessity. Bernard de Mandeville (2017) was the first to see benefits of the aggregate of immoral actions for the system as a whole. In his book *The Fable of the Bees: or, Private Vices, Public Benefits*, he compared people acting on the market with bees in a beehive, each individual strictly minding his own business, but producing general welfare. The premise of this book has been further developed by Adam Smith in *The Wealth of Nations* (1998). Smith explained how the constellation of individuals seeking private gains will by the intervention of an 'invisible hand' generate prosperity at the aggregate level. By exchanging products for products, products for money, money for money and money for products, it becomes possible to improve one's position, while, unintentionally, also improving the general level of well-being.

The system of the market co-evolves with so-called bourgeois values that emphasize individuality, thriftiness, entrepreneurship, and above all self-love (Jacobs, 1992b; McCloskey, 2010). Nowadays, self-love would be understood as self-interest, which, according to Albert Hirschman, resolves the capriciousness of the passionate nature of self-love (Hirschman, 1977). The market increasingly becomes understood as a predictable system that mechanistically relies on fixed motivations. The need for control emerged out of the desire to make society more moderate and more peaceful. *Douce commerce* could replace the violence-ridden constellation of fighting royal families for whom securing pride and power was more important than maintaining peace.

Economic liberalism is a reductionist expression of political liberalism, in the sense that it identifies individuals strictly as economic agents and the private sphere strictly as the market. Like in political liberalism, the state should be kept as small as possible, not for ideological reasons, but in order to facilitate prosperity. With the reduction of the private sphere to the market, the domestic domain is excluded from the public sphere. This creates a new situation as the market originally was connected to the domestic domain, in which the production of goods was a joint effort by the whole family. So, on the one hand the emergence of the modern market confirmed the hierarchy of the public realm to serve the needs of the private realm; on the other hand, it meant that women were firmly delegated to sphere of privacy (Pateman & Phillips, 1987). They had no role to play outside of the house. In sum, the market brought peace, but also brought along a stripped version of liberalism that, in a disguised fashion, enforced a new structure upon domestic life.

15.2.3 The Public Sphere as a Space for Speech

In opposition to the liberal conception of the public/private divide, another conception emerges in which public space is explicitly substantiated as a realm that acknowledges the intrinsic value of politics, namely that of democracy. In this space, individuals, in their role as citizens, aim to define a public interest outside of the realm of the state. The two conceptualizations, as contrastive as they may be, came to coexist and influenced and often implied each other.

The genealogy of this democratic conception of the public space knows its paradoxes, as it can be said to evolve from the private sphere. Jürgen Habermas (1999) notably located the roots of this public sphere in coffeehouses of London. But before that, the clandestine churches of Holland seemed to have played a constitutive role, as these churches were among the first places in which a social collective came together to decide upon its identity by organizing an ongoing debate. A similar type of secretive proto-public spaces were provided by masonic lodges (Kaplan, 2002). The activity of judging public matters spilled over to debates about culture, building on developments such as the commodification of literature, the commercial popularity of theaters, and the emergence of parks and squares that invited people to stroll around, to meet up and have a conversation about life in the city (Van Horn Melton, 2001). These new forms of urban public space and sociability also became embedded in a culture of print, with the rise of newspapers, political journalism, novels and criticism. The result was a culture that was engaged with the creation of a body of shared meanings that was not only independent from the patchwork of private spaces, but also from the public space of the state. As such, an authentic bourgeois culture emerges in the eighteenth century.

This new public space can be seen as a culmination of the set of separations that characterizes the modern city, at the background of an even wider range of cultural achievements. First, the separation of an autonomous society from the authoritarian state, the pursuit of individual and collective goods became independent from the mechanisms of state coercion. This newly found autonomy also had its expression in the free market and in the emerging domain of science (Van Horn Melton, 2001). Instead of obeying authority, these realms were increasingly subject to self-contained laws. Second, the separation between the household as the sphere of intimacy and secrecy, and the street as the sphere of sociability and openness in which everything, including politics, could be subjected to criticism.

The propensity of members of an autonomous civil society to discuss political matters cannot be fitted into the liberal model well, which after all does not acknowledge the existence of a substantive political realm in the first place. The historical analysis of Habermas (1999) has inspired the articulation of an alternative model that tries to identify the conditions for a politically engaged society that is able to decide on its shared norms, values and goals by means of deliberation. Inspiration was not only found in the coffeehouses of the eighteenth century, but also in the classic works of Aristotle. Deliberative models that have been developed in which the value of participation and civic empowerment came to be stressed (Bohman, 1998; Dryzek, 1994; Fischer, 1999). An important ingredient of this deliberative sphere is that it can build on the plurality of views, moral principles and ideological starting points that are nurtured in the multitude of private spaces; a debate based on a heterogeneity of claims can be seen as productively binding people, while emphasizing their autonomy.

15.2.4 Accommodating Paradoxes

It is useful to make a short expansion here to the origins of 'public' and 'private'. The liberal conceptualizations that have been shown in previous sections have Roman sources and have become part of Medieval jurisprudence via the *Corpus Juris Civilis*, emperor Justinian's compilation of Roman law from the sixth century AD (Weintraub, 1997). These concepts were mainly used to differentiate public law from private law, where public law referred to concerns of the empire and private law referred to the interests of individuals (Geuss, 2001). Over time, more meanings have been bestowed upon these terms. Most important of these originate from ancient Greek philosophy. The common translations of public and private are, respectively, *to koinon* and *to idiom* (Saxonhouse, 1983). The first term refers to something that is held in common, something that is shared, especially connected to the sphere of a family that shares ancestors, space, religion, meals, etc. The second term refers to something that differentiates or separates.

The public sphere of democracy relates very strongly to the Greek understanding of the polis, which was most basically seen as a context that resembled the sphere of the family, in which what we share defined us as members of a certain collective. The public sphere that revolved around deliberation and critical assessment of the affairs of the collective is one that defines its commonality on the basis of speech and exchange of arguments. With this, societies can actively establish their own identity and its own collective course.

The role of the family is completely reversed in this deliberative account, instead of being the core of the private sphere, it now becomes symbolic of the public sphere. This ambiguity characterizes liberal democracy (Berlin, 1969; Mouffe, 2000; Pesch, 2005). The addition of a sphere of commonality is necessary for allowing a functional democracy, and it stresses the values of inclusivity and a common identity (Taylor, 1995). However, like any notion of kinship it also brings about a fundamental opposition between those who are members of a community and who are not. Exclusionary tendencies are intrinsically tied to inclusion—even while acknowledging the need for plurality.

In all, the liberal public space embraces contradictions and paradoxes. It allows the establishment of connections and separations between different societal groups; it includes plurality and heterogeneity, while maintaining commonality; it juxtaposes fixed rights and ongoing deliberation. The public space is continuously subjected to the confrontations of these contrasts and finds ways to reconcile them in practice.

The accommodation of contradictions and paradoxes appears to depend on the separation of our physical and mental world into different contexts. The distinction between spheres of sociability has spilled over in different roles, different institutions, and different activities. Indeed, our moral universe has come to be defined by the boundaries between different spheres in which contrastive morals may reign (Rawls, 1997; Walzer, 1983, 1984). The plurality of conceptions of distinctions between the public and the private as well as the indeterminacy of these conceptions recruit continuous processes of negotiation and renegotiation about what exactly is to be considered public or private. As such, public space is fundamentally open to take on new concerns and new issues (Dewey, 1927; Marres, 2007).

15.3 The Sanitary City

The contours of the sanitary city can be found in the work of Edwin Chadwick, an attorney from London who was concerned with the circumstances of the poor. His *Report on the Sanitary Condition of the Labouring Population of Great Britain* (Chadwick, 1842) became a bestseller, and defined the thinking of the hygienic movement: the supposed lack of civilization was no longer used as an explanation, as it became clear that the lack of hygienic infrastructure was the culprit for epidemics. This awareness brought along a shift in cultures of care and charity. Instead of relying on individual benefactors, economic circumstances and the lack of public hygiene had to be addressed.

Not only was there a moral demand for improving the environment of poor classes, there also was the conviction that a science-based approach could effectively resolve sanitary problems. By creating waterworks, paved streets, ventilated buildings, viaducts, water towers, filtration techniques, pumps, and so on, the city could be turned into a body in which infrastructures played the role of arteries that, according to Chadwick, had to allow the flowing of vital streams. The city came to be seen as a living metabolism, which was marked by *movement* and in which all houses were organically connected to *streams* of waste, air and water (Melosi, 2000; Sennett, 1996). Chadwick's idea were turned into reality after the 'Great Stink' that occurred in the dry, hot summer of 1858 when feces and other forms of organic waste created an intolerable smell in the city center of London. Members of Parliament, who were in the middle of the stench, saw firsthand that change was necessary. London was given underground sewers, increasingly governed by authorities.

The developments of London were not idiosyncratic—sanitary systems were also constructed in other European and in North American cities. For instance, Georges-Eugène Haussmann considered Paris to be a sick body from which tumors had to be removed and its blood circulation had to be started again (De Waal, 2012). For this he created his famous boulevards, but above all he developed a sewerage system that allowed water to clean the city without people even noticing.

> These underground galleries would be the organs of the metropolis and function like those of the human body without ever seeing the light of day. Pure and fresh water, along with light and heat, would circulate like the diverse fluids whose movement and replenishment sustain life itself. These liquids would work unseen and maintain public health without disrupting the smooth running of the city and without spoiling its exterior beauty (Hausmann (1854), quoted in Gandy, 1999, p. 24)

The new waste infrastructures, and to a significant extent also further infrastructures, were based on the way organic waste was dealt with. Our idea of pollution was, and basically still is, based on the way we deal with human feces. These have the role of a taboo, something that has to remain hidden (Douglas, 1966; Jewitt, 2011; Moore, 1985; Pesch, 2015b). The infrastructures developed in the nineteenth century allow cities to remove human excretion from public eyesight, by transporting it by underground pipelines to places that are not visited. Other forms of waste management follow this template; for example, streams of dirt have to be transported in such a way that it minimizes visual impact. It is not surprising that around the turn of the twentieth century, the role of local authorities was compared with that of housewives who, just like the municipal services, swept up the dirt in their homes and collected it in garbage cans (Melosi, 1981).

To a large extent, the sanitary city builds on the liberal city. It perfects the distinction between the roles that we play in visible life and the secretive intimacy of our genuine selves. Playing a role boils down to self-control, will power overrides our direct urges. Biological necessities had to be suppressed in public life. Physical processes had to take place in sheltered spaces. The establishment of sanitary infrastructures meant that bodily control could be pursued even more effectively, it allowed everyone to retreat from eyesight in order to relieve oneself.

The industrialization of cities posed new challenges for cities, adding new conceptions of publicness to the already existing repertoire. Most notably, these conceptions follow the development of the sanitary city: a city in which infrastructures take care of the flow of streams composed out of waste, goods, traffic and people. Three dimensions of publicness can be connected this new type of city. On the one hand, the constellation of infrastructures that characterized urban contexts gives rise to the metaphor of the metabolic city (Gandy, 2004; Melosi, 2000), turning the public space of city streets into canals for moving streams instead of spaces for meeting other people. The agencies necessitated for managing urban infrastructures resurrects the role of government as the dominant bearer of publicness. On the other hand, the establishment of these infrastructures, especially regarding waste, can be related to the growing awareness of the interconnectedness between socio-economic strata. Civil society that in earlier times could be secluded from the concerns of the poor and of women now had to give in to the fact that the urban fabric of the industrial city is deeply interwoven.

15.3.1 A Space for Government

It was not a self-evident outcome that governmental authorities assumed the management for the sewage systems. The earliest initiatives for public hygiene had a private character, carried out by groups of citizens and charity organizations (Houwaart, 1991; Lintsen, 1992; Van Zon, 2002). Also there was no instantaneous development of systems that were based on the principle of circulation: many different technologies for disposing human waste were tested (Dekkers, 2014; Noort, 1990). However, over the course of time, the image of a city being controlled by local authorities became dominant. In classic liberalism, the state presents the conditions for the market, but is no economic actor itself. However, in many cases the state is the only agency that can develop economic initiatives that are beneficial to the general welfare of to the market itself (Pesch, 2005; Wright, 2000).

Cleaning human and urban bodies above all seems to boil down to making feces invisible. Human hygiene has increasingly been based on the control of bodily excretion. Urban hygiene basically applies this control dogma to the level of the city; the urban metabolism that comes into existence comes to include an increasing range of waste, which, in turn, has to become subjected to control. More and more stuff is seen as wrongly located, as things that need to be hidden (Douglas, 1966; Gastelaars, 1994; Pesch, 2015b).

As the range of pollution increases, the government is given more tasks. While waste at first consisted out of rags, leftovers, and old iron, at later stages, dirt from households and industry were added to this repertoire. The management of these forms of waste became the designated job of a corps of professionals that were increasingly on the payroll of public organizations (Van Zon, 1990). These professionals collected waste, and came to control its flow towards reserved locations that allowed for invisibility (Melosi, 1981, 2000).

It can be said that the model of waste management has been created to liberate the city streets from excretion, and over time this model has come to serve other forms of waste. Scientifically trained experts together with public professionals were designated to take care of a new form of public hygiene: the aim was no longer to battle epidemics that were the result of the discharge of feces on the streets, but to protect the 'health' of the living environment itself. Instead of dealing with waste, 'pollution' had to be targeted; everywhere around us, there could be stuff that made us ill. A local approach to manage such pollution would no longer do, and around the turn of the twentieth century, Western states centralized the science-based care for public hygiene.

With local and national governments administering flows in urban infrastructures, public space was overtaken. It now became the space supervised by the public sector, up to the point they coincide, especially in Western European states. Little is left of a neutral public space in which free citizens met each other. The responsibility for the public space, both above and beneath the pavement, was given to the authorities and professionals; their job was to facilitate public space as a space of movement, so it remained clean.

Sanitary infrastructures brought along a new substantiation of public space. Streets were seen less as meeting places, but became increasingly understood in terms of the transport they allowed. The physical public space that created societal coherence became invisible, existing out of underground pipes and cables that connected all individual households to the greater urban context.

The management of public space also had implications for the private space. The liberal city had been based on a clear separation between the public space of the street and the private space of the home. In the sanitary city, the city becomes a body, an autonomous subject controlling itself. Many of its subterranean bifurcations, pipes for water, gas and excretion, cables for electricity and information, effortlessly crossed the boundary between public and private. The authorities got access to the area behind the front door, as far as to the toilet bowl itself, providing essential household services.

15.3.2 A Space for Emancipation

The hygienist movements brought along a physical and an ideological idea of community. Physical in the sense that the proximity of poor people can affect the general level of health, as such it is simply beneficent to establish encompassing infrastructures that serve the city as a whole (De Swaan, 1990). Ideological as it creates a common fate; the hygienists felt solidarity with those less well off (Van Zon, 2002). Sanitary initiatives were soon followed by other initiatives to relieve poor classes. At the same time, such initiatives had strong disciplinary intentions and with that severe stigmatizing ramifications. Just like the sewers allowed for dirt to be controlled, welfare reforms aimed for the control of disorderly behavior by the lower classes (Gandy, 1999).

In contrast with these initiatives that clearly had a top-down character, the nineteenth century also witnessed the reverse dynamic. Up from the bottom came a range of emancipatory claims that can be seen as demands for access to public space. In spite of the emphasis on individuals, the liberal city only allowed a minority of people to act as the kind of liberal individuals that could alternate the public sphere of politics and equality with the private sphere of the home (Pateman & Phillips, 1987). Women were confined to the private sphere of the house, and as such they were effectively denied citizenship. A professional career, contributing to public opinion, and participation in civil society were all out of reach. Also large segments of the male population, such as laborers, minorities, and immigrants, were excluded from public and political life. The right to vote was usually based on income, so that the democratic system was above all the effective guardian of the interests of the richer classes.

Increasingly, however, groups deprived of a public role came to organize themselves, appealing for more equality. One may, admittedly anachronistically, describe the rise of such emancipatory movements as the formation of 'counterpublics' (Fraser, 1990; Warner, 2002). These counterpublics challenge the unitary character and exclusionary tendencies of the liberal public sphere. Different groups may develop their own common discourse and shared identity, and claim their right for such self-expression. The paradox of these counterpublics is that they aim to create a protected, almost semi-private, realm that allows the development of an identity that is, in the end, comprehensively accepted as a legitimate identity. Emancipation means both becoming equal and becoming autonomous.

In the account of the liberal city, it has been shown how any political community has exclusionary features. A political community is not the same as the social collective that is contained by the rules and arrangements that are consented with by the citizens. Suffrage is not universal, groups like children and tourists are not given citizenship rights. The emergence of counterpublics can be seen as the appeal of social groups to gain access to the political community. This raises some paradoxes with regards to the liberal creed, because these groups often present their claim for access in terms of their identity as a group, which might contradict the individualist starting points of liberalism, while these at the same time are justified by the liberal ideal of equality. Another upshot is that the public sphere of deliberative politics does not only relate to the outcome of a certain debate, but also to the struggle about which groups are allowed to participate in this debate. The ambiguity of public space that has been described earlier allows such contradictions to be discussed and resolved. Expanding circles of emancipation and empowerment are based on the possibility of renegotiating what belongs to which side of the public/private divide.

The development of the sanitary city did not only bring centralized management of infrastructures and flows, it also contributed to the homogenization and emancipation of society. The infrastructures allowed societal groups to become connected, not only in spatial terms, but also in terms of culture and morality. These processes have, in many cases, involved the emancipatory struggles of counterpublics. At the same time, we may observe the emergence of a split between the social *production* of the physical creation of public space and social *construction* of the symbolic experience of public space (Lefebvre, 1991; Low, 1996). The 'hardware' of the sanitary city composed out of urban arteries came to be opposed to the 'software' of publics and counterpublics that used urban space as the arena for creating both shared and contesting meanings. The conceptual distinction between the social construction and production of public space has become a common understanding in urban studies and it is often used to oppose the capacity of communities to create their own understandings about a place with the meanings and norms that authorities aim to maintain in the physical design of that space. As such, this distinction figures as a major factor in the contemporary understanding of public space.

15.4 The Rational City

In this section, I will introduce an intellectual infrastructure that is added to the social and technical infrastructures presented earlier and which presents a description of the world in terms of objective, universalist rules. Such a description allows for mechanistic manipulations in order to achieve certain preset goals. This worldview in which goals and means are clearly separated and in which mechanistic rules motivate actions, utilizes Max Weber's thesis on the rationalization of the life-world.

While the liberal and the sanitary city are also manifestations of this process of rationalization, as these pursue societal progress by deliberate design, this new phase of rationalization also pertains to these processes of design themselves. In other words, the social and technical arrangements that had been implemented earlier also become subject to objectification.

The historical developments that underlie the rational city are firstly formed by the processes of industrialization that were furthered in the nineteenth century, and secondly by the emergence of modern business organizations. The industrialization of cities in Europe and North America is deeply connected to the development of the sanitary city sketched in the previous section: the introduction of controllable energy sources created the possibility of concentrated forms of production. The rhythm of life came to be determined by the steam engine and the assembly line. It has been described by Thorstein Veblen (1994) how the machine came to govern the production process. The machine necessitates standardization and regularity, while the laborer needs to adjust his actions, choices and skills to these necessities, and as such his physique becomes servile to the demands of the machine.

The steam machine can be seen as the first appliance that incorporated cybernetics. The centrifugal governor invented by James Watt controls the speed of the machine by figuring as an internal feedback system. The constant speed that was enabled by this system came to directly control the functioning of the laborers. Gradually, the machine also came to determine the shape of the city, concentrating laborers in factories and their families in nearby urban settlements.

The rational city as it is described here not only pertains to the rationalization of mechanical force and labor capacity, but it also pertains to the rationalization of public space. The key for this rationalization, as will be contended here, emerges from the development of mobility infrastructures. First, railroads came not only to connect cities, harmonizing different regions of a country, but also gave rise to a new way of coordination, namely the modern business organization. Second, the car came to define not only mobility, but also the city of the twentieth century. The emphasis on movement as was introduced in the sanitary city came to denote not just dirt that needed to be flushed or swept, but also people that transported themselves, necessitating urban planners to introduce the infrastructures that facilitated these movements. What we see is how technological and social infrastructures start to figure as the physical landscape determining the ways in which people act and interact. At the same time, infrastructures create their own dependencies, they create a lock-in of technologies and related practices that very much provide a fixed platform of values and concerns that are to be shared (Correljé & Groenewegen, 2009).

15.4.1 A Space for Functionality

The liberal city that embraced both a public and a private sphere has been contested in the twentieth century by modernist architects, most notably by Frank Lloyd Wright and Le Corbusier. These renowned designers share their 'metropolophobia'; seeing the city as an unnatural, impure living environment. Instead of sanitizing the city, their cure was to eradicate the city. Frank Lloyd Wright (1932) claimed: "To look at the plan of any great city is to look at a cross-section of some fibrous tumor." It was the task of the architect to destroy this tumor, by redesigning the built environment so that people could live in a 'humane' way. The city they loathed was the disorganized and wasteful nineteenth century city. Architects and urban planners gave themselves the task to rationalize these industrial sites of chaos (Fishman, 1977).

Frank Lloyd Wright spent a great part of his life working on to the never realized utopia *Broadacre City*. The degenerated big city needed to be conquered by a planned city, which was based on new technologies such as cars, telephones, and radios. Every family was to be given one acre, and all transport was to be done by car. With sixty miles per hour, you were able to visit any of your 30,000 neighbors within 15 min. At gas stations on the edge of the city, standardized consumer products could be purchased. Broadacre City was a city without a recognizable center, without having a clear transition from city into nature. No urban districts, just hundreds of uniformly separated houses in which people could be occupied only with their own concerns. The industrial, centralized city belonged to the age of the machine. Such cities were monsters, aberrations build on greed, destroying efficient production and human values. Only the return to the natural habitat of humans, the rural context, could revive this natural balance.

Le Corbusier based his utopian city on an idealized version of Paris: a center of art, culture and industry in which elites and avant-garde could meet. He found the real Paris horrible: it was crowded and loud, houses were uninhabitable and roads were congested. For Le Corbusier such an organically grown city belonged to an age of artisans, not to an age of the machine. Modern times allows for a drastic top-down planning that could reintroduce harmony. Houses, districts, and cities had to be approached as 'living machines': like a factory in which first the functions had to be analyzed before they were assembled into an effective process, a city should also be based on functional separations. The city should allow for speed and efficiency, which could only be done by building the city from scratch. Highways, subways, bicycle roads, pedestrian lanes all had to be designed to contribute to great speed, so that an optimal exchange of information, talent and pleasure could take place.

In hindsight, the modernist architects have made a number of impressive buildings, but never created great streets or great cities (Kelbaugh, 1996). Their legacy is formed by utopian cities such as Chandigarh and Brasilia that continue to struggle for coherence. However, they have been hugely influential for urban design in the twentieth century. Most followers of the modernist utopists did not make new cities, but added new districts at the periphery of existing cities, taking transport by car as a starting point—creating a profound lock-in situation in which it has become unavoidable to drive a car.

The consequence for public space is that it has been hollowed out, it had no intrinsic value apart from facilitating people to go from one place to another (Sennett, 1992). Meeting strangers, the essence of the liberal public sphere, was discouraged by the mobility infrastructures that were to be developed in the twentieth century, these infrastructures were above all designed to cater traffic by car. The ideal situation being a trip from garage to garage, not having to engage in the pluralism that was intrinsic to the public sphere of the liberal city (Duany, Plater-Zyberk, & Speck, 2001).

15.4.2 A Space for Efficiency

Factories absorbed capital, labor, and also power. They did not bring about the modern organization, which may have more to do with another application of the steam engine, namely the construction of an extensive railroad. Alfred Chandler (1977) describes how this construction in the United States brought about a new type of business enterprise. Instead of an owner that controlled the daily affairs of a company, the vast distances that had to be covered by the new railroads necessitated the division into multiple organizational units that had to be managed separately. The multi-unit enterprise came to be run by a class of professional managers that coordinated decisions inside of and between organizational units. With that, the multi-unit organization created a new kind of strategic behavior on the one hand and bureaucratic routines on the other hand.

Organizations are a new phenomenon to the liberal templates that differentiated individuals (and families) from the encompassing domains of the state, the market and civil society. It is here that we recognize Roald Coase's notion of transaction costs, which explains the existence of organizations in the supposedly self-ruling system of the market, because there is a cost of using the price system that may be reduced by control over information (Coase, 1937).

The behavior inside of the organization became an object of academic studies, which had the aim to rationally optimize the processes taking place inside of the organization. These processes were to be described in terms of resources that had to be deployed as efficiently as possible in order to achieve the preset set of organizational goals. Labor was one of the main resources, and one that needed the most drastic revision. Ever since the *Scientific Management* of Frederick Taylor (1914), the goal of hierarchical structures related to questions about how to control human resources by aligning activities of employees with organizational demands: the division of tasks and responsibilities and the motivation of individuals by stick, carrot or sermon came to be objects of systematic study (e.g. Simon, 1997).

For the objectified study of organizational behavior the public or private status of an organization does not play a structural role. It is a contingent feature that might impact the conditions that need to be managed, but are of no further intrinsic relevance (Pesch, 2005, 2008). At the same time, the template of the modern organization originates in the private sector, which is generally assumed to create more efficient forms of management—efficiency being the holy grail of rationalization. Especially in the 1990s, the neo-liberal creed inspired the modelling of state organizations on market organization. The idea of new public management was that government 'should steer not row', this creed implied that the public status of an organization has no distinct features (Osborne & Gaebler, 1992). Privatization of state organizations, but also the stimulation of entrepreneurial activities of state officials could be observed throughout the world.

These developments suggest that the responsibility for managing vital infrastructures could be taken away from the government. Subsequently, the provision of many urban infrastructures has been taken over by private companies and all aspects of public life have fallen prey to the dictates of the private sector, which are managerialism and efficiency. There are many degrees of privatization—a service can be fully or partially taken over by a for-profit organization, or a government can act entrepreneurially—but the outcome is the same: public services become products that obey the same logic as market products. Moreover, following globalization and the dominance of neo-liberal policies, liberalism is increasingly understood exclusively in terms of *economic* liberalism.

The upshot of the developments sketched in this section is a significant *hollowing out* of the public space. In modernist architecture the space that is not of private, commercial or cultural relevance loses any intrinsic value. What is left is a *functional* space that allows people to travel as effectively as possible from one point to another, and in which public management is seen as no different than private management. What is missing in the rational city is the attention to empowerment and emancipation, which have played a role in the previous two sections. It may be that these values become harder to pursue. Not only do technical infrastructures take away a share of the adaptiveness necessary to allow negotiations about the content of public space, the increasing porosity of the boundaries between the public and the private also make it hard to distinguish an identifiable public space in the first place (Pesch, 2014). With that, it becomes unclear where and how to make appeals for inclusion and participation.

15.5 The Experimental City

According to Manuel Castells (1993) the age of the industrial city has given way to the *information society*. Industrial societies were characterized by immobile factories, in which employers could easily spend their working life. It was, in the words of Zygmunt Bauman (2000), the age of 'heavy modernity', in which everything was based on steadiness. Now we have entered the age of 'fluid modernity' in which everything is in flux. The market is now a global phenomenon based on real-time transactions. Cities have become nodes in socio-economic networks in which capital and knowledge have been integrated. The public space of the city has become occupied only temporarily by a caste of business nomads.

Public space has become diffuse, and just like the virtual space of the internet, it knows no clear spatial qualities. Public space is subjected to commercialization and commodification and citizens are increasingly seen as mere consumers. Without a clear idea of publicness, public space becomes further privatized. As government retreats from the public space, it will be left empty. The same can be said about the urban landscape which has come to exist out of an endlessly repeating series of shopping malls, megastores, fast-food chains and other 'non-places' (Loukaitou-Sideris & Banerjee, 1998).

The privatized nature of public space can also be connected to a changed relation between sanitation and urban metabolism. According to Gandy (2004), our current metabolic system can be characterized as 'antibiotic'; diseases are addressed as separate entities, instead of addressing them as epidemics that could affect a collective and that had to be managed by developing a 'bacteriological' form of urbanism. The shared destiny that was the result of the equalizing threat to public hygiene is not needed any longer; one can conveniently retreat into individualized patterns of healthcare. With the withering of boundaries and categories due to processes of globalization and digitalization, the metabolic character of the city can be more aptly described as 'rhizomatic' (cf. Deleuze & Guattari, 1987), in the sense that the underground city acts like a fungus which can be traced by seemingly random patterns of popped-up mushrooms.

Earlier we have seen how the categories of citizens and consumers emerged in the context of the institutional domains of the democratic state and the free market. It is not that these categories have become defunct, however, what can be observed is that the boundaries between them have become fundamentally porous, giving rise to hybrid activities and orientations (Pesch, 2014, 2015a). At the same time, to grant legitimacy to these activities and orientations, existing dimensions of public space are still appealed to. The different dimensions of publicness continue to play a role in contemporary cities. This will be shown first by exploring how private companies take over public responsibilities with regards to infrastructures that are increasingly integrated into 'smart systems' in order to pursue the efficient management of collective goods and services (Kitchin, 2014; Townsend, 2013; Van Timmeren & Henriquez, 2015). This development has ramifications for conceptualizations of public space that go much further than the mere osmosis between the sphere of the market and the sphere of the government, as it bestows public space with a dual function, namely that of a testing platform and of context in which people live (Evans, Karvonen, & Raven, 2016; Jackson, Gillespie, & Payette, 2014). Second, I would like to point at a wide range of activities instigated by collectives of non-institutional actors, such as local groups and communities, that deploy economic and political activities (Bulkeley & Castán Broto, 2013). With that, individuals as citizens and users try to define their own public space to an unprecedented degree, while building strongly on liberal traditions. Though these developments are contrastive in many regards, they build on the same socio-economic and technological basis, namely the availability of information flows that minimizes the costs of coordinating collective behavior.

15.5.1 The Hybrid Publicness of Smart Systems

New ICT systems come with a seemingly infinite capacity to establish and control flows of data. As such, the existing infrastructures can be managed, new infrastructures can be added, and the full repertoire of infrastructures can be integrated into one seamless web that has been given the label 'smart' (Kitchin, 2014; Sadowski & Pasquale, 2015; Pesch & Ishmaev, 2019). The integration of infrastructures allows the creation of circular processes that may revert the tendencies of the pollution management model to control waste by making it invisible. Instead, by connecting different industrial streams, waste may be regarded as a resource (Pesch, 2015b). Ideas such as cradle-to-cradle, industrial ecology and circular economy are all based on the capacity to change waste from output to input. Making such circular approaches possible demands detailed knowledge about production and consumption processes as well as detailed monitoring of material flows.

Governmental authorities can hardly neglect the promises with which they are confronted: after all, how could they withstand the lure of more efficiency? Then again, the developers of these smart systems are market-based companies that come with their own definition of what a smart city should be, a definition that becomes embedded in the technological artefacts they produce. In many cases, these companies also take over the actual control of these systems, tapping into the development noted in the previous section that the management of public infrastructures does not have to be the responsibility of public authorities. Private actors may present themselves as equally, or even better, equipped to fulfil this job.

The work of companies like Siemens, IBM and Cisco resembles that of Frank Lloyd Wright and Le Corbusier: they present a template of how the city should function and with that comes an imaginary of the public space that has strong normative connotations. Also this imaginary is remarkably similar to the above discussion: it portrays the public space as a space that needs to facilitate flows, be it of cars or of information. The actual management of these flows should be laid down in the hands of experts, not of politically motivated authorities.

Just like the modernist architects, the developers of smart systems have built working prototypes of their design (Van Timmeren & Henriquez, 2015). New cities like Songdo in South-Korea and Masdar City in the United Arab Emirates act as showcases for what a smart city could entail, stacked with sensors that monitor flows of traffic, waste, energy, water, air, and people. Smart systems producers also intervene in existing cities and by that actively change the already prevailing public space. To a significant extent, these interventionist activities can be characterized analogously to activities that internet companies deploy. New systems are not thoroughly tested before they are released in society, but they are tested by being introduced in society. In the jargon of internet companies: 'beta-versions' of new technologies are introduced so that user feedback can be used to improve the product (Jackson et al., 2014). This means that a community of consumers acts as both as a user platform and a testing site.

In the context of a city, the legitimacy of releasing beta-versions might be problematic. Cities are not just conglomerations of consumers or users, they are also conglomerations of *citizens* whose capacity of collective self-determination might be compromised by being subjected to experiments run by commercial enterprises. But this is not the only challenge to liberal starting points. As the conditions for ending the experiment are not given, it becomes almost impossible to hold public and private authorities accountable for eventual mishaps. After all, the possibility of failure is one of the key characteristics of experiments—not something to be prevented. Moreover, we have seen that the liberal city has been based on the interaction with strangers. By designing efficient infrastructures that allowed private forms of transport, the modernist architects had tried to reduce the possibility of being confronted by strangers. Smart cities can be said to go a step further: not only do efficient infrastructures reduce the possibility of being confronted by strangers, the presence of digital forms of surveillance also exert control over anyone who is considered to be out of bounds.

The eradication of strangership hints at mechanisms of exclusion that characterize smart cities, by externally defining public space, smart systems also define the conditions for accessing public space. People that are not able to fulfill these conditions, because of poverty, personal history, physical or mental qualities are effectively disabled to partake in public life. As such, it can be said that the ramifications of these processes increase the ongoing fragmentation of public space.

15.5.2 Grassroots Experiments by Community Initiatives

The reduction of the costs of controlling information flows not only allows more efficient management of institutional processes. It may also be taken up by individuals or societal groups to organize themselves to pursue certain collective goals (Hajer, 2011). With that, new citizen-based initiatives emerge that can be said to regain or at least develop new conceptualizations of public space. In this, we may recognize ideas and ideals that have originated in the liberal city. These initiatives emphasize the democratic quality of a community to actively shape their shared identity by speaking out in participatory settings (Arendt, 1958; Dryzek, 1994).

Community based initiatives are not new to society, there have been traditional forms of civic engagement as described for instance by Robert Putnam (2001). These traditional forms of engagement usually involved activities related to churches, schools, or local politics, or related to groups such as professional organizations (including labor unions), sports clubs, and literary societies. There also have been social movements that have tried to empower minority groups or criticize dominant political ideas and practices. As social movements, community initiatives mobilize resistance to existing power structures, which can be categorized according to their identity, adversary and societal goal (Castells, 1997, p. 71).

Nevertheless, we may now observe a new form of such initiatives, as these revolve around the use and development of sociotechnical systems that are implemented next to or instead of existing parts of the metabolic infrastructures that constituted the sanitary city (Smith, Hargreaves, Hielscher, Martiskainen, & Seyfang, 2016). This novel form of community initiatives can be recognized in services that allow the sharing of products such as tools and cars (Schor, 2014), community gardens (Holland, 2004; Naylor, 2012), energy cooperatives that produce electricity locally (Walker & Devine-Wright, 2008), and holistic initiatives like transition towns (Feola & Nunes, 2014; Amanda Smith, 2011) and eco-villages (Boyer, 2015; Höflehner, 2011). Civic engagement takes place against the background of a globalizing economic system that is partly driven by neo-liberal paradigms, and the tendency of sustainable civic engagement to revolve around the uptake of technology, instead of taking social and cultural elements as their starting point. The result of these features is that community initiatives are more directly connected to the provision of certain public goods, and as such bear a more functional character than conventional initiatives (De Moor, 2013). This aspect of civic engagement can be characterized as the *substitution* of the production of goods and services.

In these initiatives, citizens are given the opportunity to create insight over the way that they relate themselves to these sociotechnical arrangements—basically incorporating the concern for innovation as part of a deliberative space (Pesch, Spekkink, & Quist, 2019). With that, community initiatives can be seen as opportunities to 'cultivate' more democratic innovation processes as is expressed by Smith and Stirling (2016), as these initiatives may contribute directly and concretely to the 'democratization of innovation'.

Actors from institutions, such as policy, industry and science, have become increasingly interested in engaging with the 'energetic society', labels like co-creation, urban living labs, and social innovation are used by these actors to create legitimacy on one hand and to learn from these societal initiatives on the other hand. At least to a certain extent, these accommodation efforts are complicated by the emphasis on efficiency and the dominance of neo-liberal policy dogmas. It seems important however to take these initiatives seriously, as they are public space experiments that may counterweigh activities instigated in institutional contexts (cf. Bulkeley & Castán Broto, 2013). The activities of citizens testify to the need for collective self-determination, as opposed to being the uncritical consumers of goods and services. Policymakers have to set aside their inclination to see such initiatives as testbeds for new institutional arrangements and engage in a symmetrical relation in which citizens have an equal opportunity to explore alternative substantiations of public space.

15.6 Conclusions and Discussion

In this concluding section, I will return to the assumptions that have been stated in the introduction. The first of these assumptions was that public space could be clearly demarcated from the myriad of private spaces. This assumption does not hold: throughout this paper it has been shown how different, sometimes even contrastive, conceptualizations of publicness have been connected to urban space. Moreover, the boundaries between public and private spaces have always been the subject of ambiguity. The interpretative flexibility allows public space to be reinvented and to be fundamentally open for new actors and new substantiations.

In sum, the different periods of urban development have led to different normative dimensions of public space that, in turn, embody values that can be considered essential for our society. Table 15.1 gives an overview of the periods, characterizes these in terms of their contribution to the understanding of public space, and connects the key values to these periods.

Given this analysis, it is key for the experimental city to accommodate the identified dimensions and values in order to further the moral assets that over time came to be associated with public space. The endorsement of the values has served as the basis for further development of concrete public spaces, but they also motivate moral demands of a heterogeneity of publics that pursue access to these spaces. The

Table 15.1 Periods of urban development

Period	Characterization	Key values
The liberal city	In this period the *liberal conception of the political realm* was introduced, which can be seen as a derivative sphere that allowed for conceptualizing public space as empty in itself	Equality Welfare Empowerment
	This period also introduces a *substantive public sphere* that was added to the liberal conception, and which allowed for social empowerment and inclusion	
The sanitary city	Here, public space was transformed into an *infrastructural system* that facilitates metabolic flows	Hygiene Emancipation
	Infrastructures brought along the homogenization of society, raising demands for further social and political equality	
The rational city	The control of this metabolic system by *efficiently managed organizations*	Functionality Efficiency
The experimental city	The increased digitalization of infrastructures and management	Smartness Engagement
	Civic initiatives developing outside of existing institutional and political contexts	

analysis presents the perhaps vexing finding that the ongoing privatization of public space and the increased significance of neoliberal policies are features that are intrinsic to the dimensions of publicness. While the discourses of privatization and neo-liberalization contribute to the hollowing out of public space, the value of efficiency that they embody is not by definition external to public space. The same may be said about the role of technology, which functions as an important facilitator of the metabolic flows that provide hygiene; moreover, as has been shown above, technologies may also be utilized by a heterogeneity of publics to bestow meanings upon public space in an independent, that is democratic, way.

These insights compel us to rearticulate the threats to public space that are presented by the experimental city. The threats may not so much lie in neoliberal discourses and privatized technologies themselves, but in the way they might exclude other dimensions to be effectuated. More specifically we may observe the following challenges. First, there is the increasing reliance on technical infrastructures, giving rise to configurations that are hardly adaptive to societal and ideological changes. This rigidity goes to the extent of the character of public space in the liberal city as a space that alluded to contrastive conceptualizations of publicness, as such they were spaces that accommodated paradox—allowing social and political tensions to be addressed and resolved. Second, there is the increased privatization of the lifeworld. The pursuit of societal values and the accommodation of contradictions becomes problematic in the lack of a public space that can be recognized and appealed to, as such it seems desirable to reinvigorate public space as a domain that is fundamentally visible, open, and accessible—which is to say genuinely public (cf. Benn & Gaus, 1983; Weintraub, 1997).

This implies that, above all, we need to democratize infrastructural innovation. This need is especially urgent, because the metabolic system not only includes an increase of the types of flows, such as people, data, energy, and waste, but these flows are increasingly integrated into connected infrastructures. Network technologies like infrastructures create sociotechnical interdependencies that become unreceptive to demands for change, they become locked-in as self-reproductive entities. Creating a network of networks will only multiply these interdependencies (Arthur, 1989). At the same time, as these integrated infrastructures provide the physical shape of urban contexts, it comes to define the lifeworld of people, and as such they have a strong public character: they forward values, practices and relationships to the people that use them (Correljé, Cuppen, Dignum, Pesch, & Taebi, 2015).

To a significant extent we have come to live in a time of 'cyborg urbanization' in which the technical and the social are seamlessly connected to each other (Gandy, 2004; Sadowski & Pasquale, 2015). This suggests that there is a need to find ways for urban populations to become empowered with regards to the development of sociotechnical systems. Above I have introduced community initiatives as explorative efforts in that direction, but this experiment has not yet given rise to further diffusion of methods and policies. It seems paramount that such diffusion take place, as it cannot only be part of society itself, because the state has to ensure the liberal values of equality and inclusivity—values that cannot be easily secured by bottom-up initiatives. In the end, the physical character of public space has strong and direct political and social ramifications, and active connections have to be established to effectuate the full potential of public space.

References

Arendt, H. (1958). *The human condition*. Chicago and London: University of Chicago Press.

Ariès, P. (1962). *Centuries of childhood. A social history of the family*. New York: Vintage Books.

Arthur, W. B. (1989). Competing technologies, increasing returns, and lock-in by historical events. *The Economic Journal, 99*(394), 116–131.

Barnett, C. (2007). Convening publics: The parasitical spaces of public action. In *The handbook of political geography* (pp. 403–417). London: Routledge.

Bauman, Z. (2000). *Liquid modernity*. Cambridge: Polity.

Benn, S. I., & Gaus, G. F. (1983). The public and the private: Concepts and actions. In S. I. Benn & G. F. Gaus (Eds.), *Public and private in social life* (pp. 3–27). London & Canberra: Croom Helm.

Berlin, I. (1969). *Two concepts of liberty* (pp. 118–172). Oxford: Clarendon Press.

Bohman, J. (1998). Survey article: The coming of age of deliberative democracy. *Journal of Political Philosophy, 6*(4), 400–425.

Boyer, R. H. W. (2015). Grassroots innovation for urban sustainability: comparing the diffusion pathways of three ecovillage projects. *Environment and Planning, 47*(2), 320–337. http://www.envplan.com/abstract.cgi?id=a140250p

Braudel, F. (1982). *Civilization and capitalism, 15th–18th century: The perspective of the world*. Berkeley, CA: University of California Press.

Bulkeley, H., & Castán Broto, V. (2013). Government by experiment? Global cities and the governing of climate change. *Transactions of the Institute of British Geographers, 38*(3), 361–375.

Callon, M. (1998). Introduction: The embeddedness of economic markets in economics. *The Sociological Review, 46*(S1), 1–57.

Castells, M. (1993). European cities, the informational society, and the global economy. *Tijdschrift voor Economische en Sociale Geografie, 84*(4), 247–257.

Castells, M. (1997). The information age: Economy, society and culture. In *The power of identity* (Vol. 2). Chichester: Blackwell.

Chadwick, E. (1842). *Report on the sanitary condition of the labouring population of Great Britain: Supplementary report on the results of special inquiry into the practice of interment in towns*. Richmond: HM Stationery Office.

Chandler, A. D. (1977). *The visible hand: The managerial revolution in American business*. Cambridge, MA: Harvard University Press.

Coase, R. H. (1937). The nature of the firm. *Economica, 4*(16), 386–405.

Correljé, A. F., Cuppen, E., Dignum, M., Pesch, U., & Taebi, B. (2015). Responsible innovation in energy projects: Values in the design of technologies, institutions and stakeholder interactions. In *Responsible innovation* (pp. 183–200). London: Springer.

Correljé, A. F., & Groenewegen, J. P. (2009). Public values in the energy sector: Economic perspectives. *International Journal of Public Policy, 4*(5), 395–413.

De Moor, T. (2013). De herontdekking van het Collectief. *Samenleving en politiek: tijdschrift voor een democratisch socialisme (Sampol), 20*(4), 29–38.

De Swaan, A. (1990). De Zorg en de Staat. In *Welzijn, onderwijs, en gezondheidszorg in Europa en de Verenigde Staten in een nieuwe tijd*. Amsterdam: Bert Bakker.

De Waal, B. G. M. (2012). *De stad als interface: Digitale media en stedelijke openbaarheid*. Groningen: University Library.

Dekkers, M. (2014). *De kleine verlossing of de lust van het ontlasten*. Amsterdam: Dwarsligger.

Deleuze, G., & Guattari, F. (1987). *A thousand plateaus*. Minneapolis: University of Minnesota Press.
Dewey, J. (1927). *The public and its problems*. New York: Holt.
Douglas, M. (1966). *Purity and danger; an analysis of concepts of pollution and taboo*. New York: Praeger.
Dryzek, J. S. (1994). *Discursive democracy: Politics, policy, and political science*. Cambridge: Cambridge University Press.
Duany, A., Plater-Zyberk, E., & Speck, J. (2001). *Suburban nation: The rise of sprawl and the decline of the American dream*. New York: Macmillan.
Dumont, L. (1977). *From Mandeville to Marx. The genesis and triumph of economic ideology*. Chicago and London: Chicago University Press.
Evans, J., Karvonen, A., & Raven, R. (2016). *The experimental city*. New York: Routledge.
Feola, G., & Nunes, R. (2014). Success and failure of grassroots innovations for addressing climate change: The case of the transition movement. *Global Environmental Change, 24*, 232–250.
Fischer, F. (1999). Technological deliberation in a democratic society: The case for participatory inquiry. *Science and Public Policy, 26*(5), 294–302.
Fishman, R. (1977). *Urban utopias in the twentieth century*. Cambridge, MA: MIT.
Fraser, N. (1990). Rethinking the public sphere: A contribution to the critique of actually existing democracy. *Social Text, 25–26*, 56–80.
Gandy, M. (1999). The Paris sewers and the rationalization of urban space. *Transactions of the Institute of British Geographers, 24*(1), 23–44.
Gandy, M. (2004). Rethinking urban metabolism: water, space and the modern city. *City, 8*(3), 363–379.
Gastelaars, M. (1994). Publiek private aangelegenheden. Een essay over de wc. *Kennis en Methode, 1*, 69–89.
Geuss, R. (2001). *Public goods. Private goods*. Princeton & Woodstock: Princeton University Press.
Goffman, E. (1963). *Behavior in public places. Notes of the social organization of gatherings*. London: The Free Press of Glencoe.
Habermas, J. (1999). *The structural transformation of the public sphere. An inquiry into a category of Borugeois society*. Cambridge, MA: MIT Press.
Hajer, M. A. (2011). De energieke samenleving. In *Op zoek naar een sturingsfilosofie voor een schone economie*. Den Haag: Planbureau voor de Leefomgeving.
Hirschman, A. O. (1977). *The passion and the interests. Political arguments for capitalism before its triumph*. Princeton: Princeton University Press.
Höflehner, T. (2011). *Ecovillages as pioneers for sustainable development in communities and regions*. Graz: Karl–Franzens University.
Holland, L. (2004). Diversity and connections in community gardens: a contribution to local sustainability. *Local Environment, 9*(3), 285–305.
Houwaart, E. S. (1991). *De hygiënisten*. Groningen: Historische Uitgeverij.
Jackson, S. J., Gillespie, T., & Payette, S. (2014). The policy knot: Re-integrating policy, practice and design in CSCW studies of social computing. In *Proceedings of the 17th ACM conference on computer supported cooperative work & social computing* (pp. 588–602). New York: ACM.
Jacobs, J. (1992a). *The death and life of great American cities. The failure of town planning*. Hammondsworth & Ringwood: Penguin Books.
Jacobs, J. (1992b). *Systems of survival. A dialogue on the moral foundations of commerce and politics*. New York: Random House.
Jewitt, S. (2011). Geographies of shit: Spatial and temporal variations in attitudes towards human waste. *Progress in Human Geography, 35*(5), 608–626. https://doi.org/10.1177/0309132510394704
Kaplan, B. J. (2002). Fictions of privacy: House chapels and the spatial accommodation of religious dissent in early modern Europe. *The American Historical Review, 107*(4), 1031–1064.

Kelbaugh, D. (1996). Typology—An architecture of limits. *Architectural Theory Review, 1*(2), 33–52.
Kitchin, R. (2014). The real-time city? Big data and smart urbanism. *GeoJournal, 79*(1), 1–14.
Kohn, M. (2013). Privatization and protest: Occupy Wall Street, occupy Toronto, and the occupation of public space in a democracy. *Perspectives on Politics, 11*(1), 99–110.
Lefebvre, H. (1991). *The production of space*. Oxford: Blackwell.
Lintsen, H. W. (1992). *Wat is techniek? Een geschiedenis van menselijke secreten en discrete technieken*. Eindhoven: University of Eindhoven.
Loukaitou-Sideris, A., & Banerjee, T. (1998). *Urban design downtown: Poetics and politics of form*. California: University of California Press.
Low, S. M. (1996). Spatializing culture: The social production and social construction of public space in Costa Rica. *American Ethnologist, 23*(4), 861–879.
Low, S. M. (2006). The erosion of public space and the public realm: paranoia, surveillance and privatization in New York City. *City & Society, 18*(1), 43–49.
Low, S. M., & Iveson, K. (2016). Propositions for more just urban public spaces. *City, 20*(1), 10–31.
Macpherson, C. B. (1962). *The theory of possessive individualism*. Oxford: Oxford University Press.
Mandeville, B. (2017). *The fable of the bees: Or, private vices, publick benefits*. Wellington: The Floating Press.
Marres, N. (2007). The issues deserve more credit pragmatist contributions to the study of public involvement in controversy. *Social Studies of Science, 37*(5), 759–780.
McCloskey, D. N. (2010). *The bourgeois virtues: Ethics for an age of commerce*. Chicago: University of Chicago Press.
Melosi, M. V. (1981). *Garbage in the cities: refuse, reform, and the environment*. Pittsburgh, PA: University of Pittsburgh Press.
Melosi, M. V. (2000). *The sanitary city. Urban infrastructure in America from colonial times to the present*. Baltimore: John Hopkins University Press.
Mill, J. S. (1985). *On liberty*. London: Penguin Books.
Minogue, K. (1963). The moral character of liberalism. In *The Liberal Mind*. Indianapolis: Liberty Fund.
Mitchell, D. (1995). The end of public space? People's park, definitions of the public, and democracy. *Annals of the Association of American Geographers, 85*(1), 108–133.
Moore, B. (1985). *Privacy. Studies in social and cultural history*. Armonk & London: M.E. Sharp.
Mouffe, C. (2000). *The democratic paradox*. London: Verso.
Naylor, L. (2012). Hired gardens and the question of transgression: lawns, food gardens and the business of 'alternative' food practice. *Cultural Geographies, 19*(4), 483–504. https://doi.org/10.1177/1474474012451543
Nijman, J. (1999). Cultural globalization and the identity of place: The reconstruction of Amsterdam. *Cultural Geographies, 6*(2), 146–164. https://doi.org/10.1177/096746089900600202
Noort, J. v. d. (1990). *Pion of pionier*. Stichting PK: Gemeentelijke bedrijvigheid in de negentiende eeuw.
Osborne, D., & Gaebler, T. (1992). *Reinventing government. How the entrepreneurial spirit is transforming the public sector*. Reading: Addison-Wesley.
Pateman, C., & Phillips, A. (1987). *Feminist critiques of the public/private dichotomy*. Oxford: Basil Blackwell.
Pesch, U. (2005). *The predicaments of publicness. An inquiry into the conceptual ambiguity of public administration*. Delft: Eburon.
Pesch, U. (2008). The publicness of public administration. *Administration & Society, 40*(2), 170–193. https://doi.org/10.1177/0095399707312828
Pesch, U. (2014). Sustainable development and institutional boundaries. *Journal of Integrative Environmental Sciences, 11*(1), 39–54.

Pesch, U. (2015a). Engineers and active responsibility. *Science and Engineering Ethics, 21*(4), 925–939. https://doi.org/10.1007/s11948-014-9571-7
Pesch, U. (2015b). Publicness, privateness, and the management of pollution. *Ethics, Policy & Environment, 18*(1), 79–95.
Pesch, U., & Ishmaev, G. (2019). Fictions and frictions: Promises, transaction costs and the innovation of network technologies. *Social Studies of Science, 49*(2), 264–277. https://doi.org/10.1177/0306312719838339
Pesch, U., Spekkink, W., & Quist, J. (2019). Local sustainability initiatives: innovation and civic engagement in societal experiments. *European Planning Studies, 27*(2), 300–317. https://doi.org/10.1080/09654313.2018.1464549
Polanyi, K. (2001). *The great transformation. The political and economic origins of our time*. Boston: Beacon Press.
Putnam, R. D. (2001). *Bowling alone: The collapse and revival ofAmerican community*. New York: Simon and Schuster.
Rawls, J. (1997). The idea of public reason revisited. *The University of Chicago Law Review, 64*, 765–807.
Ryan, A. (1983). Private Selves and Public Parts. In S. I. Benn & G. F. Gaus (Eds.), *Public and private in social life* (pp. 135–154). London & Canberra: Croom Helm.
Sadowski, J., & Pasquale, F. A. (2015). The spectrum of control: A social theory of the smart city. *First Monday, 20*, 7.
Saxonhouse, A. W. (1983). Classical greek conceptions of public and private. In *Public and private in social life* (pp. 363–384). London: Croom Helm.
Schama, S. (1988). *The embarrassment of riches: An interpretation of Dutch culture in the Golden Age*. Berkeley, CA: University of California Press.
Schor, J. (2014). Debating the sharing economy. *Great transition initiative*. (October 2014), http://greattransition.org/publication/debating-the-sharing-economy.
Sedlacek, T. (2011). *Economics of good and evil: The quest for economic meaning from Gilgamesh to Wall Street*. Oxford: Oxford University Press.
Sennett, R. (1992). *The conscience of the eye: The design and social life of cities*. New York: WW Norton & Company.
Sennett, R. (1993). *The fall of public man*. London and Boston: Faber & Faber.
Sennett, R. (1996). *Flesh and stone: The body and the city in Western civilization*. New York: WW Norton & Company.
Shaw, J., & Graham, M. (2017). An informational right to the city? Code, content, control, and the urbanization of information. *Antipode, 49*, 907–927.
Shorto, R. (2013). *Amsterdam: A history of the world's most liberal city*. New York: Random House.
Silver, A. (1997). Two different sorts of commerce—Friendship and strangership in civil society. In J. Weintraub & K. Kumar (Eds.), *Public and private in thought and practice. Perspectives on a grand dichotomy* (pp. 43–74). Chicago & London: University of Chicago Press.
Simon, H. A. (1997). *Administrative behavior. A study of decision-making processes in administrative organization*. New York and London: The Free Press.
Smith, A. (1998). *An inquiry into the nature and causes of the wealth of nations*. Oxford and New York: Oxford University Press.
Smith, A. (2011). The transition town network: A review of current evolutions and renaissance. *Social Movement Studies, 10*(1), 99–105. https://doi.org/10.1080/14742837.2011.545229
Smith, A., Hargreaves, T., Hielscher, S., Martiskainen, M., & Seyfang, G. (2016). Making the most of community energies: Three perspectives on grassroots innovation. *Environment and Planning A, 48*(2), 407–432.
Smith, A., & Stirling, A. (2016). Grassroots innovations and democracy.
Springer, S. (2011). Public space as emancipation: meditations on anarchism, radical democracy, neoliberalism and violence. *Antipode, 43*(2), 525–562.
Taylor, C. (1995). *Philosophical arguments*. Cambridge, MA: Harvard University Press.

Taylor, F. W. (1914). *The principles of scientific management*. New York: Harper.

Townsend, A. M. (2013). *Smart cities: Big data, civic hackers, and the quest for a new utopia*. New York: WW Norton & Company.

Valentine, G. (2008). Living with difference: reflections on geographies of encounter. *Progress in Human Geography, 32*(3), 323–337. https://doi.org/10.1177/0309133308089372

Van Horn Melton, J. (2001). *The rise of the public in enlightenment Europe*. Cambridge: Cambridge University Press.

Van Timmeren, A., & Henriquez, L. (2015). *Ubikquity & the illuminated city. From smart to intelligent urban environments*. Delft: Delft University Press.

Van Zon, H. (1990). *Tachtig jaar RIVM*. Assen: Van Gorcum.

Van Zon, H. (2002). *Nederlandse Hygienisten* (p. 29). Groningen: Groniek.

Veblen, T. (1994). *Absentee ownership: Business enterprise in recent times: The case of America*. Piscataway: Transaction Publishers.

Walker, G., & Devine-Wright, P. (2008). Community renewable energy: What should it mean? *Energy Policy, 36*(2), 497–500. https://doi.org/10.1016/j.enpol.2007.10.019

Walzer, M. (1983). *Spheres of justice. A defense of pluralism and equality*. Philadelphia: Basic books.

Walzer, M. (1984). Liberalism and the art of separation. *Political Theory, 12*(3), 315–330. http://www.jstor.org/stable/191512

Warner, M. (2002). Publics and counterpublics. *Public Culture, 14*(1), 49–90.

Weintraub, J. (1997). The theory and politics of the public/private distinction. In J. Weintraub & K. Kumar (Eds.), *Public and private in thought and practice. Perspectives on a grand dichotomy* (pp. 1–42). Chicago & London: University of Chicago Press.

Wright, F. L. (1932). *The disappearing city*. New York: WF Payson.

Wright, V. (2000). Blurring the public-private divide. In *Governance in the twenty-first century: Revitalizing the public service* (pp. 155–174). Montreal: McGill-Queen's University Press.

Udo Pesch is associate professor at Delft University of Technology. He is affiliated to the department of Values, Technology and Innovation of the Faculty of Technology, Policy and Management. His disciplinary interests include responsible innovation, participatory decision-making, environmental politics, public policy, and ethics. Recently published work includes articles on energy justice, sustainable innovation and moral emotions. He has published a wide range of articles and book chapters, among others in *Social Studies of Science*, *Science, Technology & Human Values*, the *Journal of Responsible Innovation*, *Science and Engineering Ethics*, and *Technological Forecasting and Social Change*.

Chapter 16
The Continued Transformation of the Public Sphere: On the Road to Smart Cities, Living Labs and a New Understanding of Society

Bart Van Der Sloot and Marjolein Lanzing

Abstract This chapter explores the transformation of the public sphere by Smart Cities and Living Labs. It explores this transformation by analysing three data-driven projects in cities around the world. These projects raise significant legal and ethical concerns because they transform and challenge valuable elements of the public sphere. The first section provides the reader with a theoretical lens for this chapter by briefly describing features of a meaningful public sphere as proposed by Jürgen Habermas and discussing the concepts of Living Labs and Smart Cities. In section two, three and four, three cases are analysed, namely the smart nation project of Singapore, Google's Living Lab 'Sidewalk' in Toronto, Canada, and the Living Lab 'Stratums Eind 2.0' in Eindhoven, the Netherlands. These cases will be used to revisit the concept of the public sphere and its transformations as proposed by Jürgen Habermas. We propose that smart Cities and Living Labs challenge important features of an open, neutral and democratic public sphere by raising several ethical and legal concerns.

Keywords Smart cities · Living labs · Public sphere · Surveillance · Behavioral steering · Democratic legitimacy · Privacy · Autonomy

---

This chapter uses insights from previously published research in the Nederlands Juristenblad, 2017 (6) p. 374–382.

B. Van Der Sloot (✉)
Institute for Law, Technology and Society, Tilburg University, Tilburg, Netherlands
e-mail: b.vdrsloot@uvt.nl

M. Lanzing
University of Amsterdam, Amsterdam, Netherlands

M. Nagenborg et al. (eds.), *Technology and the City*, Philosophy of Engineering and Technology 36, https://doi.org/10.1007/978-3-030-52313-8_16

16.1 Introduction

Smart Cities and Living Labs are increasingly used as real life laboratories in order to pave the way for the implementation of smart technology in the physical environment. A public environment that is embedded with sensors and equipped with cameras that are always 'ON' may transform our understanding of the public sphere. This chapter explores this transformation by identifying the ethical and legal concerns that are raised by the development of smart cities and living labs. To this end, three projects are analysed that are currently up and running in different places of the world. Smart cities and Living Labs raise significant legal and ethical concerns because they transform and challenge valuable elements of the public sphere. In the first section we provide the theoretical lens for this chapter. It describes the features of a meaningful public sphere as proposed by Jürgen Habermas and the concepts of Living Labs and Smart Cities. In section two, three and four, three cases are presented. The first case, which is predominantly descriptive, discusses the smart nation project of Singapore and identifies three trends in smart city design.[1] This case is mainly aimed at providing the reader with an insight in the amount of data processed, an understanding of the various data flows between the different actors involved in these types of projects, and an idea of how data-driven applications are integrated into everyday life. The next two cases engage critically with these trends: Google's Living Lab 'Sidewalk' in Toronto, Canada,[2] and the Living Lab 'Stratums Eind 2.0' in Eindhoven, the Netherlands.[3] These two cases serve to discuss the implications of Smart Cities and Living Labs for the ideal of the public sphere in terms of neutrality, democracy and respect for fundamental rights.

These three projects have been selected because they represent the various aspects of Smart Cities and Living Labs. First, these cases differ in size. Singapore is already a smart nation; 'Sidewalk' is a Living Lab limited to a neighbourhood in Toronto and Stratums Eind 2.0 is a Living Lab on a street in the city of Eindhoven. Second, they differ in terms of implementation. Singapore has already implemented many aspects in concrete technologies and applications to the extent that it may have exceeded the stage of 'Living Lab' (although many aspects and practices of this smart nation are experimental); Eindhoven is experimenting and testing; and in Toronto, the plans have not yet been implemented and tested. Third, they all take place in nations that identify as democracies, but are located in different parts of the world: Asia, Europe and North-America. Although it is clear that this chapter cannot give a full and comprehensive overview of the various Smart Cities and Living Labs around the world nor of their potential legal, ethical and social implications, these three cases should enable the reader to get the general gist of new data-driven applications in cities of the future and their potential implications for the public sphere. As a final step, these cases will be used to revisit the concept of the public

[1] https://www.smartnation.sg/.

[2] https://www.sidewalklabs.com/.

[3] http://www.openlivinglabs.eu/livinglab/eindhoven-living-lab.

sphere and its transformations as proposed by Jürgen Habermas. Living Labs and Smart Cities raise legal and ethical concerns with regard to maintaining and fostering the important features of a public sphere. This is significant for monitoring and evaluating (the development of) Smart Cities and Living Labs in the future.

16.2 Theoretical Lens

16.2.1 Structural Transformations of the Public Sphere

In *The Structural Transformation of the Public Sphere* (1962), Jürgen Habermas describes the rise and transformation of the public sphere. In 1439, the goldsmith Johannes Gutenberg invented a printing press that enabled the production of affordable and large quantities of prints. By the sixteenth century, the spread of printing presses through Europe had resulted in the production of millions of copies. European citizens were increasingly able to easily access prints. Moreover, they started to congregate to discuss them. Habermas explains how these meetings, especially the salons of the eighteenth century, gave rise to the notion of the public sphere.

Within liberal democratic theory, the public sphere is traditionally distinguished from the private sphere.[4] The traditional account maintains that in the private domain, citizens can form their personality and develop their ideas and beliefs, while in the public realm, they can exchange those ideas and beliefs. Through discussion and the exchange of ideas that are originally formed in the private domain, public opinion is formed and on the basis thereof, laws are enacted. This is the ideal of the deliberative democracy. Hence, while they are separated, there exists an important and continued dialogical movement between the private and the public domain. Habermas describes this as follows:

> '(...) it was private people who related to each other in [the public sphere] as a public. The public's understanding of the public use of reason was guided specifically by such private experience as grew out of the audience-oriented subjectivity of the conjugal family's intimate domain. Historically, the latter was the source of privateness in the modern sense of a saturated and free interiority' (Habermas, 1962/1989, p. 28).

---

[4] The traditional liberal private-public divide has been notoriously criticized by feminist theory. For a long time the 'private' sphere was equated with the 'home', which was apolitical and 'free' from government interference. Philosophers like Carole Pateman (1989) and MacKinnon (1989) have criticized the private sphere of the 'home' for being a place of necessity and domination and the public sphere as a space only accessible to (white) men. For that reason, they argued for erasing the distinction between the public and private spheres: the private is the political. We do not assume the existence of a public-private dichotomy nor do we propose the (re)instalment of such a distinction. The distinction merely features within the context of Habermas' explanation about the structural transformation of the public sphere. We recognize a myriad of social contexts with different privacy expectations (Nissenbaum 2009). Nevertheless, we use the (ideal) notions of the public sphere and the private sphere to raise concerns with regard to the transformation of the public sphere along with the rise of smart cities.

The public sphere is described as an ideal, virtual and essentially social space accessible to *all* citizens, (regardless of race, class or gender) in which citizens can meet each other and can freely communicate and deliberate by exchanging ideas about different conceptions of the good life (Habermas, 1962/1991, p. 1). It is also the place where they can act according to their opinion and their conception of the good (Habermas, 1996, pp. 25–26, 28). In so far as there are restrictions on the exercise of freedom, these are based on public laws that have democratic legitimacy: the result of public deliberation and an exchange of ideas (Habermas, 1996). Furthermore, this conception of freedom within Habermas' deliberative democracy implies there must be room for minority views within a discussion and not one conception of the good imposed on its citizens. This also means that citizens are free to act according to their own conception of the good life.

Habermas describes how this conception of the public sphere originated from the eighteenth century coffeehouses, salons and tischengesellschaften. Although they differed in size and composition, the style of proceedings, the climate of debate and their topical organization, they all organized discussions among private people that tended to be ongoing and had several common features that resonate with the ideal of the deliberative democracy: equality, critique and inclusivity.

> 'First, they preserved a kind of social intercourse that, far from presupposing the equality of status, disregarded status altogether. The tendency replaced the celebration of rand with a tact befitting equals. [] Secondly, discussion within such a public presupposed the problematization of areas that until then had not been questions. The domain of "common concern" which was the object of public critical attention remained a preserve in which church and state authorities had the monopoly of interpretation not just from the pulpit, but in philosophy, literature, and art, even just at a time when, for specific social categories, the development of capitalism already demanded a behavior whose rational orientation required ever more information. To the degree, however, to which philosophical and literary works and works of art in general were produced for the market and distributed through it, these culture products became similar to that type of information: as commodities they became in principle generally accessible. [] Thirdly, the same process that converted culture into a commodity (and in this fashion constituted it as a culture that could become an object of discussion to begin with) established the public as in principle inclusive. However exclusive the public might be in any given instance, it could never close itself off entirely and become consolidated as a clique; for it always understood and found itself immersed within a more inclusive public of all private people, persons who—insofar as they were propertied and educated—as readers, listeners, and spectators could avail themselves via the market of the objects that were subject to discussion' (Habermas, 1962/1989, pp. 36–37).

Although these characteristics of equality, critique and inclusivity (through the market) have become deeply engrained aspirational values for deliberation in modern day democracies, Habermas describes how gradually, the core characteristics of the public and the private sphere transformed to such an extent that the values of both the private and the public sphere became strained.

One of the reasons for this transformation was the commercialization of important actors and institutions during the twentieth century within the public sphere such as the media (Habermas, 1962/1991, p. 181). Newspapers developed into capitalist enterprises. While commercialization is not necessarily problematic, Habermas takes a critical stance. The commercialization of the media rendered them susceptible to manipulation. Once it becomes too dominant, the commercialization of the public sphere becomes at tension with the potential of this sphere for communication and deliberation. For instance, the commercialization of the media entailed that it became a 'gate through which privileged private interests invaded the public sphere' (Habermas, 1962/1991, p. 185).

Habermas describes two structural changes that dissolved the boundary between the public and the private. On the one hand, the private and the personal lives of citizens became increasingly more public and subjected to public reason, laws and general justice. On the other hand, the public realm became increasingly privatized by organizations and companies. This problematized the ideal to have a private sphere in which one could formulate their own conception of the good life. Moreover, it problematized the ideal of the public sphere and its ideals of openness, neutrality, freedom and respectful deliberation. As a result, Habermas argues, the debate became exclusive and run by professionals who either represent the people or by organizations that stood above and outside the debate (Habermas, 1962/1989, pp. 175–176).

So how does Habermas' analysis of the structural transformations of the public sphere relate to new technological developments? What kinds of new transformations do new technologies such as Smart Cities cause for the way we understand the public sphere? In contemporary political democratic theory, the idealised version of a meaningful public space has been attributed a number of assets allowing for an open debate and interactive dialogue. Habermas' previously mentioned description of the features of an ideal public sphere is a description of ideal ingredients for the ideal of a deliberative democracy (Habermas, 1962/1991, p. 1; Habermas, 1996, pp. 25–26, 28; Habermas, 1996). Based on that description, without aiming to provide an exhaustive list, it can be roughly stated that: (1) the public sphere should be accessible to *all* people (regardless of race, class or gender), (2) should be neutral in the sense that no vision of the good life is presupposed or excluded, (3) in so far as there are restrictions on the exercise of freedom, these should be based on public laws that have democratic legitimacy, (4) that such laws should have democratic legitimation, (5) that such restrictions may never go so far as to integrally impose one idea of the good on society, there must be room for minority views, and that (6) as far as possible, human freedom should remain intact. The latter also entails that unpredictability and uncontrollability, as aspects of individual freedom, are part and parcel of the public realm.

This chapter will discuss how Living Labs and Smart Cities challenge valuable elements of the public sphere in new ways by transforming the design of the public sphere. Of course, the claim is not that an ideal public sphere as described above ever existed (or will ever exist), not in the eighteenth century, not in contemporary society and not in future communities. The idea proposed in this chapter should also not be misunderstood as arguing that all transformations of the public sphere brought about by new technologies are problematic. For instance, the spread of the printing press spurred the physical public sphere of the salons and coffeehouses where people were able to discuss ideas obtained from its prints. The development of the Internet created a virtual parallel. Connected to and intertwined with the physical public sphere, online platforms allow citizens to voice their political opinions and to participate as well as manage and foster various other social relations.

At the same time, technologies support features that reform important elements of the public sphere. For example, online platforms allow (mostly private) parties to filter or block certain voices, options and actions (Vaidhyanathan, 2011, pp. 135–138). Moreover, private parties structure most online platforms by determining the rules through their terms and conditions. Furthermore, the interest of Internet providers does not lie in providing citizens with maximum freedom and to create zones of experimentation. Rather, their interest lies in maximizing their profit, which can be facilitated through controlling and influencing citizens' behaviour (Zuboff, 2015). Tensions are created in the process of the transformation of the public sphere as a result of new technologies that reshape the design of these spheres. This chapter will engage with Smart Cities and Living Labs to discuss their specific impact on the public sphere.

16.2.2 Living Labs and Smart Cities

A recent development in the transformation of the public sphere is the rise of 'Smart Cities' and 'Living Labs' in which the 'physical' public sphere will be transformed in such a way that it will always be 'online'.

Smart cities can be roughly understood as the technological promise to redesign cities into smart environments that are governed by algorithmic decision-making processes that rely on real-time Big Data. Smart cities can respond more efficiently to the needs of their inhabitants through the use of new technologies, the Internet of Things and various 'smart' applications. The 'smartness' then refers to the fact that these environments can learn from and adapt real-time to new circumstances based on new information. They can influence the creatures living and acting in those environments to optimize the safety, mobility, efficiency, sustainability or public health of the city:

> 'Smart cities are places where information technology is wielded to address problems old and new. In the past, buildings and infrastructure shunted the flow of people and goods in rigid, predetermined ways. But smart cities can adapt on the fly, by pulling readings from vast arrays of sensors, feeding the data into software that can see the big picture, and taking action. They optimize heating and cooling in buildings, balance the flow of electricity through the power grid, and keep transportation networks moving. Sometimes, these interventions on our behalf will go unnoticed by humans, behind the scenes within the wires and walls of the city. (…) All the while, they will maintain a vigilant watch over our health and safety, scanning for miscreants and microbes alike' (Townsend, 2013, p. xii–xiii).

It is no longer the behaviour of citizens that changes and adapts to the environment, it is the environment that adapts to the citizens, in order to steer their behaviour in such a way that it becomes increasingly standardised. Actors behind smart cities aim to steer or 'nudge' citizens towards a particular conceptualisation of the good and a particular way of achieving this good. Common values that are mentioned in that respect are, for instance, 'sustainability', 'public health' and 'safety'.

A Living Lab can be described as '(…) an open innovation environment in real-life settings in which user-driven innovation is the co-creation process for new services, products, and societal infrastructures. Living Labs encompass societal and technological dimensions simultaneously in a business-citizens-government-academia partnership.'[5]

'Living Labs' serve as a testing ground for various 'smart' products and technologies in the context of the development of Smart Cities. The Living Lab is characterized by the experiments that take place within an existing social and territorial context, so that the research can be carried out in a 'real' environment beyond the walls of the 'artificial' world of the laboratory. In other words, specific zones within the public sphere or entire cities become designated areas of experimentation. Let us now look at what Smart Cities and Living Labs currently amount to across the world by exploring several cases. What do Smart Cities look like, what do they require and what ethical and legal concerns are raised with regard to the public sphere?

16.3 Singapore: Smart Nation

One of the most advanced Smart Cities in the world is the Smart Nation project in Singapore. The entire country of Singapore is conceptualised as a smart nation. The project aims at making Singapore 'an economically competitive global city and a liveable home. It is a whole-of-nation movement to harness digital technologies to build a future Singapore, to improve living and build a closer community, empower citizens to achieve their aspirations through good jobs and opportunities, and encourage businesses to innovate and grow.'[6]

This section briefly discusses three trends that show how the public sphere is transformed. First, there is the trend of datafication, meaning that increasingly, reality, people and objects are mirrored in data images. These data images are used to make decisions and predictions about reality. Second, as a consequence, there is a hybridisation of the public sphere. Physical reality is combined and integrated with smart products that are in constant connection with the Internet and that react autonomously to changes in their environment. Third, the public space is privatised. Citizens are encouraged to gather data for the police and prevent and deter crime and commercial parties are allowed to run projects and experiments in the public sphere.

16.3.1 Datafication

The Smart Nation project is one of the most open, transparent and well-documented smart cities around the world.[7] This endeavour consists of several projects. There are the National Digital Identity project, which aims to allow citizens and businesses to

---

[5] http://www.diva-portal.org/smash/get/diva2:979171/FULLTEXT01.pdf.

[6] https://www.smartnation.sg/about/Smart-Nation.

[7] Most of what is described in this section comes directly from the project's website or is a paraphrase thereof.

transact digitally in a convenient and secure manner and e-Payments project, which aims to allow everyone to make simple, swift, seamless, and safe payments. Furthermore, there are the Smart Nation Sensor Platform, in which sensors and other Internet of Things (IoT) devices are deployed in order to make Singapore liveable and secure and the Smart Urban Mobility project, in which data and digital technologies, including artificial intelligence and autonomous vehicles, are used to enhance public transport. Finally, there is the Moments of Life initiative, in which government services, across different agencies, offer personalised services to citizens at key moments of their lives in which they will have to adapt their lifestyle -such as the birth of a child. In order to facilitate these projects, Singapore promotes Open Data initiatives and various Living Labs.

There are 14 apps that can be downloaded from the government site. These include apps for health and lifestyle, biodiversity, mobility, suspicious sightings, taxes, official statistics, the weather, municipal issues and immigration services.[8] There are also various initiatives to make Singapore the centre of technological innovation and digital creation. Huge investments are made in that respect:

'Singapore is transforming to become a Smart Nation, where citizens live meaningful and fulfilled lives empowered by digital technology, where digital connectivity leads to stronger community bonds and many more opportunities for Singaporeans to pursue their aspirations and contribute to Singapore's future. This is a whole-of-nation journey that Singapore is embarking on, enabled by digital technologies. Digital technologies will impact how we live our daily lives, open up new possibilities for the way we manufacture goods and deliver services, expand healthcare options, and revolutionise the way we plan and run our city.'[9]

For instance, the National Electronic Health Record (NEHR) is a patient data–sharing platform that enables healthcare professionals to access their patients' healthcare history.[10] But, the data of such platforms may also be used for other purposes like congestion control.[11] Mention can be made of no less than eight open data platforms (between government agencies, between businesses and the government) on transportation, economic and social demographics, financial institutions, statistics, press releases, and local, governmental services and information.[12] The

---

[8] https://www.smartnation.sg/apps; https://www.sgsecure.sg/; https://www.police.gov.sg/sgsecure; http://www.nea.gov.sg/; http://www.singstat.gov.sg/services/singstat-mobile-app; https://www.scdf.gov.sg/content/scdf_internet/en/community-and-volunteers/mobile_phone_technology.html; http://www.mnd.gov.sg/mso/mobile-about.htm.

[9] https://www.nrf.gov.sg/rie2020.

[10] https://www.scientificamerican.com/products/singapore-a-smart-nation/singapore-a-smart-living-laboratory/.

[11] http://brandinsider.straitstimes.com/hitachi/implementing-the-smart-city/—COE likely stands for Certificate of Entitlement https://www.lta.gov.sg/content/ltaweb/en/roads-and-motoring/owning-a-vehicle/vehicle-quota-system/overview-of-vehicle-quota-system.html.

[12] https://www.smartnation.sg/resources/open-data.

promises of an 'open-data society' lie in the realm of public health and the improvement of city governance.[13,14]

This description of Singapore's Smart Nation project, however brief, indicates a trend towards datafication. This means that increasingly large amounts and varieties of data are gathered about personal lives and the affairs in both public and private sphere. Data and the data image about reality are becoming increasingly important. Profiles and data-analytics become increasingly dominant. Within the Smart Nation project, this is viewed as a positive development. Data-driven applications are described by the Smart Nation project as more open and transparent towards citizens and more efficient and effective in terms of operation. Through data-driven applications, citizens and the community can share information and together build the nation they strive for. Although Singapore's Smart Nation project may be an extreme example, datafication is a general element in most smart cities and living labs. A second element general to such project is hybridisation, that is, the amalgamation of offline and online. This trend will be central to the next sub-section.

16.3.2 Hybridisation

Real and virtual, offline and online, physical and digital are increasingly merged in Singapore's Smart Nation project.[15] To begin with, the physical city is studded with cameras, sensors and other objects that can gather data.

> 'In Singapore we started building our own network of sensors, especially CCTV cameras, some time ago. The Police installed CCTVs at HDB void decks and [o]ther agencies have different sensors. PUB has sensors to detect water levels in drains to know whether it is likely to flood or not. LTA has cameras to monitor traffic conditions and deter illegal parking. [] We are building an integrated national sensor network. We are making "every lamp-post a smart lamp-post", meaning it can mount different types of sensors on any of the lampposts. We are installing more CCTV cameras in public places. We are combining inputs from different sources—police, LTA, hotels and commercial buildings, even handphones, which are effectively sensors on the ground. [] So if I have 10,000 cameras, I do not need 1,000 people watching those cameras. I just need maybe just 10 people. Each person can watch 1,000 cameras and if the AI detects that something funny is happening, it will pop up and the man can pay attention and a response can be direct.'[16]

The second step is to integrate data-driven devices in the physical world and to make objects and environments contextual, meaning that they correspond to activities or the state of affairs in which they play a role. This entails that these objects can communicate with other devices over the internet and that they can adapt and make

[13] https://www.scientificamerican.com/products/singapore-a-smart-nation/singapore-a-smart-living-laboratory/.

[14] https://www.nrf.gov.sg/programmes/virtual-singapore.

[15] https://www.tech.gov.sg/-/media/GovTech/Media-Room/Speeches/2017/5/Factsheet-Smart-Nation-Sensor-Platform.pdf.

[16] http://www.pmo.gov.sg/national-day-rally-2017.

decisions based on profiles and patterns distilled from previous data collection and analytics. Singapore has several relevant projects in this sense in the area of health,[17] urban life,[18] mobility[19] and public services.[20]

Health projects include various initiatives. One initiative promotes the usage of robotics in healthcare, which are believed to help seniors and those with disabilities through ‘automated bathing machines and autonomous droids that can interact and monitor patient progress’.[21] Another initiative enables users to access their medical records and useful health information and share it with their caregivers.[22] In addition, it offers self-tracking apps and virtual elderly monitoring services that alerts caregivers when abnormalities are detected.[23,24] In the category living, one particular app that stands out is the smart homes project.[25] With approximately 80% of Singapore households living in public housing, the government promotes smart homes equipped with sensors. Moreover, a number of private sector-driven smart home trials have been launched.[26]

Mobility is also increasingly becoming smart. In public transport, autonomous technology and public transport data support mobility on demand. [27,28] Another project uses public data sets and data analytics to assist the government and Singaporeans to cooperate on urban transport planning and transport solutions.[29] Finally, there is a project of self-driving vehicles.[30]

---

[17] https://www.smartnation.sg/initiatives/Health.

[18] https://www.smartnation.sg/initiatives/Living.

[19] https://www.smartnation.sg/initiatives/Mobility.

[20] https://www.smartnation.sg/initiatives/Services.

[21] https://www.smartnation.sg/initiatives/Health/assistive-technology-analytics-and-robotics-for-aging-and-healthcare.

[22] https://www.smartnation.sg/initiatives/Health/healthhub-portal%2D%2Da-digital-healthcare-solution.

[23] https://www.smartnation.sg/initiatives/Health/national-steps-challenge%2D%2Dan-app-towards-healthy-and-active-lifestyle.

[24] https://www.smartnation.sg/initiatives/Health/telehealth%2D%2Dintegrated-and-seamless-healthcare-services-at-home.

[25] https://www.smartnation.sg/initiatives/Living/smart-homes%2D%2Dtech-enabled-solutions-for-homes-in-singapore.

[26] http://www20.hdb.gov.sg/fi10/fi10296p.nsf/PressReleases/F93B15F80588397748257D500009CE6C?OpenDocument.

[27] https://www.smartnation.sg/initiatives/Mobility/contactless-fare-payment-for-public-transport-in-singapore.

[28] https://www.smartnation.sg/initiatives/Mobility/mobility-on-demand%2D%2Dreal-time-demand-driven-transport-through-apps.

[29] https://www.smartnation.sg/initiatives/Mobility/open-data-and-analytics-for-urban-transportation.

[30] https://www.smartnation.sg/initiatives/Mobility/self-driving-vehicles-sdvs%2D%2Dfuture-of-mobility-in-singapore. See also: https://www.smartnation.sg/initiatives/Mobility/spearheading-research-in-standards-for-self-driving-vehicles-sdvs.

Most public services are digitalized. Digital government services include public service chatbots,[31] the Moments of Life initiative, that was previously mentioned,[32] and various financial technology projects.[33] Consequently, the city as well as the home and the public as well as the private sphere, are increasingly hybrid environments. Objects, sensors and cameras constantly gather data and react to the activities and state of affairs within those environments, contributing to smart governance with automated decisions.

16.3.3 Privatisation

A final trend that can be witnessed in the Smart Nation project in Singapore is the privatisation of the public sphere. The public sphere is increasingly owned and controlled by private parties, which can be either commercial organisations that commercialise and exploit data or private citizens that are granted special rights, positions or privileges to act in the public sphere. 'Public-private partnerships are essential for the successful development of a smart city.'[34] One of the projects that received funding is the Corporate Laboratory@University Scheme, which is described as the establishment of key laboratories by industries in universities. [35] Research and Development is conducted in collaboration between universities and companies, which ensures that universities achieve impact by developing cutting edge solutions for problems faced by industry.[36]

Also in other sectors, public-private partnerships are encouraged. The underlying rationale is that the public and public institutions benefit from such collaborations because private organisations help in promoting public goals, produce knowledge and introduce services that facilitate citizens in their everyday life. Public officials in Singapore state that:

> 'We strongly believe in the value of public-private partnership because in the space of smart cities (...) government policies and support can strongly influence how companies innovate in those spaces.'[37]

---

[31] https://www.smartnation.sg/initiatives/Services/digital-government%2D%2Dpublic-services-made-more-seamless-with-technology.

[32] https://www.smartnation.sg/initiatives/Services/moments-of-life-initiative-begins-with-supporting-every-young-child.

[33] https://www.smartnation.sg/initiatives/Services/regulatory-sandbox-for-innovative-fintech-experimentation. See also: https://www.smartnation.sg/initiatives/Services/towards-a-smart-cashless-society-with-contactless-payment.

[34] http://brandinsider.straitstimes.com/hitachi/implementing-the-smart-city/.

[35] https://www.nrf.gov.sg/programmes/corporate-laboratory@university-scheme.

[36] One of such collaborations is in the area of promoting cyber security. https://www.nrf.gov.sg/Data/PressRelease/Files/201610241232595116-2016-1022%20NUSSingtel%20Corp%20Lab%20-%20Press%20Release%20(final).pdf

[37] https://www.opengovasia.com/articles/7182-exclusive-fostering-public-private-sector-partnerships-in-singapore.

In their turn, private organisations are allowed to enter the public sphere and use powers traditionally in the hand of governmental institutions in order to gather data and to control public life. The National Research Foundation of the prime minister's office of Singapore states on their website: 'Businesses can tap on the wealth of data and information within Virtual Singapore for business analytics, resource planning and management and specialised services.'[38]

16.3.4 Conclusion

Three essential characteristics of smart cities have been described on the basis of the Singapore Smart Nation project. First, they rely heavily on datafication. Reality is mimicked in data profiles; these data profiles subsequently replace reality in decision-making processes. Second, smart cities transform the physical reality into a hybrid space, in which objects are equipped with sensors and camera's in order to gather data about citizens and their affairs and in which these objects are connected to the internet, making them smart objects that resonate with the context within which they function. Third, smart cities allow for a privatization of the public sphere. Private organizations are allowed to gather data in public and commercialize part of that sphere. In the next two sections, discussing the cases of Sidewalk Toronto and Stratums Eind 2.0, these characteristics are taken as a given when reflecting on the implications in terms of transforming the meaning of the public sphere.

16.4 Toronto: Sidewalk Toronto

On the eastern side of the Toronto Waterfront lies Quayside. Quayside is the projected site for Sidewalk Toronto: a joint urban innovation venture between the local community, urban development non-profit corporation Toronto Waterfront and Google's Sidewalk Labs. It aims to turn this area into an 'affordable, eco-friendly, smart neighbourhood' (Sauter, 2018). Already $50 million has been committed to a one-year process and Sidewalk's design plans extend beyond the initial Quayside location, stretching along the entire Waterfront. This case will be used to highlight how the goal of improving the quality of life relates to a particular conception of the good life and how a dominant private party may shape this conception. Moreover, it discusses how this relates to the transformation of the public sphere from the perspective of a deliberative democracy. First, Sidewalk Labs exemplifies that within Living Labs as public-private partnerships, it becomes possible for powerful third parties to 'run' the city without having been part of a democratic procedure involving elections. Secondly, the case study shows that within these partnerships, the

---

[38] https://www.nrf.gov.sg/programmes/virtual-singapore.

neutrality of the public sphere is challenged because a particular (commercially driven) conception of the good life may permeate every corner of the public sphere. Finally, this also means that the private sphere is affected too, because the boundary between these spheres is increasingly dissolved.

16.4.1 Neutrality

Ideally, the public sphere of a democracy should be 'neutral'. In a liberal democracy, the idea of state neutrality applies. This means that not one dominant perspective on the 'good' life should be imposed on the citizens by either a government, company or other powerful actor(s). Without this proviso, citizens will have a difficult time to freely discuss different versions of the good life or to experiment with dissenting ideas. In the case of Sidewalk Labs, the neutrality of the public space is challenged. While the government always tries to influence the behavior of citizens, it uses laws and regulations that leave the choice to the public as much as possible. Citizens should be free to pursue an alternative version of the good life.

In the case of Sidewalk Labs, there exists a significant power imbalance between the actors within the public-private partnership. This means that it is likely that the city will be shaped according to the vision of the most powerful actor, which is in this case a commercial actor. Google's vision with regard to the quality of life reflects an idea about what it means to live the good life. In other words, their suggestions with regard to the quality of life are not neutral, but normative. Sidewalk Toronto is the materialization of a dream that has been held by the founders of Google for a long time: to have a city and to be in charge (Dingman, 2017). The wish to 'be in charge' relates to Google's ambition to accelerate urban innovation in order to improve quality of life. Sidewalk Labs is the pilot for many cities to come (Crawford, 2018). It aims to benefit its citizens by improving the sustainability, safe mobility, economic opportunity, affordability of housing, community building and reclaiming public space for social connection of the area. Sidewalk Labs explicitly states this on their homepage accompanied by attractive visuals of a green neighbourhood alive with social interaction.[39] Google's interventions with regard to shaping the city and designing it to respond to the needs of its citizens are designed according to a certain norm associated with a particular conception of the good life. For example, altering the light on the smoker's patio to an annoying electric blue, based on the value of public health, constitutes a normative interference with a person's desire to enjoy a cigarette. The same holds true for the protection of public order and safety; although safety is one of the most commonly shared ideals and generally seen as a precondition of a 'good life', there is still a part of the population for which a good night life experience certainly does not exclude a bar fight. And even if everybody on the planet would agree that values like climate protection,

[39] https://www.sidewalklabs.com/.

public health, safety and responsible energy use are worth pursuing, there might still be disagreement about the realisation of these values in the form of concrete normative interferences.

Smart environments increasingly dissolve the boundary between the public and private sphere. This means that conceptualizations of the good life as proposed by Google also infiltrate what used to be called the 'private' sphere. This also has an impact on the development of and, as a result, on the diversity of conceptualizations of the good life that are available for discussion in the public sphere. Minority, including dissenting, views or behaviour may not be accommodated or recognized by the smart systems embedded within the environment due to a normative blind spot. In addition, the normative interferences may be economic or state incentives that do not necessary have the best interest and well being of the public at heart.

16.4.2 Democracy

Of course what usually legitimates governments to surveil public spaces and promote a certain conception of the good life is that it has democratic legitimation. Sidewalk's CEO Dan Doctoroff expressed that Google's systems, which require a large amount of sensors and camera's in the public sphere, could actually improve on democracy (Crawford, 2018). While Doctoroff does not explain how democracy could be improved, some suggestions can be found that may resonate with this goal on the Sidewalk Labs website. The website identifies with a quote about the relationship between open data and democracy: 'Democracies provide pathways for government to learn from their citizens. Open data makes those pathways so much more powerful.'[40] In this reading of a democracy, the government can use democracy, which presupposes the outcome of deliberations and discussions in the public sphere, to learn about what their citizens want in terms of leading their lives. By collecting data about their citizens, the government is able to respond more effectively to the needs of their citizens that they can deduce from patterns in the data. Democracy itself is reduced to an instrument of the government to effectively organize society.

This view on the role of the public sphere differs from the conceptualization of the public sphere within a deliberative democracy. Within a deliberative democracy, a public sphere entails active discussion between citizens about a multitude of different versions of the good life and how this should be realized in society. Argumentation and citizen participation is key for a discussion and for persuading other citizens whilst respecting their autonomy. Conceptualizing the public sphere as a space that can be mined, in which the collection of data and the deduction of dominant 'patterns' are considered as reflective and representative of the needs of

[40] https://www.sidewalklabs.com/blog/the-tremendous-untapped-power-of-data-to-reinvent-city-services/.

citizens, changes its meaning. One of the implications of such a transformed conception of the public sphere is that the question 'why' matters less and less. Data, however valuable, does not inherently provide an explanation or justification (Boyd & Crawford, 2012; Mayer-Schönberger & Cukier, 2013).

Sidewalk Labs more or less circumvents this concern by forming a partnership with the local community. It has developed a civic engagement plan that features meetings, workshops and events in which citizens can discuss and offer ideas about issues related to the city. These do not only encompass their views on realizing the values of mobility or sustainability, but also values and issues such as social inclusion. There is even a Sidewalk summer camp for children. Sidewalk Labs also plans to launch innovation trials across town, not only in Quayside to familiarize people with the concept of a smart environment.

The problem remains that, despite the effort to include public participation and co-creation, the values for quality of life that arise from these meetings are not safeguarded by a formal democratic procedure. Regardless of the well-meant intentions, the power imbalance between the municipality, its citizens and Google is too great to ensure that the input from citizens is heard and acted upon (in the future). In that sense, its dominant and powerful status prevents it from actually being able to serve public interest.

16.4.3 Commodification

Google's ambitions are realized by extensive real-time data collection enabled by an unprecedented digital network of sensors. Sidewalk Labs will run on 'a traffic-sending network that will collect data from smartphones, embedded sensors and cameras to identify areas that could use more bike-sharing slots, or where a self-driving vehicle should be routed, or where a future pop-up store could find a market for its wares' (Dingman, 2017) in order to control anything from 'the ambient temperature of buildings to crosswalk signals to assigned uses of adaptable private and publics spaces' (Sauter, 2018). In order to experiment with different technologies synchronically, a substantial amount of different types of sensors will be installed. Sidewalk Labs has the potential to become an intensively surveilled place.

The main incentive for Sidewalk Labs is commercial. Google has pledged to invest $10 million dollars in the project, as long as the government signs up for a plan that benefits the company (Crawford, 2018). Google has offered to share the data that they collect with the city, but it remains to be seen how much of this data actually benefits the government. In the meantime, Google has access to a substantial amount of data and can learn about all sorts of aspects of urban life including (predictive) patterns with regard to social service distribution, mobility and energy use. This creates opportunities for commodification of that data and new products may be developed by Google based on these insights (Zuboff, 2015). In Living Labs, as public-private partnerships, the distinction between public and private interests becomes blurred.

Google's dominant market position and its large financial resources, make Google an interesting partner for governments with limited resources, but also a powerful and dominant actor in shaping the terms and conditions of a project like Sidewalk Labs. In the case of Sidewalk Labs, Google is effectively allowed to run the city according to their vision on improving the quality of life, which presupposes a certain conception of the good life. This entails a shift away from the public sphere and democratic decision-making that stems from its discussion. Living Labs provide an opportunity for powerful private actors to shape the city without being elected through a formal democratic procedure as the representative that acts on behalf of the public interests.

Moreover, Google's plan for civic engagement has been criticized as a way to normalize societal experimentation and the idea of being a 'guinea pig' in Sidewalk's experiments so as to ensure a smooth transition and familiarity with these technologies so that when these pilots are later integrated within a built-environment, citizens will have no qualms inhabiting smart housing and neighbourhoods (Dingman, 2017). This concern relates to considerations regarding privacy and fairness. The people who will inhabit the affordable Sidewalk Lab are likely to be vulnerable individuals such as children and people who need affordable housing (Sauter, 2018). They may not agree with Google's terms and conditions but find themselves forced to accept them due to their socio-economic circumstances. While the question whether the commodification of personal data and the steering of behaviour should be allowed is already a controversial debate, circumstances of inequality may indicate a particular form of harmful and inappropriate commodification (Roessler, 2015, p. 149; Satz, 2010, p. 98). While Google presents it is a trade-off of interests, this frame represents a very particular view on information that is, first and foremost, in the interest of a powerful commercial entity.

16.4.4 Conclusion

This section discussed Sidewalk Toronto, the Living Lab run by Google in cooperation with the local government and its citizens. This case has been examined from the perspective of a deliberative democracy and highlights how state neutrality with regard to conceptualizing the meaning of 'the good life' becomes strained in public-private endeavours such as Living Labs and Smart Cities in which they aim to improve quality of life. This is largely due to a power imbalance between powerful private actors, citizens and local governments. Moreover, the absence of a neutral public sphere problematizes democratic discussion amongst multiple different versions of the good life. The absence of a neutral public sphere, however well intentioned, may not recognize minority views or behaviour. The freedom to formulate alternative views or to behave alternatively might become increasingly reduced and limited, which is problematic from the viewpoint of a deliberative democracy. Powerful private actors may, due to their position, undermine the rule of law that constitutes the legitimacy of democratic procedure and increasingly compromise the neutrality of the public sphere.

16.5 Eindhoven: Stratums Eind 2.0

The city of Eindhoven aims to increase security and decrease the number of incidents (night life violence) in the nightlife area called Stratumseind through the use of a Living Lab.[41] The project, called Stratumseind 2.0,[42] is a partnership between the municipality of Eindhoven, Eindhoven University and companies such as Phillips, which is specialised in lightning, personalised healthcare and smart products.[43] In order to maximize security, the area is monitored permanently by cameras, heat sensors, object recognition is deployed and the data are combined with live Twitter feeds and social media post. In addition, those data are combined with existing databases owned by the municipality.[44] On the basis of all these data, predictions are made about citizens' behaviour and preventive measures are taken. In addition, behaviour is manipulated by various experiments performed at citizens. Will coloured light make people less aggressive than bright white light (Haans & De Kort, 2014)? Will people behave more relaxed when tangerine scent is sprayed in the streets?[45]

There are three aspects of this case that are at tension with our initial description of a meaningful public sphere. First, the projects undermine the fundamental rights of citizens, such as the right to privacy and the right to data protection. These rights are not only bestowed on citizens by democratic means, they are described as preconditions for developing one's character and personality and hence for active and engaged citizenship. Second, there is often no or very limited legal basis for the experiments conducted or allowed by the government, which challenges one of the fundamental principles of the rule of law, namely that governmental action is based on democratic rules. Third and finally, Living Labs are zones of experimentation. This means that the behaviour of citizens is tacitly influenced. This may undermine citizens' freedom and autonomy, and consequently, the preconditions for democracy. Moreover, citizens have very limited influence on the design of Living Labs (Chesbrough, 2006).

[41] http://www.openlivinglabs.eu/livinglab/eindhoven-living-lab. http://www.midpointcsi.nl/powered-by-social-innovation/wp-content/uploads/2016/07/LLTrillion2015.pdf.

[42] https://e52.nl/een-kijkje-in-het-lab-van-het-stratumseind/.

[43] http://www.smartdatacity.org/stratumseind-living-lab/.

[44] http://www.philips.nl/a-w/about/news/archive/standard/about/news/press/20131209-Stratumseind-Eindhoven-wordt-proeftuin.html.

[45] https://www.cursor.tue.nl/fileadmin/user_upload/Magazines/PDF_jes_Nederlands/57/Cursor_04__NL_spread.pdf.

16.5.1 Privacy and Data Protection

In Europe, citizens are granted the right to privacy and the right to data protection.[46] The right to the protection of personal data, is embedded in the General Data Protection Regulation (GDPR) of the European Union.[47] The right to data protection will apply to large parts of projects such as Stratumseind 2.0, data processing being a central element in most Living Labs. Several examples will be provided on how data processing within smart cities will be evaluated from the perspective of the GDPR.

First of all, the GDPR requires that data should be gathered for a specific and concrete purpose.[48] A concrete purpose exists, for example, when a pizza delivery service needs a person's home address to be able to deliver the pizza just ordered. Many living lab projects do not have such a concrete an specific purpose for gathering data, but rather pursue general and vague purposes such as 'maximising security' or 'optimising traffic flows'.

Secondly, the GDPR requires that the data collection should be necessary and effective to achieve the concrete and specific purpose.[49] However, many Living Labs, including the Stratumseind 2.0 project, are based on experimentation, as will be discussed below. It is often unsure whether and to what extend the experiments will in fact promote the goals selected. Also, the GDPR requires that when data are collected for a specific purpose, they should not be re-used for other, unrelated purposes.[50] However, in Living Labs, public-private partnerships are nourished more often than not, and data which are gathered, for example, to promote security, may be shared with private companies that want to use the data to test products.

Furthermore, the GDPR requires that data gathering and processing must be limited to the minimum[51] and that data must be deleted when the goal for which they were gathered has been achieved.[52] With respect to Living Labs, however, the trend is to gather as much as data possible and to store them for as long as deemed useful.

Finally, the GDPR requires that citizens must be informed about the fact that their data are gathered and why, by whom and how their data are processed.[53] In reality, citizens are mostly unaware of the fact that so many data are gathered about them and remain oblivious about the fact that they enter a Living Lab area. There is no information, such as a small sign, that informs citizens. This also may clarify the

[46] http://www.europarl.europa.eu/charter/pdf/text_en.pdf.

[47] Regulation 2016/679 of the European Parliament and of the Council of 27 April 2016 on the protection of natural persons with regard to the processing of personal data and on the free movement of such data, and repealing Directive 95/46/EC (General Data Protection Regulation).

[48] Article 5 paragraph 1 sub b GDPR.

[49] Article 5 paragraph 1 sub a GDPR.

[50] Article 5 paragraph 1 sub b GDPR.

[51] Article 5 paragraph 1 sub c GDPR.

[52] Article 5 paragraph 1 sub e GDPR.

[53] Article 5 paragraph 1 sub a GDPR.

lack of an intensive public debate about Stratumseind. Nevertheless, a small sign at the beginning the street, such as Stratumseind 2.0, stating 'You are now entering a Living Lab' or 'Data about you can and will be collected if you enter this street' is not enough to be GDPR compliant.[54]

More generally, the right to privacy contains the right to develop one's personality and flourish to the fullest extent (Van der Sloot, 2015). Such a right may be undermined because people will restrict their behavior if they know or fear that they might be watched in public spaces. This is called the chilling effect. The argument that citizens' can chose to avoid certain Living Lab areas, and thus avoid limitations to their privacy (or put stronger, that they have consented to privacy limitations by entering that area) has been squashed by the European Court of Human Rights. Inter alia, in a case in which the municipality of Amsterdam had designated an area as a risk zone, in which people could be randomly selected for a search, the Dutch government argued that the complainant in that case was not frisked and thus could not claim to be a victim. The Court, however, ruled: [55]

> '(...) that the applicant was engaged in lawful pursuits for which he might reasonably wish to visit the part of Amsterdam city centre designated as a security risk area. This made him liable to be subjected to search orders should these happen to coincide with his visits there. (...) It follows that the applicant can claim to be a "victim" within the meaning of Article 34 of the Convention and the Government's alternative preliminary objection must be rejected also.'

Consequently, if people have to avoid city centers in order not to be monitored or subjected to privacy infringing activities, they can be seen as victims by that very fact.

16.5.2 Legality and Legitimacy

One of the core principles of the rule of law and democratic reign is that governmental conduct should be based on a law and those laws should have democratic legitimacy. The people decide what the government should do, not the other way around. This principle is also embedded, inter alia, in the European Convention on Human Rights, which holds that the right to privacy may only be limited if necessary in the pursuit of a public interest and based on a law. The ECtHR has specified that laws must be public, accessible and understandable and should contain limits to the use of government power. Thus, laws may not provide a blanket power to the state, but must set conditions for the use of powers, mechanisms for oversight and possibilities to challenge governmental action before a court of law.[56] This principle is put under pressure by the tendency to call Living Labs 'experiments'; it is often

---

[54] Article 12, 13 and 14 GDPR.

[55] ECtHR, Colon v. the Netherlands, application no. 49458/06, 15 May 2012.

[56] https://www.echr.coe.int/Documents/Guide_Art_8_ENG.pdf.

argued that such experiments are not in need of an explicit legal basis, as is the case with Stratumseind 2.0.

In addition, the scientific standards for doing experiments on study objects, such as normally conducted in artificial surroundings, such as labs in university buildings, are often pushed aside. It is clear that experiments in Living Labs come into conflict with many of those standards (Beauchamp, 2011). Two standard principles are (1) that people are informed about the experiment and (2) that they consent to being part of that experiment (APA, 2010; BPS, 2014; Hansson, 2003). A traditional experiment on a university premises would entail a special lab or room designed to conduct an experiment and test subjects explicitly entering the lab, knowing that they will be part of an experiment (even though they may not be informed in full about what type of experiment or which type of research the test is conducted for). In Living Labs, it is often difficult to ask permission from the research subjects, as this requires a very tedious process and would undermine the 'naturalness' of the test environment. In addition, being transparent about which data are collected, how and for what purposes is often impossible, because Living Labs are often hybrid projects being in constant flux. The experiments may change constantly, as well as which data are gathered and why (Pols, 2016). Consequently, two of the most basic preconditions for legitimate experiments are often neglected in Living Lab projects such as Stratumseind 2.0.

In addition to these principles, other safeguards are included in the traditional guidelines on research ethics that aim to protect the autonomy of citizens. For example, both the American Psychological Association (APA) and the British Psychological Society (BPS) apply strict criteria with regard to the deliberate misleading of participants within an experiment. Misleading or secretly collecting data can only be justified within a research plan (a) if it is essential to obtain the intended research results—and there are no alternatives—(b) if the research objective has a very high scientific value and (c) if there is a good plan for risk management and harm reduction in case of undesirable results (BPS, 2014, p. 24; APA, 2010, p. 11). In addition, the research must be carried out in such a way that the dignity and autonomy of the subjects are safeguarded. The BPS states that studies that take place outside the controlled environment of the laboratory must respect the privacy and psychological wellbeing of the individuals studied. Observation in public spaces without consent is only acceptable where those observed can reasonably expect to be monitored (BPS, 2014, p. 25). Obviously, this is not the case in smart cities such as Eindhoven.

Consequently, Living Lab projects such as Stratumseind 2.0 challenges two fundamental conditions for the rule of law, namely legality and legitimacy. Municipalities and governmental organizations often conduct experiments, monitor public spaces and gather data about citizens without a specific legal basis. Researchers often conduct experiments on test subjects without their knowledge and without any form of informed consent. And it is unsure to what extent such experiments promote public goals. For example, Article 8 ECHR allows for limitation on the right to privacy if based on a law, if necessary in a democratic society (which is often unsure) and when one of the following public interests involved: 'national security, public safety or the economic well-being of the country, for the prevention of disorder or crime, for the protection of health or morals, or for the protection of the rights and freedoms of others'. Especially in public-private partnerships, it is unsure to what extent companies that have access to the data gathered will have the public interest at heart (Felt et al., 2007; Krohn & Weyer, 1994; Martin & Schinzinger, 2000; Van de Poel, 2009; Van de Poel, 2011).

16.5.3 Freedom and Autonomy

Finally, two principles of democracy are at stake in Living Lab experiments such as Stratumseind 2.0. First, the preconditions for democracy, namely that citizens can engage freely and autonomously in public deliberation, debates and elections. Like the right to privacy can be seen as a precondition for developing one's personality and becoming an independent and critical citizens, the ability to act and engage with others as a critical and independent citizen can be seen as a precondition for democracy. If all would follow public opinion without discussion, democracy would be an empty concept. Second, the idea of democracy is not only that governmental action should be based on democratic law, but also that citizens should have the ultimate decision on how public spaces are designed and in which way they function.

To start with the latter point, the report of the expert group 'Science and Governance' to the European Commission stated in 2007 that there is a new regime in the field of science: collective experiments of co-creation in which society becomes a laboratory. In this type of experiments, the goal of the project is not made explicit in advance, but the goals of the research arise as the experiment progresses (Felt et al., 2007). The advice was to build a European Knowledge Society on the basis of collective experiments in which innovation was distributed among various actors in society.[57]

In practice, however, citizens cannot or have very limited influence on smart city and living lab projects, and it is increasingly difficult to appoint stakeholders or representatives thereof, in particular when whole cities or provinces are designated areas for experimentation. Consequently, while public spaces are increasingly non-neutral and dynamic, citizens have increasingly less grip on such changes. Rather, private parties and companies are assigned an important role in designing the public space, as was stressed in the previous section.

In addition, there are concerns over the protection of the individual freedom and autonomy of citizens. This concern mainly concerns the use of data in Living Labs to influence the behaviour of citizens. In Eindhoven, data are collected on the relationship between light and aggression, in order to be able to adjust the lighting in nightlife areas in such a way that the behaviour of visitors becomes less aggressive. The same applies to spraying tangerine scent in the streets, which would make citizens more relaxed. In a way, it echoes the rumour that airlines reduce the amount of

[57] See also: Von Hippel, 2005.

oxygen in the airplane, so that people are less energetic and fall asleep more often. That way, customers are easier to handle by flight personnel. Citizens that are tacitly made more relaxed by influencing their behaviour in public spaces are also easier to handle by public authorities.

This is a form of nudging. A nudge is ‘a form of choice architecture that changes people’s behavior in a predictable way without forbidding other options or changing their economic motives’ (Thaler & Sunstein, 2008, p. 6). The problem of steering the subconscious is that it can undermine individual freedom, without the individual being aware of that fact or being capable or exercising any form of control over it (Hausman & Welch, 2010; Wilkinson, 2013). The person is steered in a certain direction on the basis of psychological mechanisms (Nys & Engelen, 2016). Of course, the government has always steered behaviour of citizens, for example through public campaigns (for example, to quit smoking) or through public laws, which prohibit certain conduct (for example, to smoke in bars and restaurants). The essential difference is of course that laws and public campaigns are transparent—the citizen knows he is being influenced. In addition, because he is aware of that fact, he can ultimately make a decision, even if that means breaking the law (Hausman & Welch, 2010, p. 130). In the case of smart cities, it will often be hidden to citizens that they are steered in their behaviour and decisions based on (their) data and corresponding social or psychological vulnerabilities. This phenomenon is already apparent in the online realm and referred to as ‘hypernudging’ (Lanzing, 2019; Yeung, 2018). Moreover, it may not even be for the sake of their interests, but rather for the sake of whoever is steering them, raising the even stronger moral concern of manipulation (Susser, Roessler, & Nissenbaum, 2018).

16.5.4 Conclusion

This section discussed the Living Lab project in Eindhoven, called Stratumseind 2.0, which is a public-private partnership. This case has been used to study the relationship between Living Labs on the one hand and democracy and the rule of law on the other hand. It has signalled three particular tensions. First, Living Labs can undermine the fundamental rights of citizens, such as the right to privacy and the right to data protection. These rights are not only bestowed on citizens by democratic means, they are described as preconditions for developing one’s character and personality and hence for active and engaged citizenship. Second, a legal basis for conducting the experiments and gathering the data is often lacking and the requirements for conducting valid and legitimate experiments are often ignored. This may put pressure on the principles of legality and legitimacy and the core assumption of democracy that the government is only allowed to act on the basis of a law, enacted through a democratic process. Third and finally, Living Labs are used to manipulate and nudge people, through which the behaviour of citizens is tacitly influenced. This may undermine citizens’ freedom and autonomy, and consequently, the preconditions for democracy.

16.6 Revisiting the Transformation of the Public Sphere

Section Two described how developments such as datafication and hybridization enable the 'physical' world to increasingly become an 'online' built environment that collects data and adapts based on that data. Contexts and activities that we consider private, such as one's house or communication, are becoming increasingly monitored by third parties. Vice versa, along with developments of datafication and hybridization, the public sphere is increasingly privatised. The same development of commercialisation and privatisation witnessed with the Internet can now be detected in the actual 'physical' public sphere with the rise of smart cities. Big Data has opened up new opportunities for commodification and commercialization of public sphere and the new frontier is the development of smart cities. Smart environments allow architects to access data, monitor and nudge citizens on an unprecedented level. Yet, as described in Section Four, the legitimacy of surveillance, the nudging of citizens and the use of a (unique) public space without the consent or equal input of citizens controversial from a legal and ethical viewpoint. These features of Living Labs challenge the freedom and autonomy of citizens, which are important prerequisites for a plurality of views necessary for a flourishing democracy. As Section Three argued, while these features may be used to advance public interests such as public health, safety or sustainability, powerful commercial actors are highly influential in shaping and promoting a certain (financially beneficial) perspective on improving the 'quality of life' and underlying conception of the good life without having participated in democratic procedures of decision-making. They undermine the rule of law that constitutes the legitimacy of democratic procedure and increasingly compromise the neutrality of the public sphere.

Although Habermas finalized his manuscript early sixties of last century, it is not difficult to identify some of his observations regarding the structural transformation of the public sphere. A number of transformations of the public sphere can be discerned. First of all, the public sphere is datafied; reality is expressed in data. It is quantified and hence made legible and measurable. On the basis of this data, citizens and their activities are becoming increasingly visible. Secondly, the public sphere is increasingly becoming a hybrid place, in which the physical and the digital, offline and online, are increasingly intertwined. The public sphere is used as a place to monitor public life (a source of data) and as place to experiment on the behavior of citizens through smart applications and modern technologies. Thirdly, the public sphere is no longer under the primary care of the government. Both private (commercial) parties and citizens play an increasingly big role in gathering data and imposing certain ideas of the good life. In smart cities, the public domain is governed by public-private partnerships.

These transformations however, challenge some of the features that are deemed valuable for a public sphere within a deliberative democracy. They raise several concerns. First of all, these transformations raise the concern surrounding the value of 'neutrality'. One of the core aspects of the ideal of the public domain has been that it is relatively neutral to various visions on the good life, and leaves room for minority and dissenting opinions. Smart Cities are often explicitly based on one normative approach regarding the good life. Moreover, they may also actively try to impose, through nudging, manipulation or coercion, a particular normative perspective.

Furthermore, although one of the core aspects of the public domain has been that were norms are imposed on citizens and their freedom is limited, these should have democratic legitimation. Due to developments of privatization, both citizens and especially larger tech companies are capable of imposing norms unilaterally.

Related to these companies there is the concern of harmful commodification. In Smart Cities, the public domain is not only seen as a place where parties can impose norms and ideals, it is also a place for harvesting data about individuals and their social relationships. These data can be used to develop product and services. Thus, the ideal of the public domain as a free zone for interaction and deliberation is turned into the ideal of a zone of commodification and commercialization, a domain that has maximum utility. However, the question is to what extent these market influences should be intertwined with the lives and decisions of citizens.

This point becomes even more urgent in light of the dissolution of privacy boundaries. The ideal of the public domain depends on an idealized version of the private domain, both physically and metaphorically, where people are autonomous and can develop their thoughts and beliefs. Citizens should, more or less, be able to experiment, read and think without being interfered with by parties that they do not want to be part of that process. Privacy is the social condition that allows them to do so. In Smart Cities and Living Labs, this ideal is increasingly undermined, both because the homes of citizens are connected to the web and included in the smart strategy, and because the private lives, as far as they take place in the public sphere, can become fully controlled and monitored.

Becoming vulnerable to manipulation by third parties that have their own interests rather than the interests of citizens at heart, is problematic from the perspective of autonomy. The idea of the public domain depends on the presumption that individuals are more or less autonomous. Autonomous citizens can meet in the public sphere to discuss, deliberate and develop new ideas and opinions. Smart Cities and Living Labs may weaken autonomy by blurring the privacy boundaries between different social contexts. This allows private interests to restructure the public domain with new forms of surveillance and to experiment on people by influencing their behavior or preferences.

Finally, there is the concern that public domains that are governed by powerful private parties may undermine the legality of these domains. In the ideal of the public domain, the government controlled this realm. Its actions were supposed not only to have democratic legitimation, but also a legal basis. Laws are published, so that citizens can take notice of the rules and prohibitions, imposed on them. In addition, laws set limits to how and to what extent the government can use its powers. In Living Labs, however, governmental actions are not based on laws. Rather, laws and legal principles are often circumvented or bypassed, with reference to the fact that the initiatives are only societal experiments.

The meaning of the public sphere and our ability to regulate privacy within the public (or to keep the public eye out of our houses, communication and choices) have always been subjected to change. For instance, Habermas notes that already in the twentieth century the public domain was increasingly privatized and depoliticized. Moreover, it has always been, and increasingly is, difficult to regulate one's privacy. Importantly, the developments that have been sketched are not new and the fact that the ideal of the public sphere is challenged is not shocking. The ideals of the public sphere and private sphere as separated spheres have never existed and will never exist. There are many different social spheres that may overlap, that consist of various stakeholders with different interests and are governed by changing privacy expectations.

Nevertheless, normative value is attributed to features of the ideal of the public sphere in the context of the ideal of the deliberative democracy. Features of Smart Cities and Living Labs conflict with the values that are associated with the ideal of the public domain. This means that the developments sketched do raise an interesting question about what we want the public sphere to look like. Does it make sense to apply these ideal features to smart environments or should we reconceptualize the meaning of the public sphere? Within the next 20 years, the meaning and function of the public sphere may have transformed structurally and dramatically under the influence of powerful private actors that provide the technological soft and hardware of Smart Cities. This is something that is already experienced today. In any case, along with these transformations, there should be societal debate about whether these changes contribute and serve the ideal of a deliberative democratic society. An aim could be to bolster some valued characteristics of the public sphere before private actors have absorbed it or the characteristics could be adjusted, redesigned or reconceptualized in light of technological development. Regardless of the conclusion, what values or features the new ideal of a meaningful public sphere should have, should be part of a public debate between autonomous citizens.

Acknowledgments We would like to thank the participants of the Philosophy of the City Summer Colloquium 2018 at the University of Technology Twente for their constructive comments. Furthermore, we would like to thank the anonymous referees of this contribution for their helpful remarks.

References

American Psychological Association. (2010). *Ethical principles of psychologists and code of conduct (with the 2010 amendments)*. Washington: American Psychological Association.

Beauchamp, T. L. (2011). Informed consent: Its history, meaning, and present challenges. *Cambridge Quarterly of Healthcare Ethics, 20*, 515–523.

Boyd, D., & Crawford, K. (2012). Critical questions for big data: provocations for a cultural, technological, and scholarly phenomenon. *Information, Communication, & Society, 15*(5), 662–679.

British Psychological Society. (2014). *Code of human research ethics*. Leicester: The British Psychological Society.

Chesbrough, H. (2006). *Open innovation: the new imperative for creating and profiting from technology*. Boston: Harvard Business School Press.
Crawford, S. (2018). Beware of Google's Intentions. *Wired*. Retrieved March, 3 2018, from https://www.wired.com/story/sidewalk-labs-toronto-google-risks/
Dingman, S. (2017). With Toronto, Alphabet looks to revolutionize city-building. *The Globe and Mail*. Retrieved March 3, 2017, from: https://www.theglobeandmail.com/report-on-business/with-toronto-alphabet-looks-to-revolutionize-city-building/article36634779/
Felt, U., Wynee, B., Callon, M., Gonçalves, M. E., Jasanoff, S., Jepsen, M., et al. (2007). *Taking European knowledge society seriously*. Brussels: Directorate General for Research, Science, Economy and Society.
Haans, A., & De Kort, Y. A. W. (2014). Reducing social aggression with dynamic lighting: the De-escalate project in Eindhoven. *Justitiele Verkenningen, 40*(4), 54–64.
Habermas, J. (1989[1962]). *Structural transformations of the public sphere: An inquiry into a category of Bourgeois society*. Cambridge: Polity Press
Habermas, J. (1991[1962]). *Structural transformations of the public sphere: An inquiry into a category of Bourgeois society.* Cambridge: MIT Press
Habermas, J. (1996[1992]). *Between facts and norms. Contributions to a discourse theory of law and democracy.* Cambridge: MIT Press
Hansson, S. O. (2003). Ethical criteria of risk assessment. *Erkenntnis, 59*, 291–309.
Hausman, D. M., & Welch, B. (2010). Debate: To nudge or not to nudge. *Journal of Political Philosophy, 18*(1), 123–136.
Krohn, W., & Weyer, J. (1994). Society as a Laboratory: the social risks of experimental research. *Science and Public Policy, 21*(3), 173–183.
Lanzing, M. (2019). Strongly recommended: revisiting decisional privacy to judge hypernudging in self-tracking technologies. *Philosophy & Technology., 32*(3), 549–568. https://doi.org/10.1007/s13347-018-0316-4
MacKinnon, C. A. (1989). *Toward a feminist theory of the state*. Cambridge, MA: Harvard University Press.
Martin, M. W., & Schinzinger, R. (2000). *Ethics in engineering* (2nd ed.). New York: McGraw Hill.
Mayer-Schönberger, V., & Cukier, K. (2013). *Big data: A revolution that will transform how we live, work and think*. London: John Murray Publishers.
Nissenbaum, H. (2009). *Privacy in context. Technology, policy, and the integrity of social life*. Palo Alto, CA: Stanford University Press.
Nys, T., & Engelen, B. (2016). Judging nudging: Answering the manipulation objection. *Political Studies, 65*(1), 199–214.
Pateman, C. (1989). *The disorder of women: Democracy, feminism, and political theory*. Stanford, CA: Stanford University Press.
Pols, A. (2016). May stakeholders be involved in design without informed consent? the case of hidden design. *Science and Engineering Ethics, 23*(3), 1–20.
Roessler, B. (2015). Should personal data be a tradable good? On the moral limits of markets in privacy. In B. Roessler & D. Mokrosinksa (Eds.), *Social dimensions of privacy* (pp. 141–161). Cambridge: Cambridge University Press.
Satz, D. (2010). *Why some things should not be for sale*. Oxford: Oxford University Press.
Sauter, M. (2018). Google's Guinea-Pig City: Will Toronto turn its residents into Alphabet's experiment? The answer has implications for cities everywhere. *The Atlantic*. Retrieved February 14, 2018, from https://www.theatlantic.com/technology/archive/2018/02/googles-guinea-pig-city/552932/
Susser, D., Roessler, B., & Nissenbaum, H. F. (2018). Online manipulation: Hidden influences in a digital world. *SSRN*. Retrieved December 23, 2018, from https://ssrn.com/abstract=3306006
Thaler, R., & Sunstein, C. (2008). *Nudge*. New Haven: Yale University Press.
Townsend, A. M. (2013). *Smart cities: Big data, civic hackers, and the quest for a new utopia*. New York: W.W. Norton.

Vaidhyanathan, S. (2011). *The googlization of everything (and why we should worry)*. Berkeley, CA: University of California Press.
Van de Poel, I. (2009). The introduction of nanotechnology as a societal experiment. In S. Arnaldi, A. Lorenzet, & F. Russo (Eds.), *Technoscience in progress: Managing the uncertainty of nanotechnology* (pp. 129–142). Amsterdam: IOS Press.
Van de Poel, I. (2011). Nuclear energy as a social experiment. *Ethics, Policy & Environment, 14*, 285–290.
Van der Sloot, B. (2015). Privacy as Personality Right: Why the ECtHR's Focus on Ulterior Interests Might Prove Indispensable in the Age of "Big Data". *Utrecht Journal of International and European Law, 31*(80), 25–50.
Von Hippel, E. (2005). *Democratizing innovation*. Cambridge, MA: MIT Press.
Wilkinson, T. M. (2013). Nudging and manipulation. *Political Studies, 61*(2), 341–355.
Zuboff, S. (2015). Big other: surveillance capitalism and the prospects of an information civilization. *Journal of Information Technology, 30*, 75–89.

Bart Van Der Sloot specializes in the area of Privacy and Big Data. He also publishes regularly on the liability of Internet Intermediaries, data protection and internet regulation. Key issues are the recently adopted General Data Protection Regulation (GDPR), international data flows, especially between Europe and the United States, and data leaks. Bart formerly worked for the Institute for Information Law, University of Amsterdam, and for the Dutch Scientific Council for Government Policy (WRR), and currently works at the Tilburg Institute for Law, Technology, and Society of the University of Tilburg, Netherlands.

Marjolein Lanzing is a Postdoctoral Researcher at the Department of Philosophy, Theology and Religious Studies. She conducted her PhD research 'The Transparent Self': A Normative Investigation of Changing Selves and Relationships in the Age of the Quantified Self at the 4TU. Center for Ethics and Technology at the University of Technology Eindhoven. Marjolein studies the ethical concerns related to new technologies, in particular concerns regarding privacy, autonomy, vulnerability, solidarity and commodification, and what they mean for the way we understand ourselves and our social relationships. Her Postdoc is part of the ERC-funded project "Digital Good" and carried out within the interdisciplinary Hub for Security, Privacy and Data Governance (iHub). Within this project she researches the "Googlization of Health".Marjolein is board member of Bits of Freedom, an NGO that protects online freedom and (digital) civil rights.

Chapter 17
A Philosophy of Sidewalks: Reclaiming Promiscuous Public Spaces

Germán Bula

Abstract This paper frames Jane Jacob's ideas on urban planning in a wider philosophical framework by contrasting two philosophical orientations: *control cartesianism* and *spinozist multiplicity*, and their differing attitudes towards diversity and self-organization. Both Jacob's ideas on urban planning and Spinoza's philosophy of composite bodies show an awareness of how cohesion and organization can be achieved through the spontaneous workings of diverse elements. These ideas are then re-purposed with the aim of re-defining "smart cities", not as a brand of high-tech city management, but as cities that, through self-organization, are capable of collective deliberation and agency.

Keywords Jane Jacobs · Baruch Spinoza · René Descartes · Urban planning · Smart cities

17.1 Introduction

As Cass Sunstein (2001) predicted early on in the history of the Internet, the unlimited ability of the user to select the content she consumes has led to polarization, confirmation bias, and the creation of information bubbles, to the great detriment of democracy. At the same time, the new digital environments have created a depressingly bland and tailor-made world, what Han (2017) would call an excess of positivity. At least in the digital world, there seems to be a crisis of public spaces.

In this context, it seems appropriate to re-appropriate the work of Jane Jacobs (1961) who, in 1961, published *The Death and Life of Great American Cities*, a

G. Bula (✉)
Facultad de Filosofía y Humanidades, Universidad de la Salle, Bogotá, Colombia
e-mail: gbulalo@unisalle.edu.co

M. Nagenborg et al. (eds.), *Technology and the City*, Philosophy of Engineering and Technology 36, https://doi.org/10.1007/978-3-030-52313-8_17

manifesto against the segregating and homogenizing tendencies of mainstream city planning, and in favor of the spontaneity and autonomy of citizens in creating public spaces. Whereas the mainstream city planning of the time created cities to be used for pre-established interactions, minimizing unplanned contacts, Jacobs advocated for promiscuous cities, with many simultaneous uses in which contacts between diverse peoples were made possible by casual public spaces such as sidewalks. Whereas her ideas were revolutionary at the time, they have become common currency for today's city planners, to the point that they have been "co-opted by real estate developers and politicians" (Goldberger, 2006, p. 126) who use them to give a progressive veneer to large, centrally planned, developments. In order to put Jacob's ideas to work in our current context, we must acknowledge the emergence of cyberspace as today's dominant system of proximity (Lévy, 1997). Beyond that, we must go beyond Jacob's concrete proposals for city planning and uncover the underlying philosophy of her proposals in relation to planning, self-organization and diversity, with the hope of re-radicalizing her thought.

I posit that Jacobs recognizes the power of diversity to create cohesion. It would seem that diversity and cohesion are antagonistic variables and one must choose between either the ordered homogeneity of the platoon or the disordered multiplicity of the carnival. But in the work of Jane Jacobs we can see how social cohesion can be produced in a molecular, rather than molar, fashion: one to one relations of symbiosis can create a cohesive and diverse network (see Bula, 2017, p. 239–257). In this mode of social bonding, the more diverse a community is, the more chances for productive symbiotic relations it has, therefore cohesion and diversity become synergistic variables. This idea of diversity and cohesion can be explained through a reading of the opposition between Spinoza and Descartes. The latter's orientation towards detailed planning and against emergent properties and self-organization constitute the philosophical underpinning of the kind of urban planning Jane Jacobs opposed.

The term "smart cities" is used to denote cities that make use of data collection and information technology in their management in order to operate more efficiently. I would like to propose a "smarter" use of this term to denote cities that, as a second-order cognitive system, are engineered so as to make smart decisions. To this end, I make use of Page's (2007) ideas around diverse groups being able to produce smart decisions if adequately connected. For a deterritorialized, self-organizing collective intelligence to emerge, a proper organizational structure must be put in place; to borrow from cybernetician Beer (1994), freedom must be designed; self-organization requires proper planning. The aim of this paper is to outline some ideas and general guidelines aimed at producing sidewalks, both digital and physical, that can become places of promiscuous encounters, foster the appropriation of spaces on the part of citizens, and produce the possibility of emergent collective voices in the political sphere.

17.2 The Planner's Itch

In a rhetorical effort to justify the creation of a philosophical method apart from the scholastic tradition, René Descartes proposes, as a general rule, that works performed by a single will and reason are better than those that haphazardly integrate the decrees of many:

> (...) I spent the whole day shut up in a small room, heated by a stove, in which I could converse with my own thoughts at leisure. Among the first of these was the realization that things made up of different elements and produced by the hands of several master craftsmen are often less perfect than those on which only one person has worked. This is the case with buildings which a single architect has planned and completed, that are usually more beautiful and better designed than those that several architects have tried to patch together, using old walls that had been constructed for other purposes. This is also the case with those ancient cities, that in the beginning were no more than villages and have become, through the passage of time, great conurbations; when compared to orderly towns that an engineer designs without constraints on an empty plain, they are usually so badly laid out that, even though their buildings viewed separately often display as much if not more artistic merit as those of orderly towns, yet if one takes into consideration the way they are disposed, a tall one here, a low one there, and the way they cause the streets to wind and change level, they look more like the product of chance than of the will of men applying their reason. And if one considers further that there have always been officials whose task it was to ensure that the design of private buildings should contribute to the beauty of the town as a whole, it will become clear how difficult it is to carry anything through to completion when working only with what others have produced. (Descartes, 2006, p. 12)

The philosopher, accompanied only by a stove, realizes that the impression of "men applying their reason" is only achieved when a single will animates a project, be it single buildings, or whole towns. If Descartes is to be believed, ugly and haphazard cities such as New York or Paris should be shunned in favor of more cohesive towns such as Brasilia or Canberra! While the architectural beauty that can be found in the latter cities cannot be denied, it must be distinguished from urban beauty, which is a result of the spontaneously emerging harmony between the several uses (and building styles) of a city or city sector (Kohr, 1989, p. 98); and under this heading, messy cities such as New York clearly come out ahead (specially those sectors of the city that have not come under heavy city planning).

Be that as it may, Descartes ideas on city planning are intuitively appealing, and it is important to ask why. We must distinguish between (1) exhaustive planning (which would aim for that which is planned to reflect a single will and reason), (2) planning for self organization (that is, planning aimed at facilitating the spontaneous harmony that comes from the interaction of individual initiatives, under proper conditions), and (3) sloppy planning. Conceptually, exhaustive planning would require the planner to maximally anticipate contingencies, so that any future state of the planned system is accounted for. This means that novelty, for exhaustive planning, is necessarily tantamount to a planning failure: the beautiful art that spontaneously comes to adorn the walls of a city sector *must* be painted over, even if it is beloved by a community; from the planner's point of view, graffiti can only be a sign of urban decay, a deviation from the plan (see Hicks, 1994; Cathcart-Keays,

2015). On the other hand, planning for self-organization (e.g, Beer, 1994) expects and fosters novelty. Therefore, in as much as the point of view of exhaustive planning is dominant in a community, planning for self-organization will be viewed as sloppy (and, in as much as novelty is appreciated, e.g. in the form of graffiti, it must be planned for; the planner must, at the very least, be the one to designate *where* graffiti happens). In these conditions, if the planner does not wish to seem sloppy, and wishes to be recognized and praised for his work, he will tend to adopt the point of view of exhaustive planning, so that his works will seem to reflect "the will of men applying their reason". The drive to anticipate and control every future state of a system (whether this drive is successful or not), I will call the *planner's itch*.

There is a certain philosophical orientation behind the planner's itch; for the sake of convenience, I will call this orientation *control cartesianism*. Indeed, many traits of this philosophical orientation are quite visible in the work of Descartes: (1) the drive to work exclusively with data that is taken as certain, (2) the preference for linearly deductive knowledge, and (3) an implicit rejection of emergent properties. In the following section, I wish to contrast these traits in the philosophy of Descartes with an alternative philosophical orientation in the work of a superficially similar philosopher, Baruch Spinoza. His work will provide some conceptual groundwork for an alternative philosophy of planning, which will be presented in sections 17.4 and 17.5.

17.3 Descartes Vs. Spinoza

In many ways, Spinoza is a cartesian philosopher: Descartes was a major philosophical influence in the liberal protestant circles in which Spinoza was active; Spinoza's only work published in his lifetime and under his own name was an exposition of Descartes; and Spinoza's philosophy can be viewed as a radicalization and further unfolding of Descartes' philosophy (see Israel, 2001). However, Spinoza's work can serve as a counter-example to *control cartesianism*. Let us examine the three traits outlined above in relation to both thinkers.

The drive to work exclusively with data taken as certain is a key component of Descarte's method, insisted upon in all three of his major works: The *Principles of Philosophy*, *Meditations on First Philosophy*, and the *Discourse on the Method* (Descartes, 1967): the methodological purpose of cartesian doubt is to guarantee that the philosophers works exclusively with absolutely certain data. This flows into *the preference for linearly deductive knowledge*: once premises that are deemed certain are established, new knowledge is progressively produced through deduction; these conclusions become new premises for further conclusions, always in such a way that the certainty of the first premises is carried over to new conclusions: cartesian knowledge can be compared to a chain, in that, if just one link fails, the whole becomes useless. In order to achieve this chain of certainty, the cartesian thinker is called to divide complex problems into their simplest parts: the complex whole is then cognitively reconstructed by re-assembling the whole from the parts.

This is tantamount to *rejecting emerging properties*, that is, properties that are present in the whole but not in the parts, because they are the product of the relationships between the parts: for example, whereas the parts of a traffic jam (the cars) move, say, north to south, *the traffic jam itself* moves south to north (see Resnick, 2000); but this cannot be deduced from what is known about the parts, only from their interaction.

Spinoza, on the other hand, has a more wholistic approach to knowledge (Merçon, 2012), that need not proceed deductively through linked certainties. A good example is Spinoza's idea of the body: unlike Descartes, Spinoza believes that body and mind are strictly isomorphic, which implies that the body, as such, is capable of operations that are just as complex as the operations of the mind. Spinoza does not specify just in what way the body is complex enough to perform such operations. Rather, Spinoza calls our attention to emergent properties (1981, pp. 54–59) by speaking about how bodies (understood as processes) that are more complex (that is, are composed of more, and more varied kinds of, sub-processes), have more properties, and merely pointing out the complexity of the human body, indicating that it must have many emergent properties that science has yet to discover: "... no one has hitherto determined what the body is capable of" (1981, p. 95). Even though Spinoza presents his work deductively, emulating Euclid's *Elements of Geometry*, his arguments function more like a fishing net than a chain: although a few nodes here and there may be broken, the whole can be strong enough to work; and it is in virtue of this whole that the parts make sense: indeed, at certain points in his texts, Spinoza asks the reader to suspend judgment until she has read the whole (e.g, 1981, p. 52). Spinoza's work, although of cartesian pedigree, displays an orientation quite opposed to control cartesianism.

Descartes' main argument proceeds in the same order in all three of his major works: (1) radical doubt, (2) the *cogito* argument ("I think therefore I am"), (3) proof of the existence of God, (4) possibility of knowledge of the external world. On the other hand, Spinoza's highly hypertextual writing, in which the reader may read backwards through the branching and diverse elements of the proofs of propositions, to which are added the *scholia*, which are outside the deductive structure of the text, gives the reader the possibility of reading the work through many paths. This difference can be put in architectural terms. The many-paths structure of Spinoza's writing has been compared by Rawes (2015) to the architecture of Antonio de Sangallo's *Villa Madama* (a renaissance villa in Rome), as opposed to that of a Victorian British house. Whereas in a Victorian British house "passages become corridors that separate work from leisure and patron from servant" (Rawes, 2015, p. 69–70), Italian renaissance architecture such as *Villa Madama* displays a.

> (…) highly perforated and dynamic kind of geometry as its multiple doors in rooms and walls allow for a range of passages through the various living-rooms, vestibules and even (…) water closets (…) the density of routes through the spaces and their multimodal points of access [make] it possible for occupants of all social classes to pass through the same rooms simultaneously from different directions (Rawes, 2015, p. 69).

In a manner analogous to *Villa Madama's* multiple passages, "Spinoza's scholia perforate the boundaries of the classical geometric method because they are outsider voices" (Rawes, 2015, p. 78). In both the scholia and the hypertextual links contained in the proofs, the reader is invited to use the text in her own way, to roam the text in whatever way best fits her intellect. On the other hand, Victorian architecture suggests single, pre-determined, paths for each kind of use of the building, and could, therefore, be compared to Descartes' writing and to a form of control cartesianism in which the writer traces a single path for the reader to use. We can therefore distinguish between control cartesianism and *spinozist multiplicity*.

One of the main reasons Jane Jacobs advocates for small city blocks (1961, p. 179–192) is that it allows pedestrians to take multiple paths through and from their destination; this, in turn, makes the streets safer (because pedestrians can escape dangerous situations through side streets), and city life more diverse and fruitful (as pedestrians may visit different businesses or acquaintances making variations on their usual routes). Spinoza's writing and Jacob's ideas on city design both advocate for a multiplicity of connections that allow users the choice of multiple pathways between ideas or destinations. In the following section, I will make use of Spinoza's philosophy to develop the idea of diversity with the aim of fleshing out a philosophical underpinning for Jane Jacob's ideas on urban planning.

17.4 Diversity

In the following considerations, I will take a path somewhat less travelled. The problem of diversity is often tackled from an abstractly axiological perspective, or from the point of view of human rights (e.g Dietz, 2007). I want to think about what diversity does to a system, considered as a process that strives for viability, for perpetuation over time (see Beer, 1994); and more particularly, what diversity does to cognitive systems (see Page, 2007). In any kind of system, including a city or community, diversity and cohesion are variables in tension (Bula, 2010). When a new element becomes part of a system (e.g, when a newborn arrives in a family, or a soldier in a platoon), some mixture of accommodation and assimilation on the part of the host must take place: either the newcomer must adapt (as happens with the soldier) or the larger system must change in order to accommodate it (as happens with the baby). A new element may certainly demand too much change from a host system: liberal authors such as Benjamin Barber (1992, p. 129) concede that too much diversity in a community may destroy its unity in terms of values and cultural cohesion.

The problem of the possible excesses of accommodation and assimilation is tackled in the philosophy of Spinoza (Zourabichvilli, 2002, p. 58). In Spinoza's ontology, bodies can form composite bodies that can, in turn, join to form ever more complex composite bodies of higher orders (1981, p. 54–59); this must not be understood as a matter of assembly (say, the joining of Lego play pieces, or of Epicurean atoms), but rather as a matter of processes becoming part of higher order processes. In interaction between humans or communities, this can be understood in terms of producing positive sum games (Bula, 2012): in relationships of cooperation, the several parts, though differentiated, become part of a single system, and come to have coinciding interests; in as much as A and B are part of a higher body, what is good for A is also good for B (see Spinoza, 1981, p. 174; Bula, 2014). Conversely, a community has low cohesion in as much as its parts lack common interests.

Diversity and cohesion are variables in tension, but they are not mutually exclusive. If cohesion is achieved through homogenization (as happens in a military platoon), diversity is destructive to cohesion; however, cohesion can also be achieved *through* diversity. The more a body is capable of having many states, the more it can enter into relationships of cooperation with many different bodies (Spinoza, 1981, p. 183). Diverse people (capable of many forms of cooperation) can produce diverse societies that are cohesive but internally differentiated, as opposed to communities that achieve cohesion through homogenization (Bula, 2017, 239–257); Hardt and Negri (2001), following Spinoza, have called the former *multitudes* and the latter, mere *masses*.

According to Jane Jacobs (1961), culturally diverse cities are also more prosperous, because different elements bring to them different competences. The prosperity of Spinoza's native Amsterdam in the seventeenth century is a case in point: it had a lot to do with its receiving Jewish refugees from the Iberian Peninsula (with their connections and abilities for international trade) as well as protestants chased out of France and Belgium (with their manufacturing abilities) (*cfr* Page, 2007, p. 331). It can be argued that there are also *cognitive* benefits to diversity: considered as a second order mind, a community benefits from hosting people who think differently (indeed, seventeenth century Amsterdam also enjoyed a philosophical and scientific Golden Age). As Scott Page has shown, for an important class of problems, "diversity trumps ability" (2007, 137). A diverse group, using diverse heuristics, will perform better than a group of experts: this is because experts will tend to use the same heuristics and will therefore find good solutions but tend to get stuck on local optima, whereas a diversity of approaches will have superadditive effects that produce unforeseen innovations (Page, 2007, 157). A good example is Craig Reynolds' (1987) contribution to the understanding of flocking in birds, which came not from training or experience as a biologist, but from his work in computer animation for Hollywood films.

Diversity does not trump ability in all circumstances; for problems with few local optimal solutions, experts will outdo diverse groups; but for problems with many possible good solutions (say, designing the transportation system of a city, a problem in which good environmental solutions may not be economical, and cost-efficient solutions may not be socially inclusive, etc.), experts are outperformed by diversity (Page, 2007, p. 159–162). Schumacher (2004, p. 121–123) distinguishes between *convergent* and *divergent* problems: in the former kind, the more a problem is studied, the more people will come up with the same solution (e.g, Babylonian and Mayan mathematicians came up with the same answers independently); in the latter kind, when problems are tackled by smart and rigorous minds, they come up with different solutions. Examples of divergent problems are questions like "what is a just society?", "how should we educate our children?" or "how are we to live?". The idea of a smart city I wish to put forward would delegate a part of city governance to processes of collective cognition, but it is important to distinguish between straightforward technical problems (which are best left to experts), and problems which, by their nature, are best tackled by democratic participation: *technocracy*, as an epithet, should be used when experts illegitimately take over the solution of this latter kind of problems.

Using these ideas on diversity, In the next section I will give a brief account of Jane Jacob's ideas on city planning and her understanding of planning for self-organization. In Sect. 17.7, I will outline possible applications of Jacob's ideas for cyberspace, in order to produce truly smart cities.

17.5 Jane Jacobs Vs. the Planner's Itch

Working as a journalist for *Architectural Forum* in the 1950s, and witnessing the effects that large scale urban renewal programs had on poor communities, Jane Jacobs began writing articles and public addresses that took aim at the very foundations of mainstream urban planning: cities are safe, lively and amicable when urban space is used densely (1961, p. 152–177) and diversely (1961, p. 200–221); when city blocks combine many kinds of uses (industrial, commercial, residential) and many real estate prices (which is why aged buildings are so important for city blocks, 1961, p. 187–199). Whereas mainstream urban planners assumed that contact with strangers ought to be minimized, Jacobs posited that city life is enriched by casual contact (1961, p. 56): places of casual contact, such as sidewalks, small convenience stores, or front stoops allow many diverse people (with whom one is not intimate enough to invite into one's home) to be on "good sidewalk terms" (1961, p. 62). This casual contact produces a basis of trust that makes it possible for city dwellers to feel safe intervening in each other's lives: scolding children who are misbehaving (because they are recognizable as this or that neighbour's), intervening when witnessing a crime (because I know I will be backed up by others whom I am familiar with), or keeping each other's keys and correspondence. Jacob's ideas have influenced modern city planning, for example, in a concern for permeability (Marazuela et al., 2014, p. 15), or in Jan Gehl's work on pedestrian-friendly urban planning (e.g, Barnett, 2014).

Jane Jacobs also advocated for the city's role in facilitating casual contact among diverse people: "(…) if interesting, useful, and significant contacts among the people of cities are confined to acquaintanceships suitable for private life, the city becomes stultified" (Jacobs, 1961, p. 56). How different this is from the point of view that Jane Jacobs criticized:

> Reformers have long observed city people loitering on busy corners, hanging around in candy stores and bars, and drinking soda pop on stoops, and have passed judgement, the gist

of which is: "This is deplorable! If these people had decent homes and a more private or bosky outdoor place, they wouldn't be on the street!" (1961, p. 55).

The unlabeled, unfocused, unanticipated activity of loiterers awakens the planner's itch. On weekdays, an unlicensed empanada vendor stands on the street in front of the building where I live. There forms around him a ring of people with an empanada in one hand and coffee in the other, sometimes skillfully performing the task of applying hot sauce to the empanada with both hands already occupied: there are construction workers, shopkeepers and doormen away from their posts for a few minutes, stylists from the nearby hair salons, and office workers from a large office building around the block. There are many problems with this situation from the point of view of exhaustive planning: the vendor is unlicensed, the people are invading the sidewalk, the neatness of the street is compromised: exhaustive planners would be driven to place the empanada vendor in a commercial, rather than residential, zone; in a closed off venue with tables and chairs, where people can sit in groups. But it is in these kinds of casual encounters that shopkeepers get to know their neighbors and build enough trust to give them informal store credit; where the office worker who needs work done on his home gets a tip on a good contractor from the construction worker; where it is reported that Martha's son seems to be hanging out with the wrong crowd; and where it is speculated that Manchester United may make the finals of the Champion's League this year. The empanadas are delicious, but people also come for the casual contact (see Bula, 2017, p. 254). Sidewalks are hubs for transmitting information (Jacobs, 1961, 68–69). The richness of city life can be contrasted to life in the suburbs, in which contact is restricted to those whom I allow into my home and private life: naturally, this will only happen with people that hold views, jobs, and socioeconomic positions similar to my own (Jacobs, 1961, p. 65): suburban life risks becoming autistic (García, 2013).

Whereas it is pleasant to wander around a dense and diverse city, one is more inclined to prefer a car ride to walking for more than a few minutes to get from here to there in a suburb. The distinction between castles and pavilions seems useful: buildings are either closed and secretive, or open, extroverted and inviting to exploration (Holbrook, 2015). To walk for any space of time in a suburb is to go from castle to uninviting castle. The same happens with too-large uses of street frontage, irrespective of the type of use (Jacobs, 1961, p. 234); it is hell to transit the sidewalk along a large shopping mall.

Wandering around dense and diverse cities, one can find small niche businesses that would never be profitable in a suburb. Variety is possible because of variety: small businesses draw from the city's "variety of cultural opportunities, variety of scenes and (…) great variety in (…) population" (Jacobs, 1961, p. 148). In terms of the philosophy of Spinoza, cities can be conceived as composite bodies made cohesive through a variety of relationships of cooperation: the more internally diverse a city is, the more it can accommodate even more diversity. In practice, however, this positive feedback loop is usually interrupted because diverse areas in a city become fashionable and therefore costly and able to support only high-yield economic uses

(Jacobs, 1961, p. 243): Brooklyn goes from an interesting place to explore on foot to gentrified and hipsterized predictability.

Should the fact that certain areas are pleasant to walk around in, whereas others are not, enter into the deliberations of city planners? Such data seems subjective and irrelevant. Indeed, control cartesianism would reject it, for it is neither certain nor deductively derivable from first principles. Urban beauty is an emergent property (see Fromm, 2004, p. 23), that must be distinguished from architectural beauty. Whereas the latter has more to do with such things as a good use of geometrical proportions or the artful integration to immediate surroundings (see, e.g., Weisberg 2011 for a discussion of Frank Lloyd Wright's *Fallingwater*), the former can only be seen in the context of a co-evolving community of buildings that interact with each other (and with more ephemeral occurrences such as seasons or weekly markets), that age and gather history, and that behave as members of a living human community rather than as stand-alone pieces of decoration[1] (Makower, 2016). Urban beauty arises from the relationship between parts, so that its presence in a whole cannot be additively inferred from its presence in the parts; therefore, it does not fit well with a cartesian approach to knowledge and planning, and can seem a frivolous topic.

As it turns out, according to Leopold Kohr (1989), beauty is an important consideration for such down to earth topics as city traffic. Traffic is not a matter of insufficient roads (Kohr, 1989, p. 48–51) but of people's desires: if the periphery of the city is dull and lifeless, people will flock downtown and create traffic (Kohr, 1989, p. 82), and more roads will only encourage more crowding: beauty must be decentralized. Kohr is not suggesting that city peripheries be painted in bright colors or garnished here and there with monuments: for a periphery to be attractive, it must have its own magnetic field (Kohr, 1989, p. 62). Indeed, the reason people go to public spaces is because they want to see and meet other people, and be involved in other people's doings; spaces must be attractive because they compete for people (Walljasper, 2005, p. 22).To be of interest, city sectors must have some degree of political autonomy (Kohr, 1989, p. 82) and be largely self-sufficient, not only in terms of needs (having hospitals and housing both nurses and doctors, etc.; which means that all city sectors must accommodate people of many income levels (Kohr, 1989, p. 106)) but in terms of *wants* (having their own structures of conviviality such as churches or pubs). Cities ought to be federations of independent, and independently attractive plazas and squares (Kohr, 1989, p. 42–44), which, although

[1] <Footnote ID="Fn1"><Para ID="Par33">Of course, both kinds of beauty interact. For example, when we understand that buildings must be understood in relation to their surroundings, we can distinguish buildings that harmonize with the city around them from those that impose upon it. Imposition upon space characterizes a certain type of architecture, an "(…) architecture as commanding object which correlates directly to the ego-driven thread of architects throughout the centuries (…) this aesthetic has justified architecture which ignores the existing environment around it" (Schiler, <CitationRef aid:cstyle="CitationRef" CitationID="CR42">2007</CitationRef>, 509). This kind of architecture needs a name, so that it may be pointed out and avoided; I humbly suggest *<Emphasis aid:cstyle="Italic" Type="Italic">douchebag architecture</Emphasis>* as a new technical term.</Para></Footnote>

harmonizing with the whole, should have their subtle differences and idiosyncrasies (Kohr, 1989, p. 62–63). They must include visible structures that confer identity to a city sector, and these structures must be beautiful, and inspire love towards one's neighborhood (Kohr, 1989, p. 86). Residential streets must aspire to be more than mere traffic corridors, they must have adequate structures (such as semi-private front yards that facilitate casual contact) so that they become structures where people want to "stay and play" (Gehl, 1978).

Both Jacobs and Kohr see diversity as an indispensable component of urban beauty. Homogeneity might seem orderly but (if one is attentive to the experience of walking through a city sector) it hides experiential disorder:

> If the sameness of use is shown candidly for what it is-sameness-it looks monotonous. Superficially, this monotony might be thought of as a sort of order, however dull. But esthetically, unfortunately also carries with it a deep disorder: the disorder of conveying no direction. In places stamped with the monotony and repetition of sameness you move, but in moving you seem to have gotten nowhere. North is the same as south, or east as west. Sometimes north, south, east and west are all alike, as they are when you stand within the grounds of a large project. It takes differences-many differences cropping up in different directions to keep us oriented. Scenes of thoroughgoing sameness lack these natural announcements of direction and movement, or are scantly furnished with them, and so they are deeply confusing. This is a kind of chaos. (Jacobs, 1961, p. 223–224).

The chaos of sameness is sometimes remedied by imposing superficial architectural differences on same-use structures: this is what Jacobs calls "googie architecture" (1961, p. 224): "hot dog stands shaped like hot dogs, ice-cream stands in the shape of ice-cream cones" or gherkin shaped office buildings in London's financial district. Such exhibitionist architecture cannot mask the underlying monotony of single-use city sectors, nor can it create urban beauty: what is interesting about cities is the diversity of human endeavors (1961, p. 229). This goes to the heart of Jacob's criticism of mainstream urban planning: people are interested in other people (1961, p. 70). We seek casual contact, we hang out around front stoops, candy stores and empanada carts because we want to see "what's up".

This is no trivial point. Planning that minimizes casual contact assumes that people want to formulate plans and then follow them, and that they do not want to be distracted; that they wish for conclusions to derive from premises in an uninterrupted manner. This reflects a neglect of passivity, inherited from the philosophical modernity of which Descartes is a paradigmatic example. The fact is that passivity and interruption is essential to human life (see Innerarity, 2001): many important things in our lives (such as love, or the development of a strong commitment to a cause) happen *in spite* of our plans, because we got sidetracked. Such chance occurrences are certainly not the product of "the will of men applying their reason", and yet can be central to a human life. Examples abound: from every "meet-cute" story of chance encounters that resulted in lifelong partnerships, to a life-changing book found casually in a discount bin, to the story of Paul the Apostle, whose will was interrupted by a bolt of lightning. Human life is enriched through diversity and chance.

The importance of chance encounters does not diminish the importance of planning: life-enriching self-organization is the product of planning. Once we abandon the idea that order is exclusively the product of exhaustive planning, of the will of men applying their reason, we can see that diversity appears given certain conditions:

> So long as we are content to believe that city diversity represents accident and chaos, of course its erratic generation appears to represent a mystery.
>
> However, the conditions that generate city diversity are quite easy to discover by observing places in which diversity flourishes and studying the economic reasons why it can flourish in these places. Although the results are intricate, and the ingredients producing them may vary enormously, this complexity is based on tangible economic relationships which, in principle, are much simpler than the intricate urban mixtures they make possible. (Jacobs, 1961, p. 150).

Jacobs goes on to enumerate a fairly simple recipe: mixed primary uses, small city blocks, buildings of different age and price, and dense concentrations of people (1961, p. 150). The aim of planning is to sow the seeds of self-organization:

> (...) most city diversity is the creation of incredible numbers of different people and different private organizations, with vastly differing ideas and purposes, planning and contriving outside the formal framework of public action. The main responsibility of city planning and design should be to develop-insofar as public policy and action can do so-cities that are congenial places for this great range of unofficial plans, ideas and opportunities to flourish, along with the flourishing of the public enterprises (1961, p. 241).

Planning for self-organization, designing freedom, is a matter of variety management: control should be distributed throughout a system so that the needs of the system can be met (Beer, 1994). A self-organizing community does not form a random landscape of relationships; rather (through the trust generated by small city blocks, alternative pedestrian routes, sidewalks and front stoops, etc.), public characters (such as storekeepers or bartenders) spontaneously emerge and act as organizing hubs: they transmit news and initiatives that are of interest to the neighborhood, and put people that have mutual interests in contact with each other (Jacobs, 1961, p. 68–69). These main, general-purpose, public characters make it possible for more specialized public characters to emerge: this is the case of Jane Jacobs herself, who, after organizing a petition against a roadway in New York's Greenwich Village became known "through the grapevine" as the neighborhood's "petition person", and went on to help in many other initiatives (1961, p. 70).

In this section, I have presented Jane Jacob's ideas on urban planning in the framework of some general ideas about planning and diversity. In the following, I wish to apply this "philosophy of sidewalks" to a new and dominant space in human existence: cyberspace. Here too, there are problems of homogeneity and lack of diversity. The three main concepts I wish to apply are *sidewalks* (understood as places for causal contact, in which people with convergent interests might meet), *technocracy* (understood as putting problems that ought to be resolved in democratic fashion exclusively into the hands of experts), and *planning for self-organization*, as distinct from *exhaustive planning*. I wish to re-purpose the term "smart city" to refer to cities that are cognitively diverse and self-organizing, so I must begin with a critique of the use that the term is usually given.

17.6 Smart Cities

Smart cities, as they are usually understood, are cities that maximize the efficiency of basic services through massive data collection on citizen behavior and public-private partnerships. Monzon, who defends the concept, defines a smart city as "an integrated system in which human and social capital interact, using technology-based solutions. It aims to efficiently achieve sustainable and resilient development and a high quality of life on the basis of a multistakeholder, municipality based partnership" (2015, p. 20); smart city proponents insist on a holistic approach that takes into account social, economic and environmental needs.

The idea of smart cities has been criticized because it treats the citizen as a passive consumer of the city: "It constructs an urban subject active only to the extent that he or she shoulders responsibilities the public sector has withdrawn from, and is otherwise fundamentally passive" (Greenfield, 2017, p. 25). Critics see "smart cities" as a way for large technology companies to market new technologies and gain a foothold on the public sector (Borkowska & Osborne, 2018, p. 357). Organizing agency is left to the managers of the smart city; which results in a poverty of experience: "Consider that 'optimised' urban management tends to create epistemic and experiential bubbles, acting to prospectively eliminate the daily frictions that force us to confront the other (…)" (Greenfield, 2017, p. 25). Greenfield suggests replacing the data probes of current smart cities with more opportunities for active citizen participation, so that managers hear from the citizens (rather than infer from their uses of city resources), what kind of city they want to inhabit: for example, exercises in which, through interviews of inhabitants, maps of city sectors are created that reflect "narratives of place" (Greenfield, 2017, p. 20).

The conceptual problem with smart cities goes deeper. If a city is seen as a bundle of consumer services (e.g roads, internet access, utilities, etc), and citizen participation is reduced to demand for these services (which can be tracked through information technologies in order to make supply more efficient), the political aspect of citizenship is stripped away. This is technocracy, as defined in Sect. 17.5: the divergent problem of city is folded under the rubric of efficiency, and made invisible. According to Fernandez-Güell et al., the technology-centered focus of smart cities runs the risk of assuming an oversimplified notion of what a city is; therefore "we should turn around the tech discourse and locate technology as a transversal supportive element of the whole urban functional system" (2016, p. 64). It seems desirable to maintain a political function for cities at a time where the political legitimacy of the nation state is waning: there may be "something irreducible and nontransferable, necessary but not quite sufficient, about the city's public street and square for the realization of a meaningfully democratic citizenship" (Holston & Appadurai, 1996, p. 202). The executive power of cities (closer to the people than that of nation states and unhobbled by matters of sovereignty), has shown great potential for inter-city cooperation aimed at tackling global problems such as the proliferation of illegal guns, or climate change (Barber, 2013, p. 6).

The productive density and diversity of cities can make them powerful hubs for democracy. As outlined in the previous section, such self-organization requires certain conditions that can be planned for. First, a re-definition: *a city is smart in as much as its citizens are connected in such a way that cohesive and inclusive processes of second order cognition are possible, and can amplify the power of citizens to direct their own lives*. Notice (1) that the term "smart city" is now on a sliding scale; rather than a brand of city management, it becomes a concept by which any city can be evaluated in terms of second order cognition; (2) that efficiency in the use of resources and delivery of services is not a part of a smart city's essence, but that a smart city would have efficient services as a side effect of citizen empowerment; (3) that, in principle, the definition neither precludes nor implies heavy use of information technologies. *How* technologies are used is much more important.

Information technology is often used to empower predatory capitalism or authoritarian states through massive data collection (Mulgan, 2013, p. 147); even participatory platforms such as social media can be hijacked to benefit the interests of a few, to the detriment of democracy (e.g Alcott & Gentzkow, 2017). On the other hand, IT is not a necessary condition for second-order cognition:

> In India, the Shodh Yatra organized by the HoneyBee Network uses walking as a tool for change. In 1 week, walkers (farmers, scientists, and researchers) travel hundreds of kilometers across rural India to unearth, share, and disseminate solutions to local issues including conservation, organic farming and biodiversity as well as health and nutrition. During the day, walkers pass through farming land- usually accompanied by local farmers and laborers who discuss and reflect on their farming practices. In the evening, walkers stay in villages and hold meetings with local residents to share insights and knowledge of innovations from other parts of India (Mulgan, 2013, p. 143).

The economic power of cities comes from their *connexity*, the conditions they supply for extended cooperation (Mulgan, 2013, p. 204). If this cooperation is to extend to democratic decision making, to becoming a *smart city* as I define it, a first condition is that cities, and neighborhoods within cities, have true political power. Without it, exercises in grassroots democracy will be as specious and half-hearted as exercises in high school student government. A devolution of power from central governments to cities, and from cities to neighborhoods is necessary for grassroots democracy: people will not give their time and commitment unless they can exercise some tangible measure of power. Interest in citizen participation correlates with how effective this participation can be (e.g, Adler & Blake, 1990, on Portland's citizen participation programme for land planning).

A second condition is to realize that massive participatory democracies are indeed possible, that there is, in principle, no size limit to the *polis*. Two fields of knowledge must be combined: information technology and cybernetics (the science of variety management, Beer, 1994); sheer computing power, however large, will not do unless communications are structured in such a way that citizens' input is not attenuated, averaged away. Whenever there has been a political will towards massive participation, methods for it have been developed: in Chile, with rudimentary 1970s technologies, such cybernetic tools of participation were in the process of being implemented when the military coup cut short the country's brief socialist experiment (Medina, 2014). The Occupy movement developed the very low-tech "people's microphone" as well as "hand-signaling techniques to coordinate decision-making processes among large groups" (Greenfield, 2017, p. 24); and veterans of the movement developed the Loomio app, which allows activists to propose courses of action and discuss them asynchronously, in recognition that face to face meetings are not always possible (Greenfield, 2017, p. 24). The key is to change the focus from technology to cognition: a smart city is one that is arranged in such a way that it collectively makes smart decisions. On this count, Glasgow's commitment to become a "learning city" (in which not only citizens undergo continuing education, but the city as such discusses itself and consciously forges an identity and a way forward) seems promising, although the city has a long way to go towards true social inclusion (see Borkowska & Osborne, 2018). City managers that are serious about democracy should both experiment with platforms for digital democracy and recognize the legitimacy of experiments undertaken spontaneously by citizens, in as much as they are effective and inclusive.

Finally, physical spaces, "sidewalks", must be available for reclamation. Smart cities are fostered by the creation of promiscuous public spaces in which citizens can converge around common problems and interests, and empower each other: city planners cannot cause this convergence, but they can give it a chance, planning for self-organization. Regardless of urban planning philosophy, unused spaces in a city are clearly detrimental to city life: abandoned buildings and unused fields are a sign and a cause of urban blight. Citizen initiatives have turned these abandoned spaces into vegetable gardens and social centers in cities such as Brooklyn and Berlin: Madrid's Campo de Cebada, an abandoned commercial development site, has been turned into a site of community meetings, cultural events such as film screenings and concerts, and also hosts swimming, skating and tai-chi sessions (Greenfield, 2017, p. 20–21). The management of such spaces for many uses presents a challenge for the community: if a common good such as Campo de Cebada is to subsist, neighbors must step out of their social bubbles to meet and come to agreement with each other, and organize. This management of commons presents concrete problems around common goods, and can get citizens to talk to each other, and to practice politics outside of the polarized and ideological realm of national politics. It is easier to reason and compromise on a schedule for the use of a common space than it is to do so on Brexit; and practice on the former can produce a better performance on the latter. The job of city management is to produce a viable legal and administrative framework for such reclamation projects. If the market has spurned a piece of land, let citizen action experiment in new uses for it (Greenfield, 2017, p. 22).

I have outlined three conditions for smart cities: (1) devolution of political power to cities and neighborhoods, (2) recognition of the possibility and legitimacy of experiments in massive participatory democracy and, (3) a framework that makes it possible for communities to reclaim and manage unused urban spaces. All three conditions involve trust in the communities' capabilities to self-organize, and therefore go against the planner's itch, and imply a shift in the mindset underlying traditional, cartesian, urban planning.

17.7 Concluding Remarks

At bottom, what Jane Jacobs proposed was a new attitude towards the unforeseen. The spontaneity generated by diversity interacting with diversity can either be welcomed or seen as a failure of planning. If we see our communities and cities as functioning well, perhaps novelty can be shunned; but if we see them as in need of change, we must work towards designing communities that are capable of generating novelty and change (see Bula, 2015). This implies changing a mindset that I have called *control cartesianism* and that, in my opinion, is deeply ingrained in western thought, and has informed political thinking beyond urban planning.

References

Adler, S., & Blake, G. (1990). The effects of a formal Scitizen participation Programme on involvement in the planning process: A case study of Portland, Oregon. *State & Local Government Review, 22*(1), 37–43.

Alcott, H., & Gentzkow, M. (2017). Social media and fake news in the 2016 election. *Journal of Economic Perspectives, 31*(2), 211–236.

Barber, B. (1992). *An aristocracy of everybody*. New York: Oxford University Press.

Barber, B. (2013). *If mayors ruled the world: Dysfunctional nations, rising cities*. New Haven, CT: Yale University Press.

Barnett, J. (2014). Life takes place on foot. *Planning, 2014*, 31–33.

Beer, S. (1994). *Designing Freedom*. New York: John Wiley & Sons.

Borkowska, K., & Osborne, M. (2018). Locating the fourth helix: Rethinking the role of civil society in developing smart learning cities. *International Review of Education, 64*, 355–372.

Bula, G. (2010). Diversidad y cohesión: Arte, ecosistemas, sociedades. *Polis, 6*(10), 53–61.

Bula, G. (2012). Spinoza: Empoderamiento y ética de la composición. *Universitas Philosophica, 29*(58), 197–215.

Bula, G. (2014). El valor de lo diferente: sobre la interpretación de Matheron de la *Ética* de Spinoza. *Ingenium, 8*, 51–71.

Bula, G. (2015). Towards a non-trivializing education. *Kybernetes, 44*(6/7), 913–925.

Bula, G. (2017). *Spinoza: Educación para el cambio*. Bogotá: Universidad de la Salle/Aula de Humanidades.

Cathcart-Keays, A. (2015). Is urban graffiti a force for good or evil?. *The Guardian*, https://www.theguardian.com/cities/2015/jan/07/urban-graffiti-force-good-evil.

Descartes, R. (1967). *The philosophical works of Descartes*. Cambridge: Cambridge University Press.

Descartes, R. (2006). *A discourse on the method of correctly conducting One's reason and seeking truth in the sciences*. Oxford: Oxford University Press.

Dietz, G. (2007). Keyword: Cultural Diversity. In *Zeitschrift für Erziehungswissenschaft*, 10. Jahrg., Heft 1/2007, S. pp 7-30.

Fernandez-Güell, J., Collado-Lara, M., Guzmán-Araña, S., & Fernández-Añez, V. (2016). Incorporating a systemic and foresight approach into Smart City initiatives: The case of Spanish cities. *Journal of Urban Technologies, 23*(3), 43–67.

Fromm, J. (2004). *The emergence of complexity*. Kassel: Kassel University Press.
García, H. (2013). The autistic suburbia of Madrid: A constructive approach to recent urban developments. In *In Suburbs: Transformation and Development: Papers from the NSBB Conference 17–19 September 2013 in Denmark*. Aalborg: Danish Building Research Institute.
Gehl, J. (1978). The residential street environment. *Built Environment, 6*(1), 51–61.
Goldberger, P. (2006). Uncommon sense. Remembering Jane Jacobs, the 20th century's most influential city critic. *The American Scholar, 5*(4), 122–126.
Greenfield, A. (2017). Practices of the minimum viable utopia. *Architectural Design, 87*(1), 16–25.
Han, B.-C. (2017). *Die Austreibung des Anderen*. Frankfurt: Fischer Verlag.
Hardt, M., & Negri, A. (2001). *Empire*. Harvard, MA: Harvard University Press.
Hicks, J. (1994). *Mayor announces new assault on graffiti, Citing Its Toll on City*. The New York Times, https://www.nytimes.com/1994/11/17/nyregion/mayor-announces-new-assault-on-graffiti-citing-its-toll-on-city.html
Holbrook, T. (2015). Castles and pavilions: Creating new hybrid places of exchange. *Architectural Design, 85*(3), 26–31.
Holston, J., & Appadurai, A. (1996). Cities and citizenship. *Public Culture, 8*, 187–204.
Innerarity, D. (2001). *Ética de la hospitalidad*. Barcelona: Península.
Israel, J. (2001). *Radical Enlightenement*. Oxford: Oxford University Press.
Jacobs, J. (1961). *The death and life of great American cities*. New York: Vintage Books.
Kohr, L. (1989). *The Inner City*. Talybont: Y Lolfa.
Lévy, P. (1997). *Collective intelligence*. New York: Plenum.
Makower, T. (2016). Finding Time. *Counterpoint, 239*, 136–141.
Marazuela, A., Merrill, E., Miller, C., & Yates, C. (2014). *Thriving cities endowment brief: The beautiful*. Charlottesville, VA: University of Virginia's Institute for Advanced Studies in Culture.
Medina, E. (2014). *Cybernetic revolutionaries: Technology and politics in Allende's Chile*. Cambridge, MA: MIT Press.
Merçon, J. (2012). La filosofía de Spinoza y el pensamiento sistémico contemporáneo. *Revista de Filosofía Universidad. Iberoamericana, 133*, 83–101.
Monzon, A. (2015). Smart cities concept and challenges: Bases for the assessment of Smart City projects. In M. Helfert et al. (Eds.), *Smartgreens 2015 and Vehits 2015, CCIS 579* (pp. 17–31).
Mulgan, G. (2013). *The locust and the bee: Predators and creators in Capitalism's future*. Princeton, NJ: Princeton University Press.
Page, S. (2007). *The difference*. Princeton, NJ: Princeton University Press.
Resnick, M. (2000). *Turtles, termites and Traffick jams*. Cambridge, MA: MIT Press.
Rawes, P. (2015). Spinoza's Architectural Passages and Geometric Comportments. In B. Lord (Ed.), *Spinoza Beyond Philosophy* (pp. 66–86). Edinburgh: Edinburgh University Press.
Reynolds, C. (1987). Flocks, herds and schools: a distributed behavioral model. In *SIGGRAPH '87 Proceedings of the 14th Annual Conference on Computer Graphics and Interactive Techniques* (pp. 25–34) http://dl.acm.org/citation.cfm?id=37406.
Schiler, M. (2007). Stewardship as architectural aesthetic. *Christian Scholars Review, 35*(4), 505–516.
Spinoza, B. (1981). *Ethics*. Salzburg: University of Salzburg.
Sunstein, C. (2001). *Republic.com*. Princeton, NJ: Princeton University Press.
Walljasper, J. (2005). The future of the human race depends on public spaces. They are the starting point for all community, commerce and democracy. *New Statesman, 15*, 21–23.
Weisberg, R. (2011). Frank Lloyd Wright's Fallingwater: A case study in inside-the-box creativity. *Creativity Research Journal, 23*(4), 296–312.
Zourabichvili, F. (2002). *Spinoza: Une physique de la pensée*. Paris: PUF.

Germán Bula is professor at the Philosophy and Humanities Faculty of the Universidad de la Salle (Bogota, Colombia). He holds a doctor's degree in education from the Universidad Pedagógica Nacional (Bogotá, Colombia) and a master's degree in philosophy from the Universidad Javeriana (Bogota, Colombia). He has published the following books: Ruedas Dentadas (2003, novel), Alteridad y Pertenencia: lectura ecocrítica de María y La Vorágine (2009, with R. Bermúdez); Cibercultura, capitalismo cognitivo y educación (2015, academic editor with R. Rueda and A. Ramírez), Paranoia: emociones públicas y universidad (2016, with S. González), Spinoza: Educación para el cambio (2017) and Latin American Perspectives on Global development (2018, academic editor with M. Masaeli and S.E Harrington). He is the winner of Emerald Literati Network Awards for Excellence (2016) for the paper: "Towards a non-trivializing education" published in *Kybernetes* (vol.44, 6/7, pp. 913–925).

Chapter 18
An Ontological Analysis of Cities, Smart Cities and Their Components

Stefano Borgo, Dino Borri, Domenico Camarda, and Maria Rosaria Stufano Melone

Abstract The arising of smart cities has shown the limitations of the traditional attempts to understand and characterize cities. The smart city marks a relevant step in the evolution of urban systems which is expected to have disruptive impacts in the near future. Indeed, the 'smartness' qualification of cities (and perhaps territories) points to relevant changes in the possibilities these complex systems offer mainly due to changes in the information which is made available. This chapter studies the notion of city from an ontological viewpoint, and provides the first elements of a general framework aimed to characterize and unify how to understand the city and its evolution. It starts from historical considerations to isolate the factors that drove the changes from ancient to modern cities, and the complexity that characterizes the nature of cities. Our analysis departs from the traditional views and leads to distinguish three ontological components in the city: the modified physical place, the distributed agentivity, and the knowledge present in the system. The analysis of these components and their interactions, we argue, provides a coherent reading of the city across its evolution steps, and may be used to establish the maturity of today's smart cities.

Keywords City · Smart city · Ontology · Place · Agentivity · Knowledge · Ontological analysis

---

S. Borgo (✉)
ISTC CNR, Trento, Italy
e-mail: stefano.borgo@cnr.it

D. Borri · D. Camarda · M. R. Stufano Melone
Politecnico di Bari, Bari, Italy

M. Nagenborg et al. (eds.), *Technology and the City*, Philosophy of Engineering and Technology 36, https://doi.org/10.1007/978-3-030-52313-8_18

18.1 Introduction

The exploitation of recent technological and computational developments is introducing relevant changes in the structure and dynamics of urban environments. This phenomenon is pervasive and is, little by little, changing the city service system. The cities that have more enthusiastically embraced the opportunities of the new technologies are now called *smart cities*, a term suggesting that they are providing more opportunities to their inhabitants and are better in managing themselves as complex systems. As any change that is potentially disruptive, the evolution from pre-smart city to smart city has introduced opportunities and risks which need to be assessed. To properly address this issue, we need to reach a proper understanding of what is at the core of the phenomenon and what drives its evolution. The fact that we lack a reliable characterization of the notion of city is a crucial problem for the quality of the assessment.

It seems clear today that the reading of cities in the pre-smart era, let us call this the *traditional view*, is not suitable to analyze and study the recent changes. This observation is justified by the emphasis with which in the literature the city and the smart city are distinguished, like if the distinction were to demarcate the emergence of a truly new type of system, i.e., the smart city. On the other hand, it would be naïve to believe that the expression 'traditional view' points to a precise, even though perhaps tacit, notion of city. It would perhaps be more acceptable to talk about traditional *views* but the core of the argument would not change: unless we resort to *ad hoc* classifications like population size and density, administrative rules or historical facts, we have been lacking in the past a shared understanding of what counts as a city, even leaving aside the smart city phenomenon. On top of this, one is tempted to claim, as we actually do in this chapter, that smart cities (whatever they could be) are not giving us any better grip for understanding what a city is. In short, we claim that we must investigate how to understand cities before attempting to assess the 'smartness' of smart cities.

Let us start with what we believe to be a straightforward claim: a city is an emerging complex organization that endures, changes and provides opportunities by integrating a variety of elements. We can even list, without aiming to completeness, many relevant elements like the co-presence of several agents with different capabilities, the physical adaptation of the territory, the possibility to quickly interact in different activities, the distribution of rights and duties, the organization and coordination of services, and so on. As usual for complex systems, a city is at the same time a whole and a collection of different parts. The tight interconnections among these parts and their technical, biological, economical and social nature, make the collection dependent on the granularity one takes and, even assuming this has been agreed upon, further dependent on the relationships one takes to identify the parts. Briefly put, the city expert sees the city as a whole but understands it in terms of tens of mutually disconnected models. For this reason, the extent of our success to study the city depends first of all on how we define it.

This chapter aims to develop a general framework which is suitable to analyze both the traditional and the smart views of the city. In this sense, it aims at a unifying approach that allows us to characterize the smart city phenomenon as a step in the historical evolution of cities in general. The chapter proceeds in two steps. First, Sect. 18.1 develops a reading of the historical evolution of cities based on the capacity to balance traditional and innovation services and forces. This reading is accompanied in Sect. 18.2 with a review of the literature that specifically studies the smart city. This part of the chapter aims to show how cities are presented today and which factors are taken of primary relevance in smart cities. Section 18.3 presents the smart city as a system where data-intensive techniques help to maintain and optimize the structural and dynamical complexity of cities. Finally, the chapter develops an ontological analysis of city in Sect. 18.4, which leads to the general framework on city. Section 18.5 revisits the notion of complexity within this approach. The final section adds further observations on what has been achieved and points to future steps.

Briefly put, the objectives of the chapter are:

1) to review relevant analyses of city in different epochs;
2) to introduce an ontology-based analysis of the city;
3) to discuss the impact of the ontological view on the complexity of the city as a system.

The core of the chapter is the presentation of three ontologically motivated components and of the general framework that arises from them. The components are centered on the notion of place, of agentivity (including intentionality), and of knowledge. The chapter argues that by isolating these components and modeling their interactions, one can provide a unifying perspective of what a city is, develop a comparison of previous analyses of the city, and discuss a new view about the study of complexity in the city. The study of the interactions across the three ontological components is not in the focus of the chapter but some initial observations are provided.

18.2 Understanding Cities

This section reviews a few historical understandings of cities as discussed across different historical epochs. The purpose is to show a variety of views to indicate how the focus of these studies changed across history without any claim of systematic classification nor of coverage of the different perspectives.

Perhaps the first interesting and integrated perspective of what cities are come from ancient Greece where cities are conceived as the realm of politics, with strong emphasis on societal and institutional aspects. Philosophical theories of this period include and analyze the role of the political organization of cities addressing even the problem of their optimal size (Plato, 380 BC). The essential roles of cities (in general overlapping with states) in peace and war are vividly drawn by historians and warriors (Thucydides, 430 BC; Xenophon, 370 BC). In the Roman period the building of cities has become a central issue largely for purposes of imperial colonization: their idea of the virtuous relationship existing between people and cities, and also the possible existence and resistance in human population of non-urban places is illustrated by Tacitus in describing the lifestyles of the Germans, who in essence do not know cities (Tacitus, 98 BC). The Latin juridical system clarifies the complementary roles—social, economic, and military—of cities and territories (Justinian, 530).

Coming to modern ages, a consolidated general theory conceives cities as spatial places which deal mostly with production of services (the non-material view) and residually of goods (the material view), which instead have their realm in non-urban spaces (Weber, 1986).

With the growth of the human population and of spatially concentrated forms of life, driven by markets and economic advantages offered by proximity and interaction, the distinction between urban and non-urban space becomes less meaningful for the ongoing retrenchment of the latter. This is shown today by the irresistible growing of megacities even though the increasing remote information and communication systems, markets globalization, virtual spaces as well as theories and practices of environmental conservation, introduce forces that try to challenge urbanization and to reintroduce distinctions between urban and non-urban spaces.

How do we understand cities in the modern days? A first answer points at cities as spaces of concentrated human population and, through this, of production of services and goods. These productions are in general rarer in less concentrated population already for a basic economical factor: large markets generate and pay better than small ones. Essential features of cities can be construed also by the forms (both natural and artificial at micro, meso, and macro scales) assumed by their political components. At the same time, modern theories of cities are conscious of the relevant role played by abiotic environmental matrices, in particular by the geological forms. More recently even the structural role played by the non-human biotic population, which finds relevant potentials of commensalism in cities, has been under investigation (Newman & Jennings, 2008).

Perhaps a sharper understanding comes from comparing cities to non-urban spaces. The stronger attraction that urban spaces generate, compared to non-urban spaces with dispersed population, comes from the potentials of work and freedom. Cities provide variety and resilience of work activities, compared to monotonies and recurrent fragilities of non-urban spaces; variety of coexistence of individualism and communalism, compared to what happens in small demographic groups; strong potentials of choice and of resistance for inhabitant, compared to the oppressive tendencies of family or tribe small groups in villages and in general in non-urban spaces. The variety and specialization of activities granted by cities offer distinctive productive and generative forms at both the individual and the social levels (Levy Strauss, 1955; Mumford, 1934, 1961). The mix of the natural and artificial elements in urban environments can be remarkable because of their particular and innovative combination of causal, organizational and intentional aspects, leading to non-repeatable and intriguing cities (Calvino, 1974; Mann, 1901). However strong, these advantages can be broken. Historical cycles of urban raising and decline (of cities and systems of cities) witness that particular political and organizational dynamics of human settlements can destroy urban attraction and create conflicts within cities and against cities (Bauman, 2006; Castells, 1989).

Even the reading of the traditional triad of professional activities of rulers-warriors, priests-shamans, and merchants-craftsmen (Mumford, 1934, 1961), which stands at the origin of the urban *demos,* has been enriched to explain the typical atmosphere of collective, complex, and even contradictory activities of the city (Shakun, 1999). The number of components and interactions, the number of people and activities in cities generate emerging systemic properties and with them innovation and creation typical of complex organization (von Bertalanffy, 1968; von Neumann, 1963). The integration of the city artificial and natural environments can generate artistic masteries and professionalisms (in art, architecture, craft) serving demands of goods and services beyond routine, ordinary needs and mere reproduction. As for technique, in the context of cities it changes into automaton and increasingly extends its ruling of the human, reducing the mystery of life itself (Severino, 2016). The attractive and aggressive features of cities generate theoretical consensus including an expectation of cognitive and productive virtuosity (Marx, 1861) supporting the development of intellectual creativity. As centers of production of services, cities always supply essential social forces to the development of information markets and of systems of education at all the levels. Paralleling ICT (Information and Communication Technology) in its ongoing supply of distributed, remote and interoperable virtual tools, theories-in-action of urban knowledge have been restructured in complex ways, calling to new ontological visions of the agent and multiagent integration.

At the beginning of the twentieth century, the city is seen as a socio-biological anthropotechnical system where the increasing need of knowledge of urban processes pushes to elaborate bidirectional (from inside to outside and vice versa) cognitive observations. The study of the city is carried out in a state of cognitive isolation (the paradigm of the inlook or outlook tower) to properly conceive the integrated past-present-future dynamical essence of the city (Geddes, 1915): a sort of cognitive perspective of social reality to which the theory of decision will later give mathematical models based on integrated utility, systematic and casual, rational and emotional principles (Kahneman & Tversky, 1979; McFadden, 1974). At the end of the twentieth century influences from psychology and the theory of organizational development lead to the need of critical self-reflection and at the same time fosters networking of experts in systems distributed in urban and regional spaces (Schön, 1983; Schön & Rein, 1994). Information starts to become a central focus in understanding and managing cities. Cities have always witnessed information spreading by social community interaction (in streets and squares, residential neighborhoods, taverns etc.) and information flowing through more or less formally organized systems in parallel with the evolution of communication techniques. The *Corriere Mercantile* [*Mercantile Courier*], founded in Genoa in the 1820s and probably the oldest ordinary newspaper of a commercial Italian city, is *de facto* nothing more than a sheet reporting day by day cargos' arrivals and departures. This example illuminates the starting need to collect specialized information in port cities because of their structural openness to interaction. Yet, information in cities has always been fundamental, constrained perhaps only by the capabilities to collect and elaborate it: health and epidemiology, security and war, work and commerce, spatial guide, legislation and political organization, culture, and entertainment.

Educational systems are typical of cities and are organized in levels gradually more specialized, expensive and rare, in tune with the demographic dimension of the city and its demands of knowledge (Christaller, 1933; Losch, 1940; Zipf, 1949). Basic levels include common social forms in coexistence with family and groups, as well as religious, military, and laboratory education. Advanced levels are provided by schools and academies whose objective is the complete cognitive training of human agents (Simon, 1945). More recently the system is gradually evolving to attain continuous education and educational multilevels for sophisticated networks of self-reflecting and self-learning agents (Schön, 1983; Schön & Argyris, 1996; Schön & Rein, 1994). Current techniques of interactive ubiquitous information, whose development seems more linked to information markets than to technical limitations, assume an increasing role in changing educational environments. Urban cognitive environments, based on qualifications and organizations, are attractive for the qualified services they offer and cumulatively grow on their successes always interacting with the more or less latent stimulating features of the cognitive urban milieux (Camagni, 1995; Camagni & Capello, 2005).

With the raising of remote and real time communications, information and knowledge interaction has witnessed a fast and intense development, whose outcomes remain to be evaluated. Data across multiple domains are accessible quickly and in large quantities in smart cities even if in general most information remains restricted to peer-to-peer and paying communities if not private corporations. Technological and spatial divides continue to affect relevant populations because of a variety of educational and/or instrumental gaps operating at the agent and/or multiagent level, often relative to institutional and governing systems (e.g., restrictions from bounded information to cognitive democracy). *Ceteris paribus*, for instance in terms of individual and social political freedoms, cities and regions technologically and economically powerful develop impressive systems to support information, cognition, and problem solving, and highly benefit from these.

18.3 A Review of the Literature on Smart Cities

Cities are mostly informally described, for instance as "the changing realities of evolving human settlements" (Bowen, Dunn, & Kasdan, 2010; Deslatte, 2015). This kind of characterization is vague and looks at the city in terms of an ideal and a processual perspective (Picon, 2015) with all its ambiguities, fascination, chances and risks. Cities comprise the complex and dynamic web of human life and activity, and because of this they are at the center of innovation, today taken as a pivoting characteristic of smart cities. To be more precise, smart cities focus on some relationships among parts of the city (Okner & Preston, 2017) while being themselves part of a dynamic system of correlated and transversal mechanisms.

The rise of smart city in the literature is a genuine revolution, comparable in significance to that of industrial cities in the social vision of the nineteenth century which determined the emergence of physical processes of city transformation with complex consequences (Picon, 2015). On another perspective, 'smart city' is considered an opaque label, perhaps because related to mechanisms which are not yet really clear. These mechanisms guide the changes and technological improvements in a kind of top-down process where the power comes to be once more in the hands of a few people or companies managing technologies and data (Townsend, 2013). Thus, cities are dynamic and polycentric systems (Okner & Preston, 2017) to which the smart city qualification adds the capacity of connecting and managing data as collected by distributed sensors and by people that furnish (intentionally or else) their own data and impressions, with all the risks that arise from data aggregation and meaning extraction.

When it comes to identify the smartness in smart cities, we can point to three types of approaches: (1) a techno-centered approach characterized by a strong emphasis on "hardware", new technologies and infrastructure making ICT the key factor of the smart city; (2) a human-centered approach where there is a large emphasis on social and human capital in defining the smart city; (3) an integrated approach that sees smart cities as cities possessing both the foregoing qualities (Bencardino & Greco, 2014). Also, there is some emphasis on how cyber-physical systems (CPSs) enhance smart cities (Fromhold-Eisebith, 2017) as new CPS-driven manufacturing systems may emerge according to the trends of "urban production" (Fromhold-Eisebith, 2017; Matt, Spath, Braun, Schlund, & Krause, 2014). These changes lead to new insights for policies and new suggestions on how economic and social well-being can influence the evolution of the city (Angelidou, 2014; Fromhold-Eisebith, 2017). A major objective of smart cities, in today's narrative, seems to be sustainability from the social, economic, and environmental viewpoint (Ferrero & Vesco, 2015; Fromhold-Eisebith, 2017; Hall et al., 2000). Much has been written about the changes of spatial patterns and social behaviors in the smart city as "media city"; yet, less attention has been paid to the question of "how urban new media shape the built form" (De Waal & De Lange, 2017). The expectation is that smart cities integrate technologies, systems, services, and capabilities into an organic network which is multi-sectorial and flexible for future developments as well as open-access (Albino, Berardi, & Dangelico, 2015; Fromhold-Eisebith, 2017).

The meaning of being a smart city is however still elusive. There is a continuously increasing number of smart cities all over the world which are smart in many different ways and with a number of different articulations (Cocchia, 2014). It seems like smartness is a status that each city can reach in its own way, and the scientific literature adds to this variety by proposing an impressive amount of different definitions and of possible different lines of exploration about smart cities (Cocchia, 2014; Hollands, 2008). Often the notion is characterized by reporting an (exhaustive?) list of dimensions that are themselves qualified as smart. In this view, cities inherit the quality of being smart from the way they instantiate the different dimensions. For example, Bencardino and Greco, among others, provide six dimensions: smart economy, smart mobility, smart environment, smart people, smart living and smart governance. These six dimensions connect the traditional theories of urban growth and development, with the modern aspect of sustainable development of a city (Bencardino & Greco, 2014). Another list is identifiable following the potentialities of CPS-driven processes: smart ICT infrastructure, smart energy, smart mobility, smart construction, smart security, smart metabolism, smart industries, smart economy, smart education, smart living, and smart governance (Bencardino & Greco, 2014). The 'smart city' label is thus reduced to an umbrella that talks about the different declinations (from conceptions to implementations) about all the fields of action and scientific interests. It is early to claim whether a satisfactory collection of all these descriptive dimensions in a sort of integrated system of knowledge about the city would be possible or even useful. However, the development of a global framework that provides a coherent understanding of these dimensions would be important to organize an analysis of smart cities.

One of the recurring aspects in the definition of smart city is the intensive use of ICT. The development and use of digital content with the hybridization of technology and the physical world is often labeled 'augmented reality' (Picon, 2015). The focus on these data-intensive services raises the risk to reduce the smart city concept to aspects related to technological and mobile applications leading to the chimera of high-tech control systems capable of resolving all the urban issues while leaving aside the essence of the city with its interconnected problems (Murgante & Borruso, 2015). Furthermore, in comparison with traditional cities, smart cities raise new kinds of problems like a stronger ethical commitment (in terms of policies and impact) in governing the environment, humans and non-humans that are living in it. New sustainable urban development and sustainable polices have to face needs of equity and service delivery, land use and planning, environmental protection, and economic development (Deslatte, 2015; Fiorino, 2010).

Considering an already consolidated city with its historical nucleus and subsequent edifications and stratifications, its process of transformation towards the smart paradigm corresponds at least ideally to the growth of the consciousness of the possibilities that technological devices and systems provide, as well as of the risks and possible undesired consequences at the collective and the individual levels. This raising consciousness goes in parallel with the making of new systems and networks producing important transformations in the social tissues that make the city. The process should lead to intentional communities with new internal qualifications and dynamics that need a fully democratic participation as there is a shift of power in favor of those with greater technical expertise. The use of expert knowledge reshapes the conditions for governing the group interactions and drives the exploration of new oligarchic organizational practices in contemporary communities (Sullivan, 2016). This is an opportunity to expand both expert knowledge and knowledge sharing while maintaining a diversity of views, a characteristic of modern societies where social knowledge is systematically related to particular social locations, situations, roles and identities (Howarth, Cornish, & Gillespie, 2015). This view of the smart city is inextricably linked to that of social innovation together with the availability and quality of the ICT infrastructure (Bencardino & Greco, 2014). Perhaps, the first elements to consider are the strategic, systematic, and coordinated implementation of modern ICT applications (Fromhold-Eisebith, 2017) which lead to a new challenge, namely, to "urbanize the technologies" (Fromhold-Eisebith, 2017; Sassen, 2011).

The focus on inhabitants' knowledge, learning capacities, creativity, and human capital for innovations, in line with analytical and modeling skills, pushes to qualify smart cities as "intelligent cities" (Glaeser & Berry, 2006). The intelligent city ensures integration across the technology, the human and the social capital to create the conditions for a continuous process of growth and innovation (Bencardino & Greco, 2014). This requires a serious investment in safety, education, housing, limitation of taxes, and improving redistribution; actions that should ensure that communities become "smart" and stay "smart." (Glaeser & Berry, 2006).

18.4 Complex City and Smart City

Urban systems present high degrees of complexity in terms of relationship, organization and decision making due to the communities' cognitive and operational levels. Complexity is certainly embedded in the nature of socio-technical systems that today characterize the city (Trist & Emery, 2005) and is structurally embedded in the larger environmental ecosystem.

The concept of environmental complexity has evolved in scientific and popular literature especially starting from the reflection on natural resources after World War II (Carson, Darling, & Darling, 1962; Commoner, 1971; Forrester, Meadows, Randers, & Behrens, 1972). Later a more mature reflection has emerged with the structural integration of the analysis of natural processes, traditionally considered of non-anthropic origin, and processes related to human activity and behavior (Bennett, 1996; Miller & de Roo, 1997). Decisions, which are often critical, impact the uncertain and diversified environment with its dynamics and operational difficulties in terms of effectiveness, equity and economic development (Faludi, 1987; Krumholz & Forester, 1990; Rittel & Webber, 1973).

In this evolution, cities acquire scientific and operational interest for their inherent complexity starting from the settling environment and its complementary relations – i.e., rural, mountain, marine and so on. The dimensions of urban complexity are multiple since the city system is open and extremely variegated, with differentiated agents and emerging properties driven by such multidirectional dynamism (Portugali, 2011, p. 53). Human agents are central constituents of the city and their activities generate a complex stratification of physical artifacts and of social and behavioral constructions and transformations. While the behavior of the individual agent reveals preferences, utilities and needs intimately connected to the individual, their integration in an urban organization make manifest the multiplicity of the differentiated behaviors (Simon, 1991).

Cities build opportunities out of this variety, thus making necessary to the agents to protect and actually enhance the city as a system. In the past, this was achieved via top-down centralized models, today we are increasingly interested in models of distributed decision making. In this latter perspective, the management of knowledge, values and needs of individuals or groups can be taken into a more focused and fine-tuned consideration. Inter-agent knowledge exchanges are also enhanced by moving toward a more democratic as well as operational sensitivity of the urban system as a whole (Camarda, 2010).

From a political organization standpoint, governance models should be guaranteed by a system of horizontal (peer-based) and vertical (hierarchical) relationships and connections (Bai, McAllister, Beaty, & Taylor, 2010; Wagenaar & Hajer, 2003) but in practice this is not easy. The difficulty holds even in smaller contexts like village-based settlements where native resources are managed through family and/or tribal administrations with dense horizontal interconnections. Furthermore, vertical connections are often characterized by passive acceptance of top-down rules and directives with little feedback at the cognitive and operational levels (Cleaver & Toner, 2006; Torregrosa, Vera, Kloster, & Torres, 2017). The difference of agents' cognition, needs and behaviors in the city induces a similarly complex characterization into relational bonds and interactions, especially in large cities which are socially, ethnically and attitudinally more articulated (Portugali, 2011, p. 86).

The inclusive and diffuse management of urban characters includes also natural, artificial and hybrid environments, i.e., areas that are not specifically settlements. The very foundation of the city is often related to the presence of resources for the development of the community and for establishing structural relationships like kinship, use, protection or exploitation. These areas are riverbanks, road crossings, ridges or valleys, or historic neighborhoods of castles, farms or mineral deposits. These relationships can be explicit or tacit, deliberate or unconscious, uncertain, changeable, virtuous or subtle but largely ineluctable for the development of urban communities. An effective management of cities should aim at structurally involving these natural and artificial environmental components, together with infrastructural items or elements of technology-based support and enhancement of urban dynamics.

In this framework, one can devise a multi-agent model to support knowledge relationships across individuals, environment and technologies. Such a model would be of critical relevance for managing the urban system as a learning organization (Arentze & Timmermans, 2006). Connections are key elements in the model and may involve biotic and abiotic, human and artificial, routine and intelligent, single and aggregated agents through horizontal or vertical interactions as well as hybrid multiscalar dynamics (Camarda, 2012). This framework has the merit of being explicit about the complexity of the environmental system and of its organizational intricacy (Batty, 2007).

The management of city problems via traditional models, with their deterministic and reductionist approaches, has brought formalizations of technical or technical-economic rationality which have not reached the needed reliability particularly at the social and collective levels (e.g., Simon, 1982). Today there is awareness that an intricate system like a city requires a compromise between the reduction of complexity and the preservation of content and representative quality. The science of organization is increasingly exploring new representation approaches to embed enhancements of foundational aspects and core forms of systemic complexity (Gorban & Roose, 2010). Within this view, multiagent models for complex interactions, cognitive exchanges, and decision making have been investigated as system architectures (Pereira, Prada, & Santos, 2016). If emphasis on technology seems determinant for characterizing a smart city (Geertman, Ferreira, Goodspeed, & Stillwell, 2015), technology can deal with environmental and cognitive complexity by monitoring features and relationships and by managing emerging properties (Ismail & Zhang, 2017; McClellan, Jimenez, & Koutitas, 2017). In this context, the smart-city concept points to the new city's abilities to exist, maintain itself, progress as a multi-agent entity and as a whole. An ontology-based approach can be foreseen as an opportunity for the analysis and understanding of this multidimensional cognitive assortment even though a framework is still lacking (Borri, Camarda, Grassini, & Patano, 2016).

18.5 Ontological Components and City Types

In the previous part of this chapter we have looked at several points: (a) reasons for the development and persistence of cities, (b) historical changes in understanding of cities, (c) top-down and agent-based modeling of cities, (d) attempts to describe the essence of a modern city, and (e) increasing relevance of technology. Due to today's awareness of the complexity of the city and the recent raising of the concept of smart city, the view of city has changed again pushing us to turn to a basic question: *what composes a city?*

Since most viewpoints we have reported answers, perhaps only implicitly, in a different way to the ontological question "what is a city?", we believe that to reach a consolidated view on this topic one has first to carry out an ontological analysis of the core components of the city. This question is by no means easier but it has the advantage of allowing us to reason by hypothetical considerations of the form "would it be possible to have a city without a component (of type) X?" or "are components X and Y ontologically different?". Such ontological analysis is here carried out following guidelines developed in the field of applied ontology (Guarino, 1998; Guarino & Welty, 2009) with the hope that components characterizing cities can be clearly singled out, and that a general framework eventually emerges making possible to embrace within the same framework not only previous views of city but also future refinements and evolutions of the concept.

The first step in an ontological analysis is the isolation of what one considers the essential properties that characterize the entity or phenomenon at stake. Thus, even though we are not asking what a city is (and, analogously, what a smart city is), it seems we should first decide whether city and smart city classify the same or two kinds of entities. In the previous pages we have contrasted the traditional notion of city and the new notion of smart city. As we have seen, the change has often been presented as disruptive to the point of forcing the understanding of the smart city as a new kind of city. The idea, when taken seriously, would lead to revise the traditional conceptual view(s) in order to develop suitable representation concepts for the smart city; to adapt the methodology to study what is truly at the core of these cities; and to change the focus of the analysis to concentrate on the new dynamics these cities manifest. At this point of the chapter we prefer not to take an explicit position, and decide to address our general question about city components in general. Ideally, if smart cities turn out to be ontologically different, this fact should emerge at some stage of our investigation.

In order to understand what to expect from our analysis, assume for a moment that we have successfully isolated the ontological components of cities. Should we suppose that a city is the juxtaposition of these components? That would be too naïve. Due to the complexity of cities, it is hardly believable that they can be reliably broken down into disjoint parts. Instead, we claim that the subsequent step is to explain how the city as a whole emerges from these components, that is, how they interact and how they are practically (but not ontologically) entangled, and even how they explain the co-existence of different views of city as we have reported earlier in the chapter. Finally, unless our analysis points out some ontological distinction between city and smart cities, we should be able to consider the latter as a special type of the first and to highlight the differences, perhaps in the role of the components or in the type of interactions that are manifested.

Briefly put, the general view we are aiming at is that a city is the result of interactions (a term that itself needs to be clarified) across ontologically distinct components. We insist that the simple co-presence of the right components is not sufficient for the existence of a city. The nature of the interactions across these components is as necessary as the qualification of the components themselves. Nonetheless, we need first to discriminate and classify the components and only later we will be in a position to qualify the interactions and, then, the city itself. Our hypothesis is that the level of development of the components in a city and the quality of the interactions they manifest is a reliable way to identify the type of that city.

A leitmotif in the discussion of cities in the literature is that these are complex systems and, as such, exposed to different analyses. Each methodology makes its own implicit assumptions, including a choice of the components that have a relevant role in the system. Ontologically speaking, these components have a special feature: they can be influenced by the city as a whole but cannot be explained as mere consequences of its existence. In other terms, the city need them to exist but not vice versa. By being necessary and having an independent nature, one can claim that these characterize the city's ontological (and, in turn, conceptual) structure. Of course, the co-existence of the components is just one condition among several since a city must be capable of surviving as well as of maintaining its identity across time, that is, while functioning, adapting and evolving.

Where should we look for these components? From the discussion in the previous sections, the literature has pointed out several candidates that have a relevant role in the isolation and understanding of cities, these are: population, services, environment, information, organization, infrastructure, regulation, and space. These candidates are intertwined in several ways: most services are provided by and for the population (primarily humans and non-human animals with an increasing participation of artificial agents, themselves needing dedicated services); the environment includes space, infrastructure and population; information is needed to feed, organize and make possible services; services, users, environment and organizations produce information; and so on. This network of relationships is complex and highlights the fact that a city is not only a rich system, as it comprises many parts of different nature (heterogeneity of parts), but is also highly intertwined as all these parts interact in a variety of ways and for a variety of reasons (heterogeneity of interactions). In the ontological comparison of these elements and aspects, three components emerge quite clearly.

The city has a spatial location with its physical and material characteristics, in short, it is a *place*. Of course, the place is, strictly speaking, *modified* and not anymore a natural place: the geographical characteristics and affordances of the physical region have been intentionally modified, at least in part, to adapt to the needs and evolution of the city. Nonetheless the place itself, no matter how changed, is a core component of the city that we call the *city-place*. By city-place we mean the specific physical region, with its physical features (geographical, morphological, geological, urban structure and so on) including the arrangement of the objects in it (natural, artificial, functional and so on), that manifests the physicality of the city. The city-place should not be confused with the city location, e.g., see the notion of locality in (Garbacz, Lawrynowicz, & Szady, 2018) nor with the (natural) environment in that location, nor with the instantaneous snapshot of its status at some point in time. On the contrary, the city-place is an entity that extends in space and time, and includes the previous including the artificial and natural *material* elements that are part, perhaps only temporarily, of the city. The notion of city-place recognizes that in a city there are changes that go beyond the natural world, primarily due to intentional activities, without attempting to constrain what these are, to explain why they happen, or to identify the sources of intentionality. Intentionality can sometimes be attributed to the city itself (as a complex body with some level of self-awareness) or to one of its multiple parts like administrative services and inhabitants. Yet, intentionality is irrelevant to the understanding of the city-place component. The city-place, as understood here, is the physical manifestation of the city elaborated by the continuous large and small changes that happen in the city with varying degrees of coordination and intentionality, or even completely lacking coordination and intentionality.

Some changes in the city-place are due to natural forces exercised by natural elements like river and sea dynamics, weather, geological movements and so on. There are intentional changes as well, due to agentivity forces acting in and around the city, primarily because of the population (human agents but not only) and organizations. The fact that these are agentivity forces does not mean that the changes they trigger in the city are expected, desired or even controlled even though some certainly are. Ontologically speaking, agentivity has an essence that transcends that of the city-place, the latter being essentially a physical entity. Any entity that comprises an intentional aspect escapes the boundaries of purely physical entities. The view of the city as a modified place recognizes and registers the physical changes but can only say how those changes happen from a physical viewpoint; it cannot include *what causes* those changes. To ask that, one has to move beyond and investigate the realm of agency, where agency is often, but not necessarily, intentional. The city as a material place on the one hand and the city as a source of agentivity on the other, identify two disjoint components of the city. These components are ontologically separated (materiality cannot explain agentivity even though it can constrain agentivity to some extent; and agentivity cannot explain materiality although it can constrain material layouts to some extent) and both necessary for the city to exist. They are clearly highly interconnected in the city as the presence and evolution of one cannot be understood without the other. Interestingly, these two components already cover and organize many candidates that have been used to talk and describe cities in the literature. *City-agency*, that is the agentivity component of the city, explains the dynamic evolution of the city with the interplay between the individuals and the community: it identifies the view of the city as a network of heterogeneous and interacting acts.

We have seen that the notions of city-place and of the city-agency cover the physical realm and the agentivity forces manifested by and in the city. Still, there is something else which is not there as it is of a different ontological nature. The information about the state of the city and its parts, the available models to depict and understand the city, the means to predict something to happen or a situation to occur, in short, information and knowledge present to and available in the city are necessary to the city and yet not covered by the previous physical and agentive components. They need to be dealt with as a separate component. After all, most relevant actions are the result of a rational decision-making process, and rational decision making requires some level of information (true or false) and some level of knowledge. To fill the gap, we introduce the *city-knowledge*, a component that takes into account the information a city produces (not necessarily reliable) and that is used for controlling, guiding, foreseeing and judging. The city-knowledge provides the background for decision making, i.e., for local and global actions to take place. Knowledge is distinct from the physical and the agentive components because it is about content and modeling, comprising the realm of information and management: from its production, to its storage, distribution and interpretation. The knowledge component is where data become information, and where information is elaborated and integrated into models. By making this possible, the knowledge component enriches the city with the capacity to represent itself and the external environment, the relationships between its parts as well as their possible evolutions. These representations make comparison and anticipatory analysis possible, allow to control the city characteristics and changes, introduce a level of self-awareness of the city and of its parts.

The place, agentivity and knowledge are by no means simple components. As said, places are traditionally understood as enriched spatial areas (Ballatore, 2016; Cresswell, 2014; Stufano Melone, Borri, Camarda, & Borgo, 2017) and include things like the geomorphological state, the urban structure, the environment as the collection of structural elements (both natural and artificial like plants and rivers, roads and buildings) as well as the changes they manifest in time. The agentivity component includes the physical and the mental actions, individual and group actions as well as the rich social system that coordinates and encapsulates the agents' thinking and acting (Searle, 2006). Finally, the knowledge component integrates data, information and knowledge broadly understood with their hierarchies, denotation, characteristics and reference mechanisms (Zins, 2007). Note that the analysis of the city as composed of place, agentivity and knowledge is ontological, as these components have clearly different ontological nature. They do not provide the explanatory and teleological views that are often the focus of the discussion in the literature since these views emerge at a subsequent level, namely the level in which we analyze the interactions across these components.

Is the ontological analysis helping us in modeling cities? A full answer to this question would need to develop a theory of interaction among the components. Yet, we can anticipate some interesting point relatively to the ontological components. The nature and role of the city-place component can be modeled using known approaches, for instance in terms of a theory of physical objects, capabilities and capacities, of spatial arrangements and historical evolution. Similarly, the nature and role of the agentivity component can be analyzed in terms of multi-agent theories that model performance, decision making and goals, needs and worldviews. Since the city embraces agents of a variety of types, a theory for the agentivity component has to include in its scope also cognitive, social and cultural aspects. Several approaches to multi-agent systems (in terms of logical modeling as well as simulation) have been developed in the last 30 years. Finally, the nature and role of the knowledge component is clearly understood in terms of theories of information science and the different representation and modeling techniques that are proper of applied sciences, all domains of research traditionally well studied. Finally, waiting for a complete theory of the city, we expect that technically we will need to investigate the ontology-based integration of models in the above three components via a theory of interaction. For instance, a model available at the knowledge level can be seen as a tool for some agent in the agentivity component to decide which action to perform to change the physical place, this change, in turn, will produce new data to be interpreted at the knowledge level. The possibility and potentiality of this modeling approach remains to be explored.

We claim that the three components we presented cover the candidates that we have seen as traditionally used to study and understand cities. What might be perhaps not clear is the generality of the perspective we have introduced. As we know, a city is a complex system in space and time, and our analysis leads to separate the physical and historical component, the agentive and social component, and the information and representation component of this system. This means that our ontological view encompasses the engineering perspective of technical systems, for instance as modeled by the hardware (physical), software (agentive) and cyberware (information) distinction in cyber-physical systems, CPS (Horvath & Gerritsen, 2012), as well as the approaches to organizational structures falling within the view of socio-technical systems, STS (Borgo, 2019; Trist & Emery,

2005). Our ontological components take into account further elements which are not found in CPS and STS, like the specificity of the location/environment components, and also provides the elements to study complex relationships across components like the interpretation (knowledge) of physical changes (place) obtained by the agents (agentivity) via detecting and integrating (action) heterogeneous data.

We recognize that our analysis leaves open how cities should be classified in an ontology: are they processes, as claimed by Varzi (2019)? Or are they objects? And what kind of processes? What kind of objects? Without a proper understanding of the city ontological components and interactions, which we have started investigating here, it is hard to justify a choice.

From this viewpoint, the study of interactions is the next necessary step of our analysis. Interactions are generally studied within specific systems: engineering design for functional interaction across device components (Hirtz, Stone, McAdams, Szykman, & Wood, 2002), CPS for the integration and coordination across hardware, software and cyberware components (Borgo, 2019; Horvath & Gerritsen, 2012), biology for the gene and biochemical interactions (Kitano, 2002), multi-agent systems for coordination and consensus across communities of agents (Olfati-Saber, Fax, & Murray, 2007), and so on. The interactions manifested in a city combine most of these interaction types. Unfortunately, there is no general theory of interaction that subsumes all these particular uses but some way to distinguish interaction types have been proposed. For instance, Borgo (2019) looks at the participating entities to distinguish among natural interactions, design interactions, tool interactions, homogeneous interactions and heterogeneous interactions. Interaction types like these provide a first high-level classification in the study of technical and social systems, but need to be extended and organized into a complete framework. In the city, the participants in an interaction belong to one of the three components, or are systems decomposable in elements each of which belongs to one component. The components then tell us to which type of interactions these elements can participate. The next step of this ontological analysis starts to become clear: the development of a general classification of interaction types (and their relationships) to cover the city dynamics and evolution.

18.6 Complexity and the Ontological View of Cities

Cities show the typical extreme complexity of ecological (Alberti, 2005; Folke, 2006) and sociotechnical systems (Hillier, 2012; Hodson & Marvin, 2010). This kind of complexity is manifested in the diversified agentive, situational and temporal perspectives, which often have different semantic relevance (Arentze, Dellaert, & Timmermans, 2008). With the growing awareness of this complexity, the traditional (aggregative, deterministic, analytical) urban decision making and policy making activities have shown their partiality to the point of becoming useless, and therefore dangerous and rhetorical.

Nowadays, the analysis of urban organizations is exploring systemic and disaggregated approaches, aimed at identifying and representing characters, features, relationships, behaviors in formal theories and in simulation processes (Arentze & Timmermans, 2006; Batty, 2007; Gruner, 2010; Liu & Lathrop, 2002). Models of behavioral simulation, for example, are integrated to formal rules for agent interaction, including biotic as well as abiotic, natural, artificial agents. Similarly, situational representations have moved from standard domain-codified dimensions to inter-domain, perceptive and even cognitive representations, structurally including also temporal dynamics.

This evolution of city modeling has not overcome the intrinsic difficulty of managing such complex systems, for example, to envision conceptual and relational disaggregations for decision support architectures (De Roo and Silva, 2010). Interestingly, our analysis has shown that this modeling change is in agreement with the ontological view of the city, and pushes to study and classify elements, properties and relationships while maintaining independent representative and conceptual views. In principle, applied ontology helps to manage large formalized databases (quantitative and/or qualitative) in an interconnected, shared, interactive way (Chen, Sabri, Rajabifard, & Agunbiade, 2018; Rabino & Caglioni, 2007). The ontological analysis of complex urban systems can benefit significantly at different stages, for instance by highlighting relevant semantic and conceptual levels (Stufano Melone, Borri, Camarda, & Borgo, 2018).

It remains necessary to reflect more on how the complexity of the city system translates into the complexity of the ontological components (city-place, city-agentivity and city-knowledge). Certainly, the complexity of the physical component, with its intrinsically connected network of full and empty spaces, artifacts, materials, structures, natural elements as well as physically located artificial elements, is something we are better equipped to deal with. Similarly, the agentivity component requires to conceptualize and model agents and interactions of various types, essence, dimension, hierarchical, social, cognitive and relational relevance which is still challenging. Finally, the knowledge component determining the representation and understanding capabilities of phenomena, uses, functions, meanings, interpretative models of spatial contexts, obtained through observation, experience, abstraction and activity is a homogeneous part of the information organization and management research community. These three components are themselves complex in their own way but, overall, we are better suited in facing these types of complexity than the heterogeneous complexity of the city as a whole.

Finally, ontological analysis is at the center of the vision that allows to reconstruct the complex urban system from these three components, leading to analyze the interaction across them and, thus, to model in a second phase the complexity of the whole city. It is today still unclear how the analysis of the city system we have carried out can be complemented to deliver an approach to city complexity modeling and management. A more refined and in-depth ontological approach to the analysis of the city might need to be developed. We can say at the moment that the ontological view we have elaborated is useful to discriminate between clearly separate aspects and to develop from these a solid articulation of the city complexity problem and modeling perspectives.

18.7 Conclusions

An entity can be complex in different ways: for the number of parts it includes, for the variety of its components, for the hierarchies they form, for the ways they interact and so on. The complexity of cities is such that one can analyze them along any of these perspectives. These analyses, when carried out independently, lead to disassemble the very same system in ways that are incompatible with each other, hindering the possibility of recomposing the results in a unified view or at least in interoperable views. In the attempt to reduce, if not to avoid, this problem we turned to ontological analysis to cut the city at its ontological joints. More precisely, observing that a city is a system that includes distinct ontological features, we proposed to carefully disarticulate cities following their ontological characteristics.

In ontology, there are a few starting points and we found inspiration from the notion of spatial inclusion, the distinction between agentive and non-agentive, and the distinction between materiality and information. The first led to identify the notion of physical place, the second of agentivity, the third of knowledge. These are the essential components of a city due to their nature and their interrelationships.

What can this study tell us about smart cities? Our historical analysis highlighted the different factors that pivoted the understanding of cities in different periods of time. The ontological analysis that we carried out gives a framework to synoptically read these factors as indicators of the relative dominance of a city component (or a city relationship) over the others. Smart cities are clearly the sign of today's raising role of the knowledge component. Since at the moment smart cities focus primarily on data and their use with limited integration, which is just one aspect of the knowledge component, we can foresee that the changes due to the raising of smart cities will have an even larger impact in the future. Smart cities are today focusing on data acquisition and elaboration (the sentient city) for service provision and optimization of the city quality parameters, they thus focus on the simplest elements of the knowledge component. We anticipate an evolution towards the exploitation of more sophisticated uses of these data, in particular the development of new representation systems and interpretation mechanisms which can be directly tailored to individual and organization needs. With richer exploitation of data and more mature interpretations, the relationships with the other components will adapt, leading to enriched understanding of the modified place, and to the development of new agentivity capacities and, possibly, new or stronger social relationships.

We conclude noticing that with the smart city for the first time knowledge has taken the spotlight in the history of city, and has the potentiality to remain the pivoting component in the next city evolution.

References

Alberti, M. (2005). The effects of urban patterns on ecosystem function. *International Regional Science Review, 28*(2), 168–192. https://doi.org/10.1177/0160017605275160.

Albino, V., Berardi, U., & Dangelico, R. M. (2015). Smart cities: Definitions, dimensions, performance, and initiatives. *Journal of Urban Technology, 22*(1), 3–21.

Angelidou, M. (2014). Smart city policies: A spatial approach. *Cities, 41*, S3–S11.

Arentze, T., Dellaert, B. G. C., & Timmermans, H. J. P. (2008). Modeling and measuring individuals' mental representations of complex spatio-temporal decision problems. *Environment and Behavior, 40*(6), 843–869. https://doi.org/10.1177/0013916507309994.

Arentze, T., & Timmermans, H. (2006). Multi-agent models of spatial cognition, Learning and complex choice behavior in urban environments. In J. Portugali (Ed.), *Complex artificial environments*. Berlin: Springer.

Bai, X., McAllister, R. R. J., Beaty, R. M., & Taylor, B. (2010). Urban policy and governance in a global environment: Complex systems, scale mismatches and public participation. *Current Opinion in Environmental Sustainability, 2*(3), 129–135. http://periodici.caspur.it/cgi-bin/sciserv.pl?collection=journals&journal=18773435&issue=v02i0003&article=129_upagiassmapp.

Ballatore, A. (2016). Prolegomena for an ontology of place. In H. Onsrud & W. Kuhn (Eds.), *Advancing geographic information science* (pp. 91–103). Needham: GSDI Association Press.

Batty, M. (2007). *Cities and complexity: Understanding cities with cellular automata, agent-based models, and fractals*. Cambridge: The MIT Press.

Bauman, Z. (2006). *Liquid times: Living in an age of uncertainty*. Cambridge: Polity.

Bencardino, M., & Greco, I. (2014). Smart communities. Social innovation at the service of the smart cities. *TeMa - Journal of Land Use, Mobility and Environment*, 39–51. https://doi.org/10.6092/1970-9870/2533.

Bennett, J. W. (1996). *Human ecology as human behavior: Essays in environmental and development anthropology*. New Brunswick, NJ: Transaction Publishers.

Borgo, S. (2019). An ontological view of components and interactions in behaviorally adaptive systems. *Journal of Integrated Design and Process Science, 23*, 1–19.

Borri, D., Camarda, D., Grassini, L., & Patano, M. (2016). Technological change and innovation for sustainable cities: A multiagent-based ontological approach. In R. Papa & R. Fistola (Eds.), *Smart energy in the smart city: Urban planning for a sustainable future* (pp. 61–82). Cham: Springer. https://doi.org/10.1007/978-3-319-31157-9_4.

Bowen, W. M., Dunn, R. A., & Kasdan, D. O. (2010). What is "urban studies"? Context, internal structure, and content. *Journal of Urban Affairs, 32*(2), 199–227.

Calvino, I. (1974). *Invisible cities*. New York: Harcourt Brace Jovanovich.

Camagni, R. (1995). The concept of innovative milieu and its relevance for public policies in European lagging regions. *Papers in Regional Science, 74*(4), 317–340.

Camagni, R., & Capello, R. (2005). ICTs and territorial competitiveness in the era of internet. *The Annals of Regional Science, 39*(3), 421–438.

Camarda, D. (2010). Beyond citizen participation in planning: Multi-agent systems for complex decision making. In C. Nunes Silva (Ed.), *Handbook of research on e-planning: ICTs for urban development and monitoring* (pp. 195–217). Hershey, PA: IGI Global.

Camarda, D. (2012). Multiagent systems for the governance of spatial environments: Some modelling approaches. In B. Murgante, O. Gervasi, S. Misra, N. Nedjah, A. Rocha, D. Taniar, et al. (Eds.), *Computational science and its applications* (pp. 425–438). Berlin/Heidelberg: Springer. https://doi.org/10.1007/978-3-642-31075-1_32.

Carson, R., Darling, L., & Darling, L. (1962). *Silent spring*. Boston/Cambridge, MA: Houghton Mifflin/Riverside Press.

Castells, M. (1989). *The informational city: Information technology, economic restructuring, and the urban regional process*. Cambridge, MA: Blackwell.

Chen, Y., Sabri, S., Rajabifard, A., & Agunbiade, M. E. (2018). An ontology-based spatial data harmonisation for urban analytics. *Computers, Environment and Urban Systems, 72*, 177–190. https://doi.org/10.1016/j.compenvurbsys.2018.06.009.

Christaller, W. (1933). *Die Zentralen Orte in Suddeutschland*. Jena: Fischer.

Cleaver, F., & Toner, A. (2006). The evolution of community water governance in Uchira, Tanzania: The implications for equality of access, sustainability and effectiveness. *Natural Resources Forum, 30*(3), 207–218. https://doi.org/10.1111/j.1477-8947.2006.00115.x.

Cocchia, A. (2014). Smart and digital city: A systematic literature review. In R. P. Dameri & C. Rosenthal-Sabroux (Eds.), *Smart city* (pp. 13–43). Cham: Springer.

Commoner, B. (1971). *The closing circle: Nature, man, and technology*. New York: Knopf.

Cresswell, T. (2014). *Place: An introduction*. London: Wiley.

De Roo, PG., & Silva, EA. (2010). *A Planner's Encounter with Complexity*. Farnham: Ashgate Publishing, Limited.

De Waal, M., & De Lange, M. (2017). Owning the city: New media and citizen engagement in urban design. In K. Etingoff (Ed.), *Urban land use* (pp. 108–129). Oakville, ON: Apple Academic Press.

Deslatte, A. (2015). Reassessing "city limits" in urban public policy. *The Policy Studies Journal, 43*(S1). https://doi.org/10.1111/psj.12102.

Faludi, A. (1987). A decision-centred view of environmental planning. In *Oxford [Oxfordshire]*. New York: Pergamon Press.

Ferrero, F., & Vesco, A. (2015). *Handbook of research on social, economic, and environmental sustainability in the development of smart cities*. Hershey, PA: IGI Global.

Fiorino, D. J. (2010). Sustainability as a conceptual focus for public administration. *Public Administration Review, 70*, s78–s88.

Folke, C. (2006). Resilience: The emergence of a perspective for social-ecological systems analyses. *Global Environmental Change, 16*(3), 253–267. http://www.sciencedirect.com/science/article/pii/S0959378006000379.

Forrester, J. W., Meadows, D. H., Randers, J., & Behrens, W. W. (1972). *The limits to growth: A report to the club of Rome*. New York: Universe Books.

Fromhold-Eisebith, M. (2017). Cyber-physical systems in smart cities: Mastering technological, economic, and social challenges. In H. Song, R. Srinivasan, T. Sookoor, & S. Jeschke (Eds.), *Smart cities: Foundations, principles, and applications* (pp. 1–18). London: Wiley.

Garbacz, P., Lawrynowicz, A., & Szady, B. (2018). Identity criteria for localities. In *Formal Ontology in Information Systems: Proceedings of the 10th International Conference, FOIS 2018* (pp. 47–54). Amsterdam: IOS Press.

Geddes, P. (1915). *Cities in evolution*. Edinburgh: Williams.

Geertman, S., Ferreira, J., Goodspeed, R., & Stillwell, J. (2015). *Planning support systems and smart cities*. Berlin: Springer.

Glaeser, E. L., & Berry, C. R. (2006). Why are smart places getting smarter. In *Rappaport Institute/Taubman Center Policy Brief* (Vol. 2). Cambridge, MA: John F. Kennedy School of Government.

Gorban, A. N., & Roose, D. (2010). *Coping with complexity: Model reduction and data analysis*. Berlin: Springer.

Gruner, S. (2010). Mobile agent systems and cellular automata. *Autonomous Agents and Multi-Agent Systems, 20*(2), 198–233.

Guarino, N. (1998). *Formal Ontology in Information Systems: Proceedings of the First International Conference, FOIS 1998, 6–8 June 1998, Trento*. Amsterdam: IOS Press.

Guarino, N., & Welty, C. (2009). An overview on ontoclean. In S. Staab & R. Studer (Eds.), *Handbook on ontologies* (2nd ed., pp. 201–220). Berlin/Heidelberg: Springer.

Hall, R. E., Bowerman, B., Braverman, J., Taylor, J., Todosow, H., & Von Wimmersperg, U. (2000). *The vision of a smart city*. Upton, NY: Brookhaven National Lab.

Hillier, B. (2012). The city as a socio-technical system: A spatial reformulation in the light of the levels problem and the parallel problem. In S. M. Arisona, G. Aschwanden, J. Halatsch, & P. Wonka (Eds.), *Digital urban modeling and simulation* (pp. 24–48). Berlin/Heidelberg: Springer. https://doi.org/10.1007/978-3-642-29758-8_3.

Hirtz, J., Stone, R. B., McAdams, D. A., Szykman, S., & Wood, K. L. (2002). A functional basis for engineering design: Reconciling and evolving previous efforts. *Research in Engineering Design, 13*(2), 65–82.

Hodson, M., & Marvin, S. (2010). Can cities shape socio-technical transitions and how would we know if they were? *Research Policy, 39*(4), 477–485. https://doi.org/10.1016/j.respol.2010.01.020.

Hollands, R. G. (2008). Will the real smart city please stand up? Intelligent, progressive or entrepreneurial? *City, 12*(3), 303–320.

Horvath, I., & Gerritsen, B. H. (2012). Cyber-physical systems: Concepts, technologies and implementation principles. In *Proceedings of TMCE, 1* (pp. 7–11).

Howarth, C., Cornish, F., & Gillespie, A. (2015). Making community: Diversity, movement and interdependence. In G. Sammut, E. Andreouli, & G. Gaskell (Eds.), *The Cambridge handbook of social representations* (p. 179). Cambridge: Cambridge University Press.

Ismail, L., & Zhang, L. (2017). *Information innovation technology in smart cities*. Singapore: Springer.

Justinian. (530). Corpus Iuris Civilis.

Kahneman, D., & Tversky, A. (1979). Prospect theory: An analysis of decision under risk. *Econometrica, 47*(2), 263–292.

Kitano, H. (2002). Systems biology: A brief overview. *Science, 295*(5560), 1662–1664.

Krumholz, N., & Forester, J. (1990). *Making equity planning work: Leadership in the public sector*. New York: Temple University Press.

Levy Strauss, C. (1955). *Tristes tropiques*. Paris: Plon.

Liu, X., & Lathrop, R. G. (2002). Urban change detection based on an artificial neural network. *International Journal of Remote Sensing, 23*(12), 2513–2518. https://doi.org/10.1080/01431160110097240.

Losch, A. (1940). *Die Raumliche Ordnung der Wirtschaft*. Jena: Fischer.

Mann, T. (1901). *The Buddenbrooks*. Berlin: S. Fischer Verlag.

Marx, K. (1861). *Das Kapital. Kritik der Politischen Oekonomie*. Hamburg: Otto Meissner.

Matt, D. T., Spath, D., Braun, S., Schlund, S., & Krause, D. (2014). Morgenstadt: Urban production in the city of the future. In M. F. Zaeh (Ed.), *Enabling manufacturing competitiveness and economic sustainability* (pp. 13–16). Cham: Springer.

McClellan, S., Jimenez, J. A., & Koutitas, G. (2017). *Smart cities: Applications, technologies, standards, and driving factors*. Berlin: Springer.

McFadden, D. L. (1974). The measurement of urban travel demand. *Journal of Public Economics, 3*(4), 303–328.

Miller, D., & de Roo, G. (1997). *Urban environmental planning: Policies, instruments, and methods in an international perspective*. Aldershot: Avebury.

Mumford, L. (1934). *Technics and civilization*. New York: Harcourt Brace and Company.

Mumford, L. (1961). *The city in history*. New York: Harcourt Brace and World.

Murgante, B., & Borruso, G. (2015). Smart cities in a smart world. In T. R. Stamatina & M. P. Panos (Eds.), *Future city architecture for optimal living* (pp. 13–35). Cham: Springer.

Newman, P., & Jennings, J. (2008). *Cities as sustainable ecosystems*. Melbourne: Island Press.

Okner, T., & Preston, R. (2017). Smart cities and the symbiotic relationship between smart governance and citizen engagement. In H. Song, R. Srinivasan, T. Sookoor, & S. Jeschke (Eds.), *Smart cities: Foundations, principles, and applications* (pp. 343–372). London: Wiley.

Olfati-Saber, R., Fax, J. A., & Murray, R. M. (2007). Consensus and cooperation in networked multi-agent systems. *Proceedings of the IEEE, 95*(1), 215–233.

Pereira, G., Prada, R., & Santos, P. A. (2016). Integrating social power into the decision-making of cognitive agents. *Artificial Intelligence, 241*, 1–44. https://doi.org/10.1016/j.artint.2016.08.003.

Picon, A. (2015). *Smart cities: A spatialised intelligence*. London: Wiley.

Plato. (380 BC). Republic.

Portugali, J. (2011). *Complexity, cognition and the city*. Berlin: Springer.

Rabino, G., & Caglioni, M. (2007). Theoretical approach to urban ontology: A contribution from urban system analysis. In J. Teller, J. Lee, & C. Roussey (Eds.), *Ontologies for urban development* (pp. 109–119). Berlin/Heidelberg: Springer. https://doi.org/10.1007/978-3-540-71976-2_10.

Rittel, H. W. J., & Webber, M. M. (1973). Dilemmas in a general theory of planning. *Policy Sciences, 4*(2), 155–169. https://doi.org/10.1007/bf01405730.
Sassen, S. (2011). Keynote paper: Who needs to become 'smart' in tomorrow's cities. In *LIFT Conference: The Future of Smart Cities, Geneva, 2–4 February 2011.*
Schön, D. A. (1983). *The reflexive practitioner*. New York: Basic Books.
Schön, D. A., & Argyris, C. (1996). *Organizational learning II: Theory, method and practice*. Reading, MA: Addison-Wesley.
Schön, D. A., & Rein, M. (1994). *Frame reflection: Toward the resolution of intractable policy controversies*. New York: Basic Books.
Searle, J. R. (2006). Social ontology: Some basic principles. *Anthropological Theory, 6*(1), 12–29.
Severino, E. (2016). *The essence of nihilism*. New York: Verso.
Shakun, M. F. (1999). Consciousness, spirituality and right decision/negotiation in purposeful complex adaptive systems. *Group Decision and Negotiation, 8*(1), 1–15. https://doi.org/10.1023/a:1008620409745.
Simon, H. A. (1945). *Administrative behavior: A study of decision-making processes in administrative organization*. New York: Macmillan.
Simon, H. A. (1982). *Models of bounded rationality*. Cambridge, MA: The MIT Press.
Simon, H. A. (1991). Bounded rationality and organizational learning. *Organization Science, 2*(1), 125–134.
Stufano Melone, M. R., Borri, D., Camarda, D., & Borgo, S. (2017). *Knowledge of places: An ontological analysis of places and their semantic stratifications* (Plurimondi) (Vol. 17). Bari: Plurimondi Press.
Stufano Melone, M. R., Borri, D., Camarda, D., & Borgo, S. (2018). Knowledge of places: An ontological analysis of the social level in the city. In R. Papa, R. Fistola, & L. Gargiulo (Eds.), *Smart planning: Sustainability and mobility in the age of change* (pp. 3–14). Berlin: Springer. https://doi.org/10.1007/978-3-319-77682-8_1.
Sullivan, H. (2016). Interpretivism and public policy research. In N. Turnbull (Ed.), *Interpreting governance, high politics, and public policy* (pp. 184–204). London: Routledge.
Tacitus. (98 BC). De origine et Situ Germanorum.
Thucydides. (430 BC). History of the Peloponnesian War.
Torregrosa, M. L., Vera, J., Kloster, K., & Torres, B. (2017). *Water access technologies based on traditional knowledge in Mexico: Obstacles and strengths* (Plurimondi) (Vol. 7). Bari: Plurimondi Press. http://plurimondi.poliba.it/index.php/Plurimondi/article/view/17.
Townsend, A. M. (2013). *Smart cities: Big data, civic hackers, and the quest for a new utopia*. New York: W. W. Norton.
Trist, E., & Emery, F. (2005). Socio-technical systems theory. In J. B. Miner (Ed.), *Organizational behavior 2: Essential theories of process and structure* (pp. 169–195). London: Taylor & Francis.
Varzi, A. C. (2019). What is a city? *Topoi*. https://doi.org/10.1007/s11245-019-09647-4.
von Bertalanffy, L. (1968). *General system theory: Foundations, development, applications*. New York: George Braziller.
von Neumann, J. (1963). *Collected works*. New York: Pergamon Press.
Wagenaar, H., & Hajer, M. A. (Eds.). (2003). *Deliberative policy analysis: governance in the network society*. Cambridge, MA: Cambridge University Press.
Weber, M. (1986). *The city*. New York: The Free Press.
Xenophon. (370 BC). Anabasis.
Zins, C. (2007). Conceptual approaches for defining data, information, and knowledge. *Journal of the American Society for Information Science and Technology, 58*(4), 479–493.
Zipf, G. K. (1949). *Human behavior and the principle of the least effort*. Cambridge, MA: Addison-Wesley.

Stefano Borgo is Head of the Laboratory for Applied Ontology (LOA), a unit of the Institute of Cognitive Sciences and Technologies (ISTC) of CNR. He studied mathematics and computer science at the University of Padova, Indiana University and the Free University of Bozen-Bolzano. His research focuses on applied ontology and information modeling methodologies with applications in robotics, engineering, cyber-physical and socio-technical systems. Areas of application include engineering design, product and process modeling, laboratory data, urban planning and architecture. He is in the Editorial Board of Applied Ontology and the Semantic Web Journal, and member of the Advisory Board of the International Association on Ontology and its Applications (IAOA).

Dino Borri Full Professor (retired) in Town Planning at the Polytechnic University of Bari since 1994. From 1996 to 2009 he was Head of the Department of Architecture and Town Planning also teaching Town and Country Planning and Environmental Engineering until 2016. He is awaiting the Ministry appointment as Emeritus, meanwhile still working as supervisor to master and doctorate dissertations. Main scientific interests and research works are on planning theory, interactive planning, futures research, application of information technology to spatial planning, intelligent planning agents. He has been Scientific coordinator of many EU funded research projects among European and Euro-Asiatic Institutions on planning issues. He has written several books, more than a hundred scientific papers on international journals and conferences worldwide. He is currently head of the planning department in the Municipality of Brindisi (Italy).

Domenico Camarda is Full professor at Polytechnic University of Bari. From 2004 he has been teaching subjects related to Town and regional planning and engineering. He holds a master degree in Civil engineering, a Ph.D. in Spatial planning, and a M.Sc. in Economic policy and planning. His research interests are Environmental planning, Spatial cognition models in planning, Multi-agent planning models, Decision-support systems. He has been research fellow in EU funded research projects among European and Euro-Asiatic Institutions on planning issues. From 2012 to 2016 he served as Italian member of the Council of Representatives of the Association of European School of Planning (AESoP). He published about 80 papers in international journals and conference papers, one authored book, five edited books, and book chapters and several informational articles.

Maria Rosaria Stufano Melone is a Postdoctoral Researcher at the Polytechnic of Bari, Italy. In 2002 she graduated in Architecture at the University Institute of Architecture of Venice (IUAV). As winner of scholarship, she attended and in 2009 she earned a biennial post graduated master in Regional Public Management, organized by Formez and Regione Puglia. In 2011 she earned a Ph.D. in Science and Methods for the City and the Territory of Europe at the Faculty of Engineering of Pisa. In 2019 she earned the Master in Territorial and Environmental Planning at Polytechnic University of Bari. Her research interests are spatial cognition in environmental planning, memory and creativity management in urban planning and design, decision support systems, ontological analysis applied ontologies as method to manage knowledge in designing and planning processes. She has published research papers in national and international journals and conferences.

Chapter 19
Binding the Smart City Human-Digital System with Communicative Processes

Brandt Dainow

Abstract This chapter will explore the dynamics of power underpinning ethical issues within smart cities via a new paradigm derived from Systems Theory. The smart city is an expression of technology as a socio-technical system. The vision of the smart city contains a deep fusion of many different technical systems into a single integrated "ambient intelligence" (Ikonen et al., Emerging Technologies Report (No. D.1.2.). ETICA Project, 2010, p. 102). Citizens of the smart city will not experience a succession of different technologies, but a single intelligent and responsive environment through which they move (Dainow, Orbit J 1, 2017. Retrieved from https://www.orbit-rri.org/volume-one/smart-city-transcendent/; Australas J Inf Syst 21, 2017. https://doi.org/10.3127/ajis.v21i0.1438).

Analysis of such an environment requires a framework which transcends traditional ontologically-based models in order to accommodate this deep fusion. This chapter will outline a framework based on Latour's Actor-Network Theory (ANT) and Luhmann's treatment of society as an autopoetic (self-organising) system. We shall use this framework to map the influence of relevant factors on ethical issues, irrespective of their composition or type. For example, under this treatment, both human praxis and technical design can be viewed as comparable tools of domination.

This chapter will provide a framework for the analysis of relations between any elements of the smart city, ranging from top-level urban management processes down to individual device operations. While we will illustrate the use of this schema through examination of ethical issues arising from power dynamics within the smart city, it is intended that this example will demonstrate the wider utility of the model in general.

Keywords Ethics · Systems · City · Information · Power

B. Dainow (✉)
Department of Philosophy, Maynooth University, Maynooth, Ireland
e-mail: bd@thinkmetrics.com; Brandt.Dainow@mu.ie

M. Nagenborg et al. (eds.), *Technology and the City*, Philosophy of Engineering and Technology 36, https://doi.org/10.1007/978-3-030-52313-8_19

19.1 Introduction

This chapter offers a new way of viewing the complex phenomena which is humans living in a technologically advanced city. Using Actor Network Theory (Latour, 2005), the concepts of autopoiesis (Luhmann, 1995), symbolic capital (Bourdieu, 1986), affordances (Hutchby, 2001) and fields of contention (Bourdieu, 1990), a new framework will be developed which provides a way of describing people and technology as a single unified system. We will use this framework to characterise the system's nodes and patterns of connection, identify fields of contention and how they impact technology design, and derive consequent ethical issues.

The sophistication of modern technology is what separates modern cities from those of the past. Today's cities are becoming dependent upon advanced digital systems embedded throughout them. The life of a citizen in such a city is incomprehensible without understanding the technology of their built environment. We may therefore regard modern cities as complex socio-technical systems (Taylor, 2005). The future for technology in cities is held in the image of the smart city, in which intelligent digital systems and personalised services are dominate urban existence. If we can develop an all-encompassing framework to describe the smart city, we can use it to understand trends in urban technology today. By anticipating the ethical concerns in future smart city technology, we can identify current trends which may not be of concern today, but will give rise to issues in the future. Such knowledge offers the chance to change innovation paths now in order to prevent future issues developing in the first place.

The key characteristic of the smart city is the ubiquity of digital devices and systems. Smart cities are based on the foundation of an omnipresent sensor network which provides information about the inhabitants and environment (Balakrishna, 2012). These sensors will communicate with devices and systems to provide analysis and response. Some of these communication patterns can be predesigned, but some will appear and disappear as people move about ("ad hoc networking") (Guo, Wang, Zhang, Yu, & Zhou, 2013). Digital devices will be embedded in the civic environment, the home and other personal spaces (such as the car), worn on the person and implanted within the body (Perera, Zaslavsky, Christen, & Georgakopoulos, 2014). While some devices will be embedded in traditional materials, the rise of smart materials will often render the distinction between device and material meaningless (Addington & Schodek, 2005). Data will be processed locally ("fog computing") as well as in the cloud (Bonomi, Milito, Zhu, & Addepalli, 2012). People will be tracked individually and in aggregate. The urban environment will change in response to the movements and needs of individuals and groups. Ethical exploration of technologies is usually limited to issues associated with specific technologies, such as cars (Jaisingh, El-Khatib, & Akalu, 2016) or location detection (Martínez-Ballesté, Pérez-Martínez, & Solanas, 2013). However, many ethical issues emerge from the interaction of different technologies. Furthermore, smart cities will not be like today's cities with some digital technology laid over the top, any more than the modern city is just a Victorian one with cars instead of horses. The mature smart city will be a fundamentally different type of environment from any we have previously seen (Batty et al., 2012). Through all of history, the human built environment has been a largely dumb. It has not had the ability to initiate action, nor has it had the capacity to know us. The ability of the environment to obtain data and to respond to human action represents a fundamentally new type of built environment for human living (Picon, 2015). The number, ubiquity, heterogeneity and invisibility of digital systems in the urban fabric will force humans to deal with intelligent spaces, rather than individual devices (Rolim et al., 2016). We must therefore cease to talk in terms of technical components and instead develop a vision of the smart city which encapsulates all aspects of its operation. The starting point for analysis is to identify the functional characteristics which can unite the various heterogeneous technical forms found in the smart city. These are ambient intelligence, artificial intelligence and robotics.

Ambient intelligence is technology which embeds input, processing or response throughout the environment (Ikonen, Kanerva, Kouri, Stahl, & Wakunuma, 2010). As a lived experience, people will not experience the smart city as individual components and discrete processes isolated within atomised sections of their lives. Instead, the smart city will be experienced as a seamless experience as one moves from house to car to work, from one context to another (Dainow, 2017a). Similarly, smart cities will integrate data about people from multiple sources, creating a *digital environment*, a pervasive digital ecosystem which saturates the built environment, interacts with devices carried or worn by people and embedded within their bodies (Balakrishna, 2012) to create an integrated sensing and response environment (Psyllidis, 2015), an "intelligent information infrastructure" (ITU-T, 2008, p. 2), or ambient intelligence.

Central to the vision of the smart city is algorithmic intelligence, or "soft AI" (e.g.: expert systems and software agents) (Komninos, 2006). In addition to expectations that soft AI will provide core functionality through cloud computing, it has been suggested soft AI will need to be embedded at local level to support ad hoc networking (Komninos, 2006), to provide contextual awareness (Augusto, Nakashima, & Aghajan, 2010), and simply to handle the sheer amount of data (Komninos, Schaffers, & Pallot, 2011). As a result, the smart city will be filled with many AI systems. These systems will need to interact with each other and the human inhabitants in myriad ways. These AI's will constitute a complementary, but distinct, form of consciousness to the humans within the smart city.

'Robotics' refers to ICT devices possessing the ability to move autonomously (Ikonen et al., 2010). Self-driving cars and drones are expected to exist within smart cities (Jaisingh et al., 2016). Much building construction will occur robotically (Tibbits & Cheung, 2012). The home will see mobile IoT devices, such as robot vacuum cleaners (Cirillo, Karlsson, & Saffiotti, 2012). Public spaces, such as a hospital ward or shopping mall, will see many different types of robot moving about (Mastrogiovanni, Sgorbissa, & Zaccaria, 2010). The specific details of which robotic devices will exist within the smart city need not concern us here. The important point is that the smart city environment will involve both humans and machines moving within the same space. Just as the introduction of the car eventually led to the rise of pedestrian crossings, so an environment of moving machines will require adjustments in human behaviour to take them into account. Similarly, just as the car required the development of petrol stations, so the needs of the smart city's robotic inhabitants must inevitably require the development of specific urban structures and support systems.

The arrival of an intelligent, responsive environment containing autonomously mobile agents brings changes which must affect people. In this respect, the smart city will be created by its digital systems as much as by its human inhabitants. Device activity will respond to human activity. Humans will respond to device activity. Digital devices will have their own needs, many of which will require consideration by humans. This will sometimes bring the needs of the digital agent into conflict with the needs of the individual, just as cars require roads which gives them effective rights over people in some circumstances. Accordingly we must view digital devices as causal agents in a smart city on the same level as humans. If we are to develop a comprehensive view of the smart city, we must therefore include humans and digital systems as equals. Both are active agents capable of producing changes in the other.

Systems theory, long dominant in urban planning (Taylor, 2005), permeates approaches to the smart city (Söderström, Paasche, & Klauser, 2014), treating it as a complex system of systems (Bicocchi, Leonardi, & Zambonelii, 2015). Actor-network theory (ANT) (Latour, 2005) offers a systems approach which puts the smart city's digital agents into an interactive relationship with humans by regarding both as nodes in the same system. By treating both devices and humans as the same type of node within the same system, ANT provides a way of describing the causal effects digital agents have on humans and vice versa (Tabak, 2015). A further advantage of ANT is the lack of need to specify the details of each node within the network (Law, 1992). Instead, under ANT, agents may be a single digital device, an individual person, a group of people, a group of devices, or a mixture of these. However, ANT alone is not sufficient. The range and number of devices, their mobility, the deep interaction with humans, and the dynamic patterns of interaction generated as a result, means smart cities cannot be centrally managed, but must be self-organising, or *autopoietic*.

The concept of autopoiesis originated in biology to account for living organisms as self-sustaining systems (Varela, Maturana, & Uribe, 1981). Systems theory holds that the defining characteristics of a system are not determined by the properties of the components, but by the patterns of their relationships (Boulding, 1956). Autopoietic systems are those which can maintain or regenerate these patterns in the face of changing circumstances (Maturana, 1981). Niklas Luhmann introduced the concept of autopoiesis to the social sciences by treating society as a form of autopoietic system. Under Luhmann, the atoms from which social systems are built were *communicative events* between people. Luhmann defined communicative events as being composed of a sequence of information, understanding, and utterance (Luhmann, 1986).

This communicative atom is also visible in digital devices and systems. It is the minimum necessary characteristic of any digital device that it be able to accept input ("information"), process it ("understanding") and produce output ("utterance"). All digital systems are formed from groups of digital devices acting in such a fashion. However, smart cities will not be constituted by systems in which digital devices communicate only with each other. Humans and digital devices will engage in communicative patterns across the human-digital divide, as well as with each other. Furthermore, significant portions of human-to-human communicative activity will be mediated by the digital environment, while that digital environment's communicative patterns will respond in myriad ways to every human action. Under such circumstances, few nodes can be said to be purely human or digital; the overall patterns of communicative flow will be the result of both human and digital actants—an *Integrated Domain*.

It is important to bear in mind ANT is not explanatory. ANT is more method than theory, and the appellation of 'theory' has been declared misleading (Latour, 2005). The concept of the Integrated Domain is not intended to explain relationships or trace cause and effect. ANT provides a system of describing phenomena which cannot be attributed to exclusively digital or human causes, but does not seek to explain why or how. The descriptive framework offered here is intended to provide a language for such explanations.

19.2 Integrated Domains

An Integrated Domain is an autopoietic socio-technical system comprised of two collectives integrated into a mutually dependant partnership. One collective consists of a human *smart society*. A smart society is a human society which exists within an ambient digital environment, such that human intersubjectivity is mediated by digital technology (Hartswood, Grimpe, Jirotka, & Anderson, 2014). The other (non-human) collective is a system of digital devices and networks. Operationally, neither collective possesses strict boundaries against the other. Instead, they intermingle to such a degree it is impossible to account for phenomena within the Integrated Domain without reference to both. An example is the search engine manipulation effect, which has shown that the order in which a search engine lists items strongly influences the chance of these items being read and being believed (Epstein & Robertson, 2015). Search engines are significant cognitive tools for many people, which analyse people's behaviour in order to produce search results personalised to them (Dillahunt, Brooks, & Gulati, 2015). Thus, the user's knowledge of the world is mediated by logic created by their behaviour in combination with the search engine's algorithms. The user's knowledge of the world no longer derives from their own efforts, but from the inseparable fusion of human and digital. When the human and the digital are inseparable, we have entered the Integrated Domain.

The node which binds the Integrated Domain as a system is the *Integrated Node*, a unit of operation consisting of input, processing and output. In ANT terminology, an Integrated Node is a mediator, a node in a system which transforms what flows through it (Latour, 2005). An Integrated Node may be a person, a single device, a group of devices, a group of people, or any combination thereof. An Integrated Node is not defined by what it is made of, but by its mode of operation.

As the digital develops more cognitive capabilities through AI, so it's impact on the human becomes more varied and more complex. The internal nature and logic of these digital systems, their *logos*, therefore becomes more important in understanding their outcomes. As a consequence, it becomes imperative that we consider the digital perspective—how the world looks and is understood by digital systems (Picon, 2015). The terms 'looks' and 'understood' do not indicate anthropomorphism. Digital cognition is not the same as human cognition. However, if cognition is accepted as "the action or faculty of knowing taken in its widest sense" (Cognition, n., 2018), then it is clear that digital systems do possess a form of cognition. The most important aspects of the digital perspective are time, design, logic, epistemology and energy.

Descriptions of digital systems are inherently time-bound. Digital systems only "exist" to the degree that data flows through them. Without data on which to operate, a digital device is inert, no more able to cause effects in the world than a rock. Accordingly, a digital system is not best considered as a state but as a process (Dodge & Kitchin, 2005). For example, a service commences with anticipation of user requirements, organises processes to deliver the service to the user, is processed by the user (possibly mediated by other devices) and terminates in a stream of output data comprised of user activity and the recording thereof. Integrated Nodes are regions through which data and activity flow, defined by the manner in which they change what flows through them. Digital systems cannot be understood as actants by considering their material composition, only through understanding how they operate over time.

Digital constructionism is the view that there is nothing natural, and little inevitable, in the manner any digital system accomplishes its task. For example, it is often assumed that improving online security must require reducing online privacy, but these two values are not in competition. Enhancing privacy can actually make systems more secure (Langheinrich, 2001). The assumption that security and privacy are oppositional reflects current practice, not the necessary characteristics which must adhere in any and all digital systems (Dainow, 2015b). Any system which improves one by reducing the other is a result of decisions made by developers, who chose to code in such a manner. The algorithms, interfaces and other implementations of digital design are therefore to be understood as artificial and engineered social constructions, and always open to alternatives (Dong, 2004).

All digital systems stem from the place of toolmaking in the human. It is the primary defining characteristic of the digital that digital systems are created by humans for specific purposes. We do not create any digital systems simply so that they can exist, returning nothing to us. All digital systems are therefore inherently teleological systems. Any attempt to understand them must consider the purpose for which they exist. We can only arrive at a complete understanding of other goals, processes and values within a digital system by considering them in reference to the ultimate purpose for which that system was built. We must therefore consider digital systems in terms of whom their primary beneficiaries are and the benefits they derive. Processes within digital systems are driven by their *digital logos*, the heuristics, values and other elements which govern the design of digital cognitive processes and thus determine their outcomes. The nature of the digital logos is therefore important to those who desire control. Consequently, contention and power dynamics are concentrated around the digital logos in the form of competition for dominating values and heuristics (Albrechtslund, 2007). The overarching function of the digital logos is to serve the system's ultimate purpose. The ultimate purpose may be different from the obvious purpose. The obvious purpose is the reason why users make use of the system. For example, a commercial navigation system's obvious purpose is to guide a user to their destination. However, as a commercial system, its ultimate purpose is to generate profit. During design, elements of the system will therefore be evaluated against their profit-making capacity as well as their capacity to aid navigation. Potential elements which help the user but compromise profit generation will be unwelcome, and may not be included (Bogliacino & Pianta, 2013). The user may not get the best solution if such a solution would reduce profitability. All systems, especially those operated for profit, must therefore be regarded as, at least potentially, compromising the user in favour of some other goal. We can never assume that the user is always the primary beneficiary of any digital service.

By its very nature a digital device cannot know the physical world. What a device knows is only what its sensors generate as input. For example, an air conditioning control system does not respond to actual room temperature, but to what the sensors tell it the temperature is. If the sensor gives a false reading, the air conditioning will still respond to what the sensor tells it, not what the reality is. Hence digital devices do not know the physical world, but inhabit a totally digital environment. What they know of us is determined by our representation in digital data. Smart city systems will not respond to us; they will respond to our digital representations. If those digital representations are incorrect or incomplete, then the analysis of us and our needs will be compromised (Mittelstadt, Allo, Tadeo, Floridi, & Watchter, 2016). *Digital scepticism* is the epistemological position that digital systems can only detect that limited set of reality which they are designed to receive as input. If a system is not designed to process a certain input, then the corresponding aspect of the world does not exist for the digital system. Should this aspect of the world represent an important goal to the user, their attempts to achieve that goal will be frustrated. When combined with digital constructionism and the politics of the digital logos, the digital system's knowledge of the world becomes a product of stakeholder interests, not some "objective" view of reality.

Every process uses energy. This means we can validly apply many concepts related to energy to both digital processes and human interactions with digital systems. We can talk in terms of capacity, frequency, resistance and so forth. We can apply these same concepts to the human as well as to the digital because we do not depend upon any particular material composition for the concept of energy to be applicable. Consequently, we can discuss both digital and human systems and the interactions between them in the same terms. The amount of energy required to utilise a digital service is an important factor in its usability. Energy is expended for a purpose, as a cost towards achieving a goal. Goals have a cost threshold, beyond which the cost is not worth the goal. Increasing ease of use is a way of reducing the energy required for a task. This is a significant factor in the spread of digital technology and an important driver of innovation (Fontana & Nesta, 2009). By contrast, *nudging* is a process of gradually leeching energy through micro-thresholds in order to produce a strong cumulative effect, either to discourage input, prevent or influence processing, or reduce output (Johnson et al., 2012).

19.3 Contention and Affordances

As creations of human society and as sources of value, digital systems are inevitably included in the *fields of contention* (Bourdieu, 1990) which permeate other aspects of society (Pursell, 2007). Many of the characteristics of any digital system can be sites of contention (Mansell, 2017). Any aspect of a digital system can be designed to increase domination more than to serve the user. Characteristics of digital systems should therefore be considered as potential strategies for competition. Within the Integrated Domain digital actants are not passive spaces, but support or inhibit the competitive strategies of human actants. Bourdieu's theory of multiple forms of *capital* (Bourdieu, 1986) allows us to understand human-digital interaction as being engaged in generation, exchange and transformation of capital. Digital systems can produce multiple forms of capital—economic (OECD, 2013), cultural (Shifman & Nissenbaum, 2017), social (Jiang & Carroll, 2009) and political (Epstein & Robertson, 2015).

Generation of non-economic capital in digital systems occurs through *affordances*, which bring cognitive associations to sensory input. Any technology may be understood in different ways, according to a person's education, social environment and other factors (Hutchby, 2001), all of which are constrained by the capabilities of the technology in question (Norman, 2002). These dictate a person's affordances; how they perceive a technology and how they understand what they perceive. The concept of affordances holds that perception does not consist of a physical activity onto which understanding is overlaid *post hoc*. Instead, the perceptive process itself contains cognitive elements, such as motivation and context (Gibson, 1986). We use affordances to build conceptual models of how something works (Hutchby, 2001). There are always two conceptual models involved; the model the designers held when they constructed the artefact and the user's model when they use it. These models are never the same and user error or difficulty is usually the result. There is a significant body of analysis which argues that the design and operation of digital services today is primarily tuned to the exploitation of users to their disadvantage (Andrejevic, 2012; Vaidhyanathan, 2012). Owners use marketing to manipulate values so as to attribute symbolic capital to use of their systems (Bernays, 1928), such as creating socially aspirational personalities whose use of a system makes it desirable to their followers (Lamb & Kling, 2003). Designers of digital systems use control of development to determine which forms of capital are generated through use of their systems (Dainow, 2015a) and which are repressed (Cirucci, 2015). A common example is making privacy controls hard to locate in order to stop people using them (Liu, Gummadi, Krishnamurthy, & Mislove, 2011).

19.4 The Integrated Node

As a system, we do not understand the smart city's digital environment as being composed of states so much as processes. An Integrated Node is a node in an Integrated Domain which binds the system together through the activities of input, processing and output. This is the model for mediators of any composition within the Integrated Domain, whether composed of single or multiple actants, digital or human, or any combination thereof. Integrated Nodes are recursive. They are usually made from autopoietic systems of sub-nodes, each of which is an Integrated Node itself. Each of these will usually be constituted by their own autopoietic system of Integrated Nodes, and so on. For example, a digital device will contain CPUs and other types of chips, all of which accept input, process it and produce output, just as the device of which they are part does, and so these chips can be considered Integrated Nodes themselves. Similarly a person may interact with their digital environment through a combination of their physical body and carried digital "assistants" such as a smartphone. Here the Integrated Node is composed of both the human and the digital devices, both of which can individually accept input, process it and produce output, and thus are Integrated Nodes in their own right. Hence, the Integrated Node constitutes both the atomic unit and the essential pattern of interactions between these atomic units, transcending normally distinct ontological boundaries.

Many Integrated Nodes will be concerned with internal states of operation within both digital assets and humans. Much of the communication between digital agents concerns management of digital technology. This monitoring represents a form of digital self-awareness, though 'self-awareness' does not imply sentience. Meanwhile, the human element confers on the Integrated Domain the human level of understanding and meaning-giving. An Integrated Domain is therefore a self-aware system. We are already witnessing early examples of hybrid systems which combine human and digital devices. *Social machines* are "socio-technical systems which involve the participation of human individuals and technological components…able to extend the reach of both human and machine intelligence [by] supporting capabilities that less integrated systems might find difficult to accomplish" (Smart, Simperl, & Shadbolt, 2014, pp. 55–56). Social machines are early forerunners of the Integrated Domain's Integrated Nodes which will bind together the fabric of future smart cities.

The Integrated Node is both its own field of contention and a site within other fields of contention, in which attributions of symbolic capital determine value, some of which are convertible into economic capital. The initial basis for contention is the degree to which the user's needs align with the service provider's or designer's aims. Where the user is paying directly for the service and has a choice of service provider, the interest of the service provider is primarily to maintain the satisfaction of the user. However, where the service is free to users but paid for by other means, the service provider's concern for the user is merely to keep them using the service while value is derived from that usage of the system by other means.

Each person inhabits a digital environment comprised of a variety of devices and systems. Some of those devices are "personal" in the sense that they are worn, carried, or embedded in the body. In addition to providing direct services to the user, these devices mediate between the person and their digital environment. The smart city will not interact with people directly, but with this assemblage of human-with-personal-devices. At the level of the Integrated Domain, each individual is thus located within an *Integrated Personage* comprised of themselves plus their personal digital devices. Should one subscribe to theories of the extended self (Belk, 2013), an Integrated Personage may be regarded as a self which has been extended into the personal digital devices. However, the concept of the Integrated Personage is not dependent on the concept of an extended self, but on practical necessity—many digital systems are not designed to interact with humans, only with other digital technology, and so humans must use digital tech to access these systems. *Integrated Devices* are those devices used within the Integrated Personage. They have a special relation to the human by virtue of their role as critical mediators with the Integrated Domain and within the human smart society. They may also have, under some accounts, special relation to the user as extensions of the self, especially if they are embedded within the physical body. As a result they have a special ethical status not accorded to other owned objects.

19.5 Ethics and the Three Modes of the Integrated Node

We need an account of the technological city which does not function at the level of the thousands of individual technical systems of which it is made. Not only is such a level of analysis is too complex, it is misleading because technologies do not work in isolation but are experienced as a unified digital environment. Likewise, we need to avoid separating the digital from the human because it is often impossible to understand the one without reference to the other. In similar fashion, consideration of individual ethical concerns is to be avoided. Individual ethical concerns are many and frequently specific to particular technologies, usages or historical circumstance. The approach taken here is to instead identify a single point of ethical concern which is not dependent on specific details. This single point can be used to identify features of ethical concern which could be found in any digital environment. Autonomy is just such a foundational ethical value. It forms the legal and ethical basis for human rights and provides the justification for treating people as we do under international law and democratic politics (Dworkin, 1988). All rights given under the Universal Declaration of Human Rights derive from the premise that all humans have the capacity for autonomy (Marshall, 2008). The capacity for autonomy is what gives humans dignity and the right to be treated as an end in themselves, not a means to an end (Kant, 1998). Autonomy is thus the single point of ethical interest with the widest range of implications and shall therefore be our point of ethical concern.

However, there are many different definitions of autonomy (Dainow, 2017b). The term was initially restricted to morality by Kant, who argued that all humans had the innate capacity to determine what was morally right and wrong (Kant, 1998). This version of autonomy has become known as 'moral autonomy.' Since Kant we have added political autonomy (participation in political processes) and individual, or personal, autonomy (being the final determinant of one's actions) (Christman, 2015). In addition to these spheres of application, there are many different accounts of what constitutes autonomy in any sphere. Here the focus lies on the internal processes by which one exercises autonomy. The general concerns are to distinguish the degree of external influence which can be allowed before we can say autonomy has been lost and what, if any, specific cognitive elements are required in order for self-reflection to count as autonomous (Dainow, 2017b). Even the same definition of autonomy can lead to people living incompatible lives because they hold different values. However, it is possible to work with a generic understanding of autonomy without getting bogged down in these details by allowing that it is for each person to determine how they want to define autonomy for themselves. This means we need to assess digital environments in terms of the degree to which they enable or prevent the exercise of all forms of autonomy. At the level of the Integrated Domain, the Integrated Personage occupies the point of autonomy for the individual and so our account of the operations of the Integrated Node shall focus on the ways in which autonomy is in play within the input, processing and output phases of an Integrated Personage.

19.5.1 Input

Input may be defined as change at the interface between the Integrated Node and its environment. It is through design of digital interfaces that the most significant efforts at control occur within the Input phase. Interfaces are material elements which generate energy for internal processing. Data is a cognitive form of energy and may be analog or digital. When meaning is applied to data or use made of it, data becomes information (Ackoff, 1989). Different meanings may be applied to the same data, just as different uses can be made of it. The same data is thus capable of giving rise to variable amounts of information (Floridi, 2008). Someone's capacity to control their input is both empowered and constrained by the nature of the interfaces which exist between the human user and the digital environment. Control of interface design and operation is therefore fundamental to domination and commodification (or "monetisation") of the digital environment.

One cannot process something which has not entered the system. Input therefore determines the constraints of processing. Control of input is a fundamental determinant of the exercise of autonomy. No matter how one defines it, autonomy is always dependant on one's knowledge of the external world because it is with this knowledge that one makes deliberations. One important way in which input is restricted is through *epistemic control*. Epistemic control is the strategy of restricting the delivery of data or designing its delivery so as to hide information. A common example is the creation of privacy policies which are lengthy and difficult to understand. By doing so, the privacy policy becomes too large or too complex to be accepted as input by the user (McDonald & Cranor, 2008). Because the policy cannot be input, the website's privacy practices cannot be processed, and so the user cannot generate a response (output) which the service provider wishes to avoid (such as not using the service) (Acar, Van Alsenoy, Piessens, Diaz, & Preneel, 2015). Epistemic control is also exercised by simply hiding what is being done. For example, digital systems may make use of hidden technologies, such as Flash Long Storage Objects, which bypass privacy systems (Soltani, Canty, Mayo, Thomas, & Hoofnagle, 2010). User systems which attempt to preserve privacy by blocking cookies are an example of *epistemic contention*, in which two parties attempt to gain epistemic control of each other. In effect, the user's cookie-blocking technology is an attempt by the user to limit the tracking organisation's input, and so prevent the tracking organisation processing information about them, while the tracking organisation's attempt to hide their tracking is an attempt to prevent information about their practices becoming input to the user. This exposes an on-going conflict between service provider and user, in which each seeks epistemic control over the input of the other as a means of influencing their respective processing. A more subtle strategy for epistemic control lies through association and design, as we have seen with the search engine effect (Epstein & Robertson, 2015), whereby the mode of presentation of data at the interface can influence the affordances through which it is input to the human. More complex factors emerge in concerns as systems become more sophisticated. Factors such as cultural bias may mean the input systems for ambient intelligence, affective systems (which seek to detect emotion) and voice recognition force people into behaviours which are foreign to their culture in order to use the system (Soraker & Brey, 2007). Epistemic control can also be achieved through preventing restriction of input; effectively "force feeding" data into the Integrated Node. Advertising technology which bypasses ad blocking is an example of such force feeding, as is subliminal messaging (Verwijmeren, Karremans, Stroebe, & Wigboldus, 2011). The actions of data brokers in seeking to keep their very existence hidden (Federal Trade Commission, 2014) are an extreme example of epistemic control.

The designer of a digital system may restrict input intentionally or through ignorance. Where this is intentional, it may be done either to help the user, make life easier for the developer, or exploit the user. Where it is done to exploit the user, it constitutes an intentional attack upon their autonomy. Many theories of autonomy hold that autonomy is preserved if the person would have consented had they been given the opportunity to do so at the time (Dainow, 2017b), so hidden factors do not necessarily constitute a restriction on autonomy. However, preventing choice *post hoc* does constitute a restriction of the user's autonomy no matter what the action.

Where this prevents someone taking actions they would have chosen if they had possessed full information, it also constitutes a restriction on their freedom. Where input is restricted to suit the service provider it constitutes a secondary, but still intentional, attack on the user's autonomy, in that the user is not being considered as an end in themselves, but as a means to the service provider's end. Where input is restricted to help the user, it can be described as an accidental restriction on autonomy and should be weighed against the benefit derived. This happens fairly frequently. For example, it is often easier for users to restrict input by forcing people to choose from pre-set options, rather than allowing them to type freely. A free choice opens the possibility of unprocessable input, and so requires checking which may force the user to retry. As a result, restricting user's to pre-set choices is the preferred strategy in design.

19.5.2 Processing

Processing is the bridge between input and output. Processing therefore heavily influences which connections are made at output, although output is also constrained by material limitations. Input is what is processed. We use the term 'energy' for that which is input, including both data and physical activity (such as hearing aids and traffic lights). Because affordances append attributions of meaning and implied use, processing is strongly influenced by the affordances with which the input was received.

Processing may be described as the reorganisation of input energy patterns through manipulation. The heuristics of this manipulation are driven teleologically by the goals for which they are deployed. How input is transformed is also determined by the material composition of the Integrated Node and its internal processing structure. The processing structure may be divided into a material component and a cognitive component. Examples of the material component include biology and chip design. The material component has a determining effect on many aspects of the data, such as granularity, processing speed, computational complexity and so forth. The factors affecting processing are always defined so that they can be applied to a device, human, individual or group, no matter whether digital, organic or both. Since it is attributions of value which determine the heuristics of processing, contention occurs for attributions of value in an attempt to influence the processing. For example, placing values on using digital devices in certain ways (Zhao, Grasmuck, & Martin, 2008) is an attempt to influence the way a person processes information using these devices. In an Integrated Personage processing is also instantiated in communication patterns between the Integrated Devices, and between these devices and the human. For example, a GPS device communicates location information to a mapping device which then communicates instructions to a bluetooth headphone which then communicates sound to the human, who then changes direction and thereby communicates new location data to the GPS device. It is not only the data processed and transmitted which constitutes the processing, but the devices themselves as nodes within the autopoietic system which is the Integrated Personage.

Many of our existing issues with regard to ICT ethics today are focused on processing; issues of algorithmic justice, the moral status of autonomous systems, and use of personal data, are all examples of concerns over processing. One of the most important ethical issues here is the role of previous experience. It is indubitable that individuals bring their previous experience to their processing. Physical practice to develop muscle memory and trained reflex, central to music and sports, are a way of encoding previous experience as processing structures in the physical body. Education is a way of producing previous experience which can be used in cognitive processing. Previous experience is held to be a critical factor in processing by designers of personalisation systems, which analyse our history in order to predict and influence our future actions (Schneier, 2015). The internet economy today is based on knowledge of previous experience (OECD, 2013), as are all personalisation services of the future.

Because ANT is not an explanatory theory, it brackets how processing works. Being silent as to the internal details of processing, the exact mechanism by which human processing occurs may be described according to one's theories regarding the nature of the human mind. All that is required here is an acceptance that people have internal states of some form, determine their own actions in some way, and respond to changes in their environment. For our purposes, it is sufficient to say that processing involves the production of information and determination of action, and that input comes with a use value (Hartson, 2003), which implies how it is to be processed.

It is important to bear in mind that each human exists within the Integrated Domain at the level of the Integrated Personage, an autopoietic system composed of human + digital. The Integrated Personage is therefore the point of autonomy within the Integrated Domain, not the individual. Since autonomy involves self-governance (Kant, 1998), the ability to have control over processing within the Integrated Personage is necessary for autonomy in an Integrated Domain. The individual must be able to control processing to the degree that they are able to generate output intended to produce the states of affairs that the individual desires. It is not required that they are actually *able* to produce their desired states of affairs. Autonomy is not threatened simply because the world does not give in to whatever you wish. But the individual must be able to *try* (Christman, 2004).

A person may be unable to exercise autonomy in processing due to constraints created by designers. *Internal contention* is contention for the manner in which the Integrated Personage processes information. *External contention* is competition as to whether processing occurs within the Integrated Personage or is done by an external Integrated Node. Internal contention occurs through attempts to influence how the Integrated Personage conducts its processing. Simple examples include blocking interoperability with a device from a competing manufacturer, controlling what apps can be installed on a device, or restricting devices from accessing certain files or formats. A common technique in many systems is to discourage processing by increasing the energy required, for example the use of obscure legal terms in privacy policies. One has no difficulty physically inputting the words into one's mind. However, where the terms are not understood, the affordances providing semantic understanding at input are absent. In order to process these terms, one must look them up in a dictionary first. This raises the energy required to process the term significantly, and may reach a level which exceeds user's willingness to continue (McDonald & Cranor, 2008). It has been argued that is exactly the purpose behind using such terms (Schneier, 2015). Such design elements are examples of the way in which people can be nudged into changing the manner of their processing through design (Acquisti, 2009). When an operating system provider embeds their web browser into the interface in a manner which makes it hard to avoid using, as Microsoft once did (Fisher, 2001), they are similarly seeking to take control of an element of the processing within one's Integrated Personage.

External contention is contention over whether processing will occur within the Integrated Personage or within an external Integrated Node, such as a cloud service. Systems which upload data to the cloud for processing are transferring cognitive load from the Integrated Personage to the service provider's Integrated Node. Systems of rentier capitalism for Integrated Devices, such as software-as-a-service, represent an attempt to own portions of the individual's Integrated Personage. If books and similar digital services represent offloading of personal cognitive capabilities into digital devices, then statements of ownership over content, such as we see with Kindle (Amazon, 2017), constitute claims of ownership over the individual's cognitive capabilities. The argument from tradition that digital devices are optional tools is contrary to modern circumstances. The right to hack and to repair one's own devices contends with intellectual property initiatives and rentier capitalism for control of how individuals process within their own Integrated Personage. A digital service or device which is so commonly used that not having access to it impacts one's ability to participate in society constitutes a component of the smart society. Lack of access means one cannot fully participate in the smart society. As technology spreads deeper into the built environment, so it comes to assume certain minimum technological capabilities on the part of the inhabitants (Dodge & Kitchin, 2005). Increasingly, the city's digital environment has expectations that the Integrated Personage possesses certain capabilities—certain input and output channels and certain processing capabilities. Where an important aspect of the Integrated Domain becomes dependent on such capability, that capability becomes necessary for a full life and full participation in the smart society. At that point one acquires, as a citizen, a right to that capability. To the degree that this capability is necessary for the living of a full life, it becomes a human right.

19.5.3 Output

Output connects an Integrated Node with other Integrated Node's. One Integrated Node's output is the next Integrated Node's input. We can therefore talk in terms of an Output-Input channel (or *O/I*). For the purposes of this discussion, output may be defined as energy released from an Integrated Node in the form of physical movement or data. Some output is intentional, though cases of hidden surveillance show that output need not be originated by the user, or even known to them (Federal Trade Commission, 2014; Sandoval, 2012). Output also occurs as a by-product of existence because all digital processes leave a *data shadow* (Westin, 1970) in the form of temp files and logs.

Because the function of output within the Integrated Domain is to connect with other Integrated Nodes, output must be *coherent* with its environment. Here 'coherence' means compliance with shared standards, such as languages, protocols and data formats. The requirement for coherence grants power to those who can control it. A significant portion of the contention within digital technology is for control over standards because it confers domination over systems dependent on those standards (Cusumano, 2010). Control of connectivity accords one the status of a gatekeeper, able to control who uses an O-I channel and what they can send through it. The gatekeeper role in connectivity is one of the major points of financial return in digital services. Websites act as gatekeepers for connecting advertisers to users and extract value therefrom, charging an entrance fee for a presence in the user's input stream (Deighton & Kornfield, 2012). Telecoms companies generate income from connecting other digital systems to users (Huston, 1999). Debates over net neutrality show that many believe service providers can use this position as gatekeepers to extract economic capital (Pil Choi & Kim, 2010).

As techniques such as object-oriented programming (Rumbaugh, Blaha, Premerlani, Eddy, & Lorensen, 1991) and Digital Object Architecture (York, 2016) demonstrate, internal processing need not use the same formats as are required for output. As a result, output often requires a translation layer to convert internal information into formats which can be accepted as input by external Integrated Nodes. For output to humans, this may mean displaying information in a readable format, with consequent constraints on minimum text size, display time, using a particular human language, and conforming to graphical conventions in digital design (McKay, Christian, Matey, & Tufte, 1997). Coherence means providing digital devices with data in a format they can accept, hence the need to create keyboards for human output intended for digital systems. The Graphical User Interface (GUI) and similar systems, such as HTML, provide formats for both data and affordances (Galitz, 2007). API's, networking protocols and similar systems provide shared output formats which can bind digital devices into communicative systems. Physical output is seen in the activation of motors and switches, and the movement of muscles and limbs. Much of the intelligent functionality of the smart city is location-dependent (Picon, 2015), such as controlling traffic flow or delivering location-based services to inhabitants. The movement of the Integrated Personage within the digital environment will be accompanied by a constant flow of output based around movement through an urban space whose digital characteristics change as fast as its physical characteristics (Dodge & Kitchin, 2005). The Integrated Personage will manifest within the digital perspective as a moving cloud of output.

Output currently constitutes the primary point of value extraction for digital service providers in which the recording of user behaviour is monetised through data mining (O'Neil, 2017). As a result, commercial surveillance is one of the foremost ethical concerns regarding output. The ethical status accorded to human beings on the basis of their possession of autonomy (Marshall, 2008) confers special ethical status on data output by the Integrated Personage. This concern is greater where usage of output results in restrictions on autonomy or derivative rights. Autonomy is limited when people's digital systems do not allow them to emit the particular form of output they wish. For practical purposes, we cannot consider something a limitation on autonomy which is physically impossible. If it is not technically possible to instantiate my preferences, then my autonomy is not reduced. Nonetheless, in many cases, systems are designed to make outputting particular desires or in particular ways difficult or impossible—nudging or forcing people into using approximate variations which can be commoditised (Cirucci, 2015; Dainow, 2015a).

It is in the user's interest that their output is tuned to optimal delivery of the services they desire, in the manner they desire, under the conditions they desire. It is in the interest of a commercial service provider that they offer the services in the manner which generates the greatest income. These two interests are not necessarily aligned. Where they are not aligned the risk arises of reductions to autonomy through the use of that service. Where the service is non-essential we may say that autonomy is not threatened because one has the option of simply not taking the service. However, where the service is essential, or has achieved the status of a public utility (such that participation in society becomes difficult without it), then conditions under which one is forced to accept that service risk threatening autonomy. It is possible therefore, to treat certain aspects of service delivery as public utilities and impose limitations on allowable terms of service. It seems most in accord with the capitalist model to promote competition by requiring alternative service providers offering alternative terms of service. The construction of a smart city in which the citizen has access to alternative providers for each and every service at each and every location is feasible but would require an open technical infrastructure, similar to that offered by TCP/IP and HTML. Such open platforms enable any designer to create new systems in confidence their output will be coherent with the digital environment. No participant in the TCP/IP or HTML ecosystems can prevent a competitor arising (no one can become a gatekeeper to the web). Achieving this situation in the smart city would require standardisation of many protocols, but we can have done this before with many other standards (Mattli, 2001), and indeed it may be argued from history of utilities that it is almost inevitable that we will do so in the future (Clifton, Lanthier, & Schröter, 2011).

19.6 Summary and Conclusions

We have passed the point where technology can be treated as a dumb hammer, whose use and impact on others is totally dependent upon the human who wields it. Society, now and into the future, can only be understood by considering the human and the technological in combination. This combination results in systems giving rise to emergent features not predictable from the components. We therefore need a theoretical framework, a language, specifically designed to handle that fusion as a thing-in-itself, not as an analogue of traditional concepts. Actor Network Theory, combined with the concept of autopoiesis, offers the vision of an Integrated Domain in which the mediating nodes which provide the motive power and logic to the autopoiesis of the system are created of the interactions of humans with digital systems. The human and the built environment of the city do not directly encounter each other but interact through the medium of digital systems. The nature of their interaction is such that it cannot be split into independent components of the digital and the human, but combines them to the degree that they are indistinguishable. In terms of effects, they become one cause. They also give rise to emergent properties not predictable from their individual characteristics and not present without both.

As a dynamic system, the Integrated Domain consists of the transmission of energy in various forms through connective nodes which are mediators constituted by a three-phase process of input, processing and output. These nodes are therefore, in terms of their operational characteristics, events and processes more than they are objects and states. The ontological priority of process derives from the fact that Integrated Nodes cannot be understood, and do not exist, unless they utilise processes over time. It is true that digital devices will have a material composition, just as humans do. However, the material constituents are separable and understandable without reference to the other. By definition, the Integrated Domain is the ontological level at which the digital and the human are fused. That fusing occurs only as a process. Therefore, by definition, material composition and static states are not part of the Integrated Domain. It is only when nodes produce output and pass it as input to other nodes that the Integrated Domain comes into existence. Consequently, it is always through process and connection that we see the Integrated Domain.

Analysis of ethical issues and contention within the Integrated Domain must therefore focus on connectivity and processing, seeking to find mechanisms by which competing stakeholders pursue strategies for control and value extraction. We understand what is being transmitted through the system as energy, using the term ‘energy’ generically so as to allow for both material and cognitive (or symbolic or mental) transmission and activity. In empirical terms observed phenomena are almost always a combination of the two, just as they are a combination of the digital and the human.

From an ethical point of view, and to the individual, the most important Integrated Node is their Integrated Personage. The Integrated Personage is the active field within the digital environment constituted by the human in an inseparable fusion with their personal digital devices, such that the aggregated output from the Integrated Personage is a unified product of both human and digital, including not just the individual operations and functions of the component devices, but also the patterns and content of their interaction. The Integrated Personage, like any other Integrated Node, is an element of the Integrated Domain and therefore a combined entity in which the human and the digital cannot be treated as distinct.

Input overlaps with output. The output from one Integrated Node is the input for the next one. These output-input (O/I) connections constitute the systemic relationships between the nodes maintaining autopoiesis within the Integrated Domain. Output must be coherent with the digital environment in order to connect with it. This means that control over what constitutes coherence dictates control over how, and possibly what, can be output, and to whom. Examples include network protocols, programming languages and data formats. In the absence of contrary forces, we can expect contention over standards to continue as the smart city develops. Given the wide range of functions, spaces, contexts, volume and variety of digital devices and systems within the smart city, we can anticipate many more opportunities for such competition. On the basis of the history of digital innovation, we expect that multiple stakeholders will participate in this contention, few of whom can be assumed to have the user's best interests as their primary goal. In the absence of countervailing forces, we can therefore anticipate that the individual user of the service is unlikely to receive the best solution to the problem. Rather they are more likely to receive something just good enough to tolerate, while their use of the service is exploited, both in terms of the service delivery and the design of the systems. As the Integrated Domain rises to become more sophisticated and more the dominating feature of modern urban society, so the potential for domination by those who do not have the user at heart becomes more likely and potentially more damaging. We do not know what the future will be like, but by anticipating the flows within future urban socio-technical environments, we can identify where human autonomy will be threatened and how.

References

Acar, G., Van Alsenoy, B., Piessens, F., Diaz, C., & Preneel, B. (2015). *Facebook tracking through social plugins*. Brussels: Belgian Privacy Commission.

Ackoff, R. L. (1989). From data to wisdom. *Journal of Applied Systems Analysis, 16*(1), 3–9.

Acquisti, A. (2009). Nudging privacy: The behavioral economics of personal information. *IEEE Security & Privacy, 7*(6), 82–85.

Addington, D. M., & Schodek, D. L. (2005). *Smart materials and new technologies*. Oxford/New York: Routledge.

Albrechtslund, A. (2007). Ethics and technology design. *Ethics and Information Technology, 9*(1), 63–72. https://doi.org/10.1007/s10676-006-9129-8.

Amazon. (2017). Kindle terms of use. Retrieved June 22, 2015, from http://www.amazon.com/gp/help/customer/display.html?nodeId=200506200

Andrejevic, M. B. (2012). Exploitation in the data mine. In C. Fuchs, M. Sandoval, B. Kees, & A. Albrechtslund (Eds.), *Internet and surveillance* (pp. 71–88). New York, NY: Routledge.

Augusto, J. C., Nakashima, H., & Aghajan, H. (2010). Ambient intelligence and smart environments: A state of the art. In J. C. Augusto, H. Nakashima, & H. Aghajan (Eds.), *Handbook of ambient intelligence and smart environments* (pp. 3–31). London: Springer.

Balakrishna, C. (2012). Enabling technologies for smart city services and applications. In *2012 Sixth International Conference on Next Generation Mobile Applications, Services and Technologies* (pp. 223–227). https://doi.org/10.1109/NGMAST.2012.51.

Batty, M., Axhausen, K. W., Giannotti, F., Pozdnoukhov, A., Bazzani, A., Wachowicz, M., et al. (2012). Smart cities of the future. *The European Physical Journal Special Topics, 214*(1), 481–518. https://doi.org/10.1140/epjst/e2012-01703-3.

Belk, R. W. (2013). Extended self in a digital world. *Journal of Consumer Research, 40*(3), 477–500.
Bernays, E. L. (1928). *Propaganda*. New York, NY: Ig Publishing.
Bicocchi, N., Leonardi, L., & Zambonelii, F. (2015). Software-intensive systems for smart cities: From ensembles to superorganisms. In R. De Nicola & R. Hennicker (Eds.), *Software, services, and systems: Essays dedicated to martin wirsing* (pp. 538–551). Retrieved from). https://doi.org/10.1007/978-3-319-15545-6_31.
Bogliacino, F., & Pianta, M. (2013). Profits, R&D, and innovation—A model and a test. *Industrial and Corporate Change, 22*(3), 649–678. https://doi.org/10.1093/icc/dts028.
Bonomi, F., Milito, R., Zhu, J., & Addepalli, S. (2012). Fog computing and its role in the internet of things. In *Proceedings of the First Edition of the MCC Workshop on Mobile Cloud Computing* (pp. 13–16). New York: ACM.
Boulding, K. E. (1956). General systems theory-The skeleton of science. *Management Science, 2*(3), 197–208.
Bourdieu, P. (1986). The forms of capital. In J. G. Richarson (Ed.), *Handbook of theory and research for the sociology of education* (pp. 46–58). Westport, CT: Greenwood Publishing Group.
Bourdieu, P. (1990). *The logic of practice (R. Nice, Trans.)*. Stanford: Stanford University Press.
Christman, J. (2004). Relational autonomy, liberal individualism, and the social constitution of selves. *Philosophical Studies, 117*(1), 143–164.
Christman, J. (2015). Autonomy in moral and political philosophy. In E. N. Zalta (Ed.), *The Stanford encyclopedia of philosophy (Spring 2015)*. Retrieved from http://plato.stanford.edu/archives/spr2015/entries/autonomy-moral/.
Cirillo, M., Karlsson, L., & Saffiotti, A. (2012). Human-aware planning for robots embedded in ambient ecologies. *Pervasive and Mobile Computing, 8*(4), 542–561.
Cirucci, A. M. (2015). Facebook's affordances, visible culture, and anti-anonymity. In *Proceedings of the 2015 International Conference on Social Media & Society* (pp. 11:1–11:5). https://doi.org/10.1145/2789187.2789202.
Clifton, J., Lanthier, P., & Schröter, H. (2011). Regulating and deregulating the public utilities 1830–2010. *Business History, 53*(5), 659–672.
Cognition, n. (2018). In *OED Online*. Retrieved from www.oed.com/view/Entry/35876
Cusumano, M. (2010). Technology strategy and management: The evolution of platform thinking. *Communications of the ACM, 53*(1), 32–34.
Dainow, B. (2015a). Digital alienation as the foundation of online privacy concerns. *ACM SIGCAS Computers and Society, ETHICOMP Special Issue, 45*(3), 109–117. https://doi.org/10.1145/2874239.2874255.
Dainow, B. (2015b). Key dialectics in cloud services. *ACM SIGCAS Computers and Society, ETHICOMP Special Issue, 45*(3), 52–59. https://doi.org/10.1145/2874239.2874247.
Dainow, B. (2017a). Smart city transcendent—Understanding the smart city by transcending ontology. *Orbit, 1*. Retrieved from https://www.orbit-rri.org/volume-one/smart-city-transcendent/.
Dainow, B. (2017b). Threats to autonomy from emerging ICTs. *Australasian Journal of Information Systems, 21*. https://doi.org/10.3127/ajis.v21i0.1438.
Deighton, J., & Kornfield, L. (2012). Economic value of an advertising-supported internet ecosystem. Retrieved from Internet Advertising Bureau website: http://www.iab.net/insights_research/industry_data_and_landscape/economicvalue
Dillahunt, T. R., Brooks, C. A., & Gulati, S. (2015). Detecting and visualizing filter bubbles in Google and Bing. In *Proceedings of the 33rd Annual ACM Conference Extended Abstracts on Human Factors in Computing Systems* (pp. 1851–1856). New York: ACM.
Dodge, M., & Kitchin, R. (2005). Code and the transduction of space. *Annals of the Association of American Geographers, 95*(1), 162–180.
Dong, A. (2004). Design as a socio-cultural cognitive system. In *DS 32: Proceedings of DESIGN 2004, the 8th International Design Conference, Dubrovnik*.
Dworkin, G. (1988). *The theory and practice of autonomy*. Cambridge/New York: Cambridge University Press.

Epstein, R., & Robertson, R. E. (2015). The search engine manipulation effect (SEME) and its possible impact on the outcomes of elections. *Proceedings of the National Academy of Sciences, 112*(33), E4512–E4521.

Federal Trade Commission. (2014). Data brokers: A call for transparency. Retrieved from Federal Trade Commission website: https://www.ftc.gov/system/files/documents/reports/data-brokers-call-transparency-accountability-report-federal-trade-commission-may-2014/140527databrokerreport.pdf

Fisher, F. M. (2001). Innovative industries and antitrust: comments on the Microsoft antitrust case. *Journal of Industry, Competition and Trade, 1*(1), 41–52.

Floridi, L. (2008). A defence of informational structural realism. *Synthese, 161*(2), 219–253.

Fontana, R., & Nesta, L. (2009). Product innovation and survival in a high-tech industry. *Review of Industrial Organization, 34*(4), 287–306. https://doi.org/10.1007/s11151-009-9210-7.

Galitz, W. O. (2007). *The essential guide to user interface design: An introduction to GUI design principles and techniques*. New York, NY: Wiley.

Gibson, J. J. (1986). *The ecological approach to visual perception*. New York: Psychology Press.

Guo, B., Wang, Z., Zhang, D., Yu, Z., & Zhou, X. (2013). Opportunistic IoT: Exploring the harmonious interaction between human and the internet of things. *Journal of Network and Computer Applications, 36*(6), 1531–1539.

Hartson, R. (2003). Cognitive, physical, sensory, and functional affordances in interaction design. *Behaviour & Information Technology, 22*(5), 315–338.

Hartswood, M., Grimpe, B., Jirotka, M., & Anderson, S. (2014). Towards the ethical governance of smart society. In D. Miorandi, V. Maltese, M. Rovatsos, A. Nijholt, & J. Stewart (Eds.), *Social collective intelligence* (pp. 3–30). London/New York: Springer.

Huston, G. (1999). *ISP survival guide: Strategies for running a competitive ISP*. New York, NY: Wiley.

Hutchby, I. (2001). Technologies, texts and affordances. *Sociology, 35*(2), 441–456. https://doi.org/10.1177/S0038038501000219.

Ikonen, V., Kanerva, M., Kouri, P., Stahl, B., & Wakunuma, K. (2010). *Emerging Technologies Report* (No. D.1.2.). ETICA Project.

ITU-T. (2008). *Ubiquitous Sensor Networks (USN)* (No. 4). International Telecommunications Union.

Jaisingh, K., El-Khatib, K., & Akalu, R. (2016). Paving the way for Intelligent Transport Systems (ITS): Privacy implications of vehicle infotainment and telematics systems. In *Proceedings of the 6th ACM Symposium on Development and Analysis of Intelligent Vehicular Networks and Applications* (pp. 25–31). https://doi.org/10.1145/2989275.2989283.

Jiang, H., & Carroll, J. M. (2009). Social capital, social network and identity bonds: A reconceptualization. In *Proceedings of the Fourth International Conference on Communities and Technologies* (pp. 51–60). New York: ACM.

Johnson, E. J., Shu, S. B., Dellaert, B. G., Fox, C., Goldstein, D. G., Häubl, G., et al. (2012). Beyond nudges: Tools of a choice architecture. *Marketing Letters, 23*(2), 487–504.

Kant, I. (1998). *Groundwork of the metaphysics of morals (M. J. Gregor, Trans.)*. New York: Cambridge University Press.

Komninos, N. (2006). The architecture of intelligent cities: Integrating human, collective and artificial intelligence to enhance knowledge and innovation. In *2006 2nd IET International Conference on Intelligent Environments. IE 06. 1* (pp. 13–20). Athens: IET.

Komninos, N., Schaffers, H., & Pallot, M. (2011). Developing a policy roadmap for smart cities and the future internet. In *EChallenges E-2011 Conference Proceedings*. Florence: IIMC International Information Management Corporation.

Lamb, R., & Kling, R. (2003). Reconceptualizing users as social actors in information systems research. *MIS Quarterly, 27*(2), 197–236.

Langheinrich, M. (2001). Privacy by design—Principles of privacy-aware ubiquitous systems. In G. Abowd, B. Brumitt, & S. Shafer (Eds.), *Proceedings of the Third International Conference on Ubiquitous Computing* (pp. 273–291). Retrieved from http://www.vs.inf.ethz.ch/publ/papers/privacy-principles.pdf

Latour, B. (2005). *Reassembling the social: An introduction to actor-network-theory*. Oxford/New York: Oxford University Press.
Law, J. (1992). Notes on the theory of the actor-network: Ordering, strategy, and heterogeneity. *Systemic Practice and Action Research, 5*(4), 379–393.
Liu, Y., Gummadi, K. P., Krishnamurthy, B., & Mislove, A. (2011). Analyzing facebook privacy settings: User expectations vs. reality. In *Proceedings of the 2011 ACM SIGCOMM Conference on Internet Measurement Conference* (pp. 61–70). New York: ACM.
Luhmann, N. (1986). The autopoiesis of social systems. In F. Geyer & J. van der Zouwen (Eds.), *Sociocybernetic paradoxes: Observation, control and evolution of self-steering systems* (pp. 172–192). London: SAGE Publications.
Luhmann, N. (1995). *Social systems*. Stanford, CA: Stanford University Press.
Mansell, R. (2017). Imaginaries of the digital: Ambiguity, power and the question of agency. *Communiquer. Revue de Communication Sociale et Publique, 20*, 40–48. https://doi.org/10.4000/communiquer.2261.
Marshall, J. (2008). *Personal freedom through human rights law*. Leiden/Boston: Martinus Nijhoff.
Martínez-Ballesté, A., Pérez-Martínez, P. A., & Solanas, A. (2013). The pursuit of citizens' privacy: A privacy-aware smart city is possible. *IEEE Communications Magazine, 51*(6), 136–141.
Mastrogiovanni, F., Sgorbissa, A., & Zaccaria, R. (2010). From autonomous robots to artificial ecosystems. In H. Nakashima, J. C. Augusto, & H. Aghajan (Eds.), *Handbook of ambient intelligence and smart environments* (pp. 635–668). New York: Springer.
Mattli, W. (2001). The politics and economics of international institutional standards setting: An introduction. *Journal of European Public Policy, 8*(3), 328–344.
Maturana, H. R. (1981). The organization of the living: A theory of the living organization. *Cybernetics Forum, X*(2–3), 14–23.
McDonald, A. M., & Cranor, L. F. (2008). The cost of reading privacy policies. *ISJLP, 4*, 543.
McKay, S. R., Christian, W., Matey, J. R., & Tufte, E. (1997). *Visual explanations: Images and quantities, evidence and narrative*. Cheshire, CT: Graphics Press.
Mittelstadt, B. D., Allo, P., Tadeo, M., Floridi, L., & Watchter, S. (2016). The ethics of algorithms: Mapping the debate. *Big Data & Society, 3*(2). https://doi.org/10.1177/2053951716679679.
Norman, D. A. (2002). *The design of everyday things*. New York: Basic Books.
O'Neil, C. (2017). *Weapons of math destruction*. New York, NY: Broadway Books.
OECD. (2013). *Exploring the economics of personal data* (OECD Digital Economy Papers No. 220). Retrieved from http://www.oecd-ilibrary.org/science-and-technology/exploring-the-economics-of-personal-data_5k486qtxldmq-en
Perera, C., Zaslavsky, A., Christen, P., & Georgakopoulos, D. (2014). Sensing as a service model for smart cities supported by internet of things. *Transactions on Emerging Telecommunications Technologies, 25*(1), 81–93.
Picon, A. (2015). *Smart cities: A spatialised intelligence*. Chichester/Malden, MA: Wiley.
Pil Choi, J., & Kim, B.-C. (2010). Net neutrality and investment incentives. *The Rand Journal of Economics, 41*(3), 446–471.
Psyllidis, A. (2015). Ontology-based data integration from heterogeneous urban systems: A knowledge representation framework for smart cities. In *CUPUM 2014: Proceedings of the 14th International Conference on Computers in Urban Planning and Urban Management 2015*. Cambridge, MA: MIT.
Pursell, C. (2007). *The machine in America: A social history of technology*. Baltimore: John Hopkins University Press.
Rolim, C. O., Rossetto, A. G., Leithardt, V. R. Q., Borges, G. A., Geyer, C. F. R., dos Santos, T. F. M., et al. (2016). Situation awareness and computational intelligence in opportunistic networks to support the data transmission of urban sensing applications. *Computer Networks, 111*, 55–70. https://doi.org/10.1016/j.comnet.2016.07.014.
Rumbaugh, J., Blaha, M., Premerlani, W., Eddy, F., & Lorensen, W. E. (1991). *Object-oriented modeling and design*. Englewood Cliffs, NJ: Prentice-Hall.

Sandoval, M. (2012). A critical empirical case study of consumer surveillance on web 2.0. In C. Fuchs, M. Sandoval, B. Kees, & A. Albrechtslund (Eds.), *Internet and surveillance* (pp. 147–169). New York, NY: Routledge.
Schneier, B. (2015). *Data and Goliath: The hidden battles to capture your data and control your world*. New York, NY: Norton.
Shifman, L., & Nissenbaum, A. (2017). Internet memes as contested cultural capital: The case of 4chan's/b/board. *New Media & Society, 19*(4), 483–501.
Smart, P., Simperl, E., & Shadbolt, N. (2014). A taxonomic framework for social machines. In E. Mordini, V. Maltese, M. Rovatsos, A. Nijholt, & J. Stewart (Eds.), *Social collective intelligence* (pp. 51–85). New York: Springer.
Söderström, O., Paasche, T., & Klauser, F. (2014). Smart cities as corporate storytelling. *City, 18*(3), 307–320.
Soltani, A., Canty, S., Mayo, Q., Thomas, L., & Hoofnagle, C. J. (2010). Flash cookies and privacy. In *AAAI Spring Symposium: Intelligent Information Privacy Management, 2010* (pp. 158–163).
Soraker, J. H., & Brey, P. (2007). Ambient intelligence and problems with inferring desires from behaviour. *International Review of Information Ethics, 8*, 7–12.
Tabak, E. (2015). *Information cosmopolitics: An actor-network theory approach to information practices*. Waltham, MA: Chandos Publishing.
Taylor, N. (2005). *Urban planning theory since 1945*. London: SAGE Publications.
Tibbits, S., & Cheung, K. (2012). Programmable materials for architectural assembly and automation. *Assembly Automation, 32*(3), 216–225.
Vaidhyanathan, S. (2012). *The Googlization of everything: (and why we should worry)*. Berkeley, CA: University of California Press.
Varela, F. G., Maturana, H. R., & Uribe, R. (1981). Autopoiesis: The organization of living systems, its characterization and a model. *Cybernetics Forum, X*(2–3), 7–13.
Verwijmeren, T., Karremans, J. C., Stroebe, W., & Wigboldus, D. H. (2011). The workings and limits of subliminal advertising: The role of habits. *Journal of Consumer Psychology, 21*(2), 206–213.
Westin, A. F. (1970). *Privacy and freedom*. London: Bodley Head.
York, D. (2016). *Overview of the Digital Object Architecture (DOA)*. Geneva: ISOC.
Zhao, S., Grasmuck, S., & Martin, J. (2008). Identity construction on Facebook: Digital empowerment in anchored relationships. *Computers in Human Behavior, 24*(5), 1816–1836. https://doi.org/10.1016/j.chb.2008.02.012.

Brandt Dainow is an Assistant Professor in the Department of Philosophy at Maynooth University, where he specialises in philosophy in the Digital Age. Brandt completed his Ph.D. in the Computer Science department at Maynooth University and his undergraduate studies at the National University of Ireland and the Pontifical University of Ireland. His research interests lie in those philosophical areas underpinning human rights, such as autonomy and justice. His primary contribution is the development of Integrated Domain Theory, which applies systems theory and process philosophy to the Digital Age.

Chapter 20
Technology and the City: From the Perspective of Philosophy of Organism

Qian Wang and Xue Yu

Abstract Seen from the perspective of Philosophy of Organism, technical artifacts are generated from the process of giving them the biological features of organisms, from humans to natural objects. It shows that the technological system bears a certain degree of organic features. Meanwhile, as the place for living in human society, cities are not only corporate supports of *biological*, *artificial*, *social* and *mental organisms*, but also the platform of interaction among all types of those organisms. The constant development of cities is promoted by the interaction of those organisms. In contrast, the mutual restriction among them leads to different kinds of "urban faults", which have particularly stood out in modern cities. The rapid increase of complex artifacts results in the expansion of social organisms and the anxiety of mental organisms and further affects the health of biological organisms as humans. To achieve the goal of sustainable development of modern cities, it is important for technology to play the proper role in cities in order to adjust the scale, growth rate and functional orientation of all types of organisms. Its aim is to construct a harmonious relationship among biological, artificial, social and mental organisms. In this respect, the idea that "Tao models itself after nature" from Chinese philosophy will be inspiring because "nature" is just the harmonious state of all types of organisms. To conclude, the "Tao" of the city's mode of development mode is towards the harmonious state of organisms.

Keywords Cities · Organisms · Philosophy of organism · Tao · Harmonious state

---

Q. Wang · X. Yu (✉)
Department of Philosophy, Dalian University of Technology,
Dalian, People's Republic of China
e-mail: qianwang@dlut.edu.cn; yuxue@dlut.edu.cn

M. Nagenborg et al. (eds.), *Technology and the City*, Philosophy of Engineering and Technology 36, https://doi.org/10.1007/978-3-030-52313-8_20

20.1 Introduction

Within the framework of Philosophy of Organism, *organism* is the central concept. In the history of philosophy, many philosophers, such as Leibniz, Bergson, Whitehead, and Hans Jonas, discussed the essential features of organisms from different perspectives (Jonas 1984, 1996; Whitehead 1920, 1922). However, one of the most significant features of *organism* is that it has vitality and vigor, which is given little attention in previous work within the Philosophy of Organism and is now a focal point within it. The *vitality* of organism, shown in the process of its generation and evolution, means it has the functions of metabolism, self-adjustment, evolution, reproduction, etc., so as to develop by means of all kinds of chance. Therefore, the *vigor* of an organism signifies the degree of vitality and the capability of self-developing via adapting itself to the environment. It could offer a deeper and more comprehensive understanding of the nature and relations between all kinds of organisms. In the biosphere, degrees of vitality and vigor vary from one species to another, while within the same species, degrees of vitality and vigor vary from one living thing to another. For instant, some species and some organisms are superior to others as regards some capacities, such as capacity for action, capacity for fertility and capability for adapting to severe environments, which is the difference between vitality and vigor. This results in many different complex phenomena in biological organisms, such as the struggle for existence, natural selection and biological evolution. When humans give these organic features to human-made technical artifacts, social organizations and mental things, there emerges a more complex interaction among all kinds of *organisms*. Furthermore, the relation between technology and the city is the concentrated reflection of that interaction.

To be specific, humans give their organic features to technical artifacts and technical systems, which makes them extensions and supplements of organs' functions. In this sense, technical artifacts can be termed '*artificial organisms*' because there exists an organic interplay and/or restriction within their inner parts, whose design is driven by human's intention, purpose and controlling relationship. Thus, artificial organisms also have a lifecycle. They need to be upgraded, developed and innovated because their vitality and vigor imply the constant pursuit of becoming faster, stronger, more complex and more convenient.

Humans also give their organic features to various kinds of social organizations and institutions. This makes them the extension of organs' functions in a social sense. Herbert Spencer (1898), a French philosopher, regarded all social organizations as "*social organisms*." He noted that the division of labor in society resembles the labor division of an animal's organ, so social evolution could be explained by means of biological natural selection. Social organisms include families, family clans and affinities that are bound by a blood relationship, but also associations, parties, armies and even states, where there is no blood relationship. The vitality and vigor of social organisms implies that all kinds of social organizations and institutions can have more powerful functions, more efficient performance and stronger competitiveness.

Finally, humans give their organic features to all kinds of mental objects, which makes them embody human's mental functions. Those mental things can be called '*mental organisms*' and include human language structure, knowledge hierarchy and cultural tradition, within which exists an organic connection between all parts of those mental organisms. Furthermore, the vitality and vigor of these mental organisms implies that new thoughts, new concepts and new methods constantly emerge and spread widely. The mental and psychological elements in each individual's mind are connected to an organic integrity, which is the mental organism existing in the individual's mind. Whereas the common language, knowledge and cultural tradition of human society, as existing in the sense of Popper's "world III" (Popper, 1972), represent the mental organism from the macroscopic perspective.

It would seem that understanding artificial, social and mental organisms from the perspective of Philosophy of Organism is similar to the principle of bionics but there is, in fact, a huge distinction. Artificial, social and mental organisms have their own particular characteristics and operational rules, although they have similar basic features and similar evolutionary mechanisms of vitality and vigor to biological organisms. While Philosophy of Organism focuses on the common nature, rules and relations among all kinds of organisms, it does not simply assimilate the features of biological organisms to other kinds of organisms.

In Sect. 20.1, we firstly illustrate the organic relation between technology and the city. As the place of intensive social life of human beings, cities are not only corporate supports of biological, artificial, social and mental organisms for their intimate interaction, but also the platform of competition and coordination among these organisms. Technical systems play the driving role within the platform and result in dynamic harmony among four kinds of organisms in cities. In Sect. 20.2, we analyze the origin of "urban faults" from the perspective of Philosophy of Organism. "Urban faults" refer to some phenomena in the development of the urban, such as high density populations, low per capita resources, traffic jam, high-price cost of living, fast pace of life, serious environmental pollution, and so forth. From the perspective of Philosophy of Organism, the reason why "urban faults" appear is the mutual restriction among these organisms. Furthermore, the evolution of artificial organisms determined by technical progress plays a decisive role in the development of cities. In Sect. 20.3, we provide an alternative approach to eliminating some "urban faults" during the development process of modern cities, that is the *Tao* of sustainable urban development. It aims to explore the dynamic model of coordination of various kinds of organisms in the city, one not only needs to guarantee the vigor and vitality of each type of organism, but also ensure that the development of various types of organisms are promoting, rather than constraining, each other rather. In the final section, we emphasize that pursuing the "Tao" of cities in the context of Philosophy of Organism will promote communication and dialogue in the field of urban development between Chinese and Western philosophy and help establish a harmonious relation between technology and the city in order to realize the sustainable development of cities.

20.2 The Organic Relation Between Technology and the City

Since the Industrial Revolution, technological progress has become the most active factor in cities, while the relationship between technology and cities has become more complex. Consequently, it is necessary to emphasize the driving role of technical systems from the perspective of Philosophy of Organisms in order to investigate the organic relations between technical systems and urban systems and consider the resulting dynamic harmony among four kinds of organisms in cities.

20.2.1 The Interrelations Between Technical Systems and Urban Systems

The technical system (which is determined by technological progress) is an assembly composed of different kinds of artificial organisms. As an evolving artificial system, it is a subsystem of the complex macrosystem that is the urban. As a specific assembly of a variety of artificial organisms, the technical system itself possesses certain organic features. According to Ellul, a French philosopher, the development of technology presents a characteristic of autonomy. "Technique has become autonomous, creating its own devouring world" (Ellul, 1954, p. 12), which Ellul claims is the epitome of the features of artificial organisms. In addition, as the platform of intimate interaction among four kinds of organisms, the urban system also possesses a certain degree of organic features, which are both synthetic and comprehensive. However, these organic features emerge from the interaction of four kinds of organisms in the city, while the development of each kind of organism is relatively independent. That is why there exists relations of interaction and/or restriction between these organisms, and why they need dynamic coordination. In other words, the technical system is the most active and dynamic subsystem of the urban as a complex macrosystem, and constantly interacts with other subsystems in the city by means of organic relations. The technical system is restricted by those subsystems at the same time as it affects them.

However, the traditional understanding on the notion of the city mainly focuses on the social structure and functions, while paying little attention to the organic meaning of the urban system and the role of artificial organisms therein. The technical system is usually regarded as an exogenous factor of the complex urban system; the means of promoting the development of the city but not an organic part of the city itself. With regard to the definition of the city, there exist different understandings within different contexts. From a demographic perspective, cities are defined by population characteristics of the city; concentration, size, density and heterogeneity (Berry, 1973). From the perspective of economics, cities are considered on the percent of non-agricultural work, particularly on the marketplace and trade-commercial relationships. Jacobs (1969) indicates that developed cities, as concentrated economic innovation centers, helped to stimulate the development of surrounding areas. Moreover, as political and administrative centers, cities can also be defined according to legal or administrative criteria (Marcotullio & Solecki, 2013). Likewise, from a cultural perspective, Lewis Mumford (1961) considers that "the chief function of the city is to convert power into form, energy into culture, dead matter into the living symbols of art, biological reproduction into social creativity" (Mumford, 1961, pp. 568–576). These definitions have involved biological, social and mental organisms of the complex urban system from different perspectives, but paid little attention to artificial organisms in cities. In fact, the city contains not only a large number of enterprises which manufacture and use these artificial organisms, but also a variety of infrastructural systems, such as traffic, power supply, water supply, heat supply, gas supply, drainage, cable television and telecommunications, which are themselves artificial organisms with certain specific functions. Cities, which have the features of a dissipative structure, constantly exchange sources, energy and information with surroundings. The possibility of this exchange is realized directly by artificial organisms. By virtue of regulating and controlling through social organisms, cities could keep themselves well organized and stable, which also needs the help of artificial organisms. A complex urban system could sustain itself under different conditions to reach homeostasis, which is analogous to a biological organism (Van den Dobbelsteen, Keeffe, Tillie, & Roggema, 2012). However, uncoordinated development of these four kinds of organism will challenge this homeostasis in cities. Once the technical system disordered, it will shock the balance of the urban system and increase the difficulty of regulating its stability.

20.2.2 The Driving Role of Technical Systems in Urban Development

It is noteworthy that, in modern cities, the development speed of artificial organisms is, generally speaking, much faster than other types of organisms. The technical system plays a decisive driving role in cities. This means that the development of biological organisms, social organisms and mental organisms always refers to the development of artificial organisms. Cities have to provide suitable living conditions and environments for an increasing population. It is necessary to update infrastructures utilizing technical systems for this purpose; to create more opportunities for the basic necessities of life and to improve the conditions of medical care in order to enhance the vitality and vigor of biological organisms. Furthermore, cities have to guarantee the well-ordered and efficient operation of all social organisms and therefore motivate the vitality and vigor of social organizations in the fields of politics, economics, military, education and culture. This requires improvement in the working conditions and environment of social institutions, in order to promote the convenient and efficient exchange of information so as to raise the working efficiency with appropriate technical methods; in other words, to drive the development of social organisms by means of the updating of artificial organisms. In addition, the construction of cities has to guarantee the healthy growth of all mental organisms in order to stimulate the vitality and vigor of mental organisms, which need to construct a better atmosphere for their study, communication and innovation. Humans could get the latest news by virtue of technological means and learn knowledge from remote distances, as well as enrich the mental world through all kinds of entertainments. A high-tech city not only provides a better human settlement, but also gathers high-level administrative organizations, social groups, enterprises and universities. This should include high-level exchange of thoughts and spiritual enjoyment. All of these profit from technical systems by means of their constant driving role on the changing and developing of other types of organisms.

The driving role of technical systems in the development of cities is reflected in the construction of smart cities over recent years. The core idea of smart cities is to pursue high efficiency, high-end economy, better life of people and more beautiful city environment with the contribution and benefits of information and communication technology (ICT). The expanding impact of AI (Artificial Intelligence) systems is the typical example of the technical system that exerts influence in the development of modern cities. In a comparative study on Smart Cities in Europe and China (CAICT & PDSF, 2016, p. 6) the following criteria for smart cities have been identified:

- Smart energy production and conservation
- Smart mobility
- Smart economy
- Smart living
- ICT economics
- Smart environment
- Smart governance
- Standard of living
- Smart society

The basis of formulating these criteria is a smart technical system. Only by relying on the promotion of technology can the development of smart cities be facilitated. Various elements in smart cities, such as smart economy, smart energy production and conservation, smart environment, smart living, smart mobility and smart governance, could be evaluated by the standards of ICT. Smart technological systems will accelerate the progressive development of the vitality and vigor in all kinds of organisms, which will bring about a lot of unexpected features in the landscape as modern cities develop.

20.2.3 The Dynamic Harmony Among All Kinds of Organisms in Cities

The dynamic harmony of all kinds of organisms is a constant process because these organisms are relatively independent but also have an organic connection of interaction and restriction. The increasing demands of biological organisms, social organisms and mental organisms will enlarge the demand of artificial organisms. However, if these artificial organisms in the urban center are demanding to such a degree that the developing space becomes limited, their function and efficiency will be reduced. In that case, the demand of biological organisms, social organisms and mental organisms should be adjusted. This adjustment focuses not only on slowing down the developing speed of artificial organisms, but also on resolving contradictions among all kinds of organisms by virtue of technological and institutional innovation. In the preliminary stage of Industrial Revolution, large-scale machine production brought about severe environmental pollution in cities, as well as the problem that workers couldn't adapt to the high speed of machines, both physically and psychologically, which caused a series of social contradictions and conflicts. Nonetheless, with the advance of technology and the adjustment of other types of organisms, especially the widespread use of human engineering, the primary uncoordinated state had changed and gradually turned into a relative coordinated state.

The dynamic harmony among all kinds of organisms in the city is morphologically analogous to the dynamic adjustment of biological organisms themselves. However, the dynamic adjustment of biological organisms is based on the mechanism of physiology, while the dynamic coordination of other kinds of organisms, such as artificial organisms, social organisms and mental organisms, is borne by technological, social and thinking (mental) rules. When different kinds of organisms interact with each other, some more complex relations of powering and restriction emerge, which constantly challenge the dynamic harmony of the urban life. For example, the construction of infrastructures in cities reflects the growth of artificial organisms, but the construction of those infrastructures also needs the corresponding economic input and social administration, as well as depending on whether administrators and citizens accept it. This is so complex as to be far beyond the remit of technology and engineering. Disharmonious dynamic relations between different types of organisms in the urban life, if it is protracted for a long time and rising, will lead to different kinds of faults; which is similar to the diseases caused by the malfunction of biological organisms. Some of these problems, termed as "urban faults", indicate a variety of faults that bring different forms of difficulty to human production and daily life within the urban. Analyzing the causes and the trends of "urban faults" from the perspective of Philosophy of Organism is beneficial for identifying its core.

20.3 Analysis of "Urban Faults" from Philosophy of Organisms

From the perspective of Philosophy of Organism, "urban faults" caused by the relations between technology and the city include the following elements.

20.3.1 The Expansion of Technology and the Malposition of Cities

The expansion of technology in cities does not mean the constant progress of infrastructures and technological applications, but refers to the ignorance of other aspects in social life and immoderate increase of technology application, which is mainly reflected in vast use of technical products and rapid construction of technical systems. Many cities have, or are now carrying out, the expansion of technology in order to satisfy all kinds of citizens' demands, such as intensively constructing high buildings, actively encouraging private cars, constantly promoting the update of domestic appliances and unlimitedly supplying energy sources. This involves a variety of workforces during the expansion of technology, which requires the establishment of corresponding social organizations (e.g. hospitals, communities, schools, etc.) in order to manage these laborers, as well as the establishment of corresponding communication platforms (e.g. television stations, libraries and parks) so as to enrich the citizens' cultural life. The expansion of technology and the increase in social demands results in the natural space in the city becoming increasingly tense, the pace of urban life becomes increasingly fast, the space for free action increasingly narrow and economic activities in cities increasingly frequent. In a city with a high population density, high speed of life and a narrow space of free activities, once the technical system malfunctions, there is little leeway to resolve these malfunctions. As a result, accident risks and negative effects caused by technologies will become significant dangers.

Unlimited expansion of technology will consequently lead to the malposition of cities. When the city scale is relatively small and the city space is sufficient, the expansion of technology will quickly and notably improve the situation of biological organisms, social organisms and mental organism*s*. It will significantly improve the living standard of citizens and rapidly accelerate the economic development of enterprises, as well as continuously amplify the influence of the different social organizations. More individuals and social organizations will be attracted to the city and will struggle for survival and development by means of the expansion of technology, which will give rise to the malposition of cities. The fact of narrow living space, increasing price of commodities and housing, traffic jam, air pollution, etc. lowers the quality of life of many citizens and reduces the work efficiency of many social institutions.

Variation of the vitality and vigor of the various organisms within the city is rarely a concern. When it comes to the level of modernization of a city, focus will be put on the degree of development of infrastructures, the per capita GDP in cities and the degree of socio-cultural influence. Even when the phenomenon of malposition, due to the extremely big and rapid development of the city scale, emerges, the expansion of technology in most cities is still intensified while the attraction of metropolis and megacities is not reduced. Confronted with malposition of cities, humans prefer to solve it with technological means. They expect to solve these "urban faults" by constantly developing new products and new technologies, such as constructing higher buildings, more flyovers and tridimensional parks. In fact, this causes more problems in the development of cities.

20.3.2 Mutually Restriction Among All Kinds of Organisms Caused by the Expansion of Technology in Cities

The rhythm of technical systems, as the specific assembly of artificial organisms, decides the rhythm of other kinds of organisms in cities. The restriction among these organisms will appear if the rhythm of biological organisms, social organisms and mental organisms can't match that of artificial organisms. Humans in cities might have a stronger feeling of anxiety and urgency and their mental organisms might be under a stronger stress. The reason lies not only in the stronger competition within social organisms, but also in a faster rhythm of artificial organisms. Chronic stress and excessive working pressure will definitely damage the health of biological organisms, which is why some chronic conditions, formerly confined to middle-aged and elderly people, are seen now among younger generations. Living in a tense social and technological environment, people will unconsciously quicken their steps of self-development. Once they find their steps behind others, it will generate a sense of anxiety. For example, using private cars should save people time and energy, but it means more time on the road and consumes more energy due to traffic jams. Furthermore, travelling by cars for a long time will damage the health of their bodies. Information technology provides all kinds of possibilities for communication and shortens the physical distance between people, but widens their psychological distance. It makes interpersonal communication in real world more difficult while making communication in the virtual world easier. Nowadays, with the advent of smart cities, much work that was formerly undertaken by workers is now done by intelligent machines, such that it becomes a social problem to re-employ those displaced laborers. Therefore, while the rapid development of *artificial organisms* brings a lot of convenience, at the same time it leads to some negative effects on biological organisms, social organisms and mental organisms and restricts their development.

In return, the problems brought about by these organisms will also impact the development of artificial organisms. If biological organisms in the cities are faced with serious environmental pollution and technological risks, machines that produce a lot of contaminants, and productions which bring about a lot of contaminants, should be limited, while unqualified commodities should be inhibited so as to prevent accidents caused by high-risk engineering projects. Consequently, the speed of technological development has to slow down. If social organisms go wrong and give rise to the social instability and the disequilibrium of economic development, technical systems will run out of financial support, efficient management and a social environment beneficial to technical development and will eventually restrict the improvement of technological standards. If mental organisms go wrong, humans' technological innovation abilities will decrease, updating of artificial organisms will become more difficult, and some old-fashioned machines will be rendered obsolete in such a competitive market.

The mutual restriction between different classes of organisms in cities caused by the expansion of technology will cause human, material and economic inputs involved in the expansion of technology to fail. Some metropolises that are formed rapidly are not equipped with enough economic strength and corresponding management ability. Some of those booming metropolises that are enjoying a healthy momentum of development in the initial period of construction are losing their vigor and vitality afterwards, which is a negative consequence of "urban faults".

20.3.3 The Difficulty of Governing the Formation of "Urban Faults"

Generally, "urban faults" emerge in modern cities on a large scale. In the West, some developed countries experience "urban faults" of different levels in different periods. For instance, London had been a serious fog city, New York was one of the most serious cities with traffic jams, Tokyo also faced a series of problems caused by rapidly increasing populations, such as high price of housing, polluted environment and growing cost of living (Bettencourt & West, 2010). In China, "urban faults" usually show up in metropolis, such as Beijing, Shanghai, Guangzhou, Shenzhen, etc. The common characteristics of these cities are high density populations, low per capita resources, traffic jam, high-price cost of living, fast pace of life, serious environmental pollutions, and so forth. In some senses, the "urban faults" in some developed areas have been relieved or solved. However, in many booming cities, "urban faults" are still coming into being, while some solved "urban faults" will reappear or worsen in some new circumstances. This shows it is hard to solve an "urban fault" completely once formed. Many methods to govern "urban faults" are making some cosmetic changes but falling into one fault by avoiding another. Actually, the key here is that the root of generating "urban faults" has not been solved, which involves the problems of ideas and patterns of city development.

One of the reasons why "urban faults" still exist is that, when people discuss and analyze all different kinds of "urban faults", they usually regard cities as a synthesis formed from a variety of fields, such as politics, economics, technology, education, culture, etc., according to the traditional classifications in the social arena, without considering the existence and restrictive relations between the different kinds of organisms. Because cities are not regarded as the platform for all kinds of organisms, but as independent elements that are managed separately, humans pin their hopes on the development of technology in order to drive the city and improve the quality of citizens' life. Therefore, although the "urban faults" have already shown up, people are still adding fuel to the fire. With blind trust in solving problems of the city by means of the expansion of technology, the population of the city is increasingly crowded and social institutions are increasingly growing. The city seems to be a magic container which contains endless development space and opportunities. When young rural laborers swarm into the city to find jobs, many of them don't know clearly who can provide them with opportunities and what their future will be. Many times administrators are busy considering various technical indexes of infrastructures and they adjust development under the differing demands of politics, economics, army, education and culture. They believe that the enlargement of cities will necessarily result in the progress of society and thus they deal with the developing challenge by increasingly heightened technical indexes. By overlooking the organic feature of city systems and paying little attention to the interaction between all kinds of organisms, they can only passively face the problems caused by the restriction of these organisms, and are unable to solve these problems.

Will "urban faults" continue? It depends on how accurately humans understand and adjust the relations of all organisms in the city. Properly adjusting the driving role of artificial organisms on other organisms, and making them evolve and develop with appropriate speed and scale, will benefit the dynamic harmony of all kinds of organisms.

20.4 The "Tao" of City Development from Philosophy of Organism

It is easy to accept the idea that all kinds of organisms should be coordinated, but it is difficult to reach an agreement on how to realize this coordinated state. It is hard to fundamentally solve the problem of city development if restricted to the view of regulating technical indicators or controlling the system of city management. The discussion on the relation between "*Tao*" and "*Ji*" (technology) in traditional Chinese philosophy will bring about an alternative approach.

20.4.1 The Appropriate Orientation of Technical Systems in City Development

Traditional Chinese philosophy attaches importance to the appropriate orientation of all inner elements in the structure of a system. When analyzing the ideological characteristics of traditional China, Needham Joseph pointed out that "if they did not behave in those particular ways they would lose their relational positions in the whole (which made them what they were), and turn into something other than themselves. They were thus parts in existential dependence upon the whole world-organism" (Joseph, 1956, p. 281). The development of technical systems in the city should be appropriately positioned, aiming at coordinating all kinds of organisms. Coordination here requires the proper interrelation of all kinds of organisms in order to keep the vigor and vitality of all kinds of organisms increasing rather than decreasing, which in turn keeps the vigor and vitality of the city increasing rather than decreasing. Only by emphasizing the co-evolution of biological, artificial, social and mental organisms can we promote the sustainable development of cities. In order to reach an internal stable state, biological organisms will adjust the way their organs function so as to make these organs adapted to, and coordinated with, each other. Biological organisms generally won't have an obvious growth and enlargement when they reach a stable size at maturity (West, Brown, & Enquist, 2001). In contrast, the expansion of city scale is unlimited, as is the driving role of the technical system. The technical system, which should be a tool of man's survival, now becomes an element of manipulating biological, social and mental organisms. That is the arrogation of technology in the city.

20.4.2 Active Adjustment of All Kinds of Organisms in Cities

Harmony between all kinds of organism depends on actively adjusting the developing scale and pattern of these organisms, according to whether it is beneficial to strengthen the vitality and vigor of them. The size and density of the population in cities should be matched with the developing situation of artificial, social and mental organisms. Trying to relieve the "urban faults", some modern cities have put forward a series of policies on limiting the population. For example, in New York City in the 1960s, a strategy of "planned shrinkage" intentionally removed essential services from some urban areas to prompt people to move away (Jacobs, 1961). Nowadays, some cities, such as Beijing in China or Rio de Janeiro in Brazil, are still trying to reduce the negative effect of city development by limiting the population. For example, Beijing recently tried to set up a new district in Xiongan, Hebei Province, in order to share some functions with the capital. However, this kind of policy has to consider the potential restricting role of other organisms. It is worth noting that the rapid development of present technology is unprecedented and we don't have enough experience to adjust those organisms by drawing lessons from history. Luis Bettencourt and Geoffrey West indicate that "the time between the 'Computer Age' and the 'Information and Digital Age' was some 20 years, compared to thousands of years between the Stone, Bronze and Iron Ages" (Bettencourt & West, 2010, p. 912). Even so, a continuously accelerating rate of technical systems development will not necessarily bring about the consequence of sustainable development, but might potentially lead to internal conflict within social and economical organizations.

In this regard, people should actively change some customary ideas. For instance, it is not necessary for humans to gather in the city center only for the sake of the vitality of self-development because it will reduce the possibility of development if it is too competitive in the city. It is also not necessary for all kinds of social organizations and institutions to excessively gather together in the city center as the efficiency of working and connecting can be guaranteed by modern information technology and traffic tools. In order to relieve the pressure on the transportation system in city center, it is simultaneously necessary to change the running model of private cars that pass in and out of the city center. Moreover, it is necessary to restrict the construction of high buildings (such as buildings of more than ten floors or buildings that consume more energy), which will help relieve the stress on sources of supply to the city center. In other words, the future appearance of cities should be different from the current appearance of many modern cities. It should show more organic features in the future cities - without the jungle of cement, without severe contamination, without traffic jams - but rather with vitality and vigor in all aspects. Artificial organisms shouldn't be the Leviathan in the city, but should be organic segments harmonizing nature and humans.

20.4.3 "Tao" of Sustainable Development of Cities

"*Tao*" in traditional Chinese philosophy is generally regarded as the essence or rules of the world with a similar meaning to *logos* in Ancient Greek. In reality, '*Tao*' is a very practical term meaning the most reasonable and optimal method (or approach) amongst human practices (Wang, 2009). The approaches and methods primarily employed in practical activities might have some unreasonable or sub-optimal aspects to some extent, which need to be constantly adjusted and to accord with natural rules in order to obtain the optimal economic and social benefits. This is what Chinese philosophy called the development from *Ji* ("technology") to *Tao*. The person who comprehends and grasps the *Tao* can give full play to their physiological and psychological potential, and can reasonably adjust the relations among tools, objects and other relevant elements in practical activities in order to realize the organic combination between human and artifacts with lower input and higher output. There is some difference between "*Big Tao*" and "*Small Tao*". The latter signifies the reasonable and optimized method in *specific practices* to achieve the ideal state between practitioners and their objects or tools. In contrast, the *Big Tao* involves the reasonable and optimized approach in practice *as a whole* to achieve harmony between all aspects of environment, society, interpersonal relationship, mind-body relations, etc., in practice. As far as city development is concerned '*Small Tao*' refers to the developmental indicators of labor-saving, high quality and efficiency. It especially refers to the increase of economic interest and the optimization of social welfare profiting from the progress of technology. However, there is a certain limitation in fixing *Small Tao* as the goal, and it will lead to damage of the environment and the suppression of value rationality. It is a type of unsustainable idea of city development. The harmonious relation between biological organisms, artificial organisms, social organisms and mental organisms will be achieved if we understand the city system from the perspective of *Big Tao*. The idea that "Tao models itself after nature" from Chinese philosophy should be followed in developing cities. City development should comply with natural law and realize the harmony of all kinds of organisms. The idea of "Tao models itself after nature" is advocated in traditional Chinese culture; within which 'nature' is the harmonious state of all types of organisms. The *Tao* of city development is to approach the harmonious state of organisms.

The exploration of the *Tao* of sustainable urban development, in particular, aims at stimulating the city's vigor and vitality as a whole. To explore the dynamic model of coordination of various kinds of organisms in the city, one not only needs to guarantee the vigor and vitality of each type of organism, but also ensure that the development of various types of organisms are promoting, rather than constraining, each other rather. Since the development of biological, artificial, social and mental organisms have their respective rules, it is obviously inappropriate to restrict or restrain the development of other kinds of organisms simply through the development of one. For instance, to set and artificially regulate the development scale and speed of various types of organisms through various quantitative indicators is to regard the urban system as a giant machine and replace the *Tao* of urban development with the rule of artificial organisms, which is undesirable. The sustainable model of urban development requires adequate attention to the vigor and vitality of biological organisms. This refers not only to the living state of human beings, but also to the beneficial plants and animals that coexist with humans. To guarantee this, dynamic assessment and necessary restrictions on the effects of artificial, social and mental organisms are required. The sustainable model of urban development also requires adequate attention to the vigor and vitality of social organisms and mental organisms, as well as timely dynamic assessments and necessary adjustments. In this respect, contemplation of the concept of "resilience" (Folke et al., 2010) is similar to exploration of the way of sustainable development of cities, and they could learn from each other. The *Tao* of sustainable development of the urban provides the ideological principles and methodology for such dynamic assessments and necessary adjustments. This is the best way to explore the primary system for evaluating the vigor and vitality of the various forms of organisms in the city and by which to construct the most reasonable and optimal relationship between them and is an important direction deserving of further exploration in urban studies. In the view of the *Tao* of urban development, the city is not only a place where crowds gather, a place of economic transactions, administrative management, cultural education and entertainment and sightseeing, but also a place where biological, artificial, social and mental organisms are interdependent, fervent and constrained. In this view, the vigor and vitality of the city should be displayed and the city should be regarded as a complicated giant organism which needs to be cultivated and treated. In this way, "urban faults" can be eliminated and the city become a wonderful home for humans.

20.5 Conclusions

Urbanization is still going on around the world. As recently as 1950, only 30% of the world's population was urbanized. Today, more than half live in urban centers. In the developed world, this has reached around 80% while it has been estimated that around 2 billion people will move to cities by around 2050, especially in China, India, Southeast Asia and Africa (UN-Habitat, 2010). A lot of problems are generated during urbanization and many scholars actively seek measures from the different fields of economics, sociology and politics. In this situation, philosophical methods take on a greater significance. Using the approach of Philosophy of

Organism, we can regard cities as the platform for biological, artificial, social and mental organisms. Considering the interaction among these four kinds of organisms as the major force of city development, and regarding their restriction as the source of "urban faults", offers an alternative approach. At present, some scholars put forward the concept of *cognitive cities* to replace smart cities. This concept lays more emphasis on the efficiency, sustainability and resilience of cities, rather than focusing only on the efficiency of ICT, which is the main idea of smart cites (Finger & Portmann, 2016). The idea of *cognitive cities* underlines the adaptability and coordinability of technology on constructing cities, which shows active efforts to harmonize the technological element with others. This kind of sustainable and resilient pattern of city development echoes the *Tao* of cities, which is one of the research interests of future cities. Inspired by traditional Chinese philosophy, pursuing the "Tao" of cities in the context of Philosophy of Organism will promote communication and dialogue in the field of urban development between Chinese and Western philosophy and help establish a harmonious relation between technology and the city in order to realize the sustainable development of cities.

Acknowledgements This work has been supported by the National Social Science Fund of China with the project "Human-Machine Relations in the Age of Artificial Intelligence" (agreement: No. 18CZX019).

References

Berry, B. J. L. (1973). *The human consequences of urbanization*. New York: St. Martin's Press.

Bettencourt, L., & West, G. (2010). A unified theory of urban living. *Nature, 467*(7318), 912–913.

China Academy of Information and Communications Technology (CAICT) & EU-China Policy Dialogues Support Facility II (PDSF). (2016). *Comparative study of smart cities in Europe and China 2014*. Heidelberg: Springer.

Ellul, J. (1954). *La technique ou l'enjeu du siècle*. Paris: Armand Colin.

Finger, M., & Portmann, E. (2016). What are cognitive cities? In E. Portmann & M. Finger (Eds.), *Towards cognitive cities, studies in systems, decision and control* (Vol. 63, pp. 1–11). Cham: Springer International Publishing.

Folke, C., Carpenter, S. R., Walker, B., Scheffer, M., Chapin, T., & Rockström, J. (2010). Resilience thinking: Integrating resilience, adaptability and transformability. *Ecology and Society, 15*(4), 20. Retrieved May 26, 2019, from https://www.ecologyandsociety.org/vol15/iss4/art20/

Jacobs, J. (1961). *The death and life of great American cities*. New York: Random House.

Jacobs, J. (1969). *The economy of cities*. New York: Vintage Books.

Jonas, H. (1984). *The imperative of responsibility: In search of an ethics for the technological age*. Chicago: University of Chicago Press.

Jonas, H. (1996). *The phenomenon of life: Toward a philosophical biology*. New York: Dell Publishing Co..

Joseph, N. (1956). *Science and civilization in China (vol. 2 history of scientific thought)*. London: Cambridge University Press.

Marcotullio, P. J., & Solecki, W. (2013). What is a city? An essential definition for sustainability. In C. G. Boone & M. Fragkias (Eds.), *Urbanization and sustainability: Linking urban ecology, environmental justice and global environmental change human-environment interactions* (Vol. 3, pp. 11–25). Dordrecht: Springer.

Mumford, L. (1961). *The city in history: Its origins, its transformations, and its prospects*. London: Secker and Warburg.
Popper, K. R. (1972). *Objective knowledge: An evolutionary approach*. Oxford: Oxford University Press.
Spencer, H. (1898). *The principles of sociology (vol. 1)*. New York: D. Appleton and Company.
UN-Habitat. (2010). *State of the world's cities 2010/2011—Cities for all: Bridging the urban divide*. Retrieved May 26, 2019, from http://www.unhabitat.org
Van den Dobbelsteen, A., Keeffe, G., Tillie, N., & Roggema, B. (2012). Cities as organisms. In R. Roggema (Ed.), *Swarming landscapes: The art of designing for climate adaptation, advances in global change research* (Vol. 48, pp. 195–206). Dordrecht: Springer.
Wang, Q. (2005). Chinese perspective: Research ethics. In C. Mitcham (Ed.), *Encyclopedia of science, technology, and ethics* (pp. 323–324). Farmington Hills, MI: Macmillan Reference.
Wang, Q. (2009). *Between Dao and technique: Philosophy of technology based on Chinese culture*. Beijing: People Press.
Wang, Q., & Zhu, Q. (2009). Dao and Phronesis: Comparing approaches to ethics and technology. In *Proceedings of 16th Biennial Conference of Society for Philosophy and Technology* (pp. 185–187). Twente: University of Twente.
Wang, Q. (2006). *An outline history of ethics of science and technology in China*. Beijing: People Press.
West, G. B., Brown, J. H., & Enquist, B. J. (2001). A general model for ontogenetic growth. *Nature, 413*(6856), 628–631.
Whitehead, A. N. (1920). *Concept of nature*. London: Cambridge University Press.
Whitehead, A. N. (1922). *The principle of relativity*. London: Cambridge University Press.

Qian Wang M.A. in Philosophy of Science and Technology from Northeastern Normal University, and Ph.D. in Philosophy of Science and Technology from Northeastern University. Currently Professor with the Philosophy Department of Dalian University of Technology, P.R.C. Research interests: management of intellectual resources; management of science and technology; philosophy of technology; ethics of technology.

Xue Yu Ph.D. in Philosophy of Science and Technology from Dalian University of Technology. Currently Assistant Professor with the Philosophy Department of Dalian University of Technology, P.R.C. Research interests: philosophy of technology; ethics of technology; ethics of artificial intelligence.

Chapter 21
Ghost Walks for Wireless Networks

Robert Seddon

Abstract Cities as we know them are built on layers of their own pasts. Moreover, cities remember themselves by preserving historic buildings, erecting statues, writing history into the names of streets, and otherwise conserving and commemorating local heritage. With widespread computerisation and computer networks come new and diverse layers of the city: digital geographies that overlie physical urban sprawl. The city of tomorrow will blend data deeply into its culture and administration; the day after tomorrow, such data will have joined the layers of its historic past. Digitisation promises to become so intimately involved in the city that its layers of data invite consideration as part of the city itself. This chapter traces how the cultural geography of a city will encompass its digital geography, and what the limitless layers of networked data portend for the cultural significance of public urban space.

Keywords Smart cities · Augmented reality · Geofences · Culture · Heritage

21.1 Introduction

Urban environments impress themselves upon the senses: cityscapes are associated with buildings dominating the horizon in every direction, with bustling crowds or polluting rush hour traffic. Yet connected technologies are adding layers to city life that are invisible to unaided human perception: not just unseen parts of urban infrastructure, but technologies belonging to and shaping the cultural life of the city. These in due course will belong to the city's cultural history and heritage too, for cities are built upon accumulated layers of their own pasts.

R. Seddon (✉)
Department of Philosophy, Durham University, Durham, UK
e-mail: r.f.j.seddon@durham.ac.uk

M. Nagenborg et al. (eds.), *Technology and the City*, Philosophy of Engineering and Technology 36, https://doi.org/10.1007/978-3-030-52313-8_21

In Dresden the pale walls of the Frauenkirche incorporate darker and older stones, the 'curated ruin' (Moshenska, 2015) of a church destroyed by wartime bombs. In Valencia a garden marks the former bed of a diverted river. 'Venice is a nineteenth century reproduction of itself.' (Hewison, 2016) City-dwellers inhabit 'a palimpsest of itself' and flâneurs explore 'strata of the urban space' (Vattano, 2014). Cities never take a final shape once they have sprawled across old fields and forests: development continues, endlessly but incrementally, except where places and landmarks are so prized that steps are taken to preserve them as they are. The technological city of the emerging future will be no less thoroughly involved with its own past.

How the city remembers itself, however, may change as technology enables and requires it to change. Our connected age is already the best time there has ever been for finding information about the histories of our cities. Aficionados of 'urban exploration' share photographs of abandoned interiors online. Seekers after 'ghost signs'—faded advertisements painted on the walls of buildings—can have a smartphone app as their tour guide (BBC News, 2017). A blogger uses Land Registry data to trace tunnel systems beneath the streets and shares the results with the world (Shrubsole, 2017).

What if the flow of networked data should become so completely integral to the structures and administration of the technological city and to the lives of its inhabitants that alongside information *about* the city, we must begin to think of data that are *part of* the city? (Even if not in a strict mereological sense, in no less escapable a sense for that.) That is the possibility that animates this chapter: the prospect that 'the blood that runs through the veins of the cities of the future consists of data' (de Ridder, 2015). How will the technological city remember itself? What manner of 'palimpsest' or 'strata' of its streets will the flâneurs of the future explore?

Cities as we know them acknowledge their urban heritage in a variety of familiar ways. Sometimes the material fabric of the past is preserved for present use. It might be the grandest of cathedrals; it might be an unassuming house with a modest little plaque recording which historical figure was Born Here. Sometimes the record is the entire memorial: past heroes and historic events remembered in the names of streets and squares. Sometimes the urban past is actually forgotten and obscured until some excavation uncovers it: an archaeological search for the remains of one of England's kings led to a city council car park where a friary used to be (Buckley et al., 2013).

In such ways as these, places give acknowledgement to their pasts, and the people charged with preserving and presenting cultural heritage for the public's benefit are often keen to adopt technologies that can help them to do so. What is less well understood so far, especially because the emergence of the technological city is a process happening around us in the present, is how we are to grapple with urban technology *as* culture and *as* heritage and history: as an increasingly commonplace and ever more integral part of the many-layered palimpsest which city-dwellers inhabit.

As philosophical works go this one is less a crisp argument from premises to conclusion, and more a tentative mapping of conceptual terrain. Section 21.2 introduces the theme of cities with invisible layers, within which digital technologies generate an urban sprawl of their own. It introduces the 'smart city', in which networking and automation are threaded throughout the city and its administration. Section 21.3 segues to urban history and heritage, describing some of the ways in which connected technologies are already used to conserve places of cultural significance and present them to the public. Among these is the technology of 'augmented reality' (AR).

Section 21.4 moves from AR's usefulness in presenting the past to its place in the cultural innovations of the present. It shows how AR gaming has invisibly transformed urban spaces. Section 21.5 moves to more practically minded uses of such transformation: the potential of AR to 'hack' one's environment, erecting digital statues or bringing adblocking from the Web to the physical world.

Section 21.6 continues the theme, noting its contrasts with the solidity of cities' physical walls. Section 21.7, however, suggests that the power to edit and personalise one's environment may fall some way short of having a voice to be publicly heard and remembered.

When we redesign and reconstruct our environments in AR, are there parts of the city that exist only in images of those environments? Section 21.8 explores perceiving and recording AR in comparison with the familiar role of photography in recording the past. Section 21.9 moves to the entirely unseen geographies of 'geofences' and 'geotagging'. These too, it suggests, create real and spatial features of a city. Section 21.10, drawing on the idea of 'ubiquitous computing', explores the indistinctness of a city so conceived, in which the definite outlines of buildings coexist with gradients of wireless coverage.

Section 21.11 draws the threads together to present the technological city of a possible future: a connected and computerised city full of digital geographies, personalised perceptions, and phantasmal surroundings that flicker into being for those with the right equipment and the right credentials. If these are the urban spaces and urban culture of the future, it concludes, then what they bequeath to history may be so fragmented as to cast doubt on the very idea of a shared local heritage, yet simultaneously rich with tantalising digital archaeologies, haunting the spaces we live in and the air we breathe in a city full of ghosts.

21.2 Invisible Cities

Besides and between the visible and tangible features of a city's geography, there exist the unseen layers of past events and urban legend. In places with suitably grisly pasts (I write from one of several counties with a claim to be Britain's most haunted) there is even a market for 'ghost walks' for the paranormally curious. Nothing piques interest in local history quite like the macabre. Tourists little expect to see an actual ghost, and by touring at night they ensure that they will not see much at all. They are drawn by stories that evoke the possibility of phantom presences that linger in the air.

In our age of wireless connectivity there are most assuredly invisible and intangible presences lingering in the spaces through which we pass. In their mundane way these too belong to urban geography: ghosts make good stories, but connected technologies offer tools to find nearby taxis, eateries and people to meet. Such is our culture now, at least in some places and for some people. Such therefore will be part of our heritage—and a further challenge for the preservation of our cultural memory.

The grandest vision of digital interconnectedness is of 'smart' environments that continuously monitor themselves and systematically react: in which data on everything from traffic flow to pollution levels are continuously available across the network so that other local subsystems can react efficiently. The greatest prize for heritage may be the prospect of places with digital memories: of a city that remembers the footfall on the waterfront 10 years in the past (because it used the information to adjust illumination from the streetlights, and stores the data to enable examination of long-term trends). A city full of ghosts. A city made of memory.

I leave largely unexamined what that prospect might mean for privacy and freedom from intrusive monitoring, not because that is unimportant but because it is so important that much is already written elsewhere about the dangers of surveillance states and panopticon societies. This essay approaches the potentials of urban technology with a disposition that is mostly sunny, or at its gravest cloudy with a chance of rain. I also leave aside the related topic of network security: there are important questions, already quite widely discussed in relation to the Internet of Things, about the practical risks that connected and computerised devices present when they are vulnerable to being hacked. In what follows I generally assume that systems are working as intended and are under the control of their proper administrators.

Of course, 'working as intended' can be a flexible notion. Reportage on surveillance in Glasgow, where sensors attached to streetlights monitor their surroundings for any loud noises in case these have 'anti-social' or criminal sources, notes that this is 'just one illustration of how connected cities can veer from their original purpose into entirely new territories, which may not always be ones their inhabitants will feel entirely comfortable about' (Wakefield, 2017).

It is not yet clear how centralised and monolithic the city of the future will be. Some visions of the 'smart city' present it as 'one big joined-up urban machine' with human beings moving within it like bees pollinating flowers, or even removed from its practical operations altogether (ibid). In this direction tend the sensor networks installed by city councils to monitor congestion, rainfall, pollution and so on. Public services use the resultant data to distribute resources efficiently, sometimes without human intervention.

In this essay I focus largely on humans' interaction with technology, because it is humans who have cultures. (The idea that some animals have transmissible cultures too is gaining traction, but artificial intelligence seems nowhere close.) Many interactions in the technological city, however, will be between artificial agents: signals intended to guide self-driving vehicles, for example, 'like flickering ghosts only cars can perceive, navigational dazzle imperceptible to humans.' (Manaugh, 2019).

Other accounts of smart cities emphasise open data, shared with and among actively engaged citizens: people who can query their local pollution sensors, for example, might become more aware of nearby environmental problems and better equipped to act upon them. 'Could we find ourselves texting the wildlife, following our litter online and using houseplants to control our energy use, all from our office in the public park?' (Sheeter, 2009).

It is notoriously hard to predict quite what technologies and urban lifestyles people will want (and particularly notorious when city planners have predicted wrongly in the past). A lot of experimental technologies and speculative proposals are involved in smart cities, and probably not all of them will become widely deployed. Some charming serendipity emerged in Melbourne when officials assigned e-mail address to each local tree, expecting citizens to use them to report dangerous branches and the like: people instead began writing affectionate messages to their favourite trees (LaFrance, 2015).

Melbourne's email-a-tree service is one in a litany of municipal projects aimed at leveraging personal and institutional technologies to keep cities running smoothly. In Chicago, there's a text-based pothole tracker. In Honolulu, you can adopt a tsunami siren. (Ibid).

One thing that otherwise remains to be seen is what effect smartness will have on the cultural distinctiveness of cities. The Europeana Foundation has an initiative 'to promote cultural heritage as a pillar of smart cities' (Dobran, n.d.). However, a recent call for conference papers laments that despite their diversity of approach and features, the vision of most Smart Cities is one of the conspicuously bland, generic, ahistorical, apolitical spaces whose identity is characterized by information technologies that could be applied anywhere. (Dunn and Cureton, 2018).

21.3 Historic Cities

Both overt and largely unseen purposes have prompted the addition of heritage sites to urban networks. Smart Heritage Sensors, explicitly designed 'to be concealed or barely noticeable' (Telefónica IoT Team, 2016), enable buildings and monuments to monitor themselves for signs of decay, and to send alerts over the network. Once these cultural assets have been brought online, however, it is a short extra step to let them broadcast information about their historic importance to anyone in the vicinity, 'allowing visitors and residents to know more about a particular street, square or fountain simply using their smartphone and specific apps' (ibid). Such systems offer unobtrusive ways of making information available at a particular location, unlike physical signage or QR codes for people to scan.

While some location-based technologies depend on such broadcasts from local beacons, there is much that can be done with just an Internet-connected smartphone and the Global Positioning System (GPS). The Belfast Soundwalks application, for example, offers audio creations by 'sonic artists/composers' through which those following its trails through the streets can 'discover a hidden sonic world that reveals something of the area's environment, history or culture' (Calvium, n.d.-a). In Toronto the TXTilecity app provides an interactive roadmap of the history of local textiles (Milling, 2014).

Most technologies of this nature aim to connect members of the public with local heritage by supplying them with fresh ways of experiencing their surroundings or ready access to relevant information. Some, however, encourage citizens to take a more active role in heritage conservation. In Bristol, Heritage Eye 'allows for members of the public to submit alerts when they see a potential problem with a building' (Calvium, n.d.-b).

There is much interest in the potential of 'augmented reality' (AR) to present locations to visitors in new and vivid ways. AR can involve any or all of the senses, but it is predominantly concerned with visual perception. AR works by superimposing a computer-generated overlay onto the world as seen on a smartphone screen, or through a pair of special glasses. (Google Glass and Microsoft's HoloLens are among the better known examples of the latter type.) When your device knows where nearby objects are in relation to you and your viewing angle, it can adjust the overlay appropriately.

Some AR software paints a depthless 'head-up display' in front of the scene in view: perhaps you point your smartphone's camera at an object and a label appears on the screen, tracking its location in the video feed, to give you details about it. Other kinds of AR use scaling in order to give the impression that what the software shows you, despite being realised as a two-dimensional image, is itself present at a particular point in physical space. (The term Mixed Reality is sometimes used for software that aims to suggest that digital and physical objects are co-existing in a shared space.)

In Santander, for example, the SmartSantanderRA mobile app provides information about nearby shops and amenities.

On starting the AR view, the App creates on the smartphone screen, an overlay over the camera with nearby POIs [*sc.* Points of Interest]. If a particular POI is selected, further information (title, short description, photo and distance to the POI) is displayed. (SmartSantander, 2012).

In London, using the Streetmuseum app a 'historical image can be overlaid onto the current view of a given site' (Guin, 2017). Projects elsewhere have 'enabled viewers to see a decayed Norway mural restored to its original glory [and] an in situ history of murals that once graced the Bowery' in New York (Karlin, 2013, p. 27). Another use for AR in the city is to reveal otherwise inaccessible features like underground pipes and cables (vGIS, n.d.).

AR is of particular interest to custodians of heritage sites, since 'the owners of sites are usually reticent to permit physical reconstructions in situ so that the archaeology remains undisturbed for future generations' (Bostanci et al., 2015, p. 1). 'AR is an ideal technology to re-enact historical events without compromising original architecture and landscapes' (tom Dieck and Jung, 2017, p. 115).

In cities of the Netherlands, the Urban Augmented Reality app evokes both pasts and futures:

By means of advanced 3D models, UAR shows the user what is not there: The city as it once was—for instance by showing buildings that once stood there. The city as it might have been—by showing scale models and design drawings of alternative designs that were never implemented. And the city of the future—by showing artist's impressions of buildings under construction or in the planning stage. (Dedale, n.d.)

Here the city as an urban palimpsest becomes marvellously vivid: with technology we can visualise not only a changing city that stretches into the past and future, but myriad possible cities, streets and structures all around us that could be or might have been. 'To visit an earlier time at a location, you simply revert to a previous version [...] like a Word or Photoshop file that you can keep "undoing."' (Kelly, 2019).

21.4 Fantastic Cities

The examples of urban AR in the previous section broadly tend to show how connected technologies can serve a useful role in providing public information and education. They represent the voices of experts or officials, and employ technology as a kind of artificial tour guide or information kiosk. This is a good use of AR, but not the whole story. AR has its playful and even its subversive sides. Moreover, in augmenting the urban spaces that already exist it is by no means limited to providing information about them. The introduction of AR into a location can offer subtly transformative effects.

The explosion of AR into public awareness did not come from heritage guides or route finding, but from the video game industry. To date, indeed, its fame has hinged on the popularity of one game in particular: on the 2016–2017 craze for Niantic's *Pokémon Go*, which has seen players explore physical locations in search of 'virtual' creatures which the software had generated and would display on their smartphone screens. (At the time of writing, Niantic has been developing an AR game using the *Harry Potter* licence, hoping to repeat its previous success. It has also announced plans to work with the United Nations World Tourism Organization on games 'to encourage global tourism and worldwide civic engagement' (Beuttenmüller, 2018).

There were curious and confusing incidents as people grappled with the collision and fusion of the physical and the computer-generated. In Gujarat (BBC News, 2016) and Yekaterinburg (Cresci, 2017) there were legal actions related to items which the game had generated 'inside' places of worship. (Religious sites, after all, serve as a reminder that there is nothing new in the thought that there can be important aspects of reality all around us which are not ordinarily accessible to human perception.) Faced with 'a cunning Western plot to destabilize Russia' (Bennetts, 2016), or just spotting an opportunity, Moscow City Hall responded with an app of its own, enabling roaming Muscovites to 'catch' notable figures from Russian history (Moscow Times, 2016). As one commentator observed in a glimpse of our perplexing future and its augmented environments:

Pokémon Go, as we all know, is just the beginning when it comes to Augmented Reality [and] governments haven't got a clue how to handle it. If a virtual Speero egg in a Hindu temple is enough to cause a lawsuit, what about adverts for alcohol, or dating sites, or porn? How are the courts going to deal with copyright infringement, or the safety of AR tourist attractions, or taxing products that do not actually exist in the real world? (Cunliffe, 2016).

Instead of furnishing the public with information about what physically exists, or did, or could, this kind of software lets one opt into privately transforming physical geography into the setting for a different kind of activity. Repercussions may be unexpected: when it emerged that Niantic's sources of geographical data may have included the OpenStreetMap project, which uses contributions from members of the public to compile map data, some enterprising players are thought to have added geographically false data to OpenStreetMap in order to manipulate the game (Bushman, 2017). The emergence of AR does not always serve geographic truth and historical accuracy. Often it serves overtly commercial interests, with local businesses invited to pay for greater prominence (Tsukayama, 2016).

Niantic (then still part of Google) had previously created an AR game called *Ingress*, promoted with the tagline 'The world around you is not what it seems'. In *Ingress*, players' teams are tasked with capturing 'portals' at physical landmarks, including public places of cultural significance. 'Your local town hall has probably changed hands three times today.' (Hatfield, 2014) Before that, in turn, the potential of location-based 'mobile city gaming' to 'change the way games are played in urban settings' had been explored in Grey Area's *Shadow Cities* (Reahard, 2011). This style of gaming has taken root particularly in urban environments because of their greater concentration of available landmarks (Hatfield, 2014).

Here it becomes plainer how technologies of this sort can appropriate existing cultural geographies and privately employ them for new purposes. In time they may generate their own ghosts: it is not too hard to imagine cultural historians of the future following trails laid down by long-defunct AR fads, remarking on which landmarks still survive in the physical world and which have succumbed to city planning. We may leave more traces of our movements to such digital archaeology than we do on the surfaces of our physical environments; though in Milwaukee a legal dispute has emerged following physical damage to public parks by hordes of Pokémon hunters. After ordinances restricting the use of AR were put in place locally, the makers of an AR poker game called *Texas Rope 'Em* objected on free speech grounds (Gardner, 2017), further illustrating the confusing ways in which AR blurs digital media into physical places.

21.5 Remixed Cities

The most suggestively and subversively transformative applications of AR, however, are not to be found in the popular products of corporate video game developers.

One of the most conspicuous aspects of many cities' cultural geography is the use of public statuary to commemorate notable individuals. The choices of whom to honour in this way are sometimes controversial. Monuments to the glory of former regimes have sometimes been displaced from their original positions of prominence and removed to memorial parks: several exist which collect together statues from the Soviet era, whether off the streets but within the capital city, as at Memento Park in Budapest, or away from dense urban populations altogether, as at Grūtas Park in its Lithuanian woodland. Among the more curious incidents of attempted *damnatio memoriae* (inspired by a similar movement in South Africa) was a campaign against a statue in Oxford of the colonial-era figure Cecil Rhodes, led by a student who was himself in Oxford because he was the beneficiary of a Rhodes scholarship (Espinoza, 2016).

The characteristic size and solidity of such public art, along with the legal protections which rightly punish vandals, have meant hitherto that it takes mass politics to make dramatic changes to the selection of notables thus immortalised. Editing public statuary to suit personal preferences has not been a possibility. In the purely physical world that remains the case, but *augmented* realities can be more flexible. A desire that more statues should commemorate women led to the creation of an app called The Whole Story, through which users can place their own 'virtual statues' at locations of their choosing, and view those placed by other users (Jardine, 2017).

AR can also be used to cover up what is already present in physical space. 'Left unchecked, the proliferation of outdoor advertising can consume a city.' (Mahdawi, 2015) Some municipalities have responded with legal restrictions, but technology offers other options and ones better suited to the means of individuals. In New York, an app has been employed that uses image recognition and AR 'to take back this public space and our attention' (NO AD, n.d.). 'Just point your iPhone camera at a billboard and, hey presto, you'll see it vanish and a piece of street art will seamlessly appear where there was once corporate propaganda.' (Dormehl, 2014) (One perverse consequence was that some people started hunting for adverts in order to collect all the available pieces of substitute art.) A similar project introduces its own terminology: 'Rather than [...] a form of *Augmented Reality* technology, we consider The Artvertiser an example of *Improved Reality*.' (Artvertiser, n.d.)

Even graffiti, long associated with the defacement of public spaces and consequently with the less salubrious elements of urban living, can exist within the privately shared overlays of AR. The technology potentially 'allows users to remix the real world' (Cameron, 2012), scrawling freely all over whatever objects they please and sharing the results with other participants in a social network. In Pittsburgh, disused buildings have been transformed into clandestine gallery spaces for concealed art (Ihaza, 2014).

The effect of digitising the physical world is to bring the attitudes and techniques of 'remix culture' and 'hacker culture' along for the ride: architecture and urban space come to be treated like software or digital media. Irked by intrusive adverts online, people learn to install adblocking software; they then realise that what they can do on the Web, technology also makes possible on the New York subway.

People who hack the world to make it feature more female statues are following a similar logic to the hacker who modified the video game *Donkey Kong* to switch the roles of hero and distressed damsel (Mika, 2013).

We're living in this remix culture. This appropriation time where any grade-school kid has a copy of Photoshop and can download a picture of George Bush and manipulate his face how they want and send it to their friends. And that's just what they do. (Quoted in Lessig, 2008, p. 14).

That was in 2008 (when George W. Bush was in the White House). Today 'this appropriation time' has tentatively started to encroach upon the spaces through which we walk. (In fact, it has been possible to set the world to one's own soundtrack of choice ever since the introduction of portable audio players.) After all, cities themselves are testimony to human modification of our environments: the urban is contrasted with the rural, where farmland in turn contrasts with forests and whatever still passes for wilderness. AR permits personal remodification of the already artificial.

21.6 Hackable Cities

As in so many walks of life, then, we cannot expect to find ourselves dealing with just one, coherent, happily uniform urban culture. The technological city will speak (perhaps literally) with multiple voices and remember from myriad perspectives. Sometimes we shall encounter the voices of public officials or commercial interests, as in the following use of localised FM radio signals:

"What we want to do is enable smart cities and fabrics where everyday objects in outdoor environments — whether it's posters or street signs or even the shirt you're wearing — can 'talk' to you by sending information to your phone or car," said [...] Shyam Gollakota. [...] The team demonstrated that a "singing poster" for the band Simply Three placed at a bus stop could transmit a snippet of the band's music, as well as an advertisement for the band [...]. (Langston, 2017).

Nothing heralds technological progress quite like vigorous new forms of advertising. (As on the Web, much of that advertising is also likely to be personalised; some shops have already developed capabilities to track customers by their cellphones' WiFi.) Already Arcona's Digital Land project has been auctioning off portions of the world's cities for AR use, partly in hope of creating an 'ecosystem' for AR advertisers (Blenkinsop, 2018). Yet we have already seen that AR, sometimes used to promote businesses, can also be used to bring adblocking into the physical world. The introduction of computerised technology into the workings of a city brings with it an open invitation to the playfulness and adaptiveness of the hacker mindset:

Hackers are [...] attuned not simply to the workings of technology but also seek such an intimate understanding of technology's capabilities and constraints that they are positioned to redirect it to some new, largely unforeseen plane. They collectively and individually derive pleasure in outwitting constraint. [...] Hacking is where craft and craftiness converge. (Coleman, 2013, p. 98).

Urban space as we know it is typically unyielding. Buildings are physically solid; walls constrain people's movement with implacable stone and brick. Even those who own the structures may find themselves dealing with complex legal and regulatory regimes, entanglements of zoning laws and applications for planning permission, if they wish to make alterations, or to undertake demolitions and build replacements. (Those in charge of historic sites are likely to find the legal regimes in question especially limiting, precisely because of public recognition of architectural heritage as something worthy of preservation. Cities preserve their pasts in very solid forms.) It is telling that parkour, a form of movement derived from obstacle course training, is most characteristically associated with the navigation of urban environments (Brunner, 2011).

So characteristic of urban space are such physical restrictions on movement that in *The Burglar's Guide to the City* Geoff Manaugh associates deviation from them with the ultimate subversiveness, that of crime:

Burglars don't need doors; they'll punch holes through walls or slice down through ceilings instead. [...] Burglars seem to exist in *Matrix* space, a world where—to paraphrase that film's own metaphysics—not only *is there no door*, but there are no walls, roofs, or ceilings. Burglary, in this sense, is a world of dissolving walls and pop-up entryways... (Manaugh, 2016, pp. 13–14)

AR offers a dramatically different prospect: one in which creating or dissolving aspects of a city's geography is as trivial as switching from one app to another. Physical walls remain stubbornly unyielding to the law-abiding citizen, but *augmented* urban geography is limitlessly phantasmal. Like any form of software, it invites exploration, customisation and even subversion. As a modification of the environments in which we live, itself open to modification in turn by anyone in a position to alter its software, it fosters the playful mindset of hacker culture.

To what extent that potential will be realised remains to be seen. 'Users immersing themselves in virtual worlds,' Peter Lunenfield has written, 'are more like visitors to a theme park, choosing between a limited number of prefabricated rides, than urbanites wandering the streets of the city' (quoted in Katyal, 2017, p. 1128). Sonia Katyal comments that *augmented* reality 'might allow for a slightly greater degree of unpredictability due to user choice and external context' than fully 'virtual reality'—but also that much 'depends on the design of the augmentation and how it actually adds to the user's everyday experience' (ibid). AR carries with it the prospect of bringing the city closer to the theme park.

That is no criticism in itself: people willingly go to theme parks to have fun. Yet in making a stroll through city streets a more overtly authored experience AR will be subject to whatever intentions and motivations lie behind that authorship. A theme park is a very peculiar sort of semi-public space: almost capitalism's triumphal answer to the Soviet statue parks in its total dedication of the park to the enthralling iconography of its theme. It makes plain how a place, even for the most benign of reasons, can be minutely choreographed.

Even in liberal nations, the governance of towns and cities does not always favour free and spontaneous commingling, and technology follows suit. A controversial tool in the policing of urban space is a device called the Mosquito, which emits a high-pitched buzz. The ability to hear the unpleasant sound typically decays with age, and people lose it during their twenties. The 'anti-teen' device (BBC News, 2008) is used to discourage younger people from loitering, and indeed from being in the vicinity at all. In short, urban areas already exist that are acoustically modified to cleanse them of the young. The choreography of city life is here already, and not everyone is welcome in the park.

21.7 Overlooked Cities

As ever, then, technology portends and fosters cultural changes. As ever, some voices will be amplified over others. Many current concerns in museological circles are about who gets to speak for whom: about when institutional and academically expert authority ought to yield to the communities whose stories are ostensibly being told; about the choice of which voices to emphasise over others; about what manner of publics a public museum serves.

This is now a key strand in much heritage research: the tensions between 'authorised'–'hegemonic' and 'alternative'–'subaltern' forms of heritage discourse, between universal and particularist claims to heritage, or between 'expert' and 'lay' knowledge. (Bienkowski, 2013, p. 42).

These are concerns that can readily spill out of the museums and art galleries and onto the streets, as we have already seen in the case of controversial public statuary. As we also saw in the case of the AR app for female statues, technology promises an answer: opt into the environment you want, modified to suit your wishes.

This is the kind of answer that raises further questions. Of course, *my* city is already not altogether *your* city: even if we happen to eat at the same café, only for me is it the café at which I recall dating the girl who talked about her hobby of taxidermy. Yet it is one thing to have personal recollections of places, and quite another to edit how places seem to us by technological means.

Human beings already have great capacities for selective attention: in cities one cultivates the art of not noticing beggars. Street fundraisers then develop tactics to counter this tendency, resulting in the coinage of the term 'chuggers' ('charity muggers') in the United Kingdom. Such is the civic life to which all these exciting new technologies are coming, and it will be interesting to find out whether AR adblocking of posters affects charities as well as businesses. Earlier I noted suggestions that tapping into the data flow of a smart city might help concerned citizens to deal with local problems. Another possibility is that technology will make the unsightly and unpalatable more easily ignored.

What is out of sight and out of mind is consequently out of memory too. Computers log precisely what programmers instruct them to, using whatever range of inputs they are built to support. Besides the undesired they may overlook the merely inconvenient. Permanent structures with fixed addresses readily get their place on Google Maps, noting for public posterity the locations of concert halls, restaurants and emporia, but the life of a city may also include subway buskers and transient food stalls: trickier aspects of urban culture for anyone to capture, and ones to which the data streams of a smart city could easily be blind. (The main interest in tracking either will probably come from law enforcement: there is supposedly 'a black market for mayo and fried onions' in Toronto (Krall, 2010). Dutch police have already explored the potential use of smartphone games in citizens' hands to find stolen cars and missing people (Smith, 2017).)

Whatever makes records must include some things and filter others out, be it the 'big data' infrastructure of Google or a city network, or the personal devices of portable and wearable technology. Where the former connote the centralised, regularised and official, the latter adorn the individual who navigates a city of haphazard routes with improvised shortcuts, spontaneous detours and chance encounters, and unruly throngs of other such individuals, the quotidian pedestrians who mill about in Michel de Certeau's writings on the everyday (de Certeau, 1984, p. 91ff). When so many of us carry devices that track our locations and let us video any interesting sights, more than ever will be bequeathed to some sort of posterity, but more than ever the result may be records of cultural fragmentation. The very point of AR is seeing the world differently, and the purpose of even simple earphones is to block and replace the clamour and chatter of the street.

21.8 Imaginary Cities

The technological city brings with it a further novel twist, though: the medium is not only the message but partly constitutive of the street as well. When a smartphone camera is used to generate an AR scene it both documents its surroundings photographically *and* augments them with a reality of its own.

Photography is central to the dissemination and conservation of culture and cultural heritage in modern times and modernised places. Indeed, Peter Probst asserts that photography 'has not only been part of heritage; [...] photographs helped to create it.' (Probst, 2007, p. 100) Besides this role in the public cataloguing of culture, photography has also let individuals make and share recordings from their own lives. José van Dijck further contrasts 'the analogue age', when photography was 'first and foremost a means for autobiographical remembering' and personal photographs became 'keepsakes in someone's (family) album or shoebox', with 'the digital era', in which 'functions of communication and identity formation' are no less 'vibrant' than those of memory, or in some commentators' eyes even more so (van Dijck, 2008, p. 58).

'In contrast to a written account [...] a photograph has only one language and is destined potentially for all.' (Sontag, 2004, p. 17) Photographs preserve the faces of dead people without the painterly interpretations of traditional portraiture. When news came of Islamic State's acts of destruction in Palmyra, photographic records provided evidence of what the world had lost. So powerful and pervasive is this sense of photographic authenticity, even after the popularisation of Photoshop, that the pervasiveness of camera-equipped cellphones has given us 'citizen journalism' and, less upliftingly, the 'selfie stick'. Meanwhile strollers in many modern cities perform for a silent audience of surveillance cameras.

Where unadorned photographs and video footage hold the promise of objective record-keeping (even when we know that the choice of which shots to frame may be open to bias), AR is different. You do not, unless you are some sort of Berkeleian idealist, suppose that a stone statue in a public square exists only in some potential sense unless somebody is observing it. Yet an AR statue is precisely like that.

Though not every implementation has operated in real time (see Zöllner et al., 2010 for presentation of heritage sites using AR in a static, 'snapshot' form suitable for the weaker smartphone hardware of its time), and though it is perfectly possible to save screenshots of AR for later perusal, what differentiates AR from any other form of photographic manipulation is precisely its power to conjure up quasi-physical apparitions dangling in (perceptual) space. Augmented reality is realised through the generation and immediate perception of images; save a frame from this stream of pictures, store it as a static record, and it has ceased to function as AR. This is quite unlike making a photographic record of an independently existing object like a stone statue. A visual record of an AR statue is in a sense self-referential: the making of the record itself makes the statue flicker into fleeting existence within the recorded scene.

In this respect AR is more like an event than like an installation: lasting visual records serve as a second-hand suggestion of what it was like to be there when something happened. Yet the aspect of being 'there' is important. Why trudge into town for the sake of an AR statue instead of bringing the statue into your living room? Because the meaning of a decorative statue in the home differs from the significance of a statue prominently placed in a public square. Why invite people to wander in search of AR graffiti instead of Photoshopping graffiti into an existing photograph and sharing the resulting image? Because 'tagging' physical locations with graffiti has a cultural meaning.

It may be that some kinds of AR are more intimately involved in their environments than others. Nicola Liberati suggests that AR games are restricted by their fictionality (Liberati, 2017): like children's games of make-believe, they have an unsuitable kind of meaning to be considered part of our everyday world (though they can affect human behaviour in it). Liberati leaves open the possibility that other forms of AR might achieve that status.

Even games and fictions can become part of a place's history. Whether we construe AR in terms of objects to be found at locations in a city's geography, or in terms of the rituals that play out on the streets, our understanding of city life becomes more complete once we are aware that *Ingress* players have vied for dominion of the local post office.

21.9 Walled Cities

So perhaps there are parts of the technological city which can be glimpsed only in images of the city, because it is only through images that AR can be actualised. The technological city also has walls and fences which are not meant to look like anything at all.

A geofence defines a boundary in terms of geographic mapping co-ordinates. A device capable of tracking its own position can react when it finds itself crossing the boundary.

Some retail and hospitality businesses will set up geofences around their competition, so when you approach the boundary, you'll get a push notification prompting you to visit the other establishment. Or, you might walk into a retail store and see a coupon pushed to your device. If you download a grocery app, chances are it will register when you drive by to prompt an alert, trying to get you to stop in. (White, 2017).

We are also promised further location-sensitive developments in social networking, audience engagement at festivals and other events, surveillance by employers, secure areas and anti-theft measures, and fridges that send messages reminding you the milk is running low when you pass near the shops. Not all geofences track living beings or occupied vehicles: they are also used to guide and restrict the flight paths of aerial drones. (Ibid).

A geofenced city will be full of invisible boundaries and phantom territories, discovered through the reactions and interactions which they generate. Geofences create another layer of geographic demarcation in addition to the familiar enclosure of urban spaces between solid walls. The example of businesses that create geofences around their competitors shows that this further layer of spatial division will not always track the ownership of sites as one might have expected. The physical and legal threshold of one firm's premises may become the perimeter of a usurper's geofenced zone.

This illustrates in cynical business practices what emerged by accident in the controversies over *Pokémon Go*'s appropriation of places of worship: how geographies created abstractly with mobile technologies do not always follow conventional understanding of what a site is, where its boundaries lie, and who gets to own and control and be responsible for the site and its boundaries. Physical and visible fences we understand very well. Boundaries created by convention and in law, such as those between the public space of a pavement (a sidewalk in some countries) and the private space of a front lawn or drive, are also familiarly understood and largely respected. Divisions of space within location-based data are different, and give rise to different concerns about who exercises control and to what effect. The same applies to markings and identifiers:

Local businesses on Google Maps aren't always as local as they seem. Tens of thousands of bogus listings are added to Google Maps every month, directing browsing traffic towards fraudulent schemes [or to scammers who] falsely list themselves as owners of genuine hotels or restaurants shown on Google Maps. (Revell, 2017).

Why will abusers and trolls bother spray-painting a house, when they can geotag it maliciously? Paint washes away, but data may be more difficult to erase. (Guadamuz, 2016).

Data that persist in uncorrected form may distort a historical record long after their original purpose has been served (but then, historians have always had to deal with bias and outright lies in source material). Conversely, data may simply cease to be available if the provider goes out of business or takes an unprofitable service offline.

If we consider location data related to a physical site to be akin to a Web page about the site, then nothing in the previous paragraph looks very novel. Yet a geofence redirecting people away from competing businesses, for example, looks quite significantly unlike text sitting passively on a Web page somewhere. As in the cases of AR and smart cities, data about a site, and the online services that use the data, seem also to be aspects of the geography of the site.

21.10 Pervasive Cities

In circumscribing a site, an area, a zone, one inescapably defines a setting for it, a region outside the borders to complement the region inside. The settings within which buildings are situated give rise to endless tensions in planning and conservation: in Britain the 'Nooks and Corners' column in *Private Eye* regularly publicises local battles against planning proposals that would result in incongruous skylines, stylistic clashes and other signs of architectural hubris.

When I was a student the dimensions of the World Heritage site across the road, containing Durham Cathedral and Durham Castle, were reputedly a source of frustration for the college engineer. The main college buildings—themselves 'listed' as historic buildings meriting some legal protection—formed part of the visual setting of the World Heritage site without being able to make a case for access to its maintenance funds. Originally defined in 1986, the site was expanded in 2008 (Durham University, 2008), but UNESCO notes that 'the current boundary still does not fully encompass all the attributes and features that convey the property's Outstanding Universal Value', and considers 'a minor boundary revision' 'likely' (UNESCO, n.d.).

Even the demarcation of physical sites of significance, then, turns out to be a bit of a messy affair in practice. Humans encounter environments through sensory perception: we seek out attractive views from lofty vantage points, and flee the noise pollution of road works or overhead aircraft. Some of the subtler sensory features of our surroundings are still not much explored or mapped for conservation purposes, although an artist named Kate McLean has been 'trying to preserve our olfactory memory' by mapping cities' smells (Traverso, 2017).

It is against such a backdrop that the tangled and overlapping geographies of connected technology emerge as a novel feature of the cultural landscape. Connectivity is becoming so widespread that much is now written about the Internet of Things, full of networked artefacts that communicate with each other, and even about 'ubiquitous' or 'pervasive' computing, a possible future in which computerisation has joined technologies like writing that, as Mark Weiser famously presented the idea, 'weave themselves into the fabric of everyday life until they are indistinguishable from it' (Weiser, 1991, p. 94).

A 2003 article commenting on progress towards realising Weiser's vision noted that such a future remains some way off: 'We still view computers primarily as machines that run programs in a virtual environment that we enter to perform a task and leave when we finish.' (Saha and Mukherjee, 2003, p. 25) This is plausibly true of setting a digital alarm clock, for example: one might not ordinarily *call* it 'a computer', but it is one, and fiddling with the buttons to change modes and update the screen display certainly brings to mind all the usual frustrations of computer interaction 'in a virtual environment that we enter to perform a task'. Yet the action of *glancing* at a digital alarm clock, as readily as at a mechanical clock and perhaps more easily than at a sundial, is far more suggestive of the advancing integration of computers into everyday life.

Our environments are becoming not only more computerised but fuller of connected computers. Of course, integration into networks and dependency on networked services did not begin with these technologies: before the Internet, and the telephone, buildings were being connected to water, gas, electricity and sewerage networks. This fact has not forced reassessment of what it means for a site to have a perimeter. Yet the existence of geofences shows that when connectivity is pervasively available, when numerous networks can exist invisibly in the air around oneself, it is possible for perimeters of another kind to be drawn and redrawn in 'empty' space.

The most obvious problems are practical ones. It is nowadays routine for frustrated researchers, having followed a URL only to encounter an altered or missing page, to turn to the Internet Archive's Wayback Machine in hope of finding an archived copy. Modern libraries work to preserve electronic data as well as paper, and 'software archaeology' studies legacy computer systems. Location-based online services, designed for deployment within places in the physical world, are neither purely data, like a Web page, nor fully spatial, like solid ground and buildings. Questions follow about how best to preserve them for study, and whom to task with doing so. If computing is to become ubiquitous, then perhaps the archives of the future must become ubiquitous too.

Practical challenges lead to philosophical reflection. The more computers and networks and streams of data become integrated into how places function and how human beings experience being in them, the greater the impulse to conclude that these digital layers have become simply parts of the sites and settings through which we move. So conceived, a building may seem to extend fuzzily far into its surrounding environment, in a hazy cloud of digital signage and concentric geofences. We will be able to map out our cities like sniffer dogs, discovering not only rigid boundaries but subtler gradients and interrelations.

21.11 Future Cities

In truth cities have always had a quality of indistinctness, as building after building blends into an overall sense of place, as impressions coalesce into memories, and as memories leave their traces in turn:

Consider the way cities brand themselves; how the fictions of Meyrink, Kafka and Hašek were moulded by the versions of Prague they lived in and how that same city is now moulded by their fiction. [...] On Bloomsday, Dublin celebrates its fictional recreation [...]. City begets text begets city. [...] All cities are subject to the Rashomon effect. (Anderson, 2015, pp. 21–22).

The technological city will be branded and moulded in other ways besides: in its parallel layers hanging invisibly in the air, full of lingering trails and clouds of data strewn about in our passage.

These predictions too are more cloudy than concrete. In mid-2015 I was completing a chapter on the theme of 'video games and virtual reality' (Seddon, 2017) which had been commissioned for a textbook on the philosophy of technology. Mention of AR, at that time, brought to mind 'smart glasses' and controversies about the etiquette of wearing Google Glass. By the time the *Pokémon Go* craze began, the book was well into its review and editorial phase; luckily, images and captions were added only nearer the end, and so I was narrowly able to acknowledge the significance of smartphone-based AR gaming. Still, I had written that AR might turn out to be transformative or might prove a passing fad, and as of early 2018 that cautious judgment still seems to be holding up.

Perhaps prototypical smart cities will ultimately serve merely to highlight the human inefficiencies in municipal government. Perhaps enthusiasm for the Internet of Things will diminish each time a fridge gets hacked and added to some spammer's botnet, and the world will decide that it manages perfectly well without a computer in every light bulb. Nevertheless, computerisation and computer networks have so expanded the range of possibilities that there is every reason to expect them to shape future environments in some form. Moreover, even fads and failed experiments can be of historical interest.

As new technologies are tried out and vie to supplant older ones, there is already no shortage of ghosts around us: in a fairly literal sense, with at least four ghost-based AR games on the market or in development (*Ghostbusters World*, *Ghostbusters Paranormal Blast*, *Night Terrors* and *Ghost Snap AR Horror Survival*). Technological exploration begets fragmentation. Cities have always been tinged with the loneliness of crowds, and it may turn out that historians of the future will have better records than ever about the private worlds of individuals, but that in honouring so many preferences and voices and perspectives the technological city will engender a weakened sense of shared local heritage. If increasingly everyone exists in a cloud of personalised data, if statues are an opt-in service and signage starts to give way to geotags, one might wonder what truly public spaces will exist to be remembered in the first place.

Then again, commemoration has always been an act bestowing a public significance of its own. The little plaque announcing that some famous person was Born

Here celebrates a private family event with retrospective public recognition. Perhaps a thousand commemorative flowers will now digitally bloom, free from the formal restrictions of official plaque selection schemes. Potentially the infinite layers of this new phantom geography offer not just new challenges but also new opportunities for shared culture and common heritage, complete with Share buttons and licensed under Creative Commons.

Once created those layers of shared culture will themselves belong to local history, bestowing their own tints of meaning upon the geography of the city. In time, many will no doubt fall into desuetude, a digital residue of comings and goings past, a social archaeology haunting the spaces within which we live. But then, even the bones of a king can lie hidden beneath a municipal car park.

References

Anderson, D. (2015). *Imaginary cities*. London: Influx Press.

Artvertiser. (n.d.). *The Artvertiser: improved reality*. Retrieved March 14, 2018, from https://theartvertiser.com/

BBC News. (2008). *No plan to ban 'anti-teen' device*. Retrieved March 14, 2018, from http://news.bbc.co.uk/1/hi/uk/7241527.stm

BBC News. (2016). *Pokemon Go in Indian court for 'hurting religious sentiments'*. Retrieved March 13, 2018, from http://www.bbc.co.uk/news/world-asia-india-37294286

BBC News. (2017). *New app reveals secrets of London's ghost signs*. Retrieved March 13, 2018, from http://www.bbc.co.uk/news/av/uk-england-london-39155680/new-app-reveals-secrets-of-london-s-ghost-signs

Bennetts, M. (2016). *Russians fear Pokemon Go a Western plot to destabilize nation from within*. Washington Times. Retrieved March 13, 2018, from https://www.washingtontimes.com/news/2016/jul/28/russians-fear-pokemon-go-a-western-plot-to-destabi/

Beuttenmüller, A. (2018). *Niantic partners with UNWTO's Travel.Enjoy.Respect program*. Retrieved July 13, 2019, from https://nianticlabs.com/blog/unwto2018/

Bienkowski, P. (2013). Whose past? Archaeological knowledge, community knowledge, and the embracing of conflict. In G. F. Scarre & R. Coningham (Eds.), *Appropriating the past: Philosophical perspectives on the practice of archaeology* (pp. 42–62). Cambridge: Cambridge University Press.

Blenkinsop, C. (2018). *World's capital cities to gain augmented reality layer through 'digital land'*. Cointelegraph. Retrieved July 13, 2019, from https://cointelegraph.com/news/worlds-capital-cities-to-gain-augmented-reality-layer-through-digital-land

Bostanci, E., Kanwal, N., & Clark, A. F. (2015). Augmented reality applications for cultural heritage using Kinect. *Human-centric Computing and Information Sciences, 5*(1), 20. [20 is the paper number, used by the journal in lieu of a page range. Pages in this journal appear to be numbered uniquely per paper, in this case 1–18.].

Brunner, C. (2011). Nice-looking obstacles: Parkour as urban practice of deterritorialization. *AI & SOCIETY, 26*(2), 143–152.

Buckley, R., Morris, M., Appleby, J., King, T., O'Sullivan, D., & Foxhall, L. (2013). 'The king in the car park': New light on the death and burial of Richard III in the Grey friars church, Leicester, in 1485. *Antiquity, 87*(336), 519–538. https://doi.org/10.1017/S0003598X00049103.

Bushman K. (2017). Pokemon Go trash. Retrieved March 13, 2018, https://www.openstreetmap.org/user/BushmanK/diary/40341

Calvium. (n.d.-a). *Artist-led soundwalks in Belfast*. Retrieved March 13, 2018, https://calvium.com/projects/belfast-soundwalks/

Calvium. (n.d.-b). *Helping residents to conserve Bristol's buildings 'at risk'*. Retrieved March 13, 2018, https://calvium.com/projects/heritage-eye/

Cameron, C. (2012). *Stiktu 2.0 now out and available worldwide*. Layar Blog. Retrieved March 14, 2018, from https://www.layar.com/news/blog/2012/06/12/stiktu-2-available-worldwide/

de Certeau, M. (1984). *The practice of everyday life*. (S. Rendall, Trans. Berkeley: University of California Press.

Coleman, E. G. (2013). *Coding freedom: The ethics and aesthetics of hacking*. Princeton: Princeton University Press.

Cresci, E. (2017). *Russian YouTuber convicted for playing Pokémon Go in church*. Guardian. Retrieved March 13, 2018, from https://www.theguardian.com/technology/2017/may/11/pokemon-go-russian-youtuber-convicted-playing-church-ruslan-sokolovsky

Cunliffe, R. (2016). *Delusional governments crusade against Pokemon Go*. Reaction. Retrieved March 13, 2018, from https://reaction.life/delusional-governments-crusade-against-pokemon-go/

Dedale. (n.d.). *Augmented reality: enriching culture*. Digital meets Culture. Retrieved March 13, 2018, from http://www.digitalmeetsculture.net/article/augmented-reality-enriching-culture/

Tom Dieck, M. C., & Jung, T. H. (2017). Value of augmented reality at cultural heritage sites: A stakeholder approach. *Journal of Destination Marketing & Management, 6*(2), 110–117.

van Dijck, J. (2008). Digital photography: Communication, identity, memory. *Visual Communication, 7*(1), 57–76.

Dobran, B. (n.d.). *Digital cultural heritage vital for shaping future smart cities*. Council of Europe. Retrieved March 14, 2018, from https://www.coe.int/en/web/culture-and-heritage/-/digital-cultural-heritage-vital-for-shaping-future-smart-cities

Dormehl, L. (2014). Free app swaps New York subway ads for street art. Cult of Mac. Retrieved March 14, 2018, from https://www.cultofmac.com/298127/ad-turns-new-york-subway-street-art-wonderland/.

Dunn, N., Cureton, P. (2018). Tomorrow's city today: Palimpsest of the future. Smartness? Between Discourse and Practice. Retrieved March 14, 2018, from https://www.smartness-discoursepractice.org/tomorrows-city-today.

Durham University (2008). Durham's world heritage site expands. Retrieved March 10, 2018, from https://www.dur.ac.uk/news/newsitem/?itemno=6941

Espinoza, J. (2016). Cecil Rhodes statue to remain at Oxford university after alumni threaten to withdraw millions. Telegraph. Retrieved March 10, 2018, from https://www.telegraph.co.uk/education/universityeducation/12128151/Cecil-Rhodes-statue-to-remain-at-Oxford-University-after-alumni-threatens-to-withdraw-millions.html

Gardner, E. (2017). Milwaukee, Wisconsin, sued for requiring permits on augmented reality games. Hollywood reporter. Retrieved March 13, 2018, from https://www.hollywoodreporter.com/thr-esq/milwaukee-wisconsin-sued-requiring-permits-augmented-reality-games-996320

Guadamuz, A. (2016). Pokémon go has revealed a new battleground for virtual privacy. The conversation. Retrieved March 14, 2018, from https://theconversation.com/pokemon-go-has-revealed-a-new-battleground-for-virtual-privacy-62670

Guin, J. (2017). Augmented reality tactics for heritage interpretation. Voices of the past. Retrieved March 13, 2018, from http://voicesofthepast.org/2017/04/14/augmented-reality/

Hatfield, T. (2014). Ingress: The game that reveals Google's secret war to control London. Guardian. Retrieved March 13, 2018, from https://www.theguardian.com/technology/2014/jun/04/ingress-the-game-that-reveals-googles-secret-war-to-control-london

Hewison, R. (2016). What's at stake in digitising heritage sites such as the Lascaux cave? Apollo Magazine. Retrieved March 13, 2018, from https://www.apollo-magazine.com/digitising-heritage-sites-lascaux-cave/

Ihaza, J. (2014). This Year's VIA festival was crawling with augmented reality graffiti. Vice. Retrieved March 14, 2018, from https://creators.vice.com/en_uk/article/nz4ezm/this-years-via-festival-was-crawling-with-augmented-reality-graffiti

Jardine, A. (2017). Y&R's augmented reality app creates virtual statues of women in public spaces. Creativity online. Retrieved March 13, 2018, from http://creativity-online.com/work/yr-new-york-the-whole-story-project/51687

Karlin, S. (2013). Augmented reality for public spaces: BC "heavy" Biermann creates high-tech art projects with a civic mission. IEEE Spectrum. Retrieved March 13, 2018, from http://ieeexplore.ieee.org/stamp/stamp.jsp?reload=true&arnumber=6471051

Katyal, S. K. (2017). Technoheritage. *California Law Review, 105*(4), 1111–1172.

Kelly, K. (2019). AR will spark the next big tech platform—Call it Mirrorworld. Retrieved July 13, 2019, from https://www.wired.com/story/mirrorworld-ar-next-big-tech-platform/

Krall, H. (2010). Hot dog of the week: Toronto street meat. Serious Eats. Retrieved March 14, 2018, from https://www.seriouseats.com/2010/01/hot-dog-of-the-week-toronto-street-meat.html

LaFrance, A. (2015). When you give a tree an email address. The Atlantic. Retrieved July 6, 2018, from https://www.theatlantic.com/technology/archive/2015/07/when-you-give-a-tree-an-email-address/398210/

Langston, J. (2017). Singing posters and talking shirts: UW engineers turn everyday objects into FM radio stations. UW News. Retrieved March 14, 2018, from https://www.washington.edu/news/2017/03/01/singing-posters-and-talking-shirts-uw-engineers-turn-everyday-objects-into-fm-radio-stations/

Lessig, L. (2008). *Remix: Making art and commerce thrive in the hybrid economy*. London: Bloomsbury.

Liberati, N. (2017). Phenomenology, Pokémon Go, and other augmented reality games. *Human Studies, 41*, 211–232. https://doi.org/10.1007/s10746-017-9450-8.

Mahdawi, A. (2015). *Can cities kick ads? Inside the global movement to ban urban billboards*. Guardian. Retrieved March 13, 2018, from https://www.theguardian.com/cities/2015/aug/11/can-cities-kick-ads-ban-urban-billboards

Manaugh, G. (2016). *The burglar's guide to the city*. New York: Farrar, Straus and Giroux.

Manaugh, G. (2019). *Ghosts only cars can perceive*. Retrieved July 6, 2019, from http://www.bldgblog.com/2019/07/ghosts-only-cars-can-perceive/

Mika, M. (2013). *Why I hacked donkey Kong for my daughter. Wired*. Retrieved March 14, 2018, from https://www.wired.com/2013/03/donkey-kong-pauline-hack/

Milling, K. (2014). 'TXTilecity': Mapping a cultural space. Culture in Virtual Spaces (Cultural Institute, King's College London). Retrieved March, 13, 2018, from https://culturalvirtualspaces.wordpress.com/2014/06/17/txtilecity/

Moscow Times. (2016). *Moscow authorities to launch their own version of Pokemon Go*. Moscow Times. Retrieved March, 13, 2018, from https://themoscowtimes.com/news/moscow-authorities-to-launch-their-own-version-of-pokemon-go-54716

Moshenska, G. (2015). Curated ruins and the endurance of conflict heritage. *Conservation and Management of Archaeological Sites, 17*(1), 77–90.

NO AD. (n.d.) About. Retrieved March, 14, 2018, from http://noad-app.com/about/

Probst, P. (2007). Picturing the past: Heritage, photography and the politics of appearance in a Yoruba city. In F. de Jong & M. Rowlands (Eds.), *Reclaiming heritage: Alternative imaginaries of memory in West Africa* (pp. 99–125). London: Routledge.

Reahard, J. (2011). *GDC 2011: Grey Area's shadow cities redefines mobile MMORPGs and immersion*. Engadget. Retrieved March, 13, 2018, from https://www.engadget.com/2011/03/07/gdc-2011-gray-areas-shadow-cities-redefines-mobile-mmorpgs-and/

Revell, T. (2017). *Thousands of fake companies added to Google maps every month*. New Scientist. Retrieved March 14, 2018, from https://www.newscientist.com/article/2127150-thousands-of-fake-companies-added-to-google-maps-every-month/

de Ridder, M. (2015). *Europeana: Utilizing the cultural past to create an urban future*. Business Models Inc. Retrieved March, 13, 2018, from https://web.archive.org/web/20151115190752/http://www.businessmodelsinc.com/europeana-utilizing-the-cultural-past-to-create-an-urban-future/

Saha, D., & Mukherjee, A. (2003). Pervasive computing: A paradigm for the 21st century. *Computer, 36*(3), 25–31.

Seddon, R. F. J. (2017). Video games and virtual reality. In A. F. Beavers (Ed.), *Macmillan interdisciplinary handbooks: Philosophy: Technology* (pp. 191–216). Farmington Hills, MI: Macmillan Reference USA.

Sheeter, L. (2009). *Sentient cities may answer back*. BBC News. Retrieved March 13, 2018, from http://news.bbc.co.uk/1/hi/technology/8310627.stm

Shrubsole, G. (2017). *How Land Registry data reveals London's secret tunnels. Who Owns England?* Retrieved March 13, 2018, from https://whoownsengland.org/2017/12/15/how-land-registry-data-reveals-londons-secret-tunnels/

SmartSantander. (2012). *SmartSantanderRA: Santander Augmented Reality application*. Retrieved March 13, 2018, from http://www.smartsantander.eu/index.php/blog/item/174-smartsantanderra-santander-augmented-reality-application

Smith, M. (2017). *Policing in the future involves citizen detectives and a Pokémon go-like app*. CSO Online. Retrieved July 13, 2019, from https://www.csoonline.com/article/3232367/policing-in-the-future-involves-citizen-detectives-and-a-pokemon-go-like-app.html

Sontag, S. (2004). *Regarding the pain of others*. London: Penguin.

Telefónica IoT Team. (2016). *How to protect Smart City cultural heritage*. Telefónica. Retrieved March 13, 2018, from https://iot.telefonica.com/blog/how-to-protect-smart-city-cultural-heritage

Traverso, V. (2017). *Learning about cities by mapping their smells: The English artist Kate McLean is trying to preserve our olfactory memory*. Atlas Obscura. Retrieved March 10, 2018, from https://www.atlasobscura.com/articles/art-mapping-smell-smellscapes-kate-mclean

Tsukayama, H. (2016). *Pokémon go developer says ads are coming, and shops are already luring gamers in. Washington Post*. Retrieved March 13, 2018, from https://www.washingtonpost.com/news/the-switch/wp/2016/07/14/pokemon-go-makes-a-change-to-make-even-more-money/

UNESCO (n.d.). *Durham Castle and cathedral*. Retrieved March 10, 2018, from http://whc.unesco.org/en/list/370

Vattano, S. (2014). *Smart technology for smart regeneration of cultural heritage: Italian smart cities in comparison*. Museums and the web. Retrieved March 13, 2018, from https://mwf2014.museumsandtheweb.com/paper/smart-technology-for-smart-regeneration-of-cultural-heritage-italian-smart-cities-in-comparison/

vGIS. (n.d.). *See infrastructure like never before*. Retrieved July 6, 2019, from https://www.vgis.io/esri-augmented-reality-gis-ar-for-utilities-municipalities-locate-and-municipal-service-companies/

Wakefield, J. (2017). *Tomorrow's cities - nightmare vision of the future?* BBC News. Retrieved March 13, 2018, fromhttp://www.bbc.co.uk/news/technology-37384152

Weiser, M. (1991). The computer for the 21st century. *Scientific American, 265*(3), 94–104.

White, S. K. (2017). *What is geofencing? Putting location to work. IDG Communications*. Retrieved March 5, 2018, from https://www.cio.com/article/2383123/mobile/geofencing-explained.html

Zöllner, M., Becker, M., & Keil, J. (2010). Snapshot augmented reality—Augmented photography. In A. Artusi, M. Joly-Parvex, G. Lucet, A. Ribes, & D. Pitzalis (Eds.), *11th international symposium on virtual reality, archaeology and cultural heritage (VAST 2010)* (pp. 53–56). Paris: Eurographics Association.

Robert Seddon is a moral philosopher with research interests broadly centred on culture. He undertook doctoral study at the University of Durham, where he retains an Honorary Fellowship and membership of the Centre for the Ethics of Cultural Heritage. Much of his current work involves looking at what happens to the ethics of cultural heritage when your culture is a modern one full of emerging technologies and their transformative effects on human ways of life: for example, applying insights from archaeological ethics to the curation of online data.

ne manufacturer's authorised representative in the EU is Springer
ture Customer Service Centre GmbH, Europaplatz 3, 69115 Heidelberg,
nany. If you have any concerns regarding our products, please
ct ProductSafety@springernature.com

and bound by CPI Group (UK) Ltd, Croydon, CR0 4YY
06/10/2026
02228459-0006